Power System Restructuring and Deregulation

Power System Restructuring and Deregulation

Trading, Performance and Information Technology

Edited by
Loi Lei Lai
City University, London, UK

JOHN WILEY & SONS, LTD
Chichester . New York . Weinheim . Brisbane . Singapore . Toronto

Copyright © 2001 by John Wiley & Sons Ltd
Baffins Lane, Chichester,
West Sussex, PO 19 1UD, England

National 01243 779777
International (+44) 1243 779777

e-mail (for orders and customer service enquiries): cs-books@wiley.co.uk

Visit our Home Page on http://www.wiley.co.uk
or
http://www.wiley.com

All Rights Reserved. No part of this publication may be reproduced, stored in a retrieval system, or transmitted, in any form or by any means, electronic, mechanical, photocopying, recording, scanning or otherwise, except under the terms of the Copyright Designs and Patents Act 1988 or under the terms of a licence issued by the Copyright Licensing Agency, 90 Tottenham Court Road, London, W1P 9HE, UK, without the permission in writing of the Publisher, with the exception of any material supplied specifically for the purpose of being entered and executed on a computer system, for exclusive use by the purchaser of the publication.

Neither the authors nor John Wiley & Sons Ltd accept any responsibility or liability for loss or damage occasioned to any person or property through using the material, instructions, methods or ideas contained herein, or acting or refraining from acting as a result of such use. The authors and Publisher expressly disclaim all implied warranties, including merchantability of fitness for any particular purpose.

Designations used by companies to distinguish their products are often claimed as trademarks. In all instances where John Wiley & Sons is aware of a claim, the product names appear in initial capital or capital letters. Readers, however, should contact the appropriate companies for more complete information regarding trademarks and registration.

Other Wiley Editorial Offices

John Wiley & Sons, Inc., 605 Third Avenue,
New York, NY 10158-0012, USA

Wiley-VCH Verlag GmbH
Pappelallee 3, D-69469 Weinheim, Germany

Jacaranda Wiley Ltd, 33 Park Road, Milton,
Queensland 4064, Australia

John Wiley & Sons (Canada) Ltd, 22 Worcester Road
Rexdale, Ontario, M9W 1L1, Canada

John Wiley & Sons (Asia) Pte Ltd, 2 Clementi Loop #02-01,
Jin Xing Distripark, Singapore 129809

Library of Congress Cataloging-in-Publication Data

Power system restructuring and deregulation: trading, performance, and information technology / edited by L.L. Lai.
    p. cm
  Includes bibliographical references and index.
  ISBN 0 471 49500 X
    1. Electrical power systems – Control. 2. Electric utilities – Cost control. 3. Electric Utilities - Deregulation. 4. Electric utilities – Technological innovations. I. Lai, Loi Lei

TK1007. P68. 2001
.333. 793'2 – dc21   2001045404

British Library Cataloguing in Publication Data

A catalogue record for this book is available from the British Library

ISBN 0 47149500 X

Produced from Word files supplied by the Editor
Printed and bound in Great Britain by Antony Rowe Ltd., Chippenham, Wilts
This book is printed on acid-free paper responsibly manufactured from sustainable forestry, in which at least two trees are planted for each one used for paper production.

Contents

Foreword xv

Preface xvii

Acknowledgements xxi

Contributors xxiii

1 Energy Generation under the New Environment 1
1.1 Introduction 1
1.2 Competitive Market for Generation 2
1.3 The Advantages of Competitive Generation 3
1.4 The Role of the Existing Power Industry 4
    1.4.1 Reconfiguring the Electricity System 5
    1.4.2 Trends in Conventional Electricity Generation Technologies 6
1.5 Electricity Demand Operation and Reliability 6
    1.5.1 Power Plant Operation 7
    1.5.2 Reliability Assessment 7
    1.5.3 Availability of Fuel 9
1.6 Renewables Generation Technologies 12
    1.6.1 Biomass 12
    1.6.2 Fuel Cell 13
    1.6.3 Wind 14
    1.6.4 Photovoltaics (PVs) 14
    1.6.5 Solar 14
1.7 Combined Heat and Power 15
1.8 Energy Policy and Government Intervention 16
1.9 Costs 17
    1.9.1 Capital Costs for New Plants 17
    1.9.2 Technology Advances – Clean Coal Technologies 18
    1.9.3 Environmental Considerations 18
1.10 Distributed Generation 20
    1.10.1 Market Regulation 21
    1.10.2 The Power Pool 22
    1.10.3 Ancillary Services 25
    1.10.4 Technical Issues 25
    1.10.5 Implications and Opportunities for Network Operators and Generators 26
    1.10.6 Connection and Use of System Charges 26
1.11 Case Study1: Phase Balancing for a Self-excited Induction Generator [50] 27
    1.11.1 Introduction 27

	1.11.2	Circuit Connection and Principle	28
	1.11.3	Performance Analysis	30
	1.11.4	Solution Technique	31
	1.11.5	Results and Discussion	32
	1.11.6	Simplified Phase-balancing Scheme	35
	1.11.7	Appendix	37
1.12		Case Study 2: Controlling a Solar Power Plant [59]	37
	1.12.1	Introduction	37
	1.12.2	The Solar Power Plant	38
	1.12.3	Control Structure of the Plant	39
	1.12.4	GA Formulation	40
	1.12.5	Experimental Results	42
1.13		Conclusions	45
1.14		References	46

2 Deregulation of Electric Utilities ... 50

2.1		Introduction	50
2.2		Traditional Central Utility Model	52
2.3		Reform Motivations	52
2.4		Separation of Ownership and Operation	53
	2.4.1	Central Dispatch Versus Market Solution	53
2.5		Competition and Direct Access in the Electricity Market	54
	2.5.1	Competition in the Energy Market	54
	2.5.2	Competition and Auction Mechanisms	55
	2.5.3	Direct Access/Wheeling	57
2.6		Independent System Operator	60
	2.6.1	Pricing and Market Clearing	61
	2.6.2	Risk Taking	61
2.7		Retail Electric Providers	63
2.8		Different Experiences	64
	2.8.1	England and Wales	64
	2.8.2	Norway	68
	2.8.3	California	71
	2.8.4	Scotland	72
	2.8.5	New Zealand	72
	2.8.6	The European Union and Germany	73
2.9		References	74

3 Competitive Wholesale Electricity Markets ... 76

3.1		Introduction	76
3.2		The Independent System Operator	79
3.3		Wholesale Electricity Market Characteristics	80
	3.3.1	Small Test System	81
	3.3.2	Central Auction	82

3.3.3		Bidding	83
3.3.4		Market Clearing and Pricing	84
3.3.5		Market Timing	86
3.3.6		Sequential and Simultaneous Markets	86
3.3.7		Bilateral Trading	89
3.3.8		Scheduling	90
3.3.9		Gaming	91
3.3.10		Ancillary Services	93
3.3.11		Physical and Financial Markets	94
3.4	Market models		96
3.4.1		Maximalist ISO	96
3.4.2		Minimalist ISO Model	97
3.5	Challenges		98
3.5.1		Market Power Evaluation and Mitigation	98
3.5.2		System Capacity	99
3.5.3		Reliability	100
3.5.4		Technical Issues	103
3.6	Acknowledgements		103
3.7	References		103
4 Distribution in a Deregulated Market			**110**
4.1	Introduction to the UK Environment		110
4.2	The Development of Competition		111
4.2.1		Competition in Supply	111
4.2.2		The Responsibilities of Retail and Distribution	111
4.2.3		Why Separate Distribution and Supply?	112
4.2.4		Key Issues for Distribution Businesses	112
4.2.5		Information Management	113
4.2.6		Use of System Billing	113
4.2.7		Customer Service	114
4.2.8		Competition in Metering	114
4.2.9		Scope for Demand-side Management	115
4.3	Maintaining Distribution Planning		116
4.3.1		Regulatory Incentives	116
4.3.2		Technical Issues	117
4.3.3		Planning Drivers	117
4.3.4		Long-term Planning	122
4.3.5		Network Planning Tools	124
4.3.6		Asset Replacement Planning	125
4.3.7		Risk Assessment	125
4.3.8		Skills and Resources	125
4.3.9		Network Design	126
4.3.10		Distribution Automation	127

4.3.11	Automation Case Study - Remote Control in London Electricity	129
4.4	Future Development	132
4.5	Appendix: Distribution Automation in a Deregulated Environment	133
4.5.1	Introduction	133
4.5.2	Remote Terminal Units	134
4.5.3	SCADA Master Station	134
4.5.4	Software Functionality	136
4.5.5	Operations and Maintenance (O&M)	136
4.5.6	System Integration, Design and Management	137
4.5.7	Communication Systems	140
4.6	References	151

5 Transmission Expansion in the New Environment ... 153
5.1	Introduction	153
5.2	Role of the TP	155
5.2.1	Vertically Integrated Utility	156
5.2.2	Three Models of the Electricity Market	158
5.2.3	For-profit TP	161
5.3	New Market Organisation	163
5.3.1	Incentive Rate Design – Price-cap Regulation	164
5.3.2	Priority Insurance Scheme	167
5.3.3	Transmission Expansion	169
5.4	Conclusions	170
5.5	References	171

6 Transmission Open Access ... 172
6.1	Introduction	172
6.1.1	The Traditional Power Industry	172
6.1.2	Motivations for Restructuring the Power Industry	173
6.1.3	Unbundling Generation, Transmission and Distribution	174
6.2	Components of Restructured Systems	175
6.2.1	Gencos	175
6.2.2	BOT Plant Operators and Contracted IPPs	175
6.2.3	Discos and Retailers	175
6.2.4	Transmission Owners (TOs)	175
6.2.5	Independent System Operator (ISO)	176
6.2.6	Power Exchange (PX)	176
6.2.7	Scheduling Coordinators (SCs)	176
6.3	PX and ISO: Functions and Responsibilities	176
6.3.1	PX Functions and Responsibilities	176
6.3.2	California Power Exchange	177
6.3.3	ISO Functions and Responsibilities	178
6.3.4	Classification of ISO types	179
6.4	Trading Arrangements	183

6.4.1	The Pool	183
6.4.2	Pool and Bilateral Trades	184
6.4.3	Multilateral Trades	185
6.5	Transmission Pricing in Open-access Systems	186
6.5.1	Introduction	186
6.5.2	Rolled-in Pricing Methods	187
6.5.3	Incremental (Marginal) Pricing Methods	187
6.5.4	Embedded Cost Recovery	189
6.5.5	Transmission Pricing Method in the NGC, UK	191
6.6	Open Transmission System Operation	192
6.6.1	Dispatch	192
6.6.2	Transmission Loss Compensation	192
6.6.3	System Control	193
6.6.4	Ancillary Service Provision	193
6.7	Congestion Management in Open-access Transmission Systems	195
6.7.1	Congestion Management in Normal Operation	195
6.7.2	Integrated Transmission Dispatch Strategy	198
6.7.3	Illustration Using a Small Power System	200
6.7.4	Static Security-constrained Rescheduling	202
6.7.5	Dynamic Security-constrained Rescheduling	205
6.8	Open-access Coordination Strategies	209
6.8.1	Price Elasticity as a Means to Relieve Congestion	209
6.8.2	Relieving Congestion by ISO Executed Price Signalling	210
6.8.3	Coordination between Transactions	211
6.8.4	Illustration of Transaction Coordination	213
6.8.5	Integrated Coordination Procedure	215
6.9	Conclusions	216
6.10	Acknowledgements	216
6.11	Appendix	216
6.12	References	217
7 Electric Power Industry Restructuring in China		**220**
7.1	Introduction	220
7.2	Development of Electric Power Industry in China	222
7.2.1	Successive Growth of Power Production and Installed Capacity	222
7.2.2	Further Expansion of Power Networks	222
7.2.3	Continuous Increase of Electricity Consumption	224
7.3	Management System of Electric Power Industry in China	225
7.3.1	The State Power Corporation	225
7.3.2	Philosophy and Strategy of the SP	231
7.4	Power Market in China	232
7.4.1	Motivations for Reformation	234
7.4.2	Reform Plan of the SP	235

7.4.3		Obstacles in Establishing the Power Market in China	236
7.5	Electricity Pricing		237
7.5.1		Basic Theory of Predicting Electricity Costs	238
7.5.2		Electricity Cost Derivation	240
7.5.3		Electricity Pricing of Inter-provincial Power Market	242
7.6	Transmission pricing		246
7.6.1		Current Decomposition Axioms	247
7.6.2		Mathematical Models	249
7.6.3		Methodology of Graph Theory	252
7.6.4		Algorithms and Case Studies	253
7.7	Conclusions		254
7.8	Acknowledgements		254
7.9	References		255

8 Flexible AC Transmission Systems (FACTS) 258

8.1	Introduction		258
8.1.1		Benefits of FACTS Technology	259
8.2	Transmission System Limitations		260
8.2.1		System Stability	260
8.2.2		Loop Flows	261
8.2.3		Voltage Limits	261
8.2.4		Thermal Limits	261
8.2.5		High Short-circuit Level Limits	261
8.3	FACTS Technology		262
8.3.1		Power Switching Devices and PWM Inverter	262
8.3.2		Control Methods and DSP/Microprocessor Technology	264
8.3.3		Present Status on FACTS Activities	265
8.4	Solution Options with FACTS		265
8.4.1		Fundamental Concepts of Transmission	265
8.4.2		Shunt Controllers	266
8.4.3		Series Controllers	271
8.4.4		Combined Series/Shunt Controllers	275
8.4.5		Phase Angle Controllers	276
8.4.6		HVDC Transmission Controllers	278
8.4.7		Other Controllers	281
8.5	FACTS Applications		281
8.5.1		SVC	281
8.5.2		STATCOM	282
8.5.3		TCSC	282
8.5.4		UPFC	283
8.6	Concluding Remarks		283
8.7	Acknowledgements		284
8.8	References		284

9 Asset Management 287
9.1 Introduction 287
9.2 Pre-privatisation (1990): The Public Purse 287
9.3 Post-privatisation (1990): Freedom 288
9.4 Early-mid 1990s: Getting the Same for Less 288
9.5 1994/5+: Getting More for Less 289
9.6 Late 1990s: Capital Efficiency 289
9.7 August 1999 Interim Report: All Change? 290
9.8 The 1990/2000 Regulatory Settlement and a Major Challenge 290
9.9 Asset Ownership 291
9.10 Asset Governance 291
9.11 Asset Management 294
9.12 Asset Information and the Ageing Process 294
9.13 Condition Monitoring 295
    9.13.1 Transformers 296
    9.13.2 On-load Tap Changers 297
    9.13.3 Switchgear 297
    9.13.4 Other Plant 297
    9.13.5 Understanding Long-term Asset Costs 298
    9.13.6 Underground Cables 298
    9.13.7 HV Cables 299
    9.13.8 Partial Discharge 300
    9.13.9 Zero Sequence Impedance 301
9.14 Asset Replacement Analysis 302
    9.14.1 Benchmarking 304
    9.14.2 Asset Lifecycle 305
    9.14.3 Asset Replacement Models 307
    9.14.5 Technology Strategy 311
9.15 Refurbishment and Replacements 312
9.16 Risk Management and Insurance Consequences 313
    9.16.1 Risk Control 313
    9.16.2 Major Incidents 314
    9.16.3 Type Failures 315
    9.16.4 Common Mode Failure 315
    9.16.5 Financial Risk Management 316
9.17 Asset Information Acquisition 317
    9.17.1 Asset Management Systems 319
    9.17.2 Data Gathering 319
    9.17.3 Property Rights 320
    9.17.4 Environment 320
    9.17.5 Data Cleaning 321
    9.17.6 Confidential Information 321
    9.17.7 Quality of Information 322

9.18	Conclusions	322
9.19	Appendix: Fuzzy DGA for Diagnosis of Multiple Incipient Faults	323
9.19.1	The IEC DGA Codes	323
9.19.2	The Fuzzy IEC Code – Key Gas Method	323
9.19.3	Fuzzy Diagnosis Results	325
9.19.4	Trend Analysis of Individual Faults	327
9.19.5	Comments	328
9.20	References	329

10 Power Quality 330

10.1	Introduction	330
10.1.1	A General Overview	330
10.1.2	PQ Issues During System Disturbances	333
10.1.3	Voltage Sags	334
10.2	Disturbance Assessment	336
10.2.1	The Wavelet Transform	336
10.2.2	Wavelet Analysis	338
10.2.3	Application to PQ [25]	339
10.2.4	Automated Disturbance Assessment	341
10.3	Waveform Distortion	342
10.3.1	Harmonic Sources	342
10.3.2	Characterisation of Harmonic Sources	345
10.3.3	Harmonic Flows [30]	346
10.3.4	Aperiodic Distortion	346
10.4	Need for Adequate PQ Indices and Standards	347
10.5	Need for Adequate PQ Monitoring [70,71]	348
10.6	References	349

11 Information Technology Application 353

11.1	Introduction	353
11.2	Software Agents	354
11.2.1	Types of Agents	354
11.2.2	General Issues and the Future of Agents	360
11.3	Electricity Options Markets with Agents	360
11.3.1	Electricity Markets and Options	362
11.3.2	Agent-Based Computational Economics	364
11.3.3	Valuing Options with Agents	365
11.4	Evolutionary Programming-based Optimal Power Flow Algorithm	371
11.4.1	OPF	373
11.4.2	EP	373
11.4.3	EP-OPF	374
11.4.4	Load flow Solution	377
11.4.5	Gradient Acceleration	378
11.4.6	Application Studies	379

11.5	Complex Artificial Neural Networks for Load Flow Analysis	383
11.5.1	Conventional ANN for Real Numbers	384
11.5.2	New ANN for Complex Numbers	385
11.5.3	Comparison of the two ANNs by Computer Simulation	388
11.5.4	Application of "Complex" ANN to Load Flow Analysis	390
11.6	Virtual Reality	395
11.6.1	Types of VR systems	395
11.6.2	Non-immersive (Desktop) Systems	396
11.6.3	Fully Immersive Head-mounted Display Systems	396
11.6.4	Semi-immersive Projection Systems	396
11.6.5	Comparison between Different VR Systems	397
11.6.6	Cave	398
11.6.7	Telepresence	398
11.6.8	Augmented	398
11.6.9	Applications	398
11.7	3-D Thermal Imaging for Power Equipment Monitoring	400
11.7.1	The Hardware	401
11.7.2	The Correspondence	402
11.7.3	Display with VR	405
11.7.4	Implementation Example	407
11.8	Conclusions	410
11.9	Acknowledgements	411
11.10	Appendix: System Data and Parameter Settings	411
11.11	References	411
12 Application of the Internet to Power System Monitoring and Trading		**416**
12.1	Introduction	416
12.2	The Internet	416
12.2.1	What Is the Internet?	416
12.2.2	How Does the Internet Work?	417
12.2.3	What Would Happen Without the Internet?	417
12.2.4	How Can the Power Industry Benefit from the Internet?	418
12.2.5	How Can I Find the Information I Need?	419
12.3	Usability of the Internet	419
12.3.1	Scientific Use for Researchers	419
12.3.2	Educational Use	420
12.3.3	Internet Products	420
12.3.4	Business Competition	420
12.3.5	Multimedia Access	421
12.3.6	On-line Services	421
12.3.7	Support for Professionals	422
12.3.8	The Power Industry and the Internet	422
12.3.9	Recent Improvements on the Internet	424

12.4		Internet Technology	424
	12.4.1	Access to the Internet	425
	12.4.2	Operating Platforms on the Internet	426
	12.4.3	Web Clients	426
	12.4.4	Web Servers	427
	12.4.5	Web Protocols	427
	12.4.6	E-mail	427
	12.4.7	Internet Security	428
	12.4.8	Internet Bandwidth	431
12.5		Internet Programming Languages	433
	12.5.1	HTML	433
	12.5.2	Interpreted Versus Compiled Languages	433
	12.5.3	What Is JavaScript?	434
	12.5.4	What Is Java?	435
12.6		Web Pages	436
	12.6.1	Setting up a Web Page	437
	12.6.2	Difference Between a Static and a Dynamic Web Page	437
	12.6.3	Displaying Database Content	438
	12.6.4	Web Pages with Functionality	440
	12.6.5	Web Pages with Integrated Applications	441
12.7		XML	441
	12.7.1	Why the Need for XML?	441
	12.7.2	Reasons for XML	441
	12.7.3	Separation of Content and Layout	442
	12.7.4	XML Layout Validation with DTD	444
	12.7.5	Stylesheets	444
12.8		Case Study1: Power Station Monitoring [8]	445
	12.8.1	Requirements of Airport Substation	445
	12.8.2	System Implementation	447
	12.8.3	Monitoring Power Station Equipment	454
12.9		Case Study 2: Power Trading Application	457
	12.9.1	Trading Platform Architecture	458
12.10		Conclusions	459
12.11		Acknowledgements	460
12.12		References	460

Index **461**

Foreword

The electricity power utilities in many countries have been, or are being, restructured. There are many reasons for restructuring. In some countries restructuring has been driven by the desire of government to meet increasing demands for electricity by encouraging independent power production, which relieves government of a financial obligation. In countries where ownership of assests are in private hands, restructuring has been driven by mergers and acquisitions, as companies seek to gain competitive advantage.

In the most advanced countries, restructuring is being driven by the desire to allow consumers to choose their electricity supplier on the basis of price and service provided. These dramatic changes in the organisation of electricity power utilities bring with them new challenges and opportunities, as the previous centrally designed and operated systems are dismantled and replaced by a new compctitive framework.

Companies operating in a competitive market need more sophisticated control and management systems to ensure that their business objectives can be achieved. The development and application of new technologies is also accelerated in this new environment, as companies seek to improve their effectiveness and efficiency.

This book is contributed by a group of world authorities. It explains in depth the reason for restructuring, without including superfluous detail. Examples are given from various countries. Details are provided on new strategies and technologies which are being applied in the areas of generation, transmission and supply. The implications for the environment are also reviewed. Tools being utilised for asset management and for the effective management of infrastructure are illustrated with practical examples. The analytical modelling and general analysis of competitive power markets are also illustrated.

This book provides a comprehensive review of all the many facets of change taking place in a dynamic industry. It is compulsory reading for graduates and engineers, and other professionals, who are entering or involved in the electricity power industry.

David G. Jefferies CBE, FREng, HonFIEE

IEE President 1997/8
Former Chairman, The National Grid Group plc, UK

Preface

This book was written as a result of the ongoing stimulating world-wide deregulation and restructuring of the power industry. This move away from the traditional monopolies and towards greater competition, in the form of increased numbers of independent power producers and an unbundling of the main service, started in the United Kingdom in 1989 and this change was driven by the large differences in electricity tariffs across regions, by advancements in technologies which allow small producers to compete with large ones, and by a strong belief that competition will produce an all-win situation.

The book was contributed by an international group of experts to produce a broad and detailed coverage of the main issues. The intent has been to provide the reader with an in depth treatment but without excessive specialisation, to avoid a purely qualitative treatment by including some analytical and numerical methods, and to offer, whenever possible, real case studies, worked examples and project discussions.

Since each power utility is unique, it will not be possible to present the best path to follow in the restructuring exercise. The market models, regulation and tariffs used by transmission networks, and the mechanism for maintaining a high level of reliability, will be different. Because of the advancement of communications technology and increased computing power, it is possible to consider different market structures. Without such advancement, no information could be available in time for the business operation.

Different markets have been considered in the book. In brief, they could be summarised as three types. In the completely market-driven environment, market forces seek to moderate the behaviour of various players in the market, e.g. the suppliers, consumers and statutory regulators. In the transitional market, there is a process of transition from a highly regulated environment to a deregulated environment. In the embryonic free market, the state retains ownership of the generators and some of the transmission infrastructure, but opens up the market to limited competition at the distribution level.

As there is much uncertainty in these environments, due to the structure of the market, planning over a long-term horizon is perceived as very difficult at present. Yet, without long-term planning, it is unlikely that the electricity power industry would be at great risk, as it might not be able to supply the growing demand, or to maintain the same quality of service as it is currently providing to its consumers. The recent chaos in California is an example. This could have very serious consequences to the long-term viability of the entire industry.

This book shows how new technology will allow us to change today's highly regulated market structure to one that relies on competition to set the energy price. By using new technologies, we can use less energy, resulting in lower energy bills for customers, and avoid or defer additional expensive plant construction. The addition of new participants, such as independent power producers, power marketers and brokers, has added a new dimension to the task of maintaining a reliable electric system. This book will detail methods by taking into account some of these issues.

In the new market environment, generation represents most of the cost. Chapter 1 reports on the development of new strategies and compares different technologies for electricity generation with environmental and political considerations. This includes decentralised power supplies, renewables, regulatory constraints, new technical challenges and solutions. Different mechanisms, such as the pool, have been set up for the operation of the new emerging electrical market. The market should dictate when new generation is needed and where it is located.

Since there is a large number of players in the market, it is important to work out the type of bidding, or negotiation strategies that each player can use. It is especially important to work out the information content of the bidding strategies. Chapter 2 covers experience from various countries on power utility restructuring and deregulation. Analytical tools for the modelling and analysis of competitive power markets are presented. Chapter 3 also discusses several wholesale electricity markets around the world and most of these are in a continuous process of change. This evolutionary process is being driven by the need to address some of the outstanding issues in the design and implementation of these markets. Some challenges, such as reliability, market power evaluation and mitigation, are outlined.

Chapter 4 reports on the change in distribution business in a deregulated market. Various issues such as planning, control, load forecasting, metering, customer services and risk assessment have been considered. A case study on the remote control of London Electricity is included.

Chapter 5 deals with transmission expansion. Following development of the market, the transmission provider transforms into the independent transmission company (ITC) so as to admit a highly sophisticated market. The ITC is required to make complex business decisions over a wide range of time scales, such as the long-term, short-term and near real-time. This chapter discusses future directions and modifications to the regulatory policies to make the ITC serve as both a market maker and a service provider.

Chapter 6 presents the economic issues associated with transmission open access. The chapter also provides a discussion of some important operational issues in the emerging market environment. Normal dispatch, congestion management and the effects of security considerations have all been discussed with examples from the open-access viewpoint.

Chapter 7 deals with the Chinese market. A detailed background on the change of industry is given. It also explains why the approaches adopted by the developed countries are not suitable. The chapter also proposes a new approach to calculate transmission loss. To operate the ever increasingly complex power systems with better efficiency, an accurate transmission loss model is indispensable.

In the rapidly deregulating utility environment, reactive power control to assure voltage stability and power flow control to avoid line overloading and congestion have become necessary considerations in power system operation. Flexibility AC transmission systems (FACTS), which are based on power electronics technology, have revolutionised the field of electric power engineering. Chapter 8 presents the application of FACTS to utilise the capability of existing transmission systems. The impact of FACTS for new generation entrants is discussed.

Chapter 9 deals with asset management. A comprehensive asset management model is required to support business in the deregulated electricity market. The main purpose and characteristics of the model components are described in detail. It will benefit all internal and external users in the open-access environment, resulting in realistic and transparent open-access charges, and bring long-term economic benefits to all parties. Tools for effective asset management in power industry restructuring are illustrated with practical examples.

Electricity industry restructuring has had a dramatic impact on the energy market. To gain a competitive advantage, today's energy providers need to focus on value-added products and services, such as power quality. Power quality is a critical issue for industrial customers, especially in the high-tech sector. In order to understand power quality, many customers or energy providers have installed power quality monitoring systems to record electrical system performance and/or facility equipment reactions, and the analysis of the monitored data has become a challenge. Chapter 10 reports on the techniques, methods and standards used or proposed for power quality issues.

The explosion in the use of information technology has seen the introduction of computer-based work management systems, asset management systems, and control systems to manage system operation. Information technology is making markets more efficient, resource production less speculative and costly, and the delivery and monitoring of energy more effective, while enfranchising customers to make more intelligent choices. Improvements in information technology will continue to allow economical and reliable solutions to problems facing the power industry. Chapter 11 introduces intelligent agents, genetic algorithms, evolutionary programming, artificial neural networks and virtual reality technology, and reports on their applications to load flow, valuing electrical options and power equipment diagnosis. The chapter highlights the technology behind the new market brought about by deregulation. Energy service companies will continue to make increasing demands for more sophisticated software and equipment to monitor and control various aspects of power delivery.

In just a few years, Java has taken the networked world by storm. Java combines powerful, object-oriented programming with the ability to run on any computer platform without the need for recompiling or translating. Java promises to play a yet more fundamental role in the future of on-line computing, including electronic commerce, for it can allow anyone to make use of powerful applications anywhere. One result of its platform independence is that a scrap of code called a Java applet can be embedded in a World Wide Web page. Chapter 12 deals with the application of the Internet to power station monitoring and discusses its use for energy trading. It also presents an introduction to Web technology and its applications.

This book addresses the most up-to-date problems and their solutions in the area of power system restructuring and deregulation in a cohesive manner. It will provide invaluable information for power engineers, educators, system operators, managers, planners and researchers.

Loi Lei Lai

Acknowledgements

The editor wishes to thank Mr Peter Mitchell of Wiley and his team in supporting this project.

The editor also wishes to thank all the contributors, without whose support this book could not have been completed. In particular, the editor thanks Harald Braun in managing to complete the manuscript despite great difficulties caused by software incompatibility. The editor also wishes to thank Mrs Vinay Sood and Professor Sood for their creation of the initial manuscript. The editor is very grateful to Dr David Jefferies for writing the Foreword. The permission to reproduce copyright materials by the IEEE and IEE for a number of papers mentioned in some of the chapters is most helpful. The arrangement of the index by Miss Qi Ling Lai and Chun Sing Lai is much appreciated.

Last but not least, we all thank Wiley for supporting the preparation of this book and for the extremely pleasant co-operation.

Contributors

Dr Loi Lei Lai was appointed Senior Lecturer at Staffordshire Polytechnic (now Staffordshire University) in 1984. From 1986 to 1987, he was a Royal Academy of Engineering Industrial Fellow to both GEC Alsthom Turbine Generators Ltd and its Engineering Research Centre. He is currently Head of Energy Systems Group and Reader in Electrical Engineering at City University, London. He is also an Honorary Professor at the North China Electric Power University, Beijing. Dr Lai is a Senior Member of the IEEE and a Corporate Member of the IEE. He has authored/co-authored over 100 technical papers. In 1998, he also wrote a book entitled *Intelligent System Applications in Power Engineering - Evolutionary Programming and Neural Networks* published by Wiley. Recently, he was awarded the IEEE Third Millennium Medal and 2000 IEEE Power Engineering Society UKRI Chapter Outstanding Engineer Award. In 1995, he received a high-quality paper prize from the International Association of Desalination, USA. Among his professional activities are his contributions to the organisation of several international conferences in power engineering and evolutionary computing, and he was the Conference Chairman of the International Conference on Power Utility Deregulation, Restructuring and Power Technologies 2000. Recently, he was invited by the Hong Kong Institution of Engineers to be the Chairman of an Accreditation Visit to accredit the University BEng (Hons) degree in electrical engineering. Dr Lai is also Student Recruitment Officer, IEEE UKRI Section. In 1999, he was included in *The Dictionary of Contemporary Celebrities of Worldwide Chinese*. In 2000, his biography was included in the 18th Edition of *Who's Who in the World*, Marquis, USA. His biography has also been selected for inclusion in the 2001 *Who's Who in Science and Engineering*, Marquis, USA.

Professor Jos Arrillaga obtained his MSc, PhD and DSc from UMIST, Manchester, UK. He is a Fellow of the IEEE and a Fellow of the Royal Society of New Zealand. From 1970 to 1975, he was Head of the Power Systems and High Voltage Groups, UMIST. From 1975 to 1999 he was Professor of Electrical Engineering, University of Canterbury, Christchurch, New Zealand. From 1982 to 1995, he was also the Director of Systems Software & Instrumentation (a Christchurch-based consulting company established in 1982). From 1985 to 1990, he was Head of Department, Electrical and Electronic Engineering, University of Canterbury. From 1988 to 1995, he was a Member of the CIGRE-14 Working Group on HVdc harmonics (14-03). From 1989 to 1995, he was Convenor of CIGRE Task Force 36-05/14-03-03 on AC System Harmonic Modelling for AC Filter Design. From 1990 to 1996, he was a Member of CIGRE JWG 11/14-09 on Unit Connection. From 1996 to 1999, he was Convenor of CIGRE Task Force 14.25 on Harmonic Cross-modulation in HVdc Transmission. Since 1990 and 1995 respectively, he has been Director of CHART Instruments, Christchurch and Director of Power Quality Consulting, a Christchurch-based consulting company. Professor Arrillaga has received many awards, such as John Hopkins Premium of the IEE, UK, 1975; the Best Paper Premium, IEEE Conference on Harmonics and Quality of Power, ICHQP94, 1994; the Best Electrotechnical Paper, IPENZ Annual Conference, 1996; Uno Lamm High Voltage Direct Current Award, IEEE, 1997; John Munganest International Power Quality Award of the Power Industry, 1997; President's (Gold Medal) Award, Annual Meeting of IPENZ, 1998;

the Best electrotechnical paper, IPENZ Annual Conference, 1999; Silver Medal of the New Zealand Royal Society for Innovation in Science and Technology, 2000; and CIGRE (Paris) Technical Committees Award, 2000.

Mr Harald Braun trained in the area of power electronics with Siemens, Frankfurt, Germany, from 1985 to 1989. He obtained his Diploma in Telecommunication at Friedberg-Giessen University, Germany, in 1994. He was a part-time lecturer at City University, London, from 1994 to 1996 in teaching object-oriented programming in C++. He was a Senior Programmer at A.M. BEST International Ltd, London, from 1996 to 2000. At present, he is a Senior Software Engineer with ALTIO, London, developing new Internet technology software. He is working for his PhD at City University on a part-time basis and expects to achieve it in July 2001. His research interest is the extraction of information from data using neural network technology.

Professor A. Kumar David is Chair Professor and hHad of the Department of Electrical Engineering, The Hong Kong Polytechnic University. His BE degree is from the University of Ceylon and PhD from Imperial College, London. He has previously worked in Sri Lanka, USA, Zimbabwe and Sweden and his research interests are in power system restructuring, pricing, control, HVDC, transient stability, protection and reliability. Professor David was elected an IEEE Fellow in 2000 for his outstanding contributions to electricity supply industry reform and open transmission access. He is the regional editor for Asia of the *International Journal of Electric Power Systems Research*.

Mr Robert Friel is responsible for strategy development in LPN (London Power Networks), which is the distribution company of the London Electricity Group. Mr Friel has extensive experience of the planning and development of both private and public electricity distribution systems in the UK and abroad. He joined London Electricity in 1997 and helped develop the distribution businesses, response to the last regulatory price control. Mr Friel is a Chartered Engineer and a Member of the IEE.

Professor Marija Ilic has been at MIT since 1987 as a Senior Research Scientist in the EECS Department where she conducts research and teaches graduate courses in the area of electric power systems. Since September 1999 she has had a joint appointment at the National Science Foundation as a Program Director for Control, Networks and Computational Intelligence. Prior to coming to MIT, she was a tenured faculty at the University of Illinois at Urbana-Champaign. Professor Ilic is a recipient of the First Presidential Young Investigator Award for Power Systems; she is also an IEEE Fellow and an IEEE Distinguished Lecturer. Professor Ilic has co-authored several books on the subject of large-scale electric power systems (Ilic, M. and Zaborszky, J., *Dynamics and Control of Large Electric Power Systems*, John Wiley & Sons, New York, 2000; Ilic, M., Galiana, F. and Fink, L. (Editors), *Power Systems Restructuring: Engineering and Economics*, Kluwer Academic Publishers, Second printing 2000; Allen, E. and Ilic, M., *Price-Based Commitment Decisions in the Electricity Markets*, Springer-Verlag, London, 1999; Ilic, M. and Liu, S., *Hierarchical Power Systems Control: Its Value in a Changing Industry*, Springer-Verlag, London, 1996). She is also a contributor to the edited book *Blue Print for Transmission* (PU Reports, 2000). Her interest is in control and design of large-scale systems.

Dr David G. Jefferies, CBE has been Chairman of the National Grid Company plc since 1990, when the Company was formed as part of the privatisation of the UK electricity sector. His bold and far-sighted leadership has been a key ingredient in its success of the National Grid Group plc from the performance of the transmission system during a decade of major change in the industry, through the conception and development of Energis, to the growth of the group internationally. He retired as the Chairman of the National Grid Group plc in July 1999. Dr Jefferies was previously Chairman of the London Electricity Board and of Viridian plc. He was the 1997/98 IEE President. Owing to his huge contribution made to the Institution, he is an Honorary Fellow of the IEE. He is also a Fellow of the Royal Academy of Engineering. He was a pioneer in the restructuring and deregulation of the UK electric power utility.

Professor Chen-Ching Liu received his PhD from the University of California, Berkeley, in 1983. Since then, he has been at the University of Washington, Seattle. He is currently Professor of Electrical Engineering and Associate Dean of Engineering at the University. Dr Liu is a Fellow of the IEEE and the US representative on CIGRE Study Committee 38. His areas of interest include power system economics, intelligent system applications and vulnerability assessment.

Professor K.L. Lo obtained his MSc and PhD from the University of Manchester Institute of Science and Technology. He is currently the Head of the Power Systems Research Group at the University of Strathclyde. His group specialises in energy management systems, issues concerning the electricity market and deregulation, simulation, analysis, monitoring and control of power networks. Professor Lo has been an international advisor and member of many organising committees of international conferences, consultant/visiting professor to over 12 educational institutions, and has lectured extensively in the Far East, Europe and America. He is the author of over 260 technical publications. He is a Fellow of the IEE and a Fellow of the Royal Society of Edinburgh.

Mr Kevin Morton is a member of London Electricity's Executive and is currently the Managing Director of both London Power Networks (LPN), which is the distribution business of London Electricity, and London Electricity Services (LES), which is the private networks business of London Electricity. As Head of the Public Distribution Business he led the work during 1999 which culminated in the formation of 24seven, the joint venture network management services provider formed by LE and TXU Europe (Eastern Electricity). He has been in the electricity supply industry for 25 years in a variety of both operational and strategic roles within the distribution business. He has a practical engineering background having worked in a number of operational, project manager and leadership roles in utility power distribution. Mr Morton is a Chartered Electrical Engineer and a Fellow of the IEE. He also represents the UK in the business area of distribution at EURELECTRIC, the pan-European association of electricity companies.

Professor Mark O'Malley received his BE and PhD degrees from the National University of Ireland, Dublin, in 1983 and 1987 respectively. He is currently a Professor at the National University of Ireland, Dublin, with research interests in power systems, control theory and biomedical engineering.

Professor Gerald B. Sheblé is a Professor of Electrical and Computer Engineering, Iowa State University, Ames, Iowa. He received his BS and MS degrees in electrical engineering from Purdue University and his PhD in electrical engineering from Virginia Tech. His industrial experience includes over 15 years with a public utility (Commonwealth Edison), with a research and development firm (Systems Control), with a computer vendor (Control Data Corporation) and with a consulting firm (Energy and Control Consultants). He has participated in the functional definition, analysis and design of power system applications for several energy management systems since 1971. Dr Sheblé also designed the optimisation package in use at over 50 electric utilities to schedule electrical production. He has consulted since entering the academic world with companies in North America and Europe on electric industry deregulation as well as expert witness testimony on the National Electric Code and Intellectual Property Rights. His consulting experience includes significant projects with over 40 companies. He developed and implemented one of the first electric market simulators for the Electric Power Research Institute using genetic algorithms to simulate the competing players. He conducts approximately 24 seminars each year on optimisation, artificial neural networks, genetic algorithms and genetic programming, and electric power deregulation around the world. His primary expertise is in power system optimisation, scheduling and control. Dr Sheblé has been awarded over 1 million dollars of research support over the last 10 years, primarily in the application of adaptive agents to market bidding. He has authored a review of adaptive agent market-playing algorithms for the Kluwer press release *Power Systems Restructuring: Engineering and Economics* edited by Ilic, Galiana and Fink. He has written a monograph on tools and techniques for energy deregulation entitled *Computational Auction Mechanisms for Restructured Power Industry*. He has also been an invited guest on radio talk shows and a resource for several news articles on electric power deregulation and industrial trends. His research interests include power system optimisation, scheduling and control. Professor Sheblé is an IEEE Fellow.

Professor Vijay Sood obtained his BSc from University College, Nairobi, and his MSc degree from Strathclyde University, Glasgow, in 1969. He obtained his PhD degree in power electronics from the University of Bradford, England, in 1977. From 1969 to 1976, Dr Sood was employed at the Railway Technical Centre, Derby. Since 1976, he has been employed as a Researcher at IREQ (Hydro-Québec) in Montreal. Dr Sood also has held Adjunct Professorship at Concordia University, Montreal, since 1979. Dr Sood is a Member of the Ordre des ingènieurs du Québec, a Senior Member of the IEEE, a member of the IEE and a Fellow of the Engineering Institute of Canada. He is the recipient of the 1998 Outstanding Service Award from IEEE Canada, the 1999 Meritas Award from the Ordre des Ingènieurs du Québec, and the IEEE Third Millennium Medal. Dr Sood is presently the Managing Editor of the *IEEE Canadian Review* (a quarterly journal for IEEE Canada). He is a Director and Treasurer of IEEE Montreal Conferences Inc. He has worked on the analog and digital modelling of electrical power systems and their controllers for over 25 years. His research interests are focused on the monitoring, control and protection of power systems using artificial intelligence techniques. Recently, Dr Sood has been interested in the Internet and its applications for teaching purposes and was mandated by IEEE Canada to publish the journal *IEEE Canadian Review* on the Internet (www.ieee.ca). Dr Sood has published over 70 articles and written two book chapters. He has supervised 14 postgraduate students and examined 17 PhD candidates from universities all over the world. He is well known amongst the electrical engineering community in Canada.

Mr Cliff Walton is Technical & Regulation Manager of London Power Networks (LPN). LPN is the distribution business of the London Electricity Group. In his current position, he is responsible for all technical and regulatory matters regarding the public electricity distribution system in London and particularly the quality of supply and reliability performance that sets London apart. He has previously been Strategy Manager, Asset Manager and Planning Manager for London Electricity's Public Networks Group. In his recent roles he has championed the development of an integrated technology strategy, strategic asset management, fault causation analysis, incipient fault detection and location techniques, as well as creating the strategies behind the implementation of one of the largest distribution remote control, telemetry and automation projects. Mr Walton joined LPN when it was established in April 2000; his career in electricity distribution spans 29 years. He has worked with a number of overseas utilities and has written and presented many papers on a wide variety of technical and asset governance and management issues. He is a Chartered Electrical Engineer and a Member of both the IEE and IEEE.

Professor Xifan Wang was born in May 1936. He graduated from Xi'an Jiaotong University in 1957. He has since been with the School of Electrical Engineering of the university, where he now holds the rank of Professor and is the Director of the Electric Power System Department. He is a Senior Member of the IEEE. From September 1981 to September 1983, he worked in the School of Electrical Engineering at Cornell University in Ithaca, New York, USA as a Visiting Scientist. From September 1991 to September 1993 he worked at the Kyushu Institute of Technology in Kitakyushu, Japan, as a Visiting Professor. Prof Wang has a 40-year experience of researching and teaching in electric power system analysis and planning. His main research fields include reliability evaluation, generation and transmission network planning, operation planning, system contingency analysis, dynamic and transient stability, short-circuit current calculation, optimal load flow, and probabilistic load flow. He is especially proficient in constructing mathematical models and developing application software in the above areas. He also took part in many research and planning tasks of key electric power projects in China, such as the Three Gorges Hydro-Power Station. He proposed a new transmission system, namely the fractional frequency transmission system (FFTS) which uses a lower frequency to reduce the reactance of AC transmission systems. In recent years, he has been researching the electric power market.

Dr Neville R. Watson received his BE (Hons) and PhD degrees from the University of Canterbury (New Zealand), where he is now a Senior Lecturer. Dr Watson has authored and co-authored approximately 100 technical papers and 3 books. Paper awards received include; Best Paper Award (The Sixth International Conference on Harmonics in Power Systems, 1994), the William Perry Award (IPENZ) and Finalist for the Carter Holt Harvey Packing Award for Innovative Technology (IPENZ). He has also given a number of invited lectures in Singapore, Australia and Canada.

Professor Fushuan Wen received his BEng and MEng degrees from Tianjin University, China, in 1985 and 1988, respectively, and his PhD from Zhejiang University, China, in 1991, all in electrical engineering. He was a Postdoctoral Fellow at Zhejiang University from 1991 to 1993. He joined the faculty of Zhejiang University in 1993, and has been a Professor of Electrical Engineering since 1997. He held a visiting position at the National

University of Singapore from 1995 to 1997. He is on leave from Zhejiang University and is now with Hong Kong Polytechnic University as a research fellow. Dr Wen is recipient of the National Natural Science Award of China, Zhejiang Provincial Top Young Scientist Award and several other awards from the Ministry of Education (China), Zhejiang provincial government, Zhejiang University and the National University of Singapore. He is a member of the editorial board of the *Journal of Automation of Electric Power Systems* (in Chinese) and was a guest editor of a special issue on 'Artificial intelligence applications in power systems'. His research interests are in power system restructuring and artificial intelligence applications in power systems.

Professor Kit Po Wong obtained MSc, PhD and DEng from University of Manchester Institute of Science and Technology in 1971, 1974 and 2001 respectively. Currently he is Professor in Electrical Engineering at the University of Western Australia. He has worked on power system dynamics, protection, electromagnetic transient evaluation, long-distance transmission, artificial intelligence and computational intelligence in power system operation and planning. Professor Wong has published over 160 research papers and has been awarded the Sir John Madsen Medal of the Institution of Engineers Australia. He was the Founding Chairman of the Western Australia Chapter of the IEEE Power Engineering Society and was the Chairman of the Western Australia Section of the IEEE from 1999 to 2000. He has been a member of numerous technical committees for international conferences. Professor Wong was the General Chairman of the IEEE PES/CSEE 2000 International Conference on Power Systems Technology (Powercon 2000). He is an editorial board member of the international journal *Electric Power Systems Research* and the *Australian Journal of Intelligent Information Processing Systems*. In 1999, he was awarded the Outstanding Engineer Award of the IEEE Power Engineering Society WA Chapter. He was a recipient of the IEEE Third Millennium Medal in 2000. Professor Wong is a Fellow of the Hong Kong Institution of Engineers, Fellow of Institution of Engineers Australia, Senior Member of the IEEE, and Member of the IEE.

Miss Yee Shan Cherry Yuen acquired her degree of Bachelor of Engineering in Electrical Energy Systems Engineering at The University of Hong Kong in 1996. In the same year Miss Yuen was awarded The China Light & Power Company Prize in Electrical Energy, because of the distinction of her final year project entitled 'The application of artificial neural networks on the detection of high impedance faults'. During 1996 to 1998 Miss Yuen pursued the degree of Master of Philosophy with a thesis entitled 'Fault detection and overvoltage protection in low voltage power systems'. In 1998 she was awarded the China Light & Power Co. Ltd. Electrical Energy Postgraduate Scholarship. In the same year she was awarded John Swire & Sons Ltd. James Henry Scott Scholarship for Engineering Studies at the University of Strathclyde which enabled her to pursue the degree of Doctor of Philosophy in Scotland. Miss Yuen is also an Associate Member of the IEE. Her current research interests include the analysis of international energy markets, congestion management, transmission pricing and the application of information technology in energy markets.

Mr Yong T. Yoon received SB degrees in applied mathematics and in electrical engineering and computer science and MEng degree in electrical engineering and computer science from the Massachusetts Institute of Technology (MIT), Cambridge, in 1995 and in 1997, respectively. He is currently pursuing a PhD degree (February 2001) in electrical engineering and computer science at MIT, concentrating on electric power system economics engineering. His thesis is entitled 'Electric power network economics: underlying principles for for-profit independent transmission company (ITC) and designing architecture for reliability'. His research interests include modelling of energy markets as stochastic dynamic systems, developing concepts for the ITC and designing software tools for various energy market participants. He has a strong background in control, estimation, mathematics, research design and regulatory economics.

1

Energy Generation under the New Environment

Dr Loi Lei Lai
City University, London
UK

1.1 Introduction

Restructuring of the electricity supply industries is a very complex exercise based on national energy strategies and policies, macroeconomic developments and national conditions, and its application varies from country to country. It is important to point out that there is no single solution applicable to all countries and there is a broad range of diverse trends.

Liberalisation, deregulation (or reregulation) and privatisation are all processes under the general label of market reform. Liberalisation refers to the introduction of a less restrictive regulatory framework for companies within a power sector. This could imply deregulation, which is the modification of existing regulation. It can be argued that reregulation is a more accurate term than deregulation since new laws are being imposed on the industry with regulatory watchdogs appointed to protect consumer interests. Ideally, then, a true liberalised energy market would work within a set regulatory framework, overseen by a regulator and with no external political influence upon the participants regarding plant size or fuel choice.

Privatisation is the sale of government assets to the private sector, by itself, privatisation is not sufficient to introduce competition into a reformed sector. Competition will be the result of careful regulation of the privatised entities to allow new entrants access to the market. Competition is fundamental to most market reforms and it is introduced in order to reduce costs and increase efficiency. There is considerable variation in the extent of the competition which is introduced. For example, competition could be introduced just for the addition of new generating capacity and referred to as competitive bidding where the existing generating company invites contractors to tender to build, operate and sell electric power to the monopoly at a specified price. Alternatively all licensed generators could be allowed to compete to supply wholesalers or retailers through a short-term market (spot market) or via longer term contracts; this is called competitive generation. The next level is wholesale competition, i.e. competition in the sale of electricity to wholesale companies for resale to a retail level or directly to final customers. This usually allows the large consumers to choose their own suppliers. Competition at final consumer level, including household consumers, is called retail competition. This is usually the very last step of the reforms, as it requires a complex information technology system because of the large number of small users involved. Retail competition is usually introduced gradually starting with the larger industrial consumers, then the medium consumers and finally including smaller consumers.

Many electricity markets around the world are currently in transition towards more deregulated and competitive markets. The changes were initiated by:

- a realisation that generation and distribution functions need not be monopolies;
- a feeling that public service obligations are no longer necessary;
- the cost reduction potential of competition;
- increased fuel availability and fuel supply stability; and
- the development of new technologies in power generation and information technology.

1.2 Competitive Market for Generation

The continuing growth of competition in American electricity markets is a consequence of the 1978 passage of the Public Utility Regulatory Polices Act (PURPA). Designed as a conservation measure, PURPA established the right of co-generators and independent power producers (IPPs) to sell electricity to local regulated investor-owned utilities (IOUs). Such rights were broadened substantially by the passage of the Energy Policy Act of 1992 which requires transmission line owners to wheel bulk power [1]. Thus, under current federal regulations non-utility power producers can sell electricity to any utility on the grid. Furthermore, in April 1994, the California Public Utility Commission adopted a policy establishing complete open access to all power producers. By 1996 independent generators could compete to sell electricity directly to large industrial customers, effectively bypassing traditional utilities. By 2002, the policy will permit all electricity consumers, regardless of size, to purchase electricity from *any* utility or independent generator on the grid. No longer will the consumer be restricted to buying electricity from the local utility. A competitive market for generation will have been established [2,3].

The system evolving in the USA provides increasing competition and diversity among generators. They vary from established utilities, IPPs and co-generators to small producers that use renewable fuels and other non-utility generators (NUGs). By 1990, a decade after the reform movement got under way in the USA, co-generators and unregulated IPPs were building more generating capacity than were the traditional utilities. For example, Southern California Edison buys 30% of its power from NUGs. The Midland Co-Generation Project in Michigan consists of 12 gas turbines with a generating capacity of 1343 megawatts. The Alamita Company, in Arizona, is an independent power company that sells electricity to bulk customers, namely the Tucson Electric Power Company and Southern California Edison [4]. Compared with the deregulation of IOUs, privatisation of a state-owned monopoly requires a complete and fundamental change in the structure of ownership and property rights in the electricity supply industry in order to obtain the benefits of increased efficiency and innovation. A shift from public to private ownership refocuses the goal of the producer towards profits. Pursuit of the latter provides a strong economic incentive, in a competitive environment, to improve and maintain the quality of customer services, monitor costs more closely, and invest in productivity-enhancing technologies. These incentives are blunted by state ownership. With respect to privatisation, the UK experience since 1989 seems more germane than does the regulatory reform the USA has been undergoing since 1978.

The European Community is addressing these same issues and has agreed to draft directives calling for open access in energy markets. As of January 1993, the European Commission seeks to let large users of electricity, those using 100 gigawatts or more of power per annum (aluminium, steel, chemicals, glass and fertiliser producers), to purchase electricity from any supplier in the Community.

1.3   The Advantages of Competitive Generation

Competitive generation provides a market within which independent firms compete on the basis of price to sell electricity directly to large industrial customers, and to supply electricity, via common carrier transmission, to distributors who in turn sell power to final users [5,6]. Producers may specialise or diversify by load characteristic. For example, some may prefer to compete for long-term base-load contracts. These firms are likely to own hydro and nuclear power plants. On the other hand, firms with fossil fuel plants might seek to supply base and cycling loads. Finally, producers with gas combustion turbines and co-generators could compete to meet peak loads. Other firms may diversify and be ready to compete for base, cycling and peak loads.

Prices charged for each type of service (peak and off-peak load, daily to seasonal) could be established by contract, 24 hour advance notice, and in spot markets. Unit prices could vary by the amount of electricity purchased per period. As a result, customers would face more service options and a more complex pricing scheme. There are a number of advantages to having a variety of types of generators linked to the transmission grid.

The first major advantage involves cost savings. At any given moment, power is supplied to the transmission grid by the firm with the lowest marginal costs. Dispatch according to merit saves resources and reduces the cost of generating electricity. Because the different plants may have different load characteristics, peak and load duration curves, generating capacity can be more fully utilised and additional capital resources saved.

The second advantage of competitive generation is that a spot market for electricity will develop. The ability to sell electricity on the spot market increases the generator's flexibility in scheduling production. The presence of a spot market means that less idle capacity must be maintained in order to provide a given level of service reliability. Shortfalls and emergencies can be met by purchasing power on the spot market. Demand and supply are equilibrated by flexible spot prices.

The third advantage of competitive generation is that the market will provide an array of service standards that more closely match consumer preferences. Consumers could be offered priority service with a schedule of electricity rates increasing with the level of reliability. According to reference [7], priority service offers significant efficiency gains over random rationing with fixed electricity rates. A competitive market in electricity generation would offer a much broader array of services than do state monopolies or regulated generators. Perhaps it is not surprising that 70% of USA private utilities, facing new competitive pressure at the generation stage, now offer some form of voluntary interruptible service [8].

The fourth advantage of competitive generation is innovation. Competition not only leads firms to be more responsive to consumer demands, monitor costs more closely, and compete on the basis of price, but also provides an incentive to be innovative. Developing a new consumer service, a better method of reducing costs, or a faster way of dealing with problems promises the innovator a competitive edge.

1.4   The Role of the Existing Power Industry

The nature of the existing generating plants will affect the speed of reforms. In countries where the coal industry has dominated the economy there has been opposition to restructuring the electricity industry, which usually includes a substantial amount of coal-fired capacity. Deregulation of the electricity sector meant loss of a secured market for coal which now has to compete for its share in the market.

The nuclear industry in the UK was initially excluded from competition and subsidised. The nuclear power stations bid into the power pool and were guaranteed the sale of their electricity due to the Non-Fossil Fuel Obligation (NFFO). The NFFO placed an obligation on the distribution companies to buy a set percentage of their electricity from stations using non-fossil fuels. In 1990, this was mainly nuclear power. A Fossil Fuel Levy was placed on the electricity bill of all electricity consumers (which amounted to 10% of the total bill) and over 90% of the money collected was given to Nuclear Electric to cover generating costs not recouped from sales of electricity to the pool [9]. In 1996, when British Energy was formed, the subsidy to the nuclear power industry was abolished. The levy was reduced and since then it has been used to support renewable energy projects.

Prices tend to go down as competition is introduced and are expected to fall significantly in the long-term. For example, in the UK prices have fallen since the market opening and they are expected to fall even lower. In 1995 real prices, the price of electricity for industry decreased by almost 13% and the price for households by 6.3% between 1991 and 1995. It is has been observed that industrial prices have decreased more than household prices in most of the countries where reductions have occurred [10].

One of the consequences of privatisation is the development of the international energy company concept - a company whose focus is becoming more global and more multi-purpose. For example US electricity and gas companies have been purchasing electricity assets in the UK, and Australian and UK companies have been heavily involved in setting up independent power projects in developing countries. Another change with privatisation is the focus on shareholder value. Privately owned companies have to compete for funds in the capital market and it is important to show that they operate efficiently and are expected to do well in the business environment to attract investors. That means a completely new organisational structure and strategies for companies from what were used in the highly regulated power industry.

Coal is expected to retain a strong position in power generation worldwide in the future. In 1995 solid fuel, mainly coal, accounted for almost 40% of world electricity production and is expected to retain this percentage until 2020. In 1995, 60% of total world consumption was for power generation and this is expected to grow to 65% in 2020. The power sector's demand for coal will increasingly be dominated by Asia. Asia's share of world demand is expected to increase from 25% in 1995 to 43% in 2020 [11].

There are a number of issues that will affect future use of coal and in some cases the results are quite uncertain. The International Energy Agency (IEA) points out that projections of coal use are subject to the outcome of competition between coal and gas, particularly in Europe, and to the policies adopted by governments to improve environmental performance and comply with greenhouse gas reduction commitments [11].

1.4.1   Reconfiguring the Electricity System

In the past, power systems were developed to transmit large amounts of power at high voltage from remote generating stations and to distribute power at lower voltage down to millions of small consumers. This was the favoured pattern, allowing ever-larger power stations, mostly coalfired, to be built and achieving economies of scale and high efficiency. The national grid evolved to ensure secure supplies to all consumers and centralised control and supervision was essential. In the present privatised electricity supply industry based on free trading of electricity as a commodity, central control is unwelcome. Wherever possible, electricity generation should be closely integrated with space and process heating in a diverse array of combined heat and power (CHP) systems. Renewable energy sources should be harnessed by large numbers of wind and wave machines, marine tidal-current or small-hydro plant, solar photovoltaic generators on roofs and small generating plant close to farms supplying wood fuel or to sources of combustible waste products. Generating plant will be small and dispersed and since CHP systems must be located close to their heat loads there will be a natural tendency for most electricity generation capacity to lie close to the consumer. There will be little need to transmit large amounts of electric power over long distances. The function of the power system will be to handle the fluctuations in load and in the output from the renewable power generators. High-power, long-distance transmission will be much less important.

In the current energy structure, a central power plant is the key facility providing energy for houses, factories and offices. With decentralised co-generation of heat and power and the deployment of renewables, this situation would change. The new structure would be less centralised and more dispersed. Network stability and frequency regulation would gain in importance and energy storage would become very important. Electricity generation is provided by a large number of small units rather than a small number of large units. Co-generation is the generation, on site, of your own power and at the same time taking advantage of the exhaust heat from your gas turbine or other engine to meet on-site heat needs. Heat can be used to heat buildings, heat dryers, generate steam through an HRSG (heat recovery steam generator), or to provide air-conditioning through an absorption chiller. Power and heat can be generated locally from natural gas or liquid fuel using an efficient, reliable gas turbine.

The uncertainty in the USA today is what will happen to electricity prices. The major competing factors are limited deregulation and lack of new generating stations (particularly large coal or nuclear stations). Estimates range from modest decreases in prices, to the levelling of local inequities, and significant increases driven by demand without supply. Our view is that prices over the long haul will increase slightly with some local inequities being eased. All this means that for many sites cogen (distributed power) will be a viable option for those willing to improve their competitive position through reduced energy costs.

New enabling technologies have now improved transport of electricity in high-voltage DC systems to the point where this may be cheaper, and use less energy, than transporting fossil fuels, for distances of 5000 km and above. This might make it possible to link low-CO_2 power sources where demand is low to distant regions where demand is high.

1.4.2 Trends in Conventional Electricity Generation Technologies

Conventional sources of electricity supply will maintain their central role in primary energy supply for many years to come. Further advancement of fossil fuel generation technologies will increase the options for mitigating greenhouse gas (GHG) emissions. This is particularly important for some developing countries and transitional economies with abundant, low-cost fossil fuels, where electricity demand is increasing rapidly. The large share of nuclear and hydro in the generating mix of some countries already makes a significant contribution to mitigation of GHGs.

1.5 Electricity Demand Operation and Reliability

World electricity production is expected to grow by an annual rate of 3% in the period 1995 to 2020 according to IEA projections. Coal retains a strong position in world power generation and will continue so. However, gas is expected to grow faster - at 6% - than solid fuels at 2.9% (e.g. coal) [11]. This is because, in countries where gas is available at competitive prices, gas-fired plants are cheaper to build and operate. Deregulation has played a role in opening the way for gas to compete with other fuels. Coal is still the favoured fuel in locations close to low-cost coal production (e.g. parts of North America, Australia and South Africa), in areas where gas is unavailable or expensive (as in those developing countries that have coal available, like China and India), and in areas where there are existing coal-fired units.

Prior to deregulation, utilities tried to predict the future energy demand in their area and build new capacity accordingly. In a deregulated energy market generators know what the current demand is and try to fill as much of the demand as possible using their power plants. The predicted growth in the demand for energy on a worldwide basis should provide an incentive for generators to build new plant or extend their existing capacity to take advantage of this trend. Competition rules will determine the market players. However, the only players in practice who can invest in new capacity are those who feel they can achieve a competitive advantage. In deregulated markets this should not be market access or cost of capital but a genuine advantage such as feedstock, technology, captive market of heat, extension of existing plant to take advantage of existing assets, refurbishment, etc. The possibility of having stranded costs would seem to rule out new, large expensive power plants. Most of the additional capacity is expected to come from incremental investment in extensions done as part of general improvements or maintenance. New plants are likely to be smaller, more cost effective, and close to areas of demand that can compete effectively for local market share. This means that there could be a swing away from large fossil-fuel-fired plants in the energy mix towards smaller, less intrusive plants sited close to the area of demand. The fact that industrial sites are now allowed to install their own generating capacity and export electricity to the grid could lead to an increase in smaller scale distributed generating capacity.

1.5.1 Power Plant Operation

The operation of power plants is also changing dramatically in deregulated markets. Generating companies are no longer obliged to generate electricity; they can choose to generate and sell their electricity when they think it is profitable for them. This means that most of the generators will want to operate their plants at base load where most profit can be made. There is little incentive for the generator to provide electricity for more expensive intermediate and peak demand, which make up only a small portion of the market. As deregulation proceeds an increasing number of players enter the system which is no longer centrally controlled. This makes the quality and reliability issues more difficult to manage. Experience so far shows that deregulated markets can reliably meet demand and are expected to do so in the foreseeable future. The UK system's reliability and availability actually increased between 1992 and 1997 when the transmission and distribution network was restructured [4]. It is believed that the system will work without problems of security of supply for the next 5~10 years.

Coal contracts are also affected by changes in power plant operation. There is a general move to shorter term fuel supply contracts to match the electricity sales contracts in deregulated markets. Flexibility in plant operation is an advantage in the competitive market where conditions change quickly. Distributed generation and small-scale units could also give more flexibility to the system. An advantage of coal is the fact that it can be easily stored in stockpiles, whereas storing gas is much more complicated and expensive and restricted to certain quantities. In deregulated markets demand and availability of supplies are less predictable and therefore the risk of disruption in fuel supply is more evident. Coal stockpiles can ensure security of supply for the generator.

1.5.2 Reliability Assessment

Utilities are forced to operate in a more reliable, economic and efficient manner and plan their expansion investments more accurately. There are a number of reasons promoting interconnections among utilities. These include economic interchange, firm power and energy transactions, wheeling, improved operating reliability and flexibility and reduction in installed generation reserves. Usually utilities construct new power plants to meet the increasing demand or to replace old plants, which need large investments. However, interconnected utilities may jointly install a generating unit in which the utilities may have different or similar shares or the interconnected utilities may buy a certain percentage of the output of a generating unit, which already exists in the other utility. Therefore, the failure of a jointly owned generating unit will cause a decrease in the available capacities of all the sharing utilities simultaneously. Because of this correlation, the conventional model of a generating unit cannot be used to represent a jointly owned generating unit.

The reliability modelling and evaluation methods of composite generation and transmission systems need to be extended when the system being analysed includes generating units that are jointly owned with other interconnected systems. This is because the modelling of jointly owned units causes two major problems. The first problem is that they cannot be included in the area generation model in a conventional manner because a jointly owned generator contributes generating capacity to two or more areas. Consequently, a failure or derated state of a jointly owned generator affects all the sharing areas. This condition cannot be incorporated in the traditional generation model, which has an inherent assumption of independence among generation models of various areas. The second problem is with the transmission model. In the absence of jointly owned units the transmission links are used only for emergency help and energy transactions. Since the energy contracts and the transmission capacity states are fixed, emergency help that can be given to neighbouring areas is fixed. But when jointly owned units are included in the reliability analysis of the system, common generation flows are present and vary depending on the states of jointly owned units. Consequently, the emergency help that can be given to neighbouring areas is dependent not only on the transmission capacity states and energy contracts, but also on the common generation flows which vary according to the states of the jointly owned generating units [12,13]. Further research on a detailed system representation is necessary to consider the particular operating features of jointly owned units so that their impact on the reliability performance of the respective power systems can be investigated.

It is important to understand the market response to the increased risk associated with the introduction of competition into the market for generating electricity. Typically a vertically integrated state monopoly deals with fluctuations in demand and random equipment failure by carrying excess capacity, including redundant backup capacity. It may also address predictable fluctuations in demand by offering peak-load pricing schemes, although the incentive to do so is weakened by state ownership or regulation.

Competitive generation produces at least two additional sources of risk: a more complex pricing structure, and loop flow problems when independent producers put electricity into the transmission network. Moreover, electricity flows along the path of least resistance. Thus, for example, electricity sold by Generator A to Industrial Customer B may not travel along the 'contract path' that is, the shortest line within the network that directly links the buyer and seller. Depending on circumstances, electricity introduced into the network at any point may give rise to 'loop flow' affecting all suppliers to the grid. Loop flow can disrupt the quality and reliability of service to everybody taking electricity from the grid at the moment additional power is introduced.

If decentralised markets introduce additional risk, they have to provide a broad array of ways of dealing with it. All of these sources of risk potentially influence the quality of service to the final consumer of electricity. In general, the market offers methods to reduce risk and to price risk so that it can be spread or shared optimally.

Consider how a generator faces the risk of uncertain prices for electricity. Firstly, the producer can sell power by long-term contract to large industrial customers and regional distributors. Contracts specify prices and adjustment clauses. Thus, only a small proportion of its output may even be exposed to unknown price fluctuations [14]. Secondly, selling on the spot market on a regular basis offers normal returns because prices regress towards the mean over a large number of sales. By selling regularly on the spot market, the producer is reducing risk through diversification. Thirdly, the producer can hedge spot market sales in the futures market.

1.5.3 Availability of Fuel

Fuels used to generate electricity are produced using the following fuel sources: namely, coal, nuclear, natural gas, oil, hydrogen and renewable resources. Renewable resources include hydro power, geothermal, biomass, wind, solar and photovoltaics. Coal is the predominant fuel source. Nuclear power is projected to decline further over the next 20 years owing to retirements of existing units. Generation from both natural gas and coal is projected to increase to offset these retirements and to meet the growing demand for electricity. The coal trade has been increasing and is expected to continue doing so in the future. It is expected to increase faster than coal production. Between 1992 and 2010 the coal trade is projected to grow by an annual 4.3% whereas coal production will grow by 2.3% annually [15]. Coal prices dropped during the 1990s in line with competition and with the fact that there is excess capacity for mining coal for the international market. Cheap coal is seen as being readily available in the short and medium term. The following sections summarise the discussions of issues related to the markets for coal, nuclear, natural gas, oil and renewable fuels, followed by electric power industry restructuring on fuel markets.

Coal

Power generators will attempt to pass on market risks to coal producers and carriers wherever they can. As a result, coal purchase contracts will likely become shorter in duration and lower in price.

The existing capacity of the power industry in each country will play an important role in its future fuel mix. In the EU, 17% of the conventional thermal capacity is over 30 years old, indicating that much of the plant is in need of refurbishment or replacement [16]. Where coal-fired plants already exist it is usually more economic to operate them rather than build new gas-fired capacity. Refurbishing or repowering an existing coal-fired plant can reduce costs as the entire infrastructure remains in place. Retrofit of pollution control equipment may be necessary to meet environmental standards. In cases where hydroelectricity and/or nuclear power dominate base-load generation other fuels - notably coal and gas - will compete more strongly for position in the mid-merit market for electricity.

Nuclear Power

Some nuclear power plants are expected to become uneconomical. Competitive electricity prices may be so low that nuclear power plant operators will not see enough income to enable them to recover the costs of operating and maintaining the plants and the costs of capital improvements, such as steam generator replacements. In the immediate future, some nuclear power units will be at risk of early retirement as a result of restructuring. The additional inability of plant operators to cover a plant's full costs, including capital costs, under restructuring produces 'stranded costs'. For nuclear plants, operating costs after deregulation will be driven mainly by plant size, age, capacity factors, and requirements for new capital improvements. Average fuel costs make up only about one-fourth of the operating costs for nuclear power plants, but the competitive environment created by a restructured electric power industry will encourage nuclear power plant operators to reduce all operating costs, including the costs of purchasing and managing nuclear fuel. Moreover, if early retirements of nuclear power plants result from competition in electricity markets, the demand for nuclear fuel will be reduced. To compete, suppliers in the nuclear fuel industry will be forced to reduce prices or improve efficiency. In 1996, 434 nuclear reactors in operation in 32 countries produced 2400 TWh of electricity avoiding an estimated 10% of global human-made emissions of carbon dioxide.

Natural Gas
Natural gas is primarily used during peak demand periods and is the preferred energy source for new generating capacity. The electric power and natural gas industries are both network industries, in which energy sources are connected to energy users through transmission and distribution networks. As the restructuring of electricity markets proceeds, the development of futures contract markets and electronic auction markets could lead to greater integration of the electricity and natural gas industries and the emergence of competitive energy markets. The availability of market information and public markets for natural gas and electricity will be a key to the development of an integrated energy market for those commodities.

The use of natural gas in electricity generation has been growing rapidly. According to the IEA World Energy Outlook, gas-fired electricity output will almost double between 1993 and 2010, even under an energy savings scenario. Low capital cost, short construction time and competitive fuel price make natural gas generation attractive, especially in deregulated markets. Technologies being applied in current commercial operation are gas turbines and gas engines. The rapid development of gas turbines in recent years - bringing higher efficiency, lower cost, reduced NO_x emissions and increased operational flexibility - puts natural gas electricity generation technologies in a position to make a large contribution to GHG mitigation. For large gas turbines, complex cycles (i.e. reheat, intercooled cycles, etc.) may further improve efficiency. Combined-cycle power plants attained thermal efficiencies of 40% in 1970, and are now close to 60% efficient. Gas turbines and gas engines for small-scale generation need further to improve their efficiency, price and environmental performance to gain wider application in the market. Conversion technology using electrochemical reactions, namely fuel cells, should become competitive in the near future. Natural gas-fuelled fuel cells can attain 50% efficiency (under very high-temperature operation), which would be further improved to 70% if used in combined cycle.

Oil
Oil prices have ranged between US$10 and 20 per barrel during the 1990s and there is no sign of any shortage in the short or medium term. Owing to assumptions about electricity industry restructuring prompting the construction of less capital-intensive and more efficient natural gas generation technologies, the share of coal generation will eventually decline while the natural gas share will continue to increase. With the deregulation of electricity generation and the resulting incentive for power generators to lower fuel costs, the use of relatively expensive residual fuel oil for electricity production is likely to decline even further. As a result, petroleum refiners may be faced with a growing problem: that is, how to dispose of leftover residual fuel and petroleum coke. Among other options, two possibilities are related to electricity markets: (1) selling petroleum coke to electricity generators for use as a fuel component, and (2) gasification at the refinery by using integrated gasification combined-cycle (IGCC) technology to produce steam for process heat and for electricity production.

Renewables

Because electricity generation from renewable sources generally is more expensive than power from conventional sources, unconstrained competition in electricity generation would likely result in a reduced role for renewables. As a result, a variety of proposals, schemes and policies include specific provisions which are used to support the continued development and use of renewable energy. Renewable portfolio standards and system benefits charges are among the programmes being considered. Green marketing and pricing programmes, already being implemented by electric utilities, may also provide a means to increase consumer demand for electricity from renewable fuels. The role of renewable energy sources in competitive electricity markets will also depend on the cost and performance of the individual renewable fuels. In addition, because renewable energy generating facilities generally depend on the availability of energy resources at specific sites, often at sites remote from major electricity grids, transmission issues will affect the penetration of renewable fuels in the electricity generation market.

Renewable energy will be an essential element of the climate change programme, producing little or no GHG emissions and significantly lower levels of other pollutants. To maintain and enhance public support for renewables, policies and programmes are aimed at assisting the renewable energy industry to become competitive.

The obligation on suppliers to supply a proportion of renewable power will give renewable generators confidence that there will be a market for their product. Many renewable electricity generation schemes, using established technologies, are already producing power at prices which are more or less competitive with that generated by mainstream coal and gas. Figures 1.1 and 1.2 show the changes in the market shares and the generation mix respectively in the UK.

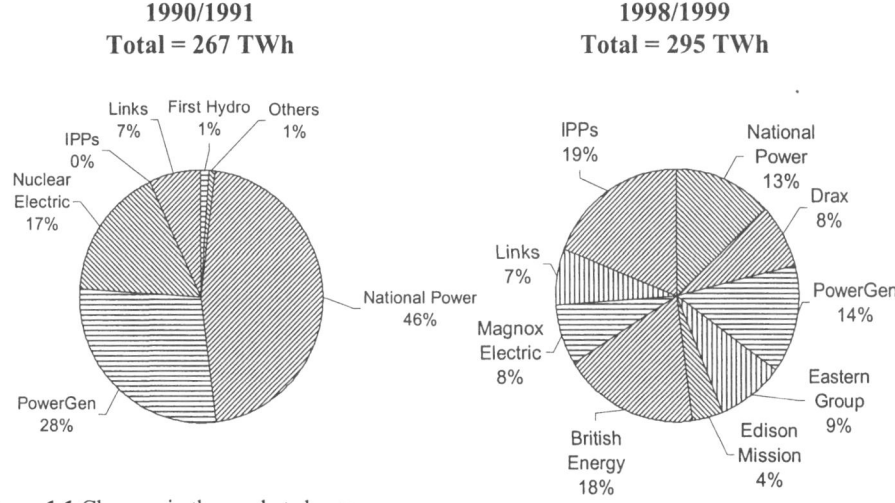

Figure 1.1 Changes in the market shares

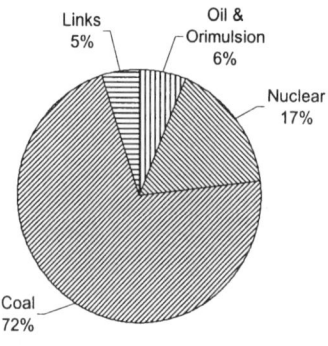
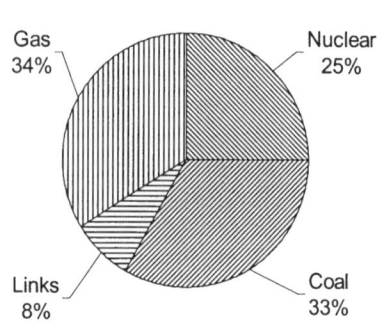

Figure 1.2 The generation mix

1.6 Renewables Generation Technologies

Although a number of the technologies are inherently small-scale compared with central station power generation, this has some distinct advantages such as increased flexibility in siting and operation. As electricity markets are restructured, distributed (embedded) generation is likely to expand and renewables will become an increasingly attractive source of supply. There is a more diverse range of technologies, although the success of any of them will depend on both their technical and economic development. These technologies include offshore wind, energy crops, photovoltaics, fuel cells, hydro, landfill gas, passive solar, biomass residues, wave power and geothermal energy. The world is changing and those changes are driven by the use of energy.

1.6.1 Biomass

Biomass power generation operates on a small to medium scale, usually under 60 MW. Integrated gasification and combined cycles can also be used with biomass and have the advantage of removing all particulates from the combustion process. In a co-generation setup it yields an efficiency of over 85%. This technology is close to commercialisation. A technology in need of further development is the fuel cell, where a biomass version may be as effective as a coal-fed version at converting chemical energy to electrical energy. Lack of concern by national authorities, difficult technical requirements and licensing arrangements, and lack of trade or promotional organisations are significant barriers to these technologies. The low tariffs for sale of biomass-generated electricity to the grid, high tariffs for standby or backup electric supplies, lack of incentives and high investment costs also constitute serious barriers.

1.6.2 Fuel Cell

A fuel cell consists of two electrodes sandwiched around an electrolyte. Oxygen passes over one electrode and hydrogen over the other, generating electricity, water and heat. Fuel cell systems will compete with other distributed generation technologies, including microturbines and reciprocating engines, available at prices competitive with existing forms of power generation. Fuel cell systems will have a competitive advantage in that they can be more easily scaled to residential size and will be more efficient in handling the load profile of residential customers. They will be quieter, environmentally cleaner, more efficient, and less expensive to install, service and maintain. Fuel cell systems will also compete with solar and wind-powered systems.

Based on regenerative fuel cell technology, National Power recently announced that it has successfully developed a new electricity storage technology, Regenesys, which could change the way power systems of the future are planned and operated [17]. Regenesys is the world's most advanced regenerative fuel cell technology. Regenerative fuel cells would be attractive as a closed-loop form of power generation. Water is separated into hydrogen and oxygen by a solar-powered electrolyser. The hydrogen and oxygen are fed into the fuel cell, which generates electricity, heat and water. The water is then recirculated back to the solar-powered electrolyser and the process begins again.

The electrochemical process, which operates like a giant rechargeable battery, has the potential to deliver commercial, operational and environmental benefits for electricity suppliers worldwide. It stores electricity when demand and costs are low and releases it when demand and prices are high, removing the need to call up more expensive power plants. The system, which can deliver power instantly, can therefore assist demand planning, improve the use of power station assets so that less capacity is needed, enhance operational control and give customers greater security of delivery. It will also offer lower lifetime costs than conventional storage. The single biggest investment advantage of Regenesys is that it will offer lower lifetime costs than either pumped storage or conventional battery plants - energy storage that could curtail peak demands. Although AC energy cannot be stored, power electronic developments offer a fast responding interface between the AC network and electrical DC energy stored in batteries. These considerations underline the potential value of energy storage in curtailing daily peak periods and that it would most effectively be located near the source of load variations, the consumers in the distribution networks [18]. Coupled with advanced power electronics, storage systems can reduce harmonic distortions and eliminate voltage sags and surges. Most distributed energy storage systems can be made multi-functional with little or no additional cost, so that, for example, both uninterrutable power supply (UPS) and energy management applications can be served by the same equipment. In combination with renewable resources, energy storage can increase the values of photovoltaic (PV) and wind-generated electricity, making supply coincident with periods of peak consumer demand. Energy storage systems can be used to follow load, stabilise frequency and manage peak loads. Regenesys has a number of distinct advantages over existing electricity storage technologies like pumped-hydro and battery storage. It offers all the benefits of pumped-hydro, but can be located anywhere on a power system thus avoiding environmental problems. Though similar to a battery storage plant, Regenesys is much more flexible. Unlike a battery, the power output and storage capacity can be specified individually. Based on fuel cell technology, Regenesys can be built in modules to the required size ranging from 5 to 500 megawatts of capacity. It is able to provide vital services to electricity grids, including frequency and voltage control. Regenesys could meet peak demand and maximise investment returns. It allows better use to be made of the cleanest generating plant by reducing the need to operate less efficient peaking plant. It can also enhance the value of renewable generators such as wind and solar power.

1.6.3 Wind

Currently some 50 countries have major wind power installations. Europe is presently the most important market but demand in Asia is growing strongly. Ease of rapid installation (six to nine months) and a free local source of power make wind an attractive technology in developing countries.

Over 1300 MW of wind-electric capacity has already been installed in Germany and more than 1000 MW is on-line in Denmark. The Danish goal is to provide 10 % of its electricity consumption through wind-electric energy by 2005 and more than 40 % by 2030. At about 4 US cents per kW of installed power, electricity from Danish turbines now costs around the same as the average cost for electricity from coalfired power plant. However, there is no such thing as a single price for wind energy as the costs depend on both wind speed and the accessibility of sites. Wind-electric energy has the potential to supply 25 % of Europe's electricity needs. Some countries could also export power to neighbouring countries.

1.6.4 Photovoltaics (PVs)

Potential applications of PVs range from basic electrification for the 2 billion people of the world without electricity to the integration of PVs in building structures in developed, urban areas. Customers need complete systems of PV modules, panels and arrays to provide electricity appropriate to their needs. Improved light-to-electricity conversion efficiency of individual cells is less important than reliable, integrated systems. The flexible thin-film amorphous silicon panel is at the forefront of PV technology. Distributed generation with PVs has been tested to relieve substation overheating and as a means to defer transmission or distribution system upgrades. Remote locations in developed countries are also practical applications for PVs. Examples include water pumping, fence electrification, and radio station power supply. PVs one of the most flexible technology supply options available for electric power production because they can supply loads from several watts to megawatts.

1.6.5 Solar

More than 350 MW of electricity are generated by commercial solar-thermal power plants in the USA. To exploit solar-thermal power fully, broader cooperation between government, electric utilities and private industry is needed. The major investments needed to develop and market solar technology must be supported by stable long-term regulatory policies, which can only be provided by government. For example, in the UK recent studies point to the need of tax equity to improve the economic competitiveness of solar-thermal plants more than technological breakthroughs.

1.7 Combined Heat and Power

World concern over carbon emissions, new domestic pollution regulations, improving small-scale technology, and the prospect of open competition for energy markets are forces that converge to demand greater efficiency in energy generation - to lower fuel costs, increase marketable products and reduce emissions. These forces argue strongly for a new paradigm of dispersed, combined heat and power (CHP) plants that have double the efficiency and produce half the pollution. Although large units will continue to operate in the short term, most will eventually be replaced by new facilities and virtually all new growth will come in the form of small units.

Readily available technologies now exist to combine the generation and supply of heat and power. By capturing unused heat energy, generators and consumers can, in effect, use the same fuel twice. Combining heat and power production reduces the net fuel demands for energy generation by supplying otherwise unused heat to residential, commercial and industrial consumers who have heating and air-conditioning needs.

CHP technologies can be widely implemented. In almost every case, such technologies will save enough money, now spent on fuel, to pay for their capital cost. By combining heat and power production and supply, 80 to 90 % of the useful energy in fuel can be put to beneficial use. When these plants extract steam from the turbines at relatively low pressure to drive industrial processes or provide heat, they lose some electricity production, but capture all of the heat, eliminating the use of other fuel to make this heat. Total efficiencies can reach 90%, depending on how well the electric and thermal needs are matched or balanced. CHP takes energy from a central electric plant and distributes it to end users as steam, hot water and chilled water using piping networks.

An increase in efficiency of 1% would result in a 2.5% reduction in CO_2 emissions. A UK study suggests that half of the CO_2 savings required up to 2010 can be met most cost-effectively with CHP. CHP can reduce fuel use, cut emissions and save money. Policy makers should take affirmative steps to encourage use of CHP. The technology is readily available, has a net economic benefit and can cut fuel consumption and pollutant emissions in the energy supply industry in half. There are many ways in which replacing separate heat and power generation with CHP systems can reduce emissions significantly. For example, producing 1 kWh of electricity, and a given amount of heat, from hard coal in a CHP system can reduce emissions by almost 30% compared with producing both separately from the same fuel. Using natural gas in the CHP system can reduce emissions by almost two-thirds compared with generating the heat and power separately from coal.

CHP meets energy needs and can save money for a wide range of energy customers - including public sector users - and also helps preserve the earth's precious energy resources, reducing the impact on the environment of harmful pollutants. The CHP shares of European power generation range from about 34% in the Netherlands to about 6% in Sweden, suggesting scope for large increases in some countries. Energy market deregulation could produce more favourable conditions for CHP, by increasing investment, innovation and market entry, and decreasing the costs of backup power and natural gas.

However, the high capital costs of these systems may deter further such investments under deregulated markets.

1.8 Energy Policy and Government Intervention

There is always going to be some government intervention in energy policy to ensure consumers' interests are protected, demand is met and competition is fair. In some European countries and the USA there are special provisions for electricity generation from renewable energy sources. In Italy, for example, new legislation requires that from 2001 all generators and importers of electricity will have to supply into the system a quota generated by renewable sources [19]. The EU directive allows member states to balance competition with public services where this is necessary in the general interest of the society, provided they comply with Community law. Examples could be an obligation for the customers to purchase a certain percentage of electricity from renewable energy sources or an obligation for distributors to supply all customers in their area at an equal price per kWh [20].

CHP has always been good value, and now it is even more so, with the UK government's decision to exempt good-quality CHP from the Climate Change Levy, which starts in April 2001. This exemption will apply to electricity generated from good-quality CHP and used on site or sold directly to other businesses. The government believes with the development of a fair and appropriate fiscal and regulatory framework, through the levy exemption and other measures such as negotiated agreements with industry, the cost-effectiveness of renewable generation and efficient CHP will be increased. This should deliver substantial increases in CHP capacity in the coming years. It should enable the government to announce, in the coming months, a new CHP target of around 10 GW by 2010 as part of the draft Climate Change Programme that would represent more than double today's CHP capacity. Action by the UK government and the EU would be essential to provide a market environment with incentives and penalties designed to ensure that the new technologies become available at competitive cost and in ample quantity. For distributed generators, there have been concerns about treatment of distributed generation by public electricity suppliers (PESs), especially distributed generation that they do not own. Under the new arrangements a distributed generator owned by a PES will be required to enter into formal arrangements with the distribution business in the same way as any other distributed generator. The same requirement to publish the agreements should minimise the risk of the PES-owned generator being treated in a more favourable way than others.

The political decisions set the economic framework in which the energy industry networks will determine success or failure in meeting the target. Private developers will install the CHP and the renewable energy plant if they see a return for their investment. If the necessary developments are to happen, unpopular measures will be required, such as taxes on emissions, incentives for the development of suitable installations, and the relaxing of restrictions imposed by planning regulations. In order to meet the new obligation a supplier can either supply the required amount of renewable electricity, or buy certificates. Any supplier who fails to meet the obligation will be required to make a payment. The UK government has recently announced the basis for its new renewable energy support mechanism. Suppliers will be able to meet their obligation either by purchasing renewable energy or by purchasing tradable green certificates. Alternatively, suppliers will be able to buy out all or part of their obligation. The buyout price will effectively cap the total cost of meeting the obligation and the associated increase in costs passed through to the end user. In addition, the provision of a guaranteed market for renewable sources at premium prices via the NFFO and also the Department of Trade and Industry's New and Renewable Energy Programme has resulted in a steady growth of renewable generation capacity.

There are a number of legislative and policy developments currently in hand that will impact on distributed generation and influence its growth. The Utility Bill is aimed at putting the customer first. The Bill will introduce important changes to the 1989 Electricity Act. These changes will include the introduction of new trading arrangements for buying and selling electricity, separation of the PES supply and distribution businesses, and a new obligation on suppliers to meet targets on renewable electricity. All of these changes affect or have implications for some if not all distributed generators. In government, each department has a 'green minister' with responsibility for ensuring that energy efficiency targets are met. Targets have been set in some departments for sourcing energy from renewable sources (such as wind) rather than conventional generation.

1.9 Costs

Deregulation has led the electricity industry to focus attention on the costs of generation and provides incentives for generators to reduce their costs and minimise their risks, e.g. by investing in smaller scale plants. Capital costs, construction time, fuel costs, operation and maintenance costs will determine the decision on what plants are built.

1.9.1 Capital Costs for New Plants

Capital costs depend on the specific site as well as the specification (size, operational flexibility, reliability, environmental performance, safety requirements, etc.). Costs will be much higher for a power plant built on the greenbelt compared with one built on an old plant site where parts of the existing infrastructure can be used. Plants close to the sea can use water for cooling purposes and avoid costs for cooling towers. It is dangerous to compare quotes from different sources as each project is site specific. Broadly speaking the capital costs of a gas-fired plant can vary from US$300 to 700/kWe. New coal-fired projects can cost from US$900/kWe for a pulverised coal combustion system to US$1400/kWe for a more advanced coal-fired system such as pressurised fluidised bed combustion [21]. Recent indicative costs as low as US$700/kWe for large pulverised coal combustion power stations have been quoted. Costs of advanced coal-fired plants have been decreasing and are expected to continue to do so as these relatively new technologies are successfully demonstrated. A gas-fired plant can be constructed and ready to operate in 18~20 months whereas a coal-fired plant would take 30~36 months. The payback period is also important for private investors. A coalfired project can fully pay back in up to 20 years whereas a gas-fired plant may take as little as 10 years [21]. From the investor's point of view gas is the more attractive option, all other factors being equal. Lifetime, however, is shorter for gas-fired plants. Most of them need significant refurbishment expenses, corresponding to the replacement of major plant components after 20~25 years, whereas coal-fired plants can reach up to 30~40 years of life.

Although gas-fired technology is cheaper in US$/kWe terms there are other factors that should be taken into account. Natural gas is not available in every country and prices are not always competitive. Moreover the infrastructure to produce and transport gas is much more capital intensive than the equivalent costs for coal. As discussed by Broadbent [22], if upstream capital costs are considered in the competitiveness of gas-fired compared with coalfired plants then the capital expenditure associated with both technologies could be the same. The high costs of the pipeline network to transport gas can outweigh the difference in capital costs for plant construction. If in place, the electricity generator can benefit from the infrastructure and build cheaper gas-fired plant. However, as demand for natural gas is growing new infrastructure will be needed. It is estimated that to cover the incremental demand of natural gas in Europe (1.7% annual growth from 1999 to 2015) investments in infrastructure of US$100~200 billion will be required [23]. Such investments could be undertaken only in the framework of long-term contracts and it is uncertain whether they could be profitable in competitive electricity and gas markets.

1.9.2 Technology Advances – Clean Coal Technologies

Clean coal technologies is a term used for technologies that achieve a higher efficiency and lower emissions for converting thermal energy to electricity than conventional pulverised coal combustion (PCC) with subcritical steam and without emissions control. The term is also used to include emission control systems such as NO_x control equipment. Clean coal technologies are the way forward for coal as they can ensure compliance with the tightening environmental standards. There has been considerable effort to develop these technologies at competitive costs. Reserves of coal are large and widely distributed and are likely to continue to be widely used, so more efficient and cleaner coal technologies (CCTs) are an important option in a future energy strategy. CCTs will enable the use of coal with higher energy efficiency and minimum environmental impacts. There are four main types of coal technologies applicable to large-scale power generation: supercritical steam PCC technologies with emissions control equipment (SPCC), atmospheric circulating fluidised bed combustion (CFBC); pressurised fluidised bed combustion (PFBC); and integrated gasification combined cycle (IGCC) technologies.

The status of these technologies today and a comparison of the costs of the different technologies with gas-fired power generation in various countries are given in [4]. Trends in planning and construction of plants using these technologies worldwide are presented in [24]. CCTs can also be used to repower existing coal-fired power stations approaching the end of their lifetime, instead of building new plant, and therefore reduce overall costs. Retrofitting pollution control equipment is also important as future and existing coal-fired plant may need to meet increasingly stringent environmental standards.

1.9.3 Environmental Considerations

Energy is one of the most critical resources for modern society. At a global level, it is expected that energy consumption will at least double in the next 50 years and grow by a factor of up to five in the next 100 years. At present levels of consumption, our use of energy poses threats to the climate, with potentially severe environmental consequences; given the levels of consumption likely in future, it will be an immense challenge to meet the global demand for energy without unsustainable long-term damage to the environment. This situation has attracted the attention of political leaders across the world, and at the Kyoto meeting of the parties to the UN Framework Convention on Climate Change in December 1997 there was agreement to tackle one aspect - the amount of greenhouse gases emitted to the atmosphere. The levels of atmospheric CO_2, for example, have increased from 285 ppm before the Industrial Revolution to about 350 ppm now. It is now generally accepted that there is a strong case for acting to mitigate the threat of drastic climate change associated with the unrestrained continuation of this trend. The Kyoto meeting produced pledges by the industrialised nations to cut their GHG emissions, by 2012, to an average of 5% below the 1990 levels.

Deregulation could play a positive role by giving flexibility to different plants or even countries to trade emissions. In this way a generator could have a portfolio of plants including some using renewable energy and therefore meet overall environmental requirements. It could also help the development of less costly pollution control technologies. In the single European electricity market, however, where electricity will be traded between member states, it is not yet clear where to allocate emissions. It could be the country where electricity is produced or where it is actually used. This is particularly important in the view of commitments to reduce GHG emissions.

US environmental regulations have caused a major shift in demand for lower sulphur coal supplies. Since the 1990 amendment to the Clean Air Act, there has been a noticeable shift in coal use by generating companies in the USA towards lower sulphur coal.

Deregulation increases the opportunities for using CHP, since the power generated can more easily be distributed and sold. CHP units can supply both electricity and heat at the same time, achieving high efficiencies and therefore reducing emissions to the atmosphere compared with separate generation of electricity and heat. In all countries CHP can be economic on industrial sites or community heating schemes where there is already demand for heat. In deregulated markets industrial users can set up a small CHP plant on their sites to supply heat and sell on any surplus electricity to the local grid. Before deregulation this practice was either not allowed, or at least not encouraged in many countries [24].

There are two ways to reduce GHG emissions. One way is to increase our reliance on nuclear power; the other is to develop a wide range of alternative methods of extracting energy from nature. The nuclear option is clean and feasible but it is hard to see that public opinion would switch from its present hostility to the acceptance of a massive programme of construction of new nuclear power stations. The role of nuclear power is expected to decrease in Europe as the perception of its environmental and economic performance has substantially changed. In the 1970s nuclear power was regarded as a source of cheap and emissions-free electricity. High costs involved in decommissioning nuclear reactors and the unresolved issue of nuclear waste have changed the image of nuclear plants. Italy has phased out nuclear generation since the early 1990s after the Chernobyl accident. Germany decided in late 1998 to phase out nuclear power and is now discussing possible ways for implementation. The UK government has plans to start phasing out nuclear power in 2010. It is clear that the construction of new nuclear plants in Europe will cause public opposition and is unlikely to materialise, particularly in deregulated markets where such investments are not competitive, as they are too expensive. The contribution from nuclear power to the fuel mix is expected to decrease and will be replaced by other sources including coal.

Without nuclear power, the power system must evolve to deliver power in a decentralised manner. This will reinforce the need to ensure diversity from other sources.

In a deregulated market the environmental image of fuels and technologies is important. It can influence decisions taken by developers and politicians for new projects as well as by consumers. Competition in retail will certainly create more possibilities for electricity consumers to influence developments. Although cost and security of supply remain the main factors affecting customer choice, as environmental concern grows this choice may be influenced by the environmental consideration. In the UK a number of electricity suppliers are planning or have launched environmental tariffs for household consumers, selling electricity from renewable energy projects. An opinion poll in the UK found that 86% of consumers would prefer to buy electricity from renewable sources, but only 21% of those would be prepared to pay more for it. In California an energy supply company has launched a green energy scheme which gives customers the option to buy a part of or all their electricity from renewable energy sources [25].

In some countries there has been opposition to the construction of new coal-fired plants for environmental reasons. The poor environmental image of coal is associated with perceptions that derive from the polluting plants of an earlier generation. This image can have a severe impact on new projects. Even state-of-the-art plant equipped with emissions control technology where the residues are reused in building materials can come under criticism. Therefore the promotion of CCTs and their excellent environmental performance is important, to ensure a role for coal plants in the future [26].

1.10 Distributed Generation

Deregulation and separation of the generating function have changed. Entrepreneurs will decide when and where to build new capacity, and, if their judgement is wrong, they will bear the losses. They will face an additional unknown that is, what competing entrepreneurs may decide to do. To minimise their risks, they are almost certain to select generation types with short construction periods and low unit capital costs. For this reason, wherever gas is available at a reasonable price, it is expected that new generation in any deregulated jurisdiction will be gas fired. Nuclear, hydro and coal or oil-fired base-load units would put too much capital at risk compared with the potential rewards [27]. Most alternative energy sources fall in the same category, but they can be subsidised.

Smaller scale generation technologies can undercut central station utility costs and have helped inspire today's competition. Several new technologies are being developed and marketed for distributed generation. They range in capacities from a few kilowatts to 100 MW and include microturbines, fuel cells, photovoltaic systems, wind energy systems, diesel engines, gas turbines and battery storage. Trends are moving towards smaller scale generation producing power near or at the customer's premises and transporting it over shorter distances at lower voltages. With stand alone generation units, customers are less susceptible to the cascading power failures that affect thousands or millions of power users, making them essentially immune from 'reliability' concerns. Distributed generation technologies can also exploit a market opportunity in the newly emerging regulatory environment, namely micro grids. A micro grid is an electrically isolated set of generators that supply all of the demand of a group of customers [28-40].

Many studies indicate that distributed generation (DG) might play a significant role in the future power system structure. A study by the Electric Power Research Institute (EPRI), for example, indicates that by 2010, 25 % of the new generation will be distributed [41]. Owing to variations in government regulations, different definitions for DG are used in different countries. In England and Wales, the term 'distributed generation' is predominantly used for power units with less than 100 MW capacity. In Sweden, DG is often defined as generation up to 1500 kW. In Australia DG is often defined as power generation with a capacity of less than 30 MW. In New Zealand, DG is often considered as generation up to 5 MW. There is no special definition of DG in the Californian and Norwegian electricity markets.

A general definition for DG could be an electric energy source connected directly to the distribution network or load centre. DG is decentralised and located closer to the point of use, making greater economic and environmental sense. Several main reasons have combined to make DG a technically, commercially, environmentally and, to an extent, politically attractive proposition.

Customers benefit from the success of DG because:

- The use of distributed energy will allow improvements in the dispatchability of resources and improve the integrity of the transmission and distribution systems.
- Identification and use of alternatives to power generation, transmission and systems controls will improve load levelling, load management and overall power quality.
- The system will become more robust in its ability to tolerate natural disasters, suffer less damage and minimise the dependence upon the need for immediate restoration of the grid system.
- Overall system reliability will improve.

To get a better understanding of the possible future development of DG in a competitive market, some examples of typical DG applications are as follows:

- Renewable energy technologies, e.g. wind power or solar power. These projects might receive certain subsidies, or customers might pay premium prices for renewable energy.
- Peak supply systems, based for example on emergency generators or on-site generators. Such systems typically sell to the power exchange for only a very short period per year to capture extremely high peak prices.
- CHP systems, e.g. district heating, whereby a high efficiency can be achieved and additional revenue from selling heat can be obtained.
- On-site generation based on microturbines or fuel cells. Electricity as well as heat are most likely to be used locally.

1.10.1 Market Regulation

In competitive power markets, DG competes with centralised power generation. Hence, market regulations should ensure that DG can act freely within power markets, similar to centralised generation. It is, however, often argued that most market regulations used worldwide have been designed with large centralised generation in mind and that, therefore, DG often faces significant barriers within the competitive market.

1.10.2 The Power Pool

The power pool is used to create an efficient marketplace for trading electricity. The power pool is usually operated by a centralised, independent organisation that defines the standards for electricity price bids and the evaluation of these bids, as well as organising the bidding and evaluation procedure. The evaluation of power pool regulations regarding the treatment of DG is a very complex issue.

The main difference between various approaches for electricity markets is that the trading of electricity through a power pool (or power exchange) is optional in some countries, e.g. in Nord Pool (Scandinavia), and mandatory in others, e.g. England and Wales as well as in the National Electricity Market in Australia. In California, the participation in the pool market is optional, except for three large private utilities. They have to trade through the power exchange until the year 2002.

The reason for a regulator to set up a mandatory pool system instead of an optional market is usually to achieve a high market transparency, e.g. to prevent some large generators from gaining market power. In general, all market participants will benefit from a transparent power market however, other options are also possible to prevent large generators getting market power, e.g. by splitting up the generators as was done in New Zealand. The disadvantage of a mandatory pool approach is that all market participants have to join the pool. That leads to various fixed costs, e.g. membership fees, and/ or energy fees. Both fees are a way to recover the cost for the operation of the power pool. The membership fee is usually a fixed annual fee and the energy fee is based on the energy actually traded via the power exchange. These costs may be a major barrier for independently owned generation companies that focus on DG to enter the electricity market. Therefore, exceptions to the mandatory rule were included in the regulations in England and Australia for small-scale generation. The exceptions depend on installed capacity (30 to 50 MW) of the DG source. However, there is no obvious reason for a DG source with a capacity of 25 MW to be treated differently from one with a capacity of 55 MW. Furthermore, technical limitations in a distribution network may be different in a rural and an urban distribution network. Hence, regulations based on a certain installed capacity influence the way certain market participants to behave.

The cost problem for participating in the pool market, however, remains, even if certain capacity limits are removed. This issue is of particular interest for DG concepts that aim at providing peak power generation, probably for only a few hours per year. To capture the high prices during extreme load/price peaks, these distributed generators must participate in the power exchange. Therefore, high annual fees can be seen as a major barrier for distributed generators to participate in a power market. As a solution, the cost recover for the operation of the pool exchange should mainly be based on energy fees. In addition, it should be mentioned that within the national electricity market in Australia distributed generators have to sell all generated power within the distribution network [42]. This kind of definition significantly reduces the market opportunities of small-scale generation. With regard to DG, the treatment of the individual imbalance of each market participant is particularly important for fluctuating power sources, such as wind or solar power. Such technologies have the disadvantage that the power output during an upcoming time period, e.g. the next 12 hours, can only be predicted with some uncertainty for each wind turbine or each wind farm.

In the UK, there are three main problems associated with the pool price:

1. Most energy contracts effectively bypass the pool.
2. Since system marginal price is paid to all, it is mathematically impossible for bid prices to be meaningful.
3. Average pool prices bear no relation to any real price parameter. The price of fuel and hence of generation has been falling steadily since 1990. Pool prices, by contrast, rose until about 1994, steadied, and now seem set on an upward path again.

Figures 1.3 and 1.4 show the pool and its problems respectively. Pool prices are related to centralised generation, but most renewables - and all DG, by definition - are injected into low (33 kV or below) voltage networks. So electricity at the retail level is being valued at wholesale prices, which is wrong. The characteristics of pool pricing are marginal pricing, complexity of bids and pool capacity payments. These lead to slow progress in change and decision making, lack of transparency in contracts for differences, concern over prices and a reduction in consumer confidence. As a result, new electricity trading arrangements (NETA) were proposed [43].

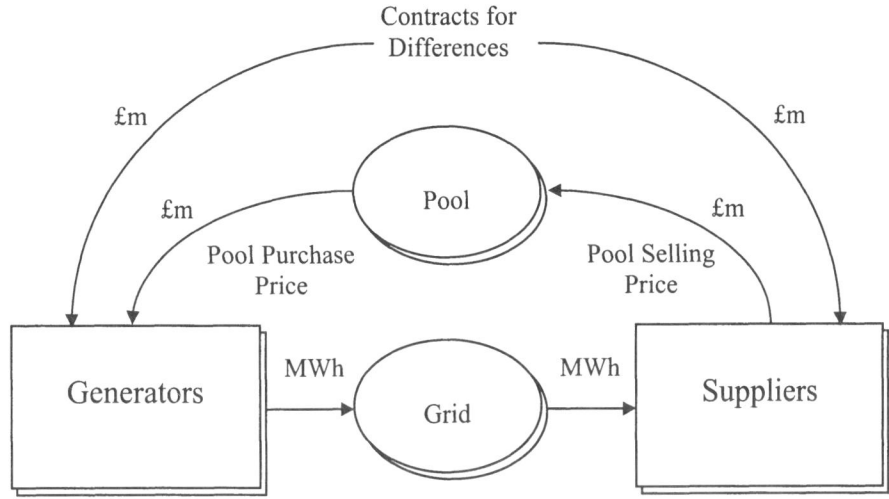

Figure 1.3 The pool

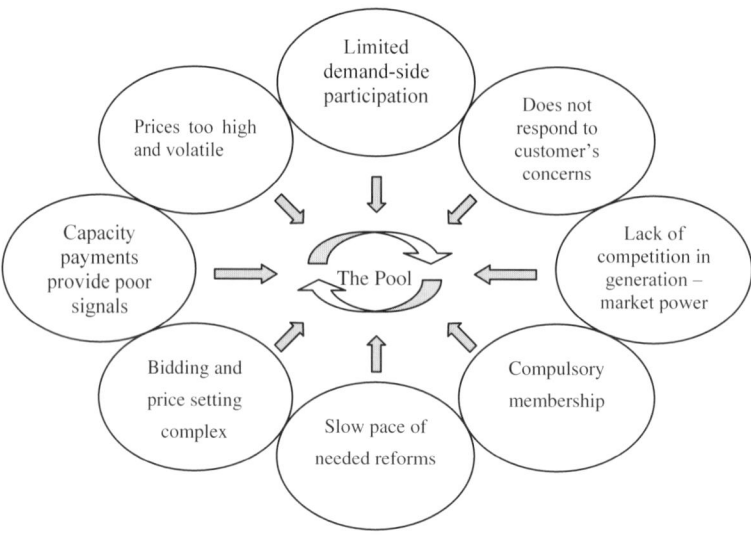

Figure 1.4 Problems with the pool

The key principles of NETA are:

- Freedom of action - 'self-despatch'
- Pay as bid and firm commitment
- Single price (no capacity element)
- Security of supply maintained
  - Long term via market economics
  - Short term via NGC balancing actions.

The expected benefits of NETA are:

- Lower prices from more efficient and competitive trading
- Focus on bilateral contracts and firm trades
- More direct demand-side participation
- Sharper incentives to manage risks
- Greater choice of markets
- Increased transparency and price discovery
- More efficient and effective governance.

The National Grid Company acts as the system operator and will operate and manage the system via the balancing mechanism and contracts for balancing services.

The government's reforms to the electricity market, of which NETA is an important part, will lead to lower electricity prices - at least 10 % lower at the wholesale level over the medium term. A fundamental change in the market will have a big impact on different buyers and sellers in the electricity market. The reformed market will encourage generators to deliver what they say they will. The introduction of the new arrangements will have different effects for different market participants. Some distributed generators, particularly renewable and CHP generators, are concerned since the arrangements will favour generators with flexible and predictable output and those generators with inflexible and unpredictable output will face exposure to imbalance charges. Therefore, not every CHP scheme will benefit from the NETA reforms. However, the position of renewables and CHP generation needs to be looked at in a wider context.

CHP, renewable and other distributed generators are likely to grow significantly in coming years and the government has, therefore, paid careful attention to aspects of the new electricity market that may adversely affect the economics of DG. It is important to ensure that distributed generators, including CHP, obtain access to the electricity distribution market on fair terms. As part of NETA, arrangements that will enable distributed generators to manage their risks and achieve fair prices have been developed. There are other proposals too, to deal with the needs of CHP and other distributed generators.

1.10.3 Ancillary Services

Ancillary services are those functions performed to support the basic services of generating capacity, energy supply and power delivery. The costs for ancillary services can be significant; for example, in the USA the total costs for ancillary services are about US$12 billion a year.

The general approach for pricing ancillary services within competitive electricity markets is based on fixed contracts for a certain time period between the independent system operator and market participants that are able to provide the required ancillary services. Thereby, the ancillary services are split up into different services, e.g. active reserves, reactive reserves and system restart. In regard to DG, active reserves and reactive reserves are of particular interest. Many DG approaches aim at participating in the ancillary service market to get an additional income just like 'traditional' generators. However, often DG is not regarded be able to maintain the 'right' level of reliability and quality and at the same time as being cost competitive.

The contracts for different ancillary services in competitive electricity markets are usually awarded in tender processes. However, new technologies, such as DG, usually face significant barriers within specific bidding processes, as the tender is often only open to traditional technologies. This assumption is confirmed by the Norwegian regulated power market, since it is only open to generators with a minimum installed capacity of 40 MW.

1.10.4 Technical Issues

The generation of electricity by distributed generators within distribution networks raises a number of technical issues concerning real and reactive power flows, voltage regulation, fault levels and power quality [17,30,44].

Load Flows
Most distribution networks operate on a radial or open-ring basis, and so have been designed broadly on the principle that load flows only in one direction and generally reduces along the length of each distributor. However, the output of a single connected distributed generator can effectively reverse the traditional direction of load flow at a given point on a distributor or interconnected network and this could affect conventional automatic voltage control schemes which cater only for conventional load profiles. Also the design of protective relaying systems is much more complicated because currents are going both ways.

Reactive Flows
Most small distributed generators, such as the majority of wind generators and small-scale CHP systems, are based on induction machines which have no steady-state reactive power generation capability. There is a need to import reactive power to provide field excitation.

Fault Levels
Injection by generators, particularly synchronous generators, can lead to localised increases in fault-level which can potentially exceed the short-time ratings and making ratings of switchgear.

Power Quality
Potential problems arise with systems using inversion (e.g. PVs and fuel cells) which can create harmonic distortion in the supply voltage waveform. Problems can also arise with systems subject to rapid torque changes within the prime mover (e.g. wind generators).

1.10.5 Implications and Opportunities for Network Operators and Generators

From a PES perspective, the effect of DG is that networks will become more active in nature and less predictable in behaviour. From a generator's perspective, although it may perceive that it is adding to overall capacity, the characteristics of generators and networks may not always be compatible, and network constraints could result in generators having to operate at less than optimum outputs at certain times [45,46].

Increased use of CHP and co-generation will result in lower usage of the network in terms of energy transportation and, therefore, potentially lower levels of income. The overall effect of DG might be to increase rather than reduce capital expenditure, whilst the effect of managing and controlling a more complex and increasingly active network might also be to increase operating costs.

Generators suitably located may also offer benefits to a distributor by, for example, offsetting the need for reinforcement or provision of other services such as voltage support. In these cases payments to generators will be substituting for other expenditure by the distributor.

1.10.6 Connection and Use of System Charges

Suppose that there is a need to replace a circuit breaker as the connection of the new generator increases the fault level. It is important to stress that this circuit breaker is installed because of all generators. The contribution of each generator can be readily computed using conventional short-circuit analysis tools. These contributions to the short-circuit current then may be used to allocate the cost of replacing the circuit breaker. It is important to emphasise that an argument such as there would not be a need to replace the breaker if the new generator did not appear, cannot be credibly used to require the new entry to recover all system reinforcement cost. In this case, the distribution network owner replaces the circuit breaker, and in the following price review period increases the use of system charges accordingly to all generators with respect to their contribution in order to recover the system investment.

The derivation of charges for assets that provide the connection of a discrete plant to the system should be differentiated from those for the use of the system. In the former case the asset is provided for a sole user and could have been financed directly, and even owned, by that user. In this instance charges should be based on the historic cost of the asset and a fair return on the cost of the capital provided by the distribution company. In the latter case the assets are used by a number of system users, past, present and future, and charges should be on the basis of a tariff differentiated by voltage. The difficulty arises as to how reinforcement costs of the infrastructure of the system should be treated when a new user joins. There is also a difficulty with the costs of stranded infrastructure assets when an existing user departs.

1.11 Case Study1: Phase Balancing for a Self-excited Induction Generator [50]

1.11.1 Introduction

In regions where renewable energy resources are abundant but usually situated in remote locations, connection to the central power grid is expensive and in many cases difficult to provide. Small-scale, autonomous generation schemes, on the other hand, are both economical and practicable. They utilise the energy resources available and supply the consumers in the local regions. The system cost can be reduced by using cage-type, self-excited induction generators (SEIGs) [47-52] since these machines are cheap and readily available.

Autonomous power systems often employ single-phase generation and distribution schemes for reasons of low cost, ease of maintenance and simplicity in protection [53]. When a three-phase SEIG is used to supply single-phase loads, however, the stator currents are seriously unbalanced, causing degradation in generator performance such as overcurrent, overvoltage, poor efficiency and machine vibration. These undesirable effects can be alleviated to a certain extent by the use of the Steinmetz connection [54] in which the excitation capacitance and load are connected across different phases. For isolated operation, however, perfect phase balance cannot be achieved when the load is purely resistive.

The objective of this case study is to introduce a modified Steinmetz connection that enables perfect phase balance to be achieved in a three-phase SEIG which supplies single-phase loads. A general performance analysis is presented and experimental results are given to validate the principles.

1.11.2 Circuit Connection and Principle

Figure 1.5 shows the modified Steinmetz connection (MSC) for a delta-connected SEIG, which supplies a single-phase load. It is assumed that the rotor is driven in such a direction that it traverses the stator winding in the sequence A-B-C, i.e. in the same direction as the positive-sequence rotating field. Hence, if *A*-phase is taken as the reference phase, *B*-phase is regarded as the lagging phase. The main excitation capacitance C_2 and the auxiliary load resistance R_{L2} are connected across *B*-phase (the lagging phase), while the auxiliary excitation capacitance C_1 and the main load resistance R_{L1} are connected across *A*-phase (the reference phase). Compared with the original Steinmetz connection [54], it is noticed that the auxiliary load resistance R_{L2} and auxiliary excitation capacitance C_1 have been introduced. These circuit elements provide additional current components that result in the flow of balanced line currents into the SEIG.

In a practical autonomous power system, the auxiliary load resistance R_{L2} can be local loads such as lighting, storage heating or battery charging. Alternatively, it could be a portion of the remote loads.

For the purpose of analysis, all the circuit parameters in Figure 1.5 have been referred to the base (rated) frequency f_{base} by introducing the per-unit frequency a and the per-unit speed b [55]. Thus, each voltage shown in Figure 1.5 has to be multiplied by a in order to give the actual value and the per-unit slip is equal to $(a - b)/a$. Besides, the motor convention has been adopted for the direction of phase and line currents.

The phase-balancing capability of the MSC for a three-phase SEIG may be studied by reference to a voltage/current phasor diagram. It is assumed that the values of C_1 and C_2 are sufficiently large so that the SEIG has built up its voltage and is supplying the loads. Figure 1.6 shows the phasor diagram for the SEIG under balanced conditions. Because the stator winding is delta connected, the line currents I_1, I_2 and I_3 lag the corresponding phase voltages V_A, V_C and V_B by $(\phi_p + \pi/6)$ rad, where ϕ_p is the positive-sequence impedance angle of the SEIG.

The line current I_2 is contributed by the current I_{C2} through C_2 and the current I_{R2} through R_{L2}. Meanwhile, the line current I_1 is contributed by $-I_{C1}$ (where I_{C1} is the current through C_1) and $-I_{R1}$ (where I_{R1} is the current through R_{L1}). It can be shown that the angle γ between I_{C2} and I_2 is equal to $(\phi_p - 2\pi/3)$ rad, while the angle δ between $-I_{R1}$ and I_1 is $(5\pi/6 - \phi_p)$ rad. The phasor diagram in Figure 1.6 can be drawn only when I_{C2} leads I_2, which implies that perfect balance can be achieved for values of ϕ_p exceeding $2\pi/3$ rad.

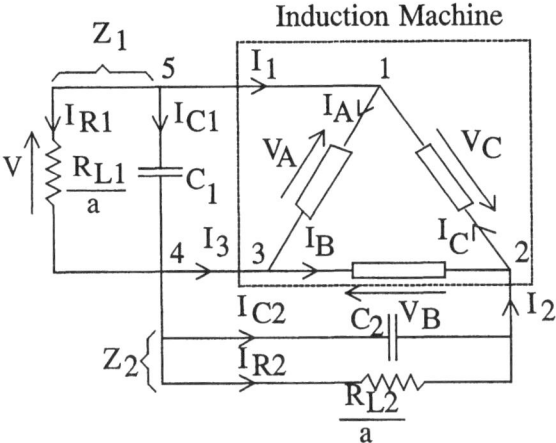

Figure 1.5 Modified Steinmetz connection for three-phase SEIG

From the current phasor triangles in Figure 1.6, the following relationships can be deduced:

$$G_1 = \sqrt{3}\,|Y_p|\cos(5\pi/6 - \phi_p) \tag{1.1}$$

$$B_1 = \sqrt{3}\,|Y_p|\sin(5\pi/6 - \phi_p) \tag{1.2}$$

$$G_2 = \sqrt{3}\,|Y_p|\sin(\phi_p - 2\pi/3) \tag{1.3}$$

$$B_2 = \sqrt{3}\,|Y_p|\cos(\phi_p - 2\pi/3) \tag{1.4}$$

where, $G_1 = a/R_{L1}$, $B_1 = a^2.2\pi f_{base}.C_1$, $G_2 = a/R_{L2}$, and $B_2 = a^2.2\pi f_{base}.C_2$.

For a given total output power, (1.1) to (1.4) can be used to determine the values of the load and phase converter elements required for perfect phase balance, provided that Y_p, ϕ_p and a of the SEIG are known.

Equation (1.2) shows that B_1 vanishes when $\phi_p = 5\pi/6$ rad, which implies that the auxiliary capacitance C_1 can be dispensed with. When ϕ_p exceeds $5\pi/6$ rad, B_1 becomes negative, implying that perfect balance can be achieved with an auxiliary inductance. In practice, however, the full-load power factor angle of an SEIG ranges from $2\pi/3$ rad to $4\pi/5$ rad, and hence it is very unlikely that an inductive element need be used.

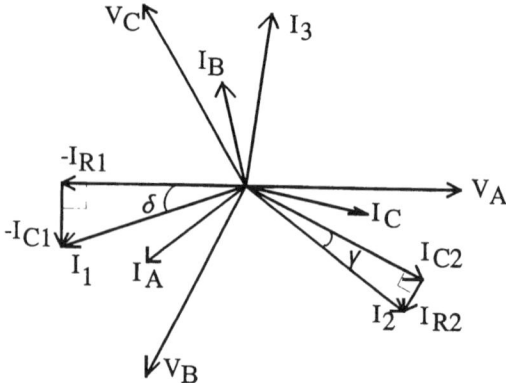

Figure 1.6 Phasor diagram of SEIG with MSC under balanced conditions

1.11.3 Performance Analysis

A general analysis of the SEIG with MSC can be carried out using the method of symmetrical components. All the equivalent circuit parameters are assumed to be constant except the magnetising reactance, which is a function of the positive-sequence air gap voltage. With reference to Fig. 1.5, the following 'inspection equations' [56] may be written:

$$V = V_A \tag{1.5}$$

$$V_A + V_B + V_C = 0 \tag{1.6}$$

$$I_1 = I_A - I_C = -\frac{V}{Z_1} = -V\,Y_1 \tag{1.7}$$

$$I_2 = \frac{V_B}{Z_2} = V_B\,Y_2 \tag{1.8}$$

where,

$$Y_1 = \frac{1}{Z_1} = G_1 + jB_1 = \frac{a}{R_{L1}} + j\,a^2 .2\pi\,f_{base} \cdot C_1 \tag{1.9}$$

and

$$Y_2 = G_2 + jB_2 = \frac{a}{R_{L2}} + j\,a^2 .2\pi\,f_{base} \cdot C_2 \tag{1.10}$$

Equation (1.6) implies that zero-sequence voltages and currents are absent in the SEIG. By solving (1.5) to (1.8) in terms of the delta system of symmetrical components [57], the positive-sequence voltage V_p and negative-sequence voltage V_n can be determined:

$$V_p = \sqrt{3}V \cdot \frac{Y_n + \frac{e^{j\pi/6}}{\sqrt{3}} Y_2}{Y_2 + Y_p + Y_n} \quad (1.11)$$

$$V_n = \sqrt{3}V \cdot \frac{Y_p + \frac{e^{-j\pi/6}}{\sqrt{3}} Y_2}{Y_2 + Y_p + Y_n} \quad (1.12)$$

where Y_p and Y_n are the positive-sequence and negative-sequence admittances of the SEIG.

The input impedance Z_{in} of the SEIG when viewed across stator terminals 1 and 3 (Figure 1.5) is given by

$$Z_{in} = R_{in} + jX_{in} = \frac{Y_2 + Y_p + Y_n}{3Y_p Y_n + Y_p Y_2 + Y_n Y_2} \quad (1.13)$$

Applying Kirchhoff's voltage law to loop 1345 in Figure 1.5,

$$I_1(Z_1 + Z_{in}) = 0 \quad (1.14)$$

For successful voltage build-up, $I_1 \neq 0$; hence

$$Z_1 + Z_{in} = 0 \quad (1.15)$$

Equation (1.15) can be solved for the excitation frequency a and magnetising reactance X_m. After a and X_m have been determined, the positive-sequence air gap voltage is found from the magnetisation curve. The generator performance can then be computed using (1.5) to (1.12).

1.11.4 Solution Technique

The input impedance Z_{in} as given by (1.13) involves the generator admittances Y_p and Y_n whose real and imaginary parts are high-order polynomials of a and X_m. As a result of the algebraic manipulations involved, both R_{in} and X_{in} in (1.13) are extremely complicated functions of the above two variables. Serious difficulties will be encountered when solving (1.15) using conventional techniques such as the Newton-Raphson method [47] owing to the lengthy mathematical derivations required. To overcome these difficulties, a function minimisation technique is employed in this case study for solving (1.15). This is based on the observation that, for given values of a and X_m, the input impedance Z_{in} can be computed readily.

The following scalar impedance function is first defined:

$$Z(a, X_m) = \sqrt{(R_1 + R_{in})^2 + (X_1 + X_{in})^2} \quad (1.16)$$

where R_1 and X_1 are respectively the equivalent series resistance and reactance of Z_1.

The solution of (1.15) is next formulated as the following optimisation problem:

For given values of load resistances, excitation capacitances and speed, determine the values of a and X_m such that the function $Z(a, X_m)$ is minimum.

It is obvious that $Z(a, X_m)$ has a minimum of zero and the corresponding values of a and X_m also satisfy (1.15).

Any optimisation algorithm that does not require the evaluation of function derivatives may be used for the above problem. In this study, the pattern search method [58] is used for function minimisation. The method employs two search strategies, namely exploratory moves and pattern moves, in order to arrive at the optimum point. A function evaluation is required each time an exploratory move or pattern move is to be made.

For normal operation of an SEIG, a is slightly less than the per-unit speed b whilst X_m is less than the unsaturated magnetising reactance X_{mu}. Accordingly, b and X_{mu} could in general be chosen as initial estimates for a and X_m for starting the search procedure. In practice, it was found that a smaller initial value for the variable a (say $0.97b$) would give more rapid convergence.

To simplify the calculations and for easy comparison, all the machine parameters are expressed in per-unit values using the rated phase voltage, rated phase current and rated power per phase of the induction machine as bases. Table 1.1 shows typical computed results for the experimental machine. The function minima obtained imply that very accurate solutions are possible. Over a wide range of load, the number of function evaluations N required to reach a solution varies from 350 to 450.

Table 1.1 Computed results for SEIG with MSC

R_{L1}	a	X_m	N	$Z(a,X_m)$
(p.u.)		(p.u.)		(p.u.)
1000	0.977193	1.2021	412	9.94e-6
10	0.975109	1.2205	402	7.73e-6
5	0.973059	1.2404	345	2.09e-6
2	0.967218	1.3084	377	3.50e-6
1	0.958454	1.4576	401	4.48e-7
0.5	0.944063	1.9230	449	1.88e-6

$b = 1.0$; $a_0 = 0.97b$; $X_{m0} = X_{mu} = 2.48$ p.u.
$C_1 = 47$ µF; $C_2 = 146$ µF; $R_{L2} = 2.3$ p.u.

1.11.5 Results and Discussion

To illustrate the phase-balancing capability of the MSC, experiments were carried out on a 2.2 kW, delta-connected induction machine whose equivalent circuit data is given in the Appendix. The speed of the SEIG was maintained at rated value ($b = 1.0$) and the values of R_{L1}, C_1, R_{L2} and C_2 were carefully adjusted until perfect phase balance was obtained. Typical results are given in Table 1.2. The good agreement between computed and experimental results confirms the principle of phase balancing for a three-phase SEIG using the MSC.

Table 1.2 Conditions for perfect phase balance in three-phase SEIG with MSC

V	I_{ph}	Y_p	ϕ_p	R_{L1}	C_1	R_{L2}	C_2
(p.u.)	(p.u.)	(p.u.)	(deg)	(p.u.)	(µF)	(p.u.)	(p.u.)
0.918	0.967	1.053	130.8	0.59	50	2.73	146
				(0.56)	(49)	(2.82)	(146)
0.835	1.037	1.214	134.7	0.51	49	1.78	168
				(0.49)	(46)	(1.87)	(167)
0.805	0.954	1.186	135.5	0.52	44	1.64	161
				(0.50)	(42)	(1.83)	(160)
0.796	0.789	0.992	133.8	0.62	41	2.26	136
				(0.61)	(39)	(2.44)	(136)

Normal: experimental values; bracketed: computed values

Figures 1.7-1.9 show the steady-state performance of the SEIG with phase converter elements fixed at the following values: C_1 = 47 µF, C_2 = 146 µF and R_{L2} = 2.3 p.u. It is seen that the SEIG is balanced at a load current (experimental value) of 1.52 p.u, which corresponds to a phase voltage of 0.86 p.u. and a phase current of 0.92 p.u. The total electrical power output is 1.63 p.u. (1940 W, or 88% of rated power), of which 80% is delivered to the main load R_{L1} while 20% is consumed by the auxiliary load R_{L2}. Under the above conditions, the performance of the SEIG is the same as if it were excited with balanced capacitances and supplying a balanced load. For loads close to the balanced operating point, an experimental efficiency of 80% can be obtained. Very good correlation between computed and experimental results is obtained; hence the validity of the symmetrical component analysis and solution technique is verified.

Figures 1.7 and 1.8 show that, when the values of the phase converter elements are fixed, the currents and voltages in A-phase and B-phase may exceed the rated values when the load is reduced, particularly when the SEIG has been balanced at heavy loads. One method to alleviate this undesirable effect is to balance the SEIG at part load (say 80% of full-load current). The performance of the SEIG will then be satisfactory between this load and full load. Another method is to balance the SEIG again at smaller loads, which involves multi-valued phase converter elements controlled by a simple switching strategy.

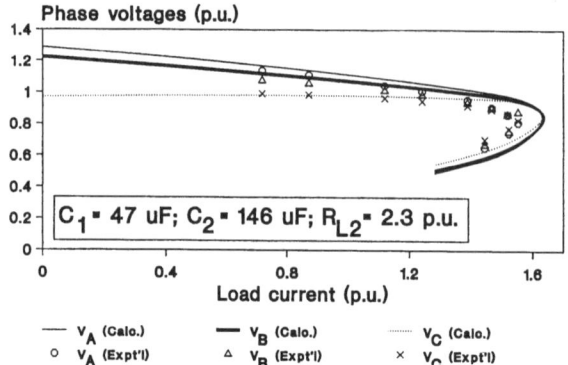

Figure 1.7 Phase voltages of three-phase SEIG with MSC

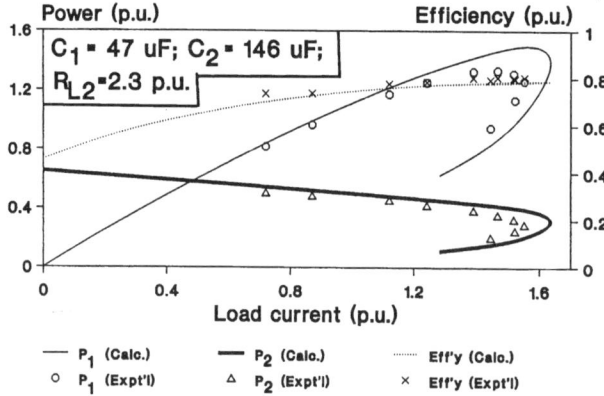

Figure 1.8 Phase currents of three-phase SEIG with MSC. P_1: output power to main load R_{L1}; P_2: output power to auxiliary load R_{L2}

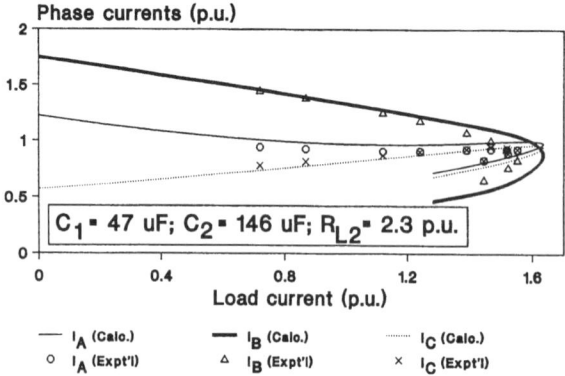

Figure 1.9 Output power and efficiency of three-phase SEIG with MSC

1.11.6 Simplified Phase-balancing Scheme

In circumstances where it is not practicable to provide auxiliary loads, or when auxiliary loads need not be supplied, the simplified Steinmetz connection (SSC) shown in Figure 1.10 may be employed. In this case, all the electrical power output of the SEIG is delivered to the single-phase load R_{L1}. The phasor diagram for the MSC (Figure 1.6) and the corresponding equations (1.1)-(1.4) may be used to identify the conditions for perfect phase balance for the SSC. Since the auxiliary load resistance R_{L2} is absent, the value of G_2 in (1.3) is forced to assume a zero value. Accordingly the positive-sequence impedance angle ϕ_p of the SEIG must be equal to $2\pi/3$ rad for (1.3) to be satisfied. From (1.1), (1.2) and (1.4), the values of the load conductance and phase-converter susceptances that result in a balanced operation of the SEIG are: $G_1 = 3Y_p/2$, $B_1 = \sqrt{3}Y_p/2$ and $B_2 = \sqrt{3}Y_p$.

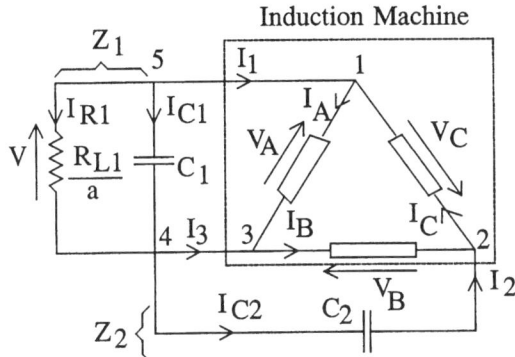

Figure 1.10 Simplified Steinmetz connection for three-phase SEIG

The auxiliary excitation capacitance C_1 is thus one-half of the main excitation capacitance C_2. By selecting proper values of C_1 and C_2, perfect phase balance can be obtained for a specific value of stator current.

Analysis of the SEIG with the SCC is similar to that for the SEIG with the MSC, except that the admittance Y_2 is now equal to $(0 + jB_2)$. Figures 1.11–1.13 show the computed and experimental performance of the SEIG with the SSC at rated speed. With main and auxiliary excitation capacitances fixed at 110 µF and 55 µF respectively, the SEIG is balanced at a load current (experimental value) of 1.13 p.u., which corresponds to a phase voltage of 0.985 p.u. and a phase current of 0.77 p.u. Under this condition, power of 1.11 p.u. (1320 W) is delivered to the load and the efficiency of the SEIG is 79.6%. Again very good agreement between the computed and experimental results is observed.

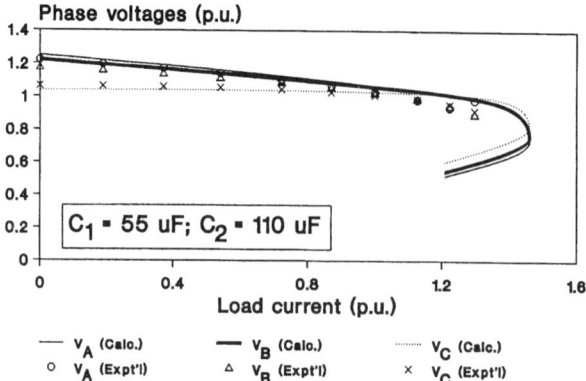

Figure 1.11 Phase voltages of three-phase SEIG with SSC

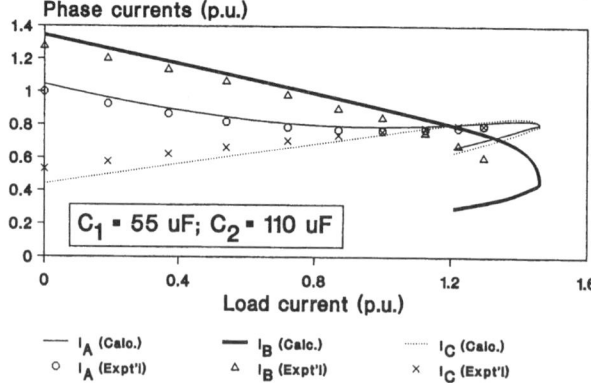

Figure 1.12 Phase currents of three-phase SEIG with SSC

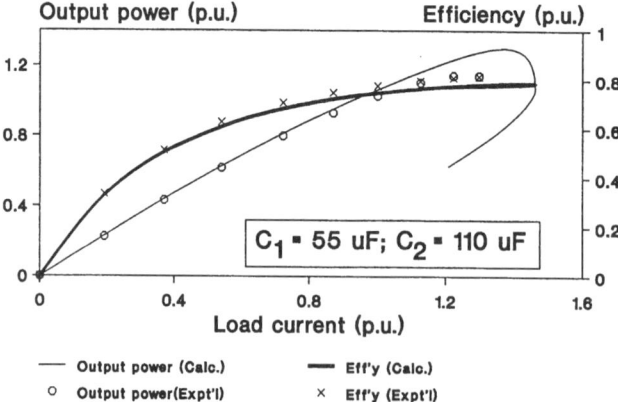

Figure 1.13 Output power and efficiency of three-phase SEIG with SSC

1.11.7 Appendix

The experimental machine has the following particulars:

2.2 kW, 220 V, 9.4 A, four-pole, 50 Hz, three-phase, delta-connected, squirrel-cage induction motor. The machine parameters (in per-unit values) are:

Stator resistance R_1	=	0.0844
Stator leakage reactance X_1	=	0.112
Positive-sequence rotor resistance R_{2p}	=	0.0621
Negative-sequence rotor resistance R_{2n}	=	0.0981
Rotor leakage reactance X_2	=	0.1
Core loss resistance R_c	=	22
Friction and windage loss P_{fw}	=	0.013

Variation of positive-sequence air gap voltage E_1 with magnetising reactance X_m is modelled by the following describing equations:

$$E_1 = \begin{cases} 1.345 - 0.203\,X_m, & X_m < 1.728 \\ 1.901 - 0.525\,X_m, & 1.728 \leq X_m < 2.259 \\ 3.156 - 1.08\,X_m, & 2.259 \leq X_m < 2.446 \\ 37.79 - 15.12\,X_m, & 2.446 \leq X_m < 2.48 \\ 0, & 2.48 \leq X_m \end{cases} \qquad (1.17)$$

1.12 Case Study 2: Controlling a Solar Power Plant [59]

1.12.1 Introduction

Whilst thermal heating technology may generally be accepted as being well developed, cost reduction is still expected to continue owing to larger scale production and advanced control techniques [60]. This is evident from the research effort devoted to improving the efficiency of solar-thermal power plants operating with distributed collectors, by means of various advanced control-methods [61-63]. These control methods, developed and tested at the solar power plant Plataforma Solar de Almería (PSA) in Spain, shown in Figure 1.14, bear a varying degree of both complexity and success. The key characteristic of a solar power plant is that its primary energy source, solar radiation, cannot be manipulated, and varies throughout the day and exhibits seasonal cycles. This results in significant variations in the dynamic characteristics of the collector field and the plant as a whole. Thus, conventional controllers such as the proportional-integral (PI) schemes which use a fixed set of system parameters optimised for a prescribed range of operations have proved to be inadequate. Recently, fuzzy logic control (FLC) schemes, which encompass the human-like approach of processing and handling of information, have been applied to the control of non-linear plants with promising results. However, early studies show that in such FLC schemes the optimisation of the 'if-then' rule base is often a cumbersome and laborious process involving 'trial and error'. Genetic algorithms based on the natural law of evolution lend themselves as an ideal optimisation tool to be used in conjunction with FLC systems. This study is one of the first of its kind to show the development of a GA-FLC scheme aimed at optimising the response time of a solar power plant to input power and temperature demand.

Figure 1.14 The solar power plant under investigation, Plataforma Solar do Almeria (PSA), in Almeria, Spain

1.12.2   The Solar Power Plant

Figure 1.15 shows the block diagram of the solar power plant. Solar energy is collected by a distributed collector field called the ACUREX field which consists of 480 distributed solar collectors arranged in 20 rows forming 10 parallel loops, with each loop being 172 m long. The oil is pumped through the receiver tube where heat transfer takes place through the receiver tube walls. The storage tank is filled with oil in the far end. The oil is heated and then introduced into the storage tank to be used for electrical energy generation or for feeding the heat exchanger of a desalination plant [64]. The system is provided with a three-way valve located in the field outlet that allows the oil to be recycled in the field until its outlet temperature is adequate for entering into the top of the storage tank. The objective of the control system in a distributed collector field is to maintain the outlet oil temperature of the loop at a desired level in spite of disturbances such as changes in the solar radiance level, cloud movement, mirror reflectivity or inlet oil temperature.

Energy Generation under the New Environment

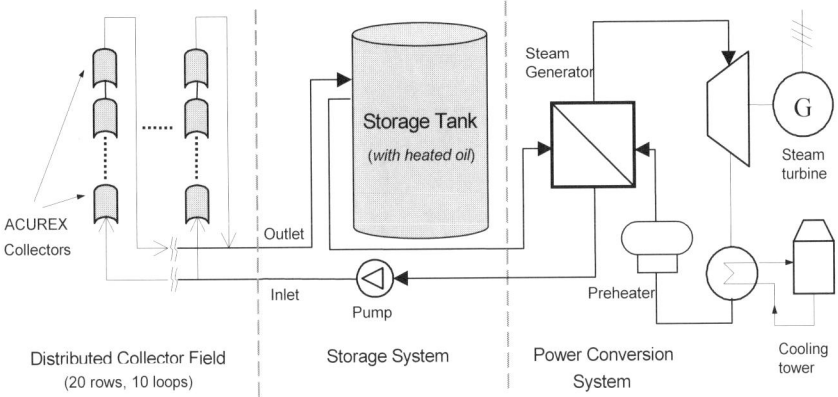

Figure 1.15 Block diagram of the solar power plant

1.12.3 Control Structure of the Plant

Figure 1.16 shows the overall control block diagram of the FLC applied to the plant with the proposed GA optimisation scheme. For this solar plant, it is noted that the outlet temperature of the field depends on other variables such as solar radiation I and the inlet temperature to the field T_{in}. Mirror reflectivity also has an influence but changes so slowly that it may be considered constant. Hence dynamically the outlet temperature T_o can be expressed as a non-linear function f of oil flow u, solar radiation I and inlet temperature T_{in}. The linearised model is based on partial derivatives (of the change in outlet temperature ΔT_o with respect to changes Δu, ΔI and ΔT_{in}):

$$T_o = f(u, I, T_{in})$$

$$\Delta T_o = \frac{\partial f}{\partial u}\Delta u + \frac{\partial f}{\partial I}\Delta I + \frac{\partial f}{\partial T_{in}}\Delta T_{in} \tag{1.18}$$

The partial derivatives can be considered as transfer functions relating the variation in outlet temperature ΔT_o to variations in oil flow Δu, solar radiation ΔI and oil inlet temperature ΔT_{in}, respectively. The mathematical model which accounts dynamically for these additional influences is complex. To approximate these effects, a feed forward element, placed in series with the FLC as shown, has been developed based on the known operating characteristics of the field:

$$u_f = \frac{0.7869I - 0.485(u - 151.5) - 80.7}{u - T_{in}} \tag{1.19}$$

where u_f is the oil flow and u is the temperature set point.

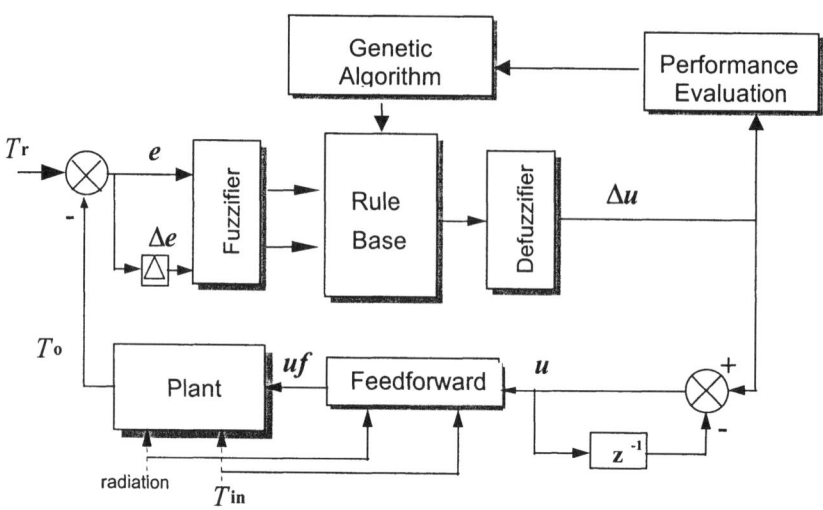

Figure 1.16 Control structure of plant

1.12.4 GA Formulation

Similar to their counterparts in nature, GAs perform the processes of natural selection, crossover and mutation, albeit at a much accelerated rate, on a pool of chromosomes [65]. The chromosomes have a format which bears encoded information representing the engineering parameters to be optimised. Here, the chromosome consists of 49 genes, linked to the 49 'if-then' rules in the rule base of the FLC as shown in Figure 1.17. The rules are typically expressed in matrix form with error (e) and change of error (de) as the antecedents. The antecedents are fuzzified into seven levels as {NL,NM,NS,ZE,PS,PM,PL}, denoting their linguistic equivalence of negative large, negative medium, negative small, zero, positive small, positive medium and positive large. After some experimentation, each gene Γ was then chosen to be represented by 5-bit data such that:

$$\Gamma_i = \{\alpha_i\ \beta_i\ \chi_i\ \delta_i\ \varepsilon_i\}$$

where $\alpha_i, \beta_i, \chi_i, \delta_i, \varepsilon_i \in [0,1]$.

Thus if Γ_1 is {00001}, Rule 1 is not significant; whereas as if Γ_{49} is {11110}, Rule 49 is very dominating. It is found that using a higher number of bits for each gene does not yield observable improvements in performance but increases the simulation time very significantly. The entire chromosome X is of the format

$$X = \{\Gamma_1, \Gamma_2, \ldots, \Gamma_{49}\}$$

Each chromosome therefore has a total of 245 bits of information. A typical chromosome is shown below:

{01010101000001011101010101010101011110101010101011110101011111110110000
00101110111000111001011110101010101000000000000101011101100111110011111100
111110001111111000110000000011111001110001001010111000111000111000111000
1010001111011101111111001010}

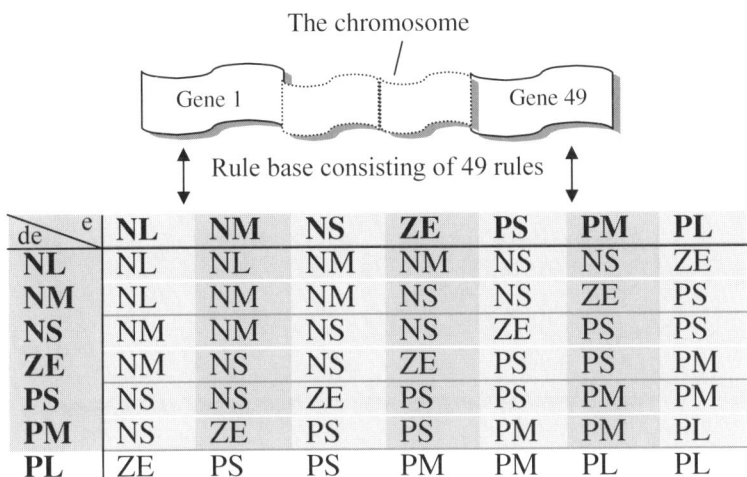

Figure 1.17 Chromosomes linked to rule base

The basic processes of initialisation, reproduction, crossover and mutation of a GA are shown in the pseudo-code of Figure 1.18. Firstly, the GA randomly generates 49 rule weights in 245 bits chromosome and copy into an initialisation pool. The initialisation will continue until the pool is full, which is set to be 30. The next process is reproduction, in which the fitness of each chromosome is first evaluated. The fitness of each chromosome is determined by monitoring the overshoot of the output of the system to the change of input. A high fitness value warrants the chromosome to be copied to the mating pool for crossover. In the mating pool, a random set of chromosomes is selected to crossover to produce offspring. Then the new chromosome is copied to the mutation pool for mutation. For mutation, a probability of 0.001 is used for 'changing the state' of 1 bit of information in the chromosome. Finally the fitness of individual members is evaluated in the reproduction and the process will repeat itself indefinitely. The present process, however, is stopped at the 100th generation as it is observed that after 80 generations the reduction of the error is negligible. This is discussed in the next section.

Step 1:	Initialising pool by generating chromosomes randomly. (size=30)
Step 2:	run = run + 1; /*run = 0 initially */
Step 3:	gen = gen + 1; /*gen = 0 initially* / /*Reproduction*/ Calculating fitness of individual chromosome from initialisation or mutation pool. Copying high fitness chromosomes to reproduction pool. (Roulette Wheel method) /*Crossover*/ Selecting chromosomes randomly from mating pool and crossover. /*Mutation*/ Copying new chromosome into mutation pool, mutation probability is 0.001.
Step 4:	If gen = no_gen Goto step 3
Step 5:	If run = no_run Goto step 2
Step 6:	End

Figure 1.18 Pseudo-code for the GA

1.12.5 Experimental Results

Experimental results on the simulator of the plant have been taken to verify the proposed GA-FLC scheme. In Figure 1.19, the effect of GA optimisation of the rule base on the performance of the plant is illustrated. The upper graph shows the error versus the number of generations. The error, an index of the fitness of the chromosome, is seen to decrease as the generations evolve. The lower left graph shows the plant output just after the 14th generation, whilst the lower right one shows the corresponding output after the 80th generation. The improvement of the dynamic response of the plant is clearly seen. Figure 1.20 compares the GA-FLC scheme on a day when solar radiation is quite fluctuating in the morning, with the conventional PI control scheme. It is shown that the GA scheme has helped to improve the plant's robustness when external changes have rendered the PI scheme very ineffective. Since there is only one solar plant and the solar radiation is uniquely different in any time interval, we cannot test the two control schemes concurrently in real-time. The validity of the comparison between the PI and GA-FLC control schemes lies in the fact that the simulator is a proven model of the plant [63] and one can capture the solar radiation in a particular period and use it as one of the inputs to the simulator. The simulator's output is then compared under different control schemes. The current investigation is based on this principle.

Energy Generation under the New Environment

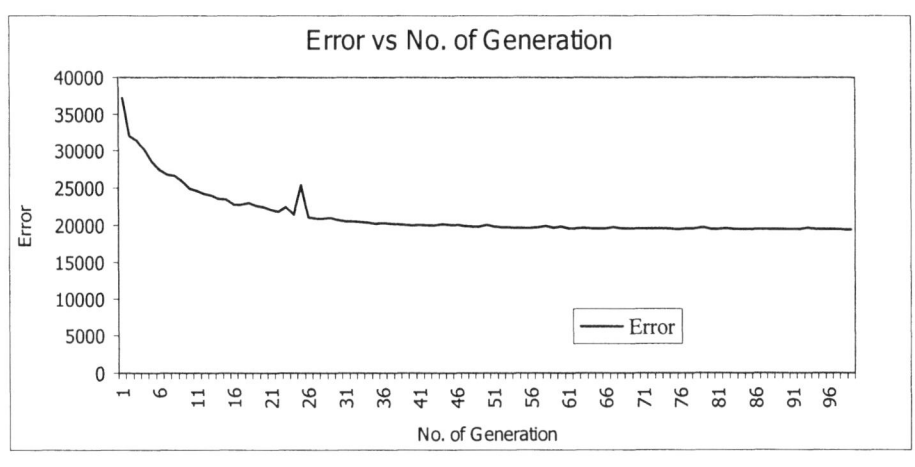

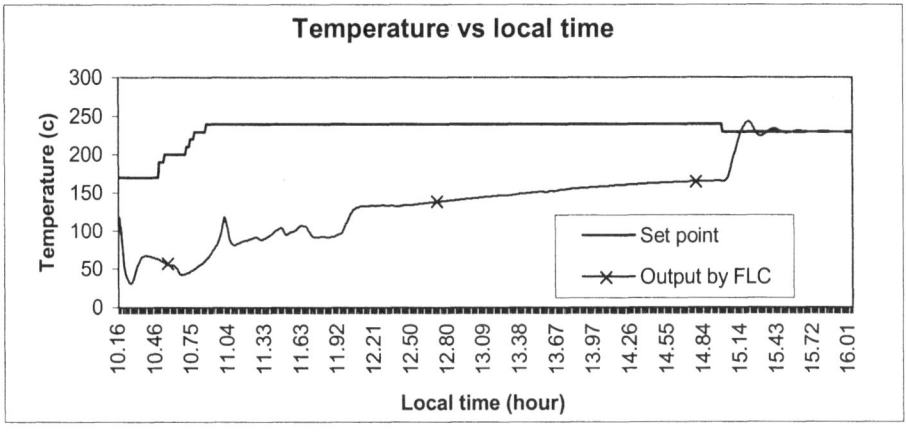

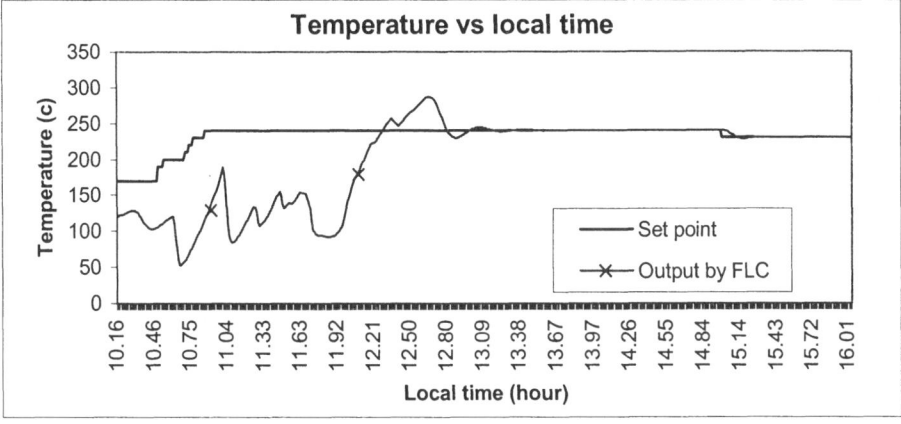

Figure 1.19 Effects on the performance of the plant by the GA optimisation

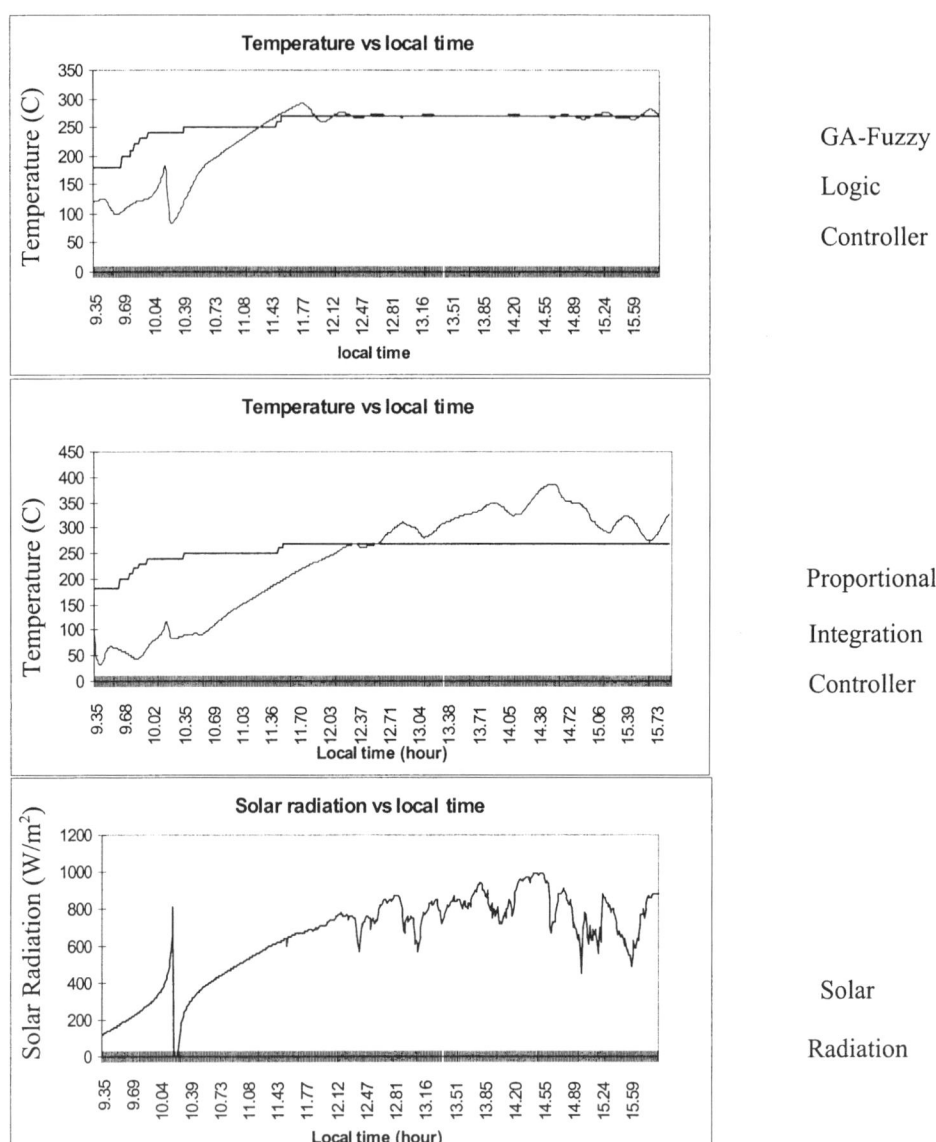

Figure 1.20 Comparison of GA-FLC with PI scheme under extreme external dynamic change

1.13 Conclusions

Deregulation leads the electricity industry to focus attention on the costs of generation and provides an incentive for generators to reduce their costs and minimise risks. It affects the way existing plants are run and operated. New plants are being built on more short-term, cost-based decisions. Long-term investments such as large-scale expensive power plants are not favoured in deregulated markets where customers cannot be secured. The rules of competition are in conflict with government intervention, which is being reduced. However, governments still intervene in deregulated markets to ensure environmental and security objectives. Most countries are concerned with security of supply and pursue a diversified fuel mix for power generation.

As environmental regulations are tightening, coal's future will depend increasingly on clean coal technologies and their competitiveness. Some of the technologies are still in the demonstration stage and it is believed that deregulation can help their commercialisation. Independent producers or industrial consumers can build their own plants and sell electricity to the grid. This means that new expertise is entering the sector. As deregulation progresses, developments are expected to be driven by the market itself. Coal makes a dominant contribution to worldwide power generation and this is forecast to continue for at least the next 10~20 years. As world electricity demand is forecast to grow steeply, coal will continue to play an important role.

Competition in generation increases load diversity, thereby reducing the inventory of standby capacity necessary to meet peak loads. In addition, competition provides an incentive to discover the demand for reliability among different customer classifications. Meeting that demand more closely by offering interruptible service saves additional capital resources. Fuel availability affects the type of co-generation system selected. The nature of the industry choosing to co-generate will often determine the fuel type, and thereby the co-generation system. Pollution concerns must be considered as well. Clean-burning fuels, either natural gas or light-grade fuel oils, will often be required in these states. Co-generation systems which most effectively utilise these fuels will probably prove to be the most economically attractive.

Unless required by government policies, the restructured electricity market is not projected to stimulate renewable energy technologies. It is foreseen that renewable resources will remain more costly than fossil fuel alternatives to at least 2015. Green power offered at or near to market rates could create a mass market demand and lead to the building of significant new capacities. Biomass, wind and geothermal would be the most likely technology choices for expanded use of renewable energy. Consumption levels increase for all fossil fuels and renewable sources, whereas consumption of nuclear electricity generation will decline as a result of retirements and the lack of new construction. Natural gas will tend to gain and coal to lose market share as the industry moves from a regulated to a competitive environment.

On the supply side, there is a need to introduce renewable energy sources along with combined heat and power facilities, advanced coal and nuclear power. Although there are some problems in the public perception and acceptability of nuclear power, it can play a proper role in mitigating global warming. Its potential contribution is too important to be ignored. Public confidence is central to the future of the nuclear enterprise. Such confidence cannot be demanded. It must be won by a mixture of openness on the part of the industry, acceptance of a degree of responsibility by the public, recognition of long-term self-interest, and clear leadership by politicians.

For distributed generation, a modified Steinmetz connection for the phase balance of a three-phase SEIG, which supplies single-phase loads, has been proposed. The feasibility of the MSC is verified by experiments on a practical induction machine, and a high generation efficiency has been obtained. Since only passive circuit elements are involved, the MSC is an economical and effective phase-balancing scheme for a three-phase SEIG used in an autonomous single-phase power system.

A new GA-based FLC scheme for the control of a solar power plant has also been proposed. The scheme optimises the strength of 49 individual 'if-then' rules in the rule base of an FLC algorithm. It uses a 5-bit word to scale the strength of each rule. The optimisation process undergoes about 100 generations until no improvement of plant performance is observed. The simulation results performed on a proven simulator of the plant show that the proposed scheme provides extra robustness for the plant under turbulent external conditions when compared with conventional PI control schemes.

Reforms in the electricity sector are progressing at different rates in different countries around the world and in many cases they have not yet even started. There are still many unresolved questions on the long-term effects, and the future development of reforms will depend on the experience gathered. New technologies suggest new market applications, imply new possibilities for market structure and so lead to pressures for regulatory reform. Regulatory and institutional changes occurring for other reasons can create opportunities for technologies, which may not yet have had a significant impact in the market.

The design of market regulations regarding fees for power exchange, the treatment of imbalances as well as the operation of the ancillary service market can have a significant impact on the economic performance of distributed generation, and hence on its future development. In this new environment, utilities win because they retain rights over their transmission system and can charge what the market will bear rather than a regulated price. They do not need to surrender control to regulators and can recover stranded costs, limited merely by the right of others to compete. They can also gain the right to invade the territory of others. Consumers win because the grid will develop naturally, since newcomers must demonstrate competence. Grid innovations become necessary and reliability is maximised.

1.14 References

[1] J.P. Walters and D.W. Smith, 'The energy policy act of 1992 - A watershed for competition in the wholesale power market', *Yale Journal of Regulation*, Vol.10, 1993, pp.447-482.

[2] M.C. Hoffman, 'The future of electricity provision', *Regulation*, Vol.3, 1994, pp.55-62.

[3] J.C. Moorhouse, 'Competitive markets for electricity generation' available from: http://www.cato.org/pubs/journal/cj14n3-3.html, Washington, DC, USA, Cato Institute, 15pp, 1998.

[4] Katerina Rousaki, Andrew Bushell and Alessandra McConville, *Liberalisation of Electricity Markets and Coal Use*, IEA, October 1999, CCC/21.

[5] R.P. Rozels, 'Competitive bidding in electric markets: A survey', *Energy Journal*, Vol.10, 1989, pp.117-138.

[6] J.B. Bushnell and S.S. Oren, 'Bidder cost revelation in electric power auctions', *Journal of Regulatory Economics*, Vol.6, 1994, pp.5-26.

[7] H. Chao and R. Wilson, 'Priority service: Pricing, investment, and market organization', *American Economic Review*, Vol.77, 1987, pp.899-916.

[8] T. Strauss, and S.S. Oren, 'Priority pricing of interruptible electric service with early notification', *Energy Journal*, Vol.14, 1993, pp.175-196.

[9] A. Midtun and S. Thomas, 'Theoretical ambiguity and the weight of historical heritage: a comparative study of the British and Norwegian electricity liberalisation', *Energy Policy*, Vol.26, 1998, pp.179-197.

[10] R. Haas, H. Auer, C. Huber, and M. Tranger, 'Limits for competition in restructured electricity markets - the European perceptive', *19th annual North American Conference*, United States Association for Energy Economics and International Association for Energy Economics, 1998, pp.103-112.

[11] IEA, *World energy outlook: 1998 edition*, Paris, France, OECD/IEA, 475pp, 1998.

[12] C. Singh and N. Gubbala, 'Reliability evaluation of interconnected power systems including jointly owned generators', *IEEE Transactions on Power Systems*, Vol.9, No.1, 1994, pp.404-412.

[13] Q. Ashan and S.F. Rahman, 'Evaluation of the reliability and production cost of interconnected systems with jointly owned units', *IEE Proceedings*, Vol.134, No.6, 1997, pp.377-382.

[14] P.L. Jaskow, 'Price adjustment in long-term contracts: The case of coal', *Journal of Law and Economics*, Vol.31, 1998, pp.47-84.

[15] IEA, *Projected costs of generating electricity, Update*, Paris, OECD/ IEA, 243pp, 1998.

[16] J. Lane, 'Sweeping the board', *Power Engineering International*, Vol.6, 1998, pp.25-29.

[17] *Proceedings of Conference on Embedded Generation*, IEE, 28 February 2000.

[18] W.R. Lachs and D. Sutanto, 'Battery storage plant within large load centres', *IEEE Transactions on Power Systems*, Vol.7, No.2, May 1992, pp.762-767.

[19] *Gazzetta Ufficiale della Republica Italiana*, Decreto legislativo 16.3.1999 n. 79, Attuazione della direttiva 96/92/CE recante norme comuni per il mercato interno dell'energia elettrica. Number 75, Rome, Italy, Instituto Poligrafico e Zecca dello Stato - S, 31 March 1999.

[20] EC, 'Guide to the electricity directive', available from: http:J/www.europa.eu.int/en/comm/dg17/elec.memor.htm, Brussels, Belgium, European Commission, Directorate-General XVII (Energy), 10pp, March 1999.

[21] P. Baruya and D. Goldsacks, 'European coal issues - European liberalisation of coal', *World Coal*, Vol. 7 (10); 1998, pp.29-34.

[22] G.A. Broadbent, 'Competitiveness of coal – the evolution of price', CS/05, London, UK, IEA Coal Research, 20pp, March 1999.

[23] UNECE, 'Security of supply in a changing European natural gas market', ENERGY/WP.3/GE.4/1996/6, Geneva, Switzerland, United Nations Economic Commission for Europe, Committee on Sustainable Energy, 17pp, June 1999.

[24] Couch G., 'OECD coal-fired power generation – trends in the 1990s', IEAPER/33, London, UK, IEA Coal Research, 83pp, April 1997.

[25] Modern Power Systems, 'World digest: Green power launched,' *Modern Power Systems*, Vol.19, February 1999.

[26] Global Private Power, 'Own coal?', *Global Private Power*, March 1999.

[27] Nuclear Energy - The Future Climate, report, The Royal Society and The Royal Academy of Engineering, UK, June 1999.

[28] T. Hoff and D.S. Shugar, 'The value of grid-support photovoltaics in reducing distribution system losses', *IEEE Transactions on Energy Conversion*, Vol.10, 1995, pp.569-576.
[29] Anna Walker, 'Developing CHP in the public sector and beyond', *Green Government*, Partnership Media Group Ltd., April 2000, pp.2-22.
[30] Nick Jenkins, Ron Allan, Peter Crossley, Daniel Kirschen and Goran Strbac, *Embedded Generation*, IEE, 2000.
[31] James Larminie and Andrew Dicks, *Fuel Cell Systems Explained*, John Wiley & Sons, Chichester, 2000.
[32] 'Big future for distributed generation', *Electricity Daily*, September 30, 1998, p.1.
[33] Thomas R. Casten, 'Electricity generation: Smaller is better', *Electricity Journal*, December 1995, p.67.
[34] Joseph F. Schuler Jr., 'Generation: Big or small?', *Public Utilities Fortnightly*, September 15, 1996, p.31.
[35] George T. Preston and Dan Rastler, 'Distributed generation: Competitive threat or opportunity', *Public Utilities Fortnightly*, August 1996, p.13.
[36] Yu-Tzu Chiu, 'Taiwan reorients energy policy to stress renewables', *IEEE Spectrum*, January 2001, page 34.
[37] Dispersed Generation; Preliminary Report of CIRED, *International Conference on Electricity Distribution*, Working Group WG04, Brussels, Belgium, June 1999.
[38] Ackermann Thomas, Andersson Göran and Söder Lennart, 'Overview of government and market driven programs for the promotion of renewable power generation', *World Renewable Energy Congress*, 1999, Perth, Australia, pp.215-219.
[39] Miles O. Jr. Bidwell, 'Structuring Markets – Determining the Optimal Amount of Regulation: A Discussion of the Changing Electricity Industry', *Pricing and Regulatory Innovations under Increasing Competition*, edited by Michael A. Crew, Kluwer Academic Publishers, 1996, pp.111 – 126.
[40] William Sweet, 'Networking assets', *IEEE Spectrum*, January 2001, pp.84-89.
[41] Thomas Ackermann, G. Andersson and L. Söder, 'Electricity market regulations and their impact on distributed generation', *Proceedings of the International Conference on Power Utility Deregulation, Restructuring and Power Technologies 2000*, IEEE, April 2000, pp.608-613.
[42] 'An introduction to Australia's national electricity market', NEMMCO (National Electricity Market Management Company Limited), 1998, available at: http://www.nemmco.com.au/nem_resources/library/publications/cr_if689.htm
[43] 'New electricity trading arrangements (NETA) program', Office of Gas and Electricity Markets, 2001.
[44] Economics of embedded generation, *IEE Colloquium*, October 1998.
[45] O.P. Malik, 'Control considerations in a deregulated electric utility environment', *IEEE Canadian Review*, Fall 2000, pp.9-11.
[46] G.C. Baker, 'The wave of deregulation, operational and design challenges', *IEEE Power Engineering Review*, Vol.19, No.11, 1999, pp.15-16.
[47] S.S. Murthy, O.P. Malik and A.K. Tandon, 'Analysis of self-excited induction generators', *IEE Proceedings - Part C*, Vol.129, No.6, November 1982, pp.260-265.
[48] G. Raina and O.P. Malik, 'Wind energy conversion using a self-excited induction generator', *IEEE Transactions on Power Apparatus and Systems*, Vol.PAS-102, No.12, December 1983, pp.3933-3936.

[49] D.B. Watson, J. Arrillaga and T. Densem, 'Controllable d.c. power supply from wind driven self-excited induction machines', *IEE Proceedings*, Vol.126, No.12, 1979, pp.1245-1248.

[50] T.F. Chan and L.L. Lai, 'Phase balancing for a self-excited induction generator', *Proceedings of the International Conference on Power Utility Deregulation, Restructuring and Power Technologies 2000 (DRPT2000)*, City University, London, IEEE, April 2000, pp.602-607.

[51] T.F. Chan and L.L. Lai, 'Steady-state analysis of a three-phase induction motor with the Smith connection', *IEEE Power Engineering Review*, Vol.20, No.10, October 2000, pp.45-46.

[52] T.F. Chan and L.L. Lai, 'A novel single-phase self-regulated self-excited induction generator using a three-phase machine', *IEEE Transactions on Energy Conversion*, Vol.16, No.2, June 2001, pp.204-208.

[53] R. Holland, 'Appropriate technology - rural electrification in developing countries', *IEE Review*, Vol.35, No.7, August 1989, pp.251-254.

[54] T.F. Chan, 'Performance Analysis of a three-phase induction generator self-excited with a single capacitance', *IEEE Power Engineering Society 1998 Winter Meeting*, Paper No. PE-028-EC-0-10-1997, February 1-5, 1998, Tampa, Florida, U.S.A.

[55] M.G. Say, *Alternating Current Machines*, London: Pitman (ELBS), 5th Ed., 1983, pp.333-336.

[56] J.E. Brown and O.I. Butler, 'A general method of analysis of 3-phase induction motors with asymmetrical primary connections', *IEE Proceedings*, Vol.100A, Pt. II, 1953, pp.25-34.

[57] J.E. Brown and C.S. Jha, 'The starting of a 3-phase induction motor connected to a single-phase supply system', *IEE Proceedings*, 1959, Vol.106A, 1959, pp.183-190.

[58] Byron S. Gottfried and Joel Weisman, *Introduction to Optimization Theory*, Prentice Hall, New Jersey, 1973.

[59] P.C.K. Luk, L.L. Lai, T.L. Tong, 'GA optimisation of rule base in a fuzzy logic control of a solar power plant', *Proceedings of the International Conference on Power Utility Deregulation, Restructuring and Power Technologies 2000*, City University, London, IEEE, April 2000, pp.221-225

[60] Brussels, COM(96)576, 'Energy for the Future: Renewable Sources of Energy', Green paper for a Community Strategy, 1996.

[61] E.F. Camacho, F.R. Rubio, and F.M. Hughes, 'Self-tuning control of a solar power plant with a distributed collector field', *IEEE Control Systems Magazine*, 1992, pp.72-78.

[62] E.F. Camacho and M. Berenguel., 'Application of generalized predictive control to a solar power plant', *Advances in Model-Based Predictive Control*, Ed. D.W. Clarke, Oxford University Press, 1994.

[63] M. Berenguel, et al., 'Incremental fuzzy PI control of a solar power plant', *IEE Proceedings - Control Theory Application*, Vol.144, No.6, 1997.

[64] A.F. Abdulbary, L.L. Lai, D.M.K. Al-Gobaisi and A. Hussain, 'Experience of using the neural network approach for identification of MSF desalination plants', *DESALINATION* 1993, Vol.92, Elsevier Science, Holland, pp.323-331.

[65] L.L. Lai, A.G. Sichanie and B.J. Gwyn, 'A comparison between evolutionary programming and genetic algorithm for fault section estimation', *IEE Proceedings - Generation, Transmission and Distribution*, Vol.145, No.5, September 1998, pp.616-620.

2

Deregulation of Electric Utilities

Prof. Kwok Lun Lo
The University of Strathclyde
Glasgow, Scotland

Miss Yee Shan Yuen
The University of Strathclyde
Glasgow, Scotland

2.1 Introduction

The success of privatisation of the airline, telecommunications industries has motivated the deregulation and restructuring of the electricity industry. In 1989, the UK became one of the pioneers in privatising its vertically integrated electricity industry. Norway and California followed in 1990 and 1996 respectively. The success of energy privatisation in the UK and Norway has encouraged other countries worldwide to follow the trend. Countries that have been undergoing energy deregulation include Argentina, Australia, Brazil, Germany, New Zealand, Spain, Taiwan and Malaysia.

Occasionally the term 'privatisation' is used to refer to what one would regard as 'deregulation' of electric utilities. While the two words are different literally, 'deregulation' often starts with the sale of state-owned utilities to the private sector. Ironically, neither is the term 'deregulation' correctly used in this context. As described in Section 2.5, none of the existing markets possess the characteristics of either perfect competition or monopoly. It is therefore necessary to have regulations to prevent the excessive exercise of market power and gaming of market participants. In any case, 'deregulation' is widely adopted to refer to the 'introduction of competition'.

Deregulation often involves 'unbundling', which refers to disaggregating an electric utility service into its basic components and offering each component separately for sale with separate rates for each component. As shown in Figure 2.1, generation, transmission and distribution could be unbundled and offered as discrete services. However, deregulation involve not only unbundling, but also the separation of ownership and operation (Section 2.4).

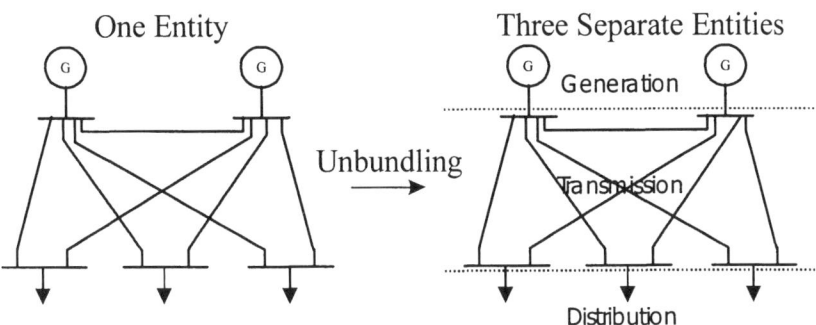

Figure 2.1 Unbundling of utilities

In many countries, a central independent body, usually called the independent system operator (frequently it is simply called the system operator), is set up to cater for the matching of supply with demand and the maintenance of system reliability and security. Sometimes the system operator is also responsible for matching the bids of generators with the demand bids to facilitate exchange, as in the England and Wales Pool. In some markets, another entity is set up just for that purpose; the so-called Power Exchange in the Californian Pool is an example. There will be more details on the roles played by the independent system operator in Section 2.6.

The establishment of the energy markets has brought a new dimension to the topic of power system economics. Economists design different forms of energy markets according to the characteristics of the electricity sector in a particular country. Most countries also learn from the lessons their precedents. The general success of energy privatisation justifies the global trend of deregulation. However, in most cases, there is room for improvement. The strengths and weaknesses of different reforms are discussed in the final section of the chapter.

In most cases, competition is introduced gradually from the wholesale market to the retail market. In most countries, retail competition is still underway; some small or residual customers have not yet benefited from the deregulation. In many countries, with existing regulations, it is difficult for new suppliers to enter the market and compete with the existing suppliers. Stringent regulations are required for mandatory third-party transmission access and the recovery of utilities' stranded costs.

Electricity, unlike many other commodities, cannot be stored easily and supply has to match demand at all times. The transportation of electricity is constrained by physical laws which need to be satisfied constantly in order to maintain the reliability and security of the power system. It soon became obvious that the transmission network is the main impediment to energy privatisation. The issue of direct access will be addressed in more detail in Section 2.5.

2.2 Traditional Central Utility Model

Before covering the details of deregulation, it is essential to have a brief summary on the traditional central utility model. In the early days economies of scale manipulated the structure of the utility structure. Later on the industry transformed into monopolies and the price of electricity became regulated. Moreover, in traditional power systems, the utilities are mostly owned by governments. Vertically integrated utilities typically own generation, transmission and distribution over a wide area. Each utility has one or more control centres that maintain security and reliability of a specific region. Since the utility has control over all generators, when a constraint becomes binding, re-dispatch can be scheduled through least cost optimisation (while satisfying all constraints) by using optimal power flow. The tariff is then the average of all the costs of the different services including generation, transmission and distribution. Under these circumstances, utilities seek to maximise profit through cost minimisation.

So far it appears that this monopolistic situation is just and sound, so why do we need deregulation? At the moment it is sufficient to consider two of the potential benefits resulting from deregulation. Firstly, the advance of technology makes low-cost power plants owned by independent power producers very efficient. These independent power producers would not emerge without the reform. Secondly, unbundling the services may result in fairer tariffs being assigned to individual services. The next section will elaborate on the motivations for reform.

2.3 Reform Motivations

Having looked into the traditional central utility structure, it is not difficult to envisage that the high prices of electricity due to monopolies could drive some societies to resort to reform. Generally speaking, the reasons for reform are complex and often political. In fact, the reformed structure is influenced by party politics in most cases. For instance, some governments favour privatisation because they wish to increase the net state revenue through asset sales and divestiture (divestiture refers to the sale, liquidation or spinoff of a corporate division or subsidiary) of fiscally draining state enterprises. However, this is beyond the scope of electrical engineering and readers who are interested can refer to the documents issued by the Energy Information Administration at its web site [1].

The UK pioneered privatisation and has been introducing full competition in its energy market. This is based on the belief that the regulated utilities know better how to make efficiency improvements when they are given the incentive to make them. Price-cap regulation is adopted as an attempt to reduce the power of natural monopolies in sectors that do not or cannot acquire full competition. The other motives of privatisation included the reduction of central government's role in economic decision making, forcing privatised companies to become more accountable to owners and encouraging the creation of a shareholder society through widespread stock ownership.

However, not all countries follow the reform as in the UK; many others prefer to introduce competition in generation with a centrally operated transmission system. The motives are obvious: efficient operation is essential to profit making and inefficiency is eradicated through competition.

The rise of independent power producers has resulted in the exploitation of different resources as the fuel of power generation in small plants. Studies done by an independent consultant company specialising in services regarding renewable energy reveal that there have been significant improvements in cost efficiency in the electricity produced by wind power [2]. As shown by these studies, there will be a continuous trend of reduction of the cost of wind energy in the near future. Nevertheless, wind energy is already competitive today, provided that it can be implemented without adding backup capacity from traditional generating technology. Taking the environmental benefits into account, it is predicted that wind power will gradually replace some of the fossil-fuel-generated power.

2.4 Separation of Ownership and Operation

One of the main issues concerning electricity reform is the unbundling of electricity assets. Competition starts with the selling of all the government's generators to several private companies. Transmission and distribution are regarded as natural monopolies, so they should remain regulated. (In some cases the distribution systems are sold to private owners.) Natural monopoly refers to a situation where one firm can produce a given level of output at a lower total cost than can any combination of multiple firms. Natural monopolies occur in industries that exhibit decreasing average long-run costs due to size, i.e. economies of scale. Economies of scale exist where the industry exhibits decreasing average long-run costs with size. According to economic theory, a public monopoly governed by regulation is justified when an industry exhibits natural monopoly characteristics.

The idea of separation of ownership and operation applies to the transmission sector. Transmission network owners and the operator are separate entities so that unjust use of the network is avoided. This central body, often called the system operator, is responsible for matching demand with supply. Generators are informed of the amount of power they should produce in different time slots. This ensures that the system is utilised in the most efficient way. Indeed, how efficient or optimal is the market under unbundling? This is addressed in the next section.

2.4.1 Central Dispatch Versus Market Solution

It is important to look at the optimality of the market solution resulting from the reformed market as compared with the traditional system with central dispatch. Without the loss of generosity, in order to maximise profit, all price-taking producers bid at incremental costs/marginal costs. One of the conditions for perfect competition in which the price of electricity is not determined by the amount generated by generators. In other words, generators do not have market power and therefore cannot control the price. They have to 'take the price' as it is set by the market mechanism. Under traditional power system operation, generation scheduling is done by economic dispatch/centralised optimisation. Now one should look at the difference between market solution and the economic dispatch solution. The results of unconstrained economic dispatch show that generators' operating cost is minimised when they operate at the same incremental cost. Therefore, theoretically, ideal market operation and centralised optimisation lead to the same solution. Nevertheless, this does not take into account system losses. Moreover, as explained in the next section, most generators can exercise market power and control the price of electricity. Market power refers to conditions where the providers of a service can consistently charge prices above those that would be established by a competitive market. Its true measure is the ratio between actual prices and the prices that would arise from true marginal cost pricing. In other words, the assumption that the generators are pricetakers is no longer valid. Market power has been a major impediment to price reduction in the England and Wales Pool and the Californian Pool. Efforts are being made to eliminate market power in both cases; in particular, there will be a dramatic reform of the energy market in the UK commencing in the year 2000.

2.5 Competition and Direct Access in the Electricity Market

Access to the transmission system is one of the main issues in energy privatisation. Rigid and sound regulations are required to facilitate transmission open access (TOA). Generally speaking, TOA refers to the regulating construct such as the rights, obligations, operational procedures, economic conditions, etc., enabling two or more parties to use a transmission network. With equal rights of access to the transmission network, it has become feasible for any generators or loads to arrange transactions with each other and hence competition is enhanced. TOA is among the key elements for facilitating competition in the energy markets. This section serves will look the details of the two issues.

2.5.1 Competition in the Energy Market

Competition is the main goal of energy privatisation. Ideally, from an economist's point of view, perfect competition is the most desirable market structure. Perfect competition is a market structure characterised most notably by a situation in which all firms in the industry are pricetakers and there is freedom of entry into and exit from the industry. It is classified according to these three criteria: independence, product substitutability and entry criterion [3]. However, in any real markets, it is rare that all of these criteria are satisfied. Considering also the technical constraints caused by the intrinsic properties of electricity, it can be deduced that perfect competition does not exist in the energy market. The performance of a market is measured by its social welfare. Social welfare is a combination of the cost of the energy and the benefit of the energy to society as measured by the society's willingness-to-pay for it. Maximum social welfare is achieved in a perfect market. Again, a real market frequently operates at a suboptimal level.

Wholesale competition has been introduced in most deregulated markets. Most of the large loads are able to choose their own supplier. Retail competition is being introduced and already some residential customers are able to select their own supplier. This competition has been complicated by the issue of direct access and the necessity for sophisticated information technology. In some countries, solid regulations are yet to be set up for wheeling. Also, it is costly for residential customers to have access to the spontaneous spot price.

In some countries, e.g. the UK, the issue of energy subsidies and the deposition of stranded costs have also complicated efforts on energy privatisation. One form of energy subsidies refers to those given to generators to purchase highly priced coal in order to sustain the local coal industry. Generators receive significantly fewer subsidies after deregulation. Stranded costs are related to the capital invested in the installation of a plant (mainly nuclear) before deregulation. Generators built at that time were supposed to produce electricity at a higher price than the market price set by a competitive market to cover, in the long-term, the cost spent on installation. Because of privatisation, this huge investment has become less worthy and investors could end up being bankrupt. Therefore, electricity reform often involves the determination of the degree of recovery of stranded costs. In some markets, e.g. the Californian Pool, the stranded costs are compensated by embedding an additional charge in the electricity bill.

2.5.2 Competition and Auction Mechanisms

The success of competition is partially determined by the auction mechanism employed in the market. This section briefly describes different potential auctions for the energy market.

Sealed-bid Uniform Price Auction
In the uniform price auction, sellers submit a price and the maximum number of units they would be willing to sell at that price. Submitted offers are ranked lowest to highest, and the lowest priced units are purchased up to the point where supply equals demand. The uniform price paid for purchased units is either the price of the last accepted offer (LAO) or the price of the first rejected offer (FRO); in either case, the price is called the reigning price. It has been suggested that FRO is in general more efficient than LAO since an LAO auction shares the strategic incentives [4]. Contrarily, existing power markets in England and Australia utilise an LAO auction. The UK Pool is an example of a non-discriminatory auction that handles the costs of transmission constraints by an 'uplift charge' and uses a uniform price for settlement purposes. It has also been shown that participants can exploit opportunities for market power in load pockets [5]. A load pocket or a load centre refers to a geographical area where large amounts of power are drawn by end users. Indeed, the two largest generators in the UK have used their market power successfully to raise prices this way [6]. In general, market power will make prices more volatile when a uniform price auction is used [7].

Discriminatory Auction
A discriminatory auction, in which participants receive their actual offers, is shown by evidence from economic theory and experimental economics not to provide incentives to reveal true costs. Despite that, it is argued that the discriminatory price auction is a better form of auction for electricity markets since the price volatility associated with errors in forecasting the load will be smaller than using a uniform price auction even when there is no appreciable market power [7].

English Auction
In an English auction, each seller initially submits an offer indicating the maximum number of units they are willing to make available from each generating facility. The auctioneer begins the auction by starting a 'clock' which sweeps down from the reservation price. Suppliers withdraw facilities whenever they wish and the clock stops when supply falls to, or below, the quantity demanded. If supply is less than demand, the clock is reset to the last price at which supply exceeded demand. If supply equals demand, the price is paid as a uniform price to all remaining sellers.

However, the English auction has a problem when implemented in the electric power framework: it is the need for the costs of all participants to be known by the system operator. The English auction fails to reveal the entire supply curve for suppliers who remain in the market, which is necessary for computing the least cost dispatch. Also, the real-time nature of the auction also imposes relatively high transaction costs compared with a sealed-bid auction.

The second price uniform and English auctions are usually selected over alternatives such as the first price uniform and discriminative auctions because of their theoretical cost revealing properties and experimental evidence about efficiency. Nevertheless, the England and Wales Pool will transform from second price uniform auction to a discriminative auction under NETA [8]. The details of the present market structure and proposed reforms will be discussed in Section 2.8.

Bilateral Models
Bilateral models promote free market competition and therefore are a good way to achieve competition in an electricity market. In this model, suppliers and consumers independently arrange trades, setting by themselves the amount of generation and consumption and the corresponding financial terms, with no involvement or interference by the system operator, e.g. Nord Pool. Economic efficiency is enhanced when consumers choose the least expensive generators. However, the unique characteristics of electric power networks make it difficult to utilise the model. Problems are caused by the lack of coordination among the independent trades, leading to the violation of transmission network constraints such as congestion and transmission system losses. In fact, to manage congestion, the system operator can use one of many different objectives. These objectives can range from a least cost formulation to one that results in the minimum possible adjustments to schedules.

Multilateral Trades
A multilateral trade is a generalisation of a bilateral trade. This operating paradigm allows suppliers and consumers primarily to seek profits on their own while the independent system operator ensures security [9]. It is postulated that it is essential to have coordinated trades involving three or more parties to relieve congestion or to ensure security of operation. This model has a minimal role for the system operator who has no data on costs or financial arrangements. The duties include verifying feasibility of trades 24 hours ahead, dispatching and monitoring trades in real-time and eliminating imbalances and charging commitment violations. Trades are arranged by brokers. A broker can be a generator or a consumer involved in the trade, but can also be an unrelated third party.

The drawback of this paradigm is that the congestion charge is shouldered by all participants and therefore it does not give any signals for locating or installing new transmission facilities. Also, it does not penalise any participants who constantly cause congestion to the network. The concern of fairness arises about how curtailment of trades is done initially. Furthermore, in reality, power system networks are complex, it will be difficult for participants to find the best trade and therefore it is unlikely that the optimal operating point will be achieved within a reasonable period.

2.5.3 Direct Access/Wheeling

Wheeling is the conveying of electric power from a seller to a buyer through a third-party-owned transmission network. In the deregulated markets, the objectives of pricing the transmission service are economic efficiency, revenue sufficiency and efficient regulation. In contrast to the traditional regulated market, the knowledge of costs is used to set prices, not just to minimise the total cost of building, operating and maintaining transmission lines. The two critical elements determining the wheeling rates are the prices determined in relation to the real-time situation and those determined through market-based competition. The components of cost are operating cost, existing system cost, opportunity cost and reinforcement cost [10].

Transmission Open Access (TOA)

Because of open access, entities that did not own transmission lines were granted the right to use the transmission system. The aim of TOA is to introduce competition into the traditional cost-of-service regulated utilities without giving up the existing regulating structure and at the same time obtain reliable and economic electric service.

Marginal cost pricing is insufficient because network revenue surplus from transmission transactions under a spot price does not cover in the long term the fixed costs of the transmission line. (But it does compensate for transmission loss.) There are two big issues concerning the implementation of TOA [11]:

1. Economic issues - marginal pricing with a supplement for revenue incentives is an economically efficient option. The different kinds of allocations are rolled-in allocation, contract path allocation, incremental cost allocation and megawatt mile allocation.
2. Operational issues - the three different considerations are before the fact, real-time and after the fact.

Transaction Cost Allocation

The pricing of transmission services plays a crucial role in the success of deregulation since it determines whether the provided transmission services are economically beneficial to both the wheelers and the customers. The rules can be divided into two main categories: embedded-cost-based approaches and marginal-cost-based approaches. The former approaches include:

- Postage stamp method: based on magnitude of the transacted power.
- Contract path method: a contract path is selected to identify the transmission facilities that are actually involved in a transaction.
- Distance-based MW-mile method: based on the magnitude of the transaction power and the aerial distance between the points of delivery and receipt.
- Power-flow-based MW-mile method: based on the extent of use of transmission facilities by the transaction. This requires a set of load flow analyses associated with each wheeling transaction when multiple wheeling transactions are considered [12].

The latter approaches are based on the solutions of optimal power flow and include short-run marginal cost methods and long-run marginal cost methods.

Variation of the power-flow-based MW-mile methods has been developed [13] which uses line utilisation factors (LUFs) and does not require a set of load flow analyses when dealing with multiple wheeling participants. However, the proposed method may not give any incentive to the wheeling participants who can alleviate any existing transmission congestion. On the other hand, it can provide the vertically integrated electric utility with the signal of each wheeling participant's actual capacity use.

One of the criticisms of embedded cost allocation methods is that they only consider the costs of the existing transmission facilities. Moreover, the existing MW-mile method requires load flow analysis whenever each wheeling transaction is exercised. The modified MW-mile allocation rule as mentioned in the previous paragraph is such that any participant that uses the system in the counter-flow direction receives an incentive in terms of smaller costs. This incentive increases as the circuit gets more loaded, up to a zero charge when the system is fully loaded; the share of each participant is proportional to its impact on the investment requirements of the system transmission. In their case, instead of evaluating the costs for a given year, they determine costs along a longer planning horizon.

Transmission pricing can also be classified as either static or dynamic [14]. Dynamic aspects can be further classified into short-term dynamic and long-term dynamic. Where and what type of new equipment should be installed concerns the static nature of the problem. When to install new pieces of equipment concerns the dynamic nature. Short-term dynamic involves the simulation of the economic agent actions of the market. Long-term dynamic is difficult to be included in transmission cost allocation methods owing to the inherent uncertainties over the future configuration of the system, hydrological conditions and the load growth. A more accurate pricing is to use long-term marginal cost (LTMC) but in practice it is difficult to implement because of the complexity of calculating the real LTMC and the constraints imposed by regulation based on the traditional cost-based pricing.

The effectiveness of the transmission charges to control the overall optimisation depends greatly on the adopted cost allocation method and the relation between generation and transmission costs. The rate of return approach and the non-negative tariff for transmission pricing could cause undesirable effects in the attempt to reach the social optimum.

Congestion
The implementation of deregulation is further complicated by the presence of congestion. Congestion refers to the binding of thermal limits of the transmission network. Congestion can be relieved by re-dispatch of generators. In the traditional power system, the utility can achieve that by re-dispatching the cheapest generator(s) available while alleviating the constraints. In the deregulated environment, generation and transmission fall into different entities. Market participants are required to pay a premium when their transactions cause congestion. Because of the parallel path flow nature of electricity (electric power flows on all interconnected parallel paths in amounts inversely proportional to each path's resistance.) in the network, a certain line could be overloaded by different transactions. A discreet charging mechanism is therefore required and there has been tremendous research in congestion management [15].

Economically speaking, congestion generally leads to higher marginal costs and reduced revenues. In some cases, congestion is the motivation for generators to exercise market power. Consider a two-area system being connected by a congested line: the flow of the line is now restrained by its thermal limit (Figure 2.2). Owing to the congestion, the perfectly inelastic (elasticity is zero) demand which is normally fed by the cheaper generator G1 is met when G1 is forced to buy some of its electricity from G2 which can theoretically raise its price as high as possible. Generator G2 is said to have acquired unlimited market power.

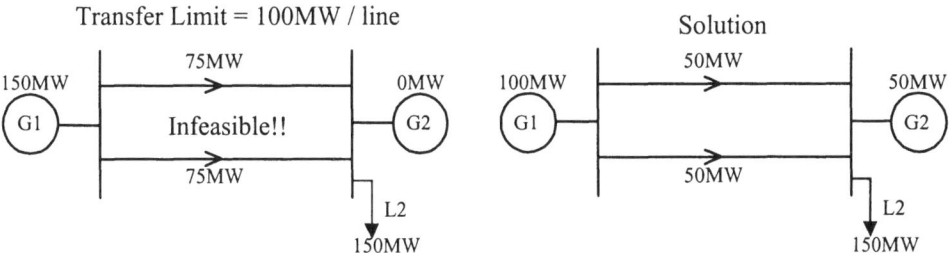

Figure 2.2 Illustration of market power caused by congestion

Figure 2.3 illustrates different congestion pricing methodologies and their classification. Privatisation across big regions is imminent (e.g. different countries in Europe and different states in the US) and proper coordination between different regions is essential to alleviate congestion 16]. Then individual regions can operate at their own congestion pricing. Nodal prices are typically calculated as dual variables or Lagrange multipliers from optimal power flow calculations. These are based on the theory of spot pricing. In places where nodal pricing is adopted, differences in nodal prices can result in congestion fees/surplus. Market participants can hedge congestion charge through the purchase of transmission congestion contracts which are also known as 'Capacity Rights', and give their holders the right to congestion charges between pairs of nodes in the transmission system. They are designed for participants who are indifferent to the difference between physical delivery and financial compensation. In the England and Wales or the Nord Pool, sharing of congestion cost is adopted and participants are credited or charged a premium on top of the system clearing price. An alternative would be congestion preventive measures. They can be subdivided into passive preventive measures and active preventive measures. Active prevention can be done by the so-called 'willingness-to-pay' method. Rather than minimising the cost functions as in traditional optimal dispatch, the new objective function is for the minimisation of curtailments on trades based on their willingness-to-pay [17]. Passive prevention is done by means of broadcasting price signals to all participants. The corresponding participants are charged a premium when the line flow is approaching the limit of the lines, i.e. before the constraints become binding [18]. The success of this method largely depends on the responsiveness of the participants and the advance of information technology. Congestion management under the multilateral trade paradigm falls under the category of both preventive measure and sharing of congestion cost. It is because all participants have to shoulder the cost of congestion implicitly that the system operator releases the technical information to all participants which seek to arrange trades without violating any transmission constraints [9,15].

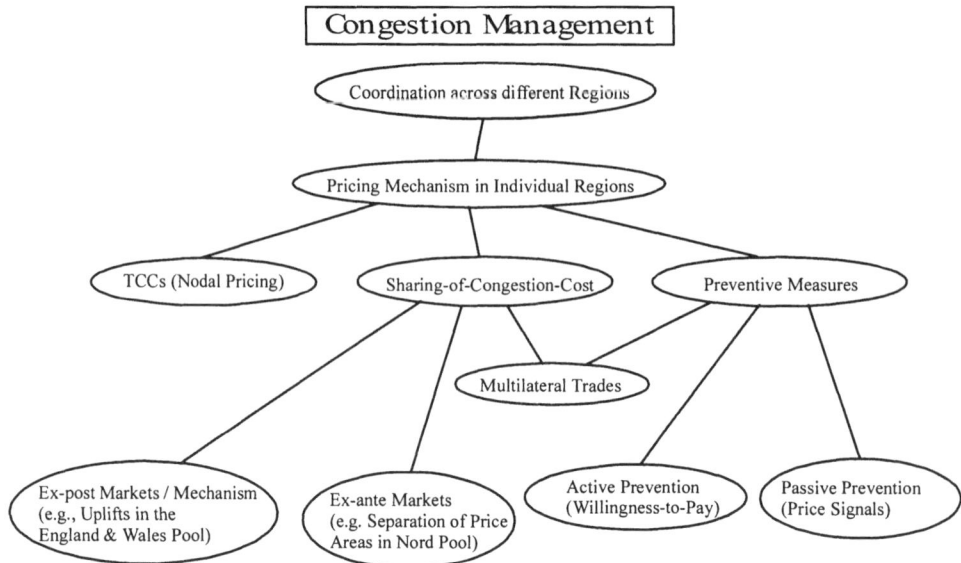

Figure 2.3 Congestion management methodologies

Transmission Losses

Transmission loss compensation is another issue of open access. Nowadays, real power transmission loss compensation is made in real-time operation as often as generation is scheduled. This is essential since the system frequency will deviate from its nominal value if the real power loss is not counterbalanced in real-time. Theoretically it is possible to work out exactly how much transmission loss a transaction can cause. Nevertheless it is a very complex process and requires sophisticated computer software. In some markets the system operator is responsible for loss compensation and participants are charged for transmission loss on a pro rata basis, e.g. 2% in the England and Wales Pool. However, transmission loss service can also be treated as a market-based service. Participants can arrange trades for transmission loss with each other and, as a matter of fact, this could promote competition.

2.6 Independent System Operator

The independent system operator (ISO) is a neutral entity responsible for maintaining the instantaneous balance of the grid system. The ISO performs its function by controlling the dispatch of flexible plants to ensure that loads match resources available to the system. Another major responsibility of the ISO is to ensure fair and impartial access to the transmission system for all generators, while maintaining reliable operation. Occasionally the duties of the ISO are integrated with power exchange (PX), i.e. act like an auction to match total demand for power with generation of power. For instance, the ISO in the England and Wales Pool creates a 'pool' or 'spot market' where price information is publicly available. It solicits bids from electricity generators and chooses the lowest bidders until it has enough supply to meet the requests to buy power. Pool prices change on

Deregulation of Electric Utilities   61

a half-hourly basis. Many customers will pay for electric power based on this price, either directly through their distribution utility or through a private power supply contract with terms that are pegged to the Pool price. The ISO can also operate markets for ancillary services, such as reactive power, spinning/non-spinning reserve and losses. The roles of the ISO and PX are shown in Figure 2.4.

2.6.1   Pricing and Market Clearing

In Section 2.5.2 different kinds of auction mechanisms were discussed. The ISO or the PX bears a duty to set the electricity price. Pricing is done essentially in either of these two fashions: *ex ante* or *ex post*. An *ex ante* market is one in which the price of the commodity is set prior to its delivery while an *ex post* market is one in which the price of the commodity is determined at the time of delivery. In an electricity market, an *ex ante* market is like a bilateral contract market in which traders/participants agree on the amount of electricity to be delivered at a certain time in the future at a certain price. In Norway, the Nord Pool combines ex ante and ex post pricing. In its spot market, system price/area prices are set up the day prior to delivery. Any difference in the forecast with the real-time delivery results in a discrepancy with the pre-set price and the spot price. This is compensated by the presence of the *ex post* mechanism. In the Nord Pool, there is a buyback market to make up for this difference. Similarly, generator bids are also submitted on a one-day-ahead basis in the England and Wales Pool, and participants are paid at the end of each day for their transactions plus compensation. The England and Wales Pool is therefore also an *ex ante* market with an *ex post* mechanism. *Ex post* markets also exist and examples are the New Zealand and Australia markets. In the New Zealand electricity market, generators and loads are allowed to change their bids until 2 hours prior to delivery and the market is cleared regularly during the bidding process. *Ex post* prices are calculated using market-clearing software with the latest offers/bids and the actual metered demand together with losses. Figure 2.5 illustrates the interaction of *ex ante* and *ex post* pricing for electricity markets.

2.6.2   Risk Taking

As in any other commodity marke, all participants have to bear a certain degree of risk in the electricity market. The system operator also has a share of the risk due to the unpredictable discrepancy of the forecast with actual demand. The degree of risk for participants or the ISO depends on the pricing mechanism of the market. For *ex ante* markets, the success of market participants in bilateral contracts depends on the accurate estimation of the system price in the future. In a hybrid market like the Nord Pool, the system price is set prior to delivery, and any real-time power imbalances are purchased by the SO in a so-called Regulating Market, which exhibits the *ex post* mechanism necessary for any ex ante markets. The Norwegian SO is therefore much more susceptible to financial loss. Congestion could cause the SO heavy financial commitment in the buyback market and it would become a serious issue when congestion is commonplace [19].

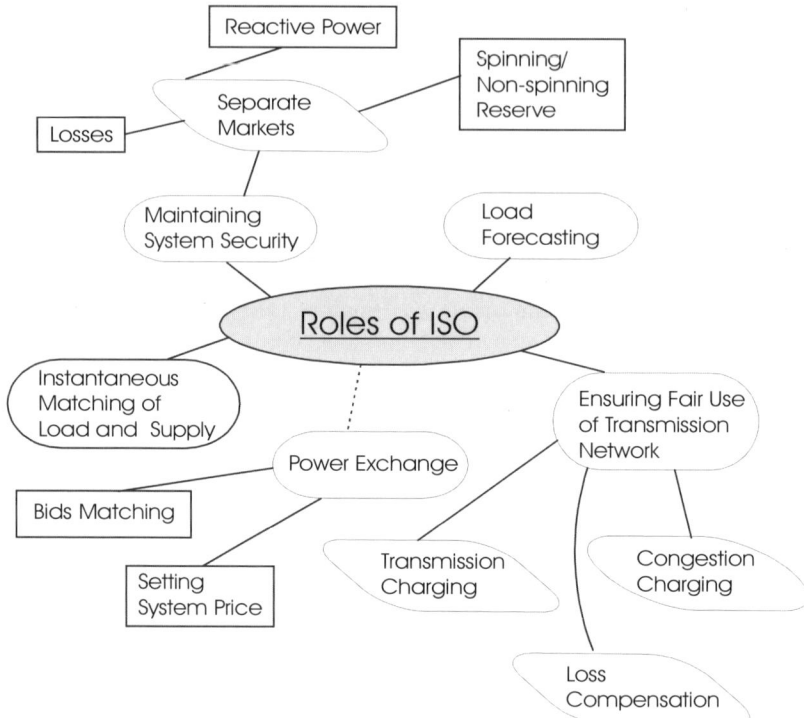

Figure 2.4 Roles of ISO

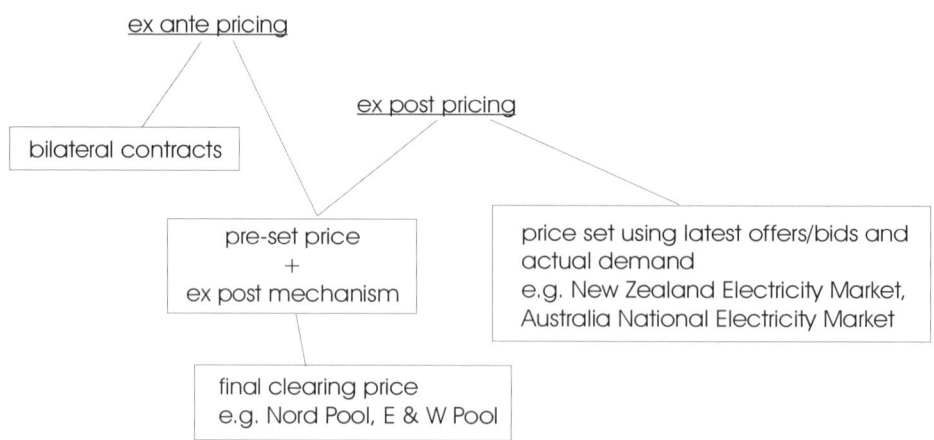

Figure 2.5 Electricity pricing

2.7 Retail Electric Providers

Retail electric providers, also known as second-tier providers, play an important role in introducing competition in the retail sector. So far not many countries have gone to the stage of retail competition. Even for the pioneer, the England and Wales Pool, retail competition did not come into practice until nine years after the establishment of a competitive wholesale market.

The main aim of retail competition is that the benefit of the average wholesale generator prices should pass to small consumers rather than be captured by monopoly distribution companies or retailers. Small consumers should have access to price signals so as to know the true cost of the amount of energy per hour. Also, competitive pressures should be exerted on energy services such as metering, billing and communications.

The difficulties of implementing retail competition are the requirement of sophisticated metering and complex settlement systems. Many residential customers do not get to benefit from energy privatisation simply because they do not have the opportunity to respond to spot prices. The need to communicate in real-time with customers makes retail competition too costly a deed to justify. Many regulators or policy makers begrudge opening the market to small consumers because they are uncertain if the benefits they gain can offset the cost required for smart metering. The complex settlement process has further exacerbated the situation. Furthermore, how much can a household consumer actually benefit by adjusting his/her energy consumption in response to the spot price? When it comes to stabilising prices, big and industrial customers with their sophisticated control can respond faster to 'spiky' prices to help clear the market before small customers can do anything about it. All in all, the implementation of retail competition is not a wise step unless its cost is reduced to the extent that competition is cost efficient. Nevertheless, no one can accurately estimate the actual benefits brought about by retail competition until it is actually implemented.

While sophisticated metering seems helpful in achieving retail competition, it is not an absolute necessity. Alternatively, retailers can offer a fixed price contract to their customers according to their individual load profiles. In this case the customers can hedge the fluctuations of the spot prices and at the same time purchase electricity at a price reasonably close to the wholesale price, plus distribution loss and/or stranded costs. The retailers can in return purchase their customers' metered quantity of electricity directly from their local distribution company through financial contracts such as the Contract for Differences (CFD) or spot purchases. CFD is a type of bilateral contract where the electricity generation seller is paid a fixed amount over time which is a combination of the short-term market price and an adjustment with the purchaser for the difference. For example, a distribution company may sell a retailer power for 1 year at P_2 cents/kilowatt-hour (kWh). The seller then gets wholesale price P_1 from the pool and the purchaser pays the distribution company the difference between the wholesale price P_1 and the contract price P_2 (or vice versa if the wholesale price should go above the contract price). Effectively, the retailer purchases power at a fixed rate of P_2-cents/kWh regardless of the changes in the wholesale price. The principles of CFDs are shown in Figure 2.6.

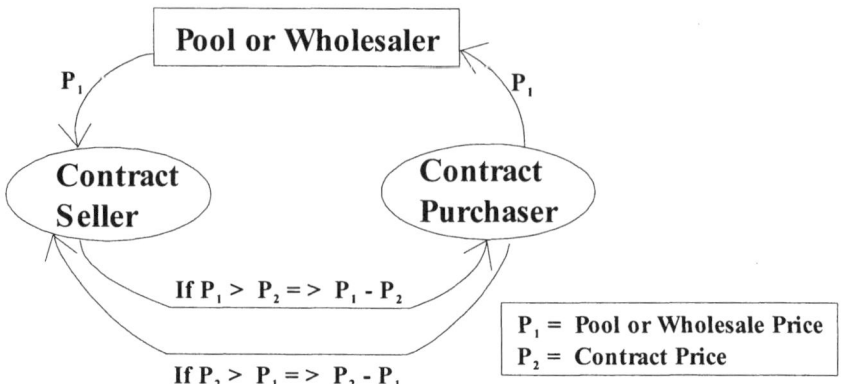

Figure 2.6 Contracts for Differences

2.8 Different Experiences

Different market structures and pool systems employed in various countries will be covered in this section. In particular there are detailed subsections on countries like England, Wales and Norway, which are among the pioneers in energy deregulation. These subsections include numerical examples to illustrate the *ex ante* and *ex post* pricing mechanisms in these countries. The relatively new Californian market will also be discussed and analysed. Each of these sections concludes with a discussion on lessons that could be learnt from each country's reforms. There is also a brief description on reforms in other countries like Scotland, New Zealand and Germany, and in the European Union.

2.8.1 England and Wales

Background

With the enforcement of the Electricity Act of 1989, the UK became one of the first nations to launch energy privatisation. The main aim was to reduce the government's role in the public sector economy. Before privatisation the market structure was a partially vertically integrated monopoly with several distributors. Generation and transmission assets were owned by the state. Nowadays the competitive electricity market in the UK comprises a number of large private generating companies that produce electricity, and 12 privately owned regional electricity companies that distribute power. The National Grid Company (NGC) owns and operates the transmission network and runs the spot market. The national grid supplies distribution companies and has interconnections with Scotland.

The UK power system features a high percentage of thermal generation (and a small percentage of nuclear generation) with an increase in the use of gas and a decrease in the use of coal after the lifting of the ban on gas for electricity generation. Owing to privatisation, independent power producers (IPPs) have emerged in the market and many high-efficiency combined-cycle gas-fired power plants are being built.

England and Wales Pool

The England and Wales Electricity Pool facilitates a competitive bidding process between generators by setting the price of electricity each half-hour of the day and establishing which generators will run to meet forecast demand. It is a mandatory pool that acts like a uniform price single-sided auction (only a very small number of demand side participants are allowed to bid) and adapts *ex ante* pricing combined with *ex post* mechanisms for power imbalances. By 10.00 a.m. every day, generators have to supply offers for each half-hour of the following day. NGC, being the system operator, is responsible for the scheduling and dispatch of generation on the day to meet actual demand. It does that by producing a forecast of demand plus reserve, taking into account the weather and demand usage patterns for each half-hour of the day. The so-called unconstrained schedule, supposedly the lowest cost generation schedule, is generated by the software generating ordering and loading (GOAL). Merit order is generated by ranking the generator bids in order to select the generators for dispatch. The system marginal price (SMP) is the price of the most expensive generator scheduled to meet the forecast demand. All the in-merit generators are paid the pool purchase price (PPP) which is SMP plus the capacity payment whose value depends on the loss of load probability (LOLP). Capacity payment is that component of the PPP which is designed to provide an incentive for generating capacity to be made available. The LOLP reflects the probability of supply being lost because generation is insufficient to meet demand.

The *ex post* mechanism that deals with real-time power imbalances augments PPP with uplifts to result in pool selling price (PSP). Uplifts represent the difference of the unconstrained schedule and the constrained schedule and account for transmission constraints, differences between forecast and actual demand, differences between offered and re-offered availability, differences between actual and instructed output, the operation of non-centrally despatched generating units, transmission losses and ancillary services. Suppliers buy electricity from the pool at the PSP. Settlement of amounts due usually takes 28 days after the trading day and any disputes can be brought up during this period.

Since it is a mandatory pool all generators have to offer bids to the pool. Trading outside the Pool is only allowed in the form of hedging contracts such as Contracts for Differences (CFDs) as mentioned in Section 2.7. The purpose of CFDs is for participants in the pool to hedge the price fluctuation of the pool prices.

Example

The following example serves to illustrate the market-clearing mechanism of the England and Wales Pool. Figure 2.7a shows the network with generators G1 and G2 having bidding cost functions $MC(PG1) = 3 + 0.02 PG1$ and $MC(PG2) = 8 + 0.07 PG2$ respectively, both having a maximum generation of 200 MW. The perfectly inelastic loads L1 and L2 are 40 MW and 150 MW respectively.

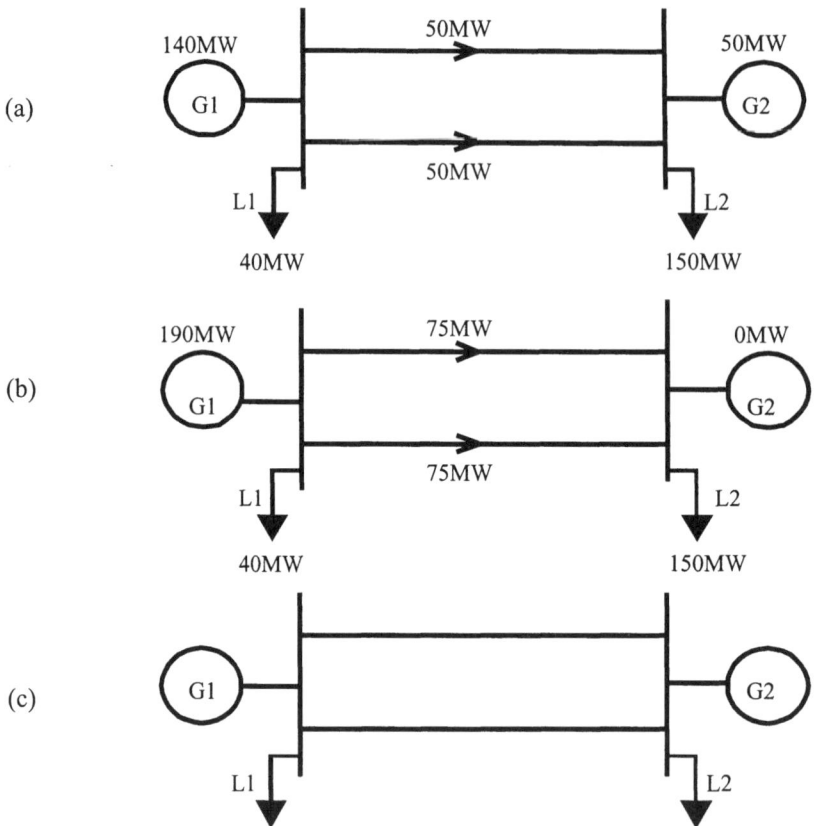

Figures 2.7 (a) The network; (b) unconstrained dispatch; (c) constrained dispatch (Source: [15])

The unconstrained dispatch is shown in Figure 2.7b and the SMP resulting from market equilibrium is 6.8 £/MWh (Figure 2.8a). Assume that the LOLP is zero and therefore PPP is equal to SMP. This sets the *ex ante* price the day prior to the actual delivery. *Ex ante* pricing comes into practice during real-time operation. For the illustration purposes congestion is chosen as one of the components of uplifts for the *ex post* pricing mechanism. The thermal limit on either of the two transmission lines is 100 MW and therefore the maximum secure loading is 50 MW on each line. The constrained dispatch is shown Figure 2.7c. Figures 2.8b and 2.8c reveal the cost functions of G1 and G2 and the necessary adjustments caused by congestion. The dark-shaded areas represent the compensation to the generators and the sum of them with the light-shaded areas results in the total payment to generators by the pool. Details of the example can be found in [15]. Uplift is embedded in the PSP paid by loads L1 and L2 and it represents the difference in the costs of unconstrained and constrained dispatch. The demand charges and generator payments are listed in Table 2.1.

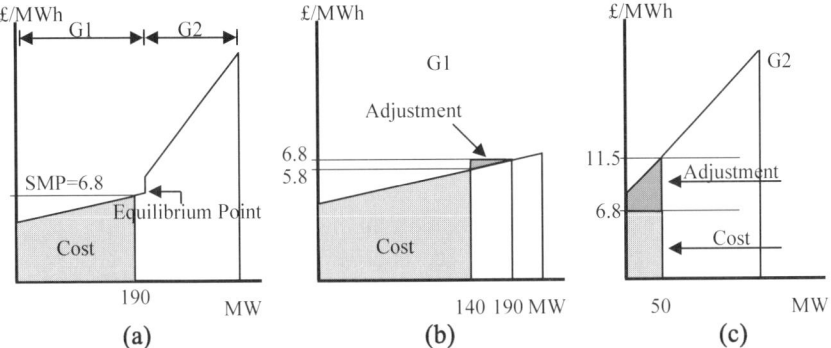

Figure 2.8 (a) SMP; (b) G1 cost function and adjustment; (c) G2 cost function and adjustment (Source: [15])

Table 2.1 Generator payments and demand charges (Source: [15])

Demand Payments	L1	L2
Demand Charges (PSP) (£/h)	308	1156
Total Charge (£/h)	1464	
Generator Payments	G1	G2
Generating Costs (Light-shaded Areas) (£/h)	616	487
Generating Payments from Pool (Output*PPP) (£/h)	952	340
Adjustments (Dark-shaded Areas) (£/h)	25	147
Total Payment (Sum of Generating Payments and Adjustments) (£/h)	1464	

Lessons Learned from Privatisation and the Proposed Reforms

Even though there has been a significant drop in electricity prices in England and Wales since privatisation, this price drop does not fully emulate the cost reduction of generation. These lower prices are not passed on to customers entirely but are partially retained by generation companies in the form of higher profit. Also, there has not yet been a significant decrease of price in the retail market. A possible reason for the inefficiency in the wholesale market is that the three largest generators could game and manipulate the wholesale market. The market lacks small IPPs which could potentially favour competition and reduce the market power of the large generators.

In view of the existing problems of the pool, the director of the Office of Gas and Electricity Markets (Ofgem) published the NETA [8], for England and Wales in 1998. The reforms should commence in 2001 and should lead to significant changes in the existing market. First of all, a discriminatory auction will replace the uniform auction. Secondly, demand-side bidding is allowed so the market will transform into a bilateral market. The reforms are designed in such a way that participants can choose over different ways they participate in the market. In a different time frame before actual delivery participants can choose to trade in the following markets:

Forwards markets: these are optional and are operated by independent organisations. Participants can sign bilateral contracts that are up to several years ahead as desired.

Short-term bilateral market: this is optional and open from 24 to 4 hours ahead of the trading period. All trades will be organised by a market operator (MO).

Balancing market: this is also optional and it is open 4 hours ahead to the end of each trading period. The SO obtains full control of the system after the close of the bilateral market. It would engage in trades to ensure that generation and demand are balanced, taking into account and resolving any constraints on the transmission network.

Real-time power imbalance charges are imposed on participants whose contracted amount is different from the actual metered amount and they could be based on the costs imposed on the SO to settle the imbalances. The reforms feature full demand-side bidding, firm offers and bids and simple offers and bids, and they aim to provide market participants with higher flexibility over different ways of trading. Nevertheless, many researchers are sceptical about the proposed reforms and believe that this is not the solution to get rid of market power and reduce prices [20,21]. Moreover, there is concern, under the proposed reforms, over the possibilities of exploitation of generators' market power in the balancing market through the increment and decrement bids. Increment bids represent the prices participants wish to be paid for an increase in output or are willing to pay for an increase in demand. Decrement bids represent the prices they are willing to pay for a decrease in output or wish to be paid for a decrease in demand.

2.8.2 Norway

Background
Market operation commenced in Norway two years after the passage of the Energy Act of June 1990. The competitive market evolved from the former 'Club-Pool', where surplus energy was traded. An optional pool was sufficient because of the large number of participants from the very beginning.

Unlike the UK power system, the Norwegian power system has a very high percentage of hydro power generation. The transmission network, which was formerly owned by Statkraft, is now owned and operated by Statnett, which is also the ISO for the regulated market. The Norwegian spot market, the Nord Pool or Elspot, is an optional market and participants are free to trade in the bilateral contract market (futures market), provided that power is in balance for generators, large customers and retailers.

Nord Pool
Similar to the England and Wales Pool, the Nord Pool utilises *ex ante* pricing to set the prices one day prior to delivery and compensates power imbalances using *ex post* mechanisms. One day ahead of actual delivery, the Nord Pool accepts generator offers and demand bids for each hour of the following day. The system price is the equilibrium point where the aggregate demand curve meets the aggregate supply curve. The Nord Pool facilitates a uniform price auction by paying all generators the last accepted bid. When there are bottlenecks between bidding areas during this process, the whole region is divided into different price areas. In the surplus area, the area price is found by the right shifting of its demand curve by an amount equal to the line capacity, whilst in the deficit area, the area price is found by the right shifting of its supply curve by the same amount. In

economic terms, the area price in the surplus area is set up in such a way that it should stimulate an extra demand which has a quantity equal to the capacity of the constrained line. On the other hand, in the deficit area, the area price is set up so that suppliers are encouraged to supply an additional amount equal to the capacity of the line. Market participants incur an additional cost and this charge is called the 'Capacity Fee' [22], which is the difference between the system price and the area price. (An example will be given to illustrate this mechanism below)

The price information is broadcast to pool participants by 2.00 p.m. on the day prior to physical delivery. Any power imbalances are compensated in a separate market, the regulating market. Generators can submit buyback bids after the day-ahead market trading is finished. These bids reveal how much a generator is willing to pay to buy back the surplus power and how much a generator costs to produce the deficit amount. Again, the system operator selects the cheapest available generators to buy or sell in case of regulation and congestion management, and all in-merit generators are paid the price set by the highest cost block. Settlement is done usually in two weeks.

Example

The network shown in Figure 2.9 is used to illustrate the *ex ante* pricing mechanism of the Nord Pool. The two generators have bidding functions $MC(PG1) = 3 + 0.02PG1$ and $MC(PG2) = 8 + 0.07PG2$ respectively and the loads L1 and L2 have demand bid cost functions $uL1(PL1) = 7 - 0.03PL1$ and $uL2(PL2) = 15 - 0.05PL2$ respectively. At first the system price P_s is calculated by finding the interception of the demand and supply curves formed by all the bids supplied by both areas (Figure 2.10a). The system price is the price at the equilibrium point, which is equal to 6.52 £/MWh. Figure 2.9a shows the unconstrained dispatch and Figure 2.9b shows the constrained dispatch.

However, because of congestion the system is split into two areas, Area 1 and Area 2. The principle is revealed graphically in Figures 2.10b and 2.10c. As shown in Figure 2.10b, in the surplus area, which is also the low-price area, the area price is found by shifting the purchase curve by 100 MW (the capacity of the constrained line) to the right (from D_0 to D_1) and finding its new intercept with the sale curve. Similarly, Figure 2.10c shows the calculation of the area price in the high-price area by shifting the sale curve to the right by 100 MW (from S_0 to S_1). Various prices are shown in Table 2.2. The settlement price for all participants is the system price and participants are charged for the use of the transmission system. L1 and G2 will be credited the Capacity Fee since they help alleviate the bottleneck, whilst G1 and L2 will be debited the Capacity Fee since they exacerbate the situation. This is reflected in the area prices. Also, because of the physical flow of 100 MW on the constrained line, there is a revenue of $100 \text{ MW} * (P_h - P_l)$ to the grid company. Table 2.2 shows how the difference of the total charge credited to L1 and G2 and the total charge debited to G1 and L2 can make up this figure.

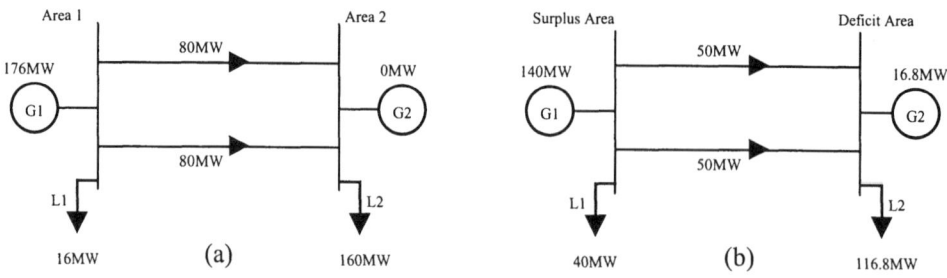

Figure 2.9 (a) Unconstrained Dispatch; (b) Constrained Dispatch (Source: [15])

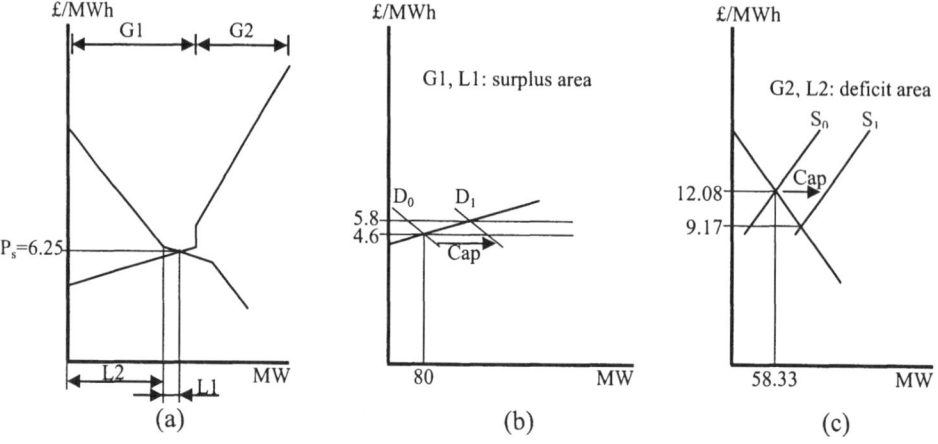

Figure 2.10 (a) System Price; (b) Surplus Area Price; (c) Deficit Area Price (Source: [15])

Table 2.2 Various Prices and Settlement

Capacity Fee in Surplus Area, C_s	$= P_s - P_l = 0.72$ £/MWh
Capacity Fee in Deficit Area, C_d	$= P_h - P_s = 2.65$ £/MWh
Settlement Price	$P_s = 6.52$ £/MWh
Charge Credited to L1 and G2 (M_c)	$= P_{L1}*C_s + P_{G2}*C_d = 73.32$ £/h
Charge Debited to G1 and L2 (M_d)	$= P_{G1}*C_s + P_{L2}*C_d = 410.32$ £/h
Net Income of Grid Company	$= M_d - M_c$
	$=$ Capacity $* (P_h - P_l) = 337$ £/h

Lessons Learned from Deregulation

The Norwegian energy markets have been a successful example of energy deregulation. Market power has not been an issue. Nevertheless the management of power imbalances has aroused concerns since it costs the SO money to resolve bottlenecks in the regulating market. Fortunately it has only contributed to a small amount of Statnett operating budget so far [19], but congestion management will be costly when congestion becomes more serious. Moreover, the selection of regulating bids using merit order, which is easily comprehensible by participants, does not necessarily result in the lowest cost to alleviate congestion.

2.8.3 California

Background

The Energy Policy Act (EPACT) of 1992 clarified the determination of the USA for a competitive energy market. It is not mandatory to implement a wholesale competitive energy market in the nation. Individual states pursuit different policies and pace of restructuring depending on their electricity prices. States with relatively high electricity prices such as California, New York, Massachusetts, etc., are more aggressive in implementing reforms. In 1998, California embarked on a four-year transitional period of deregulation.

Stranded costs have complicated deregulation in California. The state government solved this problem by issuing bonds to inflicted companies to compensate for their losses. Customers' bills include a small amount of charge (e.g. 4 cents/kWh), the so-called competition transfer charge (CTC), to account for stranded costs.

During this transitional period, participation in the pool is optional, apart from three large private utilities, which have to trade through the PX until March 2002. One distinct difference between the Californian Pool and the England and Wales Pool is that in the former case market clearing and bids matching are under a separate entity, the PX, rather than embedded in the duties of the ISO, as in the England and Wales Pool. Moreover, in California, two types of bilateral contracts exist: Contract for Differences and Direct Access Contracts. The fact that CFDs are tied to pool prices has lured some generators to game the market using their market power. The idea of Direct Access Contracts is to counteract this problem: Direct Access Contracts are not bonded to the PX and participants only have to request their transactions through the ISO.

Californian Forward Markets

The California Power Exchange (CalPX) is responsible for holding auctions for the competitive forward markets (day-ahead and day of markets). The day-ahead market is similar to its counterpart in Norway and England. Market participants provide hourly supply/demand bids to CalPX one day prior to physical delivery. MCP is actually the equivalent of system price in Norway or system marginal price in England and Wales. Uniform pricing is adopted and all participants are paid or debited the MCP. The day of market provides participants with the chance to make up for system imbalances by holding auctions at various times during the delivery day.

Zonal pricing is employed for congestion management. Market participants can submit the so-called schedule adjustment bids (SABs) which are similar in nature to the regulating bids in the Nord Pool. The SAB represents the desire of the participant to adjust its schedule if energy price varies. When there is congestion, the region is divided into zones and the ISO calculates the zonal prices using SABs. The PX uses this information to work out the final prices for participants so that upon settlement the PX remains revenue neutral [23].

Lessons Learned from Deregulation

Energy deregulation is in its early stage in California and it is premature to comment on the efficiency and operation of the various markets. However, there is concern over the operation of the spinning and non-spinning reserve markets. Generators have to reserve a certain amount of their capacity in order to bid in the reserve markets. They are not encouraged to do so unless they can make more money in the reserve markets than in the spot market. Because of that reason, generators submit very high bids to the reserve markets, resulting in non-competitive reserve prices. Non-spinning reserve has a relatively higher price than spinning reserve because there are insufficient participants in the non-spinning reserve market. For maintenance of system security, the ISO has to purchase a certain amount of both reserves. Since non-spinning reserve cannot be replaced by spinning reserve, the consequence is a higher price for a lower quality good (non-spinning reserve is not as 'worthy' to the system as spinning reserve is). These exemplify market inefficiencies caused by unapt market rules.

2.8.4 Scotland

Since the commencement of energy privatisation in 1989 in the UK, Scotland, unlike England and Wales, has not acquired a competitive and efficient wholesale market. Also, for various reasons, Scottish customers have benefited much less than their counterparts in England and Wales, despite the fact that the England and Wales Pool prices have been well in excess of the marginal generation costs in Scotland, even after taking into account the charges for transmission losses, interconnector access charges, and NGC use of system charges. Prices for transmission and distribution are regulated using the 'price-cap' control which depends on the inflation rate and electricity prices are set based on the pool prices in England and Wales with adjustments made after taking into account the differences of the markets.

Scotland is characterised by a surplus of generation capacity over demand, with generation capacity almost two times the total maximum demand [24]. It has diversified generation types including dual oil and gas, coal-fired, hydro, pumped-storage and nuclear. The two generation companies, Scottish Power and Scottish Hydro-Electric have interconnected grids and Scottish Hydro-Electric can access the grid in England via Scottish Power's transmission system. Even though these two dominant privatised generation companies, remain vertically integrated after privatisation, they are required to keep separate accounts for separate businesses, i.e. transmission, generation and distribution. Competition between the two companies is made possible through 'second-tier suppliers' who are authorised to supply electricity to customers outside their supply areas.

There exist several potential obstacles to efficient trading in Scotland. Firstly, the two biggest generators are vertically integrated. Secondly, the market is too small to be competitive. Moreover, the substantial surplus of generation capacity indicates that there is no need to build new generators in the coming future. Finally, generators from England cannot compete with Scottish Power or Scottish Hydro-Electric because of the weak interconnector between the two countries. In view of the above, Ofgem, the Office of Gas and Electricity Markets, will focus on reforms for the Scottish markets which will remove the obstacles and be consistent with the NETA [25].

2.8.5 New Zealand

The voluntary wholesale electricity market in New Zealand commenced in 1996, but before that there had already been limited competition in the supply sector. It is operated by the Marketplace Company Ltd (M-co Ltd) which has recently been sold to RMB Australia Ltd. As in the Californian Pool, market participants in New Zealand can trade outside the pool through bilateral contracts, provided that the system operator is informed of the transactions.

In New Zealand generation is dominated by hydro power, which is located in the South Island. The load concentrates on the North Island which is connected to the South Island by an HVDC interconnector. Even though the three government-owned generation companies dominate the wholesale market, the market remains transparent through the broadcast of predicted prices and load forecast. Effort was spent only on introducing competition in the retail sector between distributors and the state-owned generation company, but it was soon realised that retail competition alone was not enough to reduce electricity prices and hence the wholesale market was developed subsequently.

The New Zealand spot market is an *ex post* market featuring nodal pricing. Nodal pricing is based on the theory of spot pricing [26]. Under nodal pricing, if the market is competitive, the short-term price signals so generated should enhance the efficient operation of the market. However, there have been ongoing discussions on effective economic long-term price signals and the management of the losses and constraint surplus arising from nodal pricing [27]. *Ex post* pricing in the physical spot market is accomplished using the latest supply and demand bids and the actual measured plus losses in the system. Measurement of actual demand is vital and it is one of the main roles of MARIA, The Metering & Reconciliation Agreement. Final prices are published a few days after the actual dispatch.

2.8.6 The European Union and Germany

After years of negotiation and debate, The Council for the European Union eventually adopted Directive 96/92/EC in December 1996 to liberalise the electricity industry. According to the directive, members of the EU are required to open their markets gradually. By the year 2006 at least one-third of the EU-wide energy market will have been opened. Different European countries can liberalise their markets at their own pace, as long as the requirements set by the directive are met. Apart from introducing competition in the wholesale and retail sectors, the directive also features unbundling. Countries at the forefront of liberalisation include Spain and the Netherlands. In Spain, a pool, similar to the existing one in England and Wales, will be developed to set up pool prices based on hourly supply and demand bids, while in the Netherlands the Electricity Act in 1998 mandates a complete liberalisation of the generation section by the year 2007. However, there are also countries, like France, Italy and Belgium, which keep their liberalisation progress to the minimum level required by the directive because of domestic political reasons.

In April 1998, Germany opened its market to all suppliers and end users. As it is a country with relatively few natural resources, two-thirds of the energy consumed is imported from other countries. Effort in deregulation is therefore focused on the maintenance of security of supply. Under the Energy Law Amendment net owners are required to provide open access to facilitate competition. However, only a few out of about 700 net users have so far published the charges for using their networks [29]. At present, most net owners also operate the grid; therefore the issue of separation of ownership and operation would need to be looked into. Also, practically small customers have not been able to change their suppliers easily under the current legislation.

The German project group on the energy market is drafting a potential project sketch and it is likely that the concept for the potential energy market will be similar to the developed EEX [30] (European Energy Exchange). It is envisaged that the development of the market will be done step by step. The first step will be the development of a futures market where bilateral contracts can be traded ahead of time. Then a spot market will be founded for physical and short-term power trading. Before reaching that step, Germany has to work on the infrastructure and regulations for fast and reliable wheeling which is essential for efficient running of the spot market.

2.9 References

[1] Energy Information Administration, http://www.eia.doe.gov/emeu/pgem/electric/.
[2] BTM consult Aps, http://www.btm.dk/Articles/fed-global/fed-global.htm.
[3] Stefano Zamagni, *Microeconomic Theory: An Introduction*, Basil Blackwell, Oxford, 1987.
[4] John Bernard, Robert Ethier, Timothy Mount, William Schulze, Ray D. Zimmerman, Deqiang Gan, Carlos Murillo-Sachez, Rober J. Thomas and Richard Schuler, 'Markets for electric power: Experimental results for alternative auction institutions', available via http://www.pserc.wisc.edu/index_publications.html, *Proceedings of the Hawaii International Conference on System Sciences*, January 1997.
[5] John Bernard, Timothy Mount, William Schulze, Ray D. Zimmerman, Robert J. Thomas and Richard Schuler, 'Alternative auction institutions for purchasing electric power', available via http://www.pserc.wise.edu/psercbin/test/get/publication/, 1998.
[6] Frank A. Wolak, and R. H. Patrick, 'The impact of market rules and market structure on the price determination process in the England and Wales electricity market', selected paper presented at the *POWER Conference*, March 1997, University of California, Berkeley, Berkeley, California, February 1997.
[7] Tim Mount, 'Market power and price volatility in restructured markets for electricity', available via http://www.pserc.wisc.edu/index_publications.html, November 1998.
[8] NETA, New Electricity Trading Arrangements for England and Wales, are based on proposals published by OFFER, Office of Electricity Regulation, July 1998, available via http://www.ofgem.gov.uk/.
[9] Felix F. Wu, 'Coordinated multilateral trades for electric power networks', *12th Power Systems Computation Conference*, Dresden, August 1996.
[10] K.L. Lo and Z.Q. Mo, 'Methods for determining wheeling rates', submitted to the special issue of *International Journal of System Science on the Restructuring of the Electric Power Industry*, 2000.
[11] Ignacio J. Perez-Arriaga, Hugh Rudnick and Walter O. Stadlin, 'International power system transmission open access experience', *IEEE Transactions on Power Systems*, Vol.10, No.1, February 1995.
[12] Young-Moon Park, Jong-Bae Park, Jung-Uk Lim and Jong-Ryul Won, 'An analytical approach for transmission costs allocation in transmission system', *IEEE Transactions on Power Systems*, Vol.13, No.4, November 1998.

[13] J.W. Marangon Lima, M.V.F. Pereira and J.L.R Pereira, 'An integrated framework for cost allocation in a multi-owned transmission system', *IEEE Transactions on Power Systems*, Vol.10, No.2, May 1995.

[14] J.W. Marangon Lima and E.J. de Oliveira, 'The long-term impact of transmission pricing', *IEEE Transactions on Power Systems*, Vol.13, No.4, November 1998.

[15] K. Lo, Y.S. Yuen and L.A. Snider, 'Congestion management in deregulated electricity markets', *Proceedings of the International Conference on Power Utility Deregulation, Restructuring and Power Technologies 2000*, City University, London, IEEE, April 2000, pp.47-52.

[16] Michael D. Cadwalader, Scott M. Harvey, William W. Hogan and Susan L. Pope, 'Coordination congestion relief across multiple regions', *Harvard Energy Policy Papers*, available via www.ksg.harvard.edu/people/whogan/index.htm, October, 1999.

[17] R.S. Fang and A.K. David, 'Optimal dispatch under transmission contracts', *IEEE Transactions on Power Systems*, Vol.14, No.2, May 1999.

[18] Marija Ilic, Francisco Galiana, Lester Fink, *Power Systems Restructuring: Engineering and Economics*, Kluwer Academic Publishers, 1998.

[19] Richard D. Christie and Ivar Wangensteen, 'The energy market in Norway and Sweden: Congestion management', *IEEE Power Engineering Review*, Vol.18, No.5, May 1998

[20] John Bower and Derek Bunn, 'A model-based comparison of pool and bilateral market mechanisms for electricity trading', London Business School, Energy Market Group, May 1999, available via http://www.econ.iastate.edu/tesfatsi/epres.htm.

[21] Catherine D. Wolfram, 'Electricity markets: Should the rest of the world adopt the UK reforms?', Program on Workable Energy Regulation (POWER), University of California Energy Institute, September 1999, available via http://www.ucei.berkeley.edu/ucei.

[22] The Nordic Power Exchange, 'The spot market', available via www.nordpool.com.

[23] For details and examples refer to, 'Zonal clearing market prices: A tutorial', available via http://www.calpx.com/news/publications/index.htm.

[24] Scotland has 10,000 MW generation capacity against maximum demand of around 5,750 MW, data taken from 'Review of Scottish trading arrangements: A consultation document', The Office of Gas & Electricity Markets, October 1999, available via http://www.ofgem.gov.uk/.

[25] Details of future proposals can be found in the latest documents published by Ofgem via its web site: http://www.ofgem.gov.uk/.

[26] Fred C. Schweppe, Michael C. Caraminis, Richard D. Tablors and Roger E. Bohn, *Spot Pricing of Electricity*, Kluwer Academic Publishers, 1988.

[27] Market Pricing Working Group, 'Issues associated with a discussion of the losses and constraint surplus', The Marketplace Company Limited, July 1999, available via http://www.m-co.co.nz/H2nzem/9910.htm.

[28] Greenpeace, Germany, 'Strommarkt in Deutschland: Vom Monopol zum Kartell', November 1998, available via http://www.greenpeace.de.

[29] Information obtained in the 'Strombörsen' section at: http://www.strom.de.

3

Competitive Wholesale Electricity Markets

Prof. Mark O'Malley
University College Dublin
Ireland

Prof. Chen-Ching Liu
University of Washington
Seattle, USA

3.1 Introduction

Electricity markets throughout the world are undergoing major changes [1]. These changes are varied in their nature but the underlying trend is towards a more competitive and open environment and this results in electricity being traded as a commodity and in the creation of competitive markets to facilitate this trade. Political forces [2,3] are driving these changes. A competitive electricity market is one in which a number of suppliers (generators) are competing to sell their electricity to a number of competing customers (loads). Here we are concerned with competition in a wholesale electricity market where the customers are large consumers or a retailer who will resell the electricity to the smaller consumers.

Although electric energy can be stored in batteries it would be uneconomic to store it in large quantities and hence electricity is a real-time commodity being produced and consumed instantly. The electricity demand has significant daily, weekly and seasonal variations and also has a significant random component [4]. The main commodity being bought and sold in an electricity market is energy. There are, however, other services such as reserves, reactive power and automatic generator control (AGC) which must be provided in order that the electricity system can function reliably [5]. These ancillary services need to be provided and an electricity market needs to be structured to facilitate trading of these services [6]. The generators and their customers are typically well distributed geographically and Kirchhoff's laws determine the routes taken by the power on the transmission system. The consequence of this is that congestion can occur on this transport system and altering the supply (generator outputs) and demand (customers' consumption) alleviates this congestion [7]. These adjustments are a constraint to competition [8]. Energy, ancillary services and transmission are interdependent and this coupled with the real-time stochastic nature of the electricity demand makes designing an efficient electricity market a great challenge.

Most participants in a wholesale electricity market will be connected to the high-voltage transmission system as opposed to the low-voltage distribution system. This transmission system transports the electricity. In some markets single entities operate generating units, transmission systems and supply the customers directly. These entities are known as vertically integrated utilities (VIUs) and can be monopolies. Where monopolies exist or where a dominant market position is held in one part of the industry, particularly generation, authorities are implementing new market structures to encourage competition [2,3]. It is uniformly accepted that the transmission system is a natural monopoly and in this new environment it should be regulated to ensure a competitive and open market [9]. Here it is assumed that all other aspects of the wholesale electricity market are competitive, although it is recognised that many wholesale electricity markets have aspects that are not competitive. For example, in Norway and Sweden, amounts of reactive power in excess of some predefined limit are compensated but amounts below this limit are not [10]. Consumer demand is largely inelastic but demand-side participation in competitive markets is technically feasible and is becoming more commonplace [11].

In a monopolistic framework a regulated VIU makes planning and operational decisions based on a least cost objective, subject to constraints (generator and system) and reliability criteria [12,13]. This planning and operational process typically necessitates the deployment of a suite of scheduling algorithms, each one specialising in solving a particular optimisation problem over a distinct time frame. In real-time operation this involves economic dispatch algorithms which achieve a real-time balance between supply and demand in a least cost manner. More advanced economic dispatch algorithms such as the optimal power flow (OPF) type consider the optimal dispatch subject to security constraints including transmission line limits, voltage levels, reserve and regulation [12]. In the longer time frames unit commitment (UC) types of algorithms are used to obtain an optimal schedule, i.e. the on/off and dispatch decisions for generators, for periods of time of up to one week [14]. These UC algorithms account for intertemporal constraints such as ramping rates, minimum up and down times that are particularly important for systems with large thermal plants which are limited by these types of constraints [15]. System security issues such as the provision of reserve and transmission constraints can also be included in these UC algorithms, although typically not to the same level of detail as the OPF-type algorithms [16,17]. As the time horizon increases from the operational time frame (seconds to hours) into the planning time frame (weeks to years) the deterministic power system models are replaced by probabilistic models [18]. In this planning phase capital investment decisions are made. This planning and operational process is a continuous process. As the time for delivery approaches, the schedules and dispatch are continually refined to adapt to current circumstances.

The profits of VIUs with a monopolistic status are typically regulated. This regulation process may result, in the short term, in least cost operation of the power system. However, there is little or no motivation to reduce the long-run costs of the production of electricity. In particular investment decisions by the VIUs may be virtually risk free as they are assured of a regulated return on their investment. By creating markets with a number of competing firms the competitive forces should encourage innovation and should serve in the long term to reduce the cost of producing electricity [19]. Therefore not only should these markets result in cost minimisation in the short term but their competitive aspect should in the long run serve to reduce these costs even further. In the competitive market situation therefore a set of markets need to be developed that mimic the VIU least cost objective, subject to operational and reliability constraints. In particular, scheduling algorithms are being replaced by markets for energy, transmission and ancillary services [20,21,10]. Just as with scheduling algorithms these markets have different time frames. The real-time or balancing markets are run very frequently to maintain balance between supply and demand and to ensure system security and are similar to economic dispatch and OPF algorithms. In many markets there may be a need to run day-ahead markets that will be like the unit commitment process [22]. Long-term capacity markets may also be a feature in some systems where for reliability reasons generators are compensated for keeping available capacity [23].

Competitive electricity market design is a highly complex exercise that is influenced not only by economic and engineering considerations but also by historical, political and social constraints. Many of the current designs have recognisable flaws [24] which can be attributed to both technical and non-technical influences. Therefore existing designs need to be assessed with these factors in mind. Lessons can be learnt from existing markets but generally every market has particular circumstances which make this process difficult. Different electricity market designs in different circumstances can be equally effective in producing the desired result, an efficient and reliable electricity supply. Different market designs in the same circumstances may also produce the same desired results. There is no unique solution to the complex problem of electricity market design. While most commentators will agree that competitive electricity markets will result in broad benefits to society there are some very significant differences of opinion on some of the market design issues. These differences of opinion can be dogmatic in nature and should not be allowed to cloud the issues. Each region/country should choose a design that reaps the benefits of competition but suits their particular social, economic and political environment.

Here a broad overview of wholesale electricity markets is given starting with a description of the independent system operator in Section 3.2. The largest section in the chapter, which describes wholesale electricity market characteristics, follows in Section 3.3. These characteristics include auctions, bidding, pricing, forward and real-time markets, simultaneous and sequential markets, congestion management, bilateral trading, scheduling, gaming, ancillary services, physical and financial markets. Where appropriate simple numerical examples are given to illustrate these characteristics. In an attempt to classify wholesale electricity markets Section 3.4 describes two dominant market models. Wholesale electricity markets are still an active area of research and Section 3.5 discusses some of the challenges in the design and operation of these wholesale electricity markets. Acknowledgements are in Section 3.6 and a comprehensive list of references is given in Section 3.7.

3.2 The Independent System Operator

As more and more regions/countries open up their electricity markets to competition, the question of how to design the market in the best interests of the consumers and suppliers is of prime importance. Central to this are the energy, transmission and ancillary services markets and how they are coordinated. The competitive market principle of supply and demand functions effectively in many markets, e.g. stock markets. However, no physical delivery of a product is required by a stock market whereas in an electricity market a product must eventually be delivered instantly (i.e. no storage) and its delivery is achieved by a physical power system. The closer we get to physical delivery the more important are the operational and reliability constraints. These basic principles are well understood and practised in many functioning markets and it is universally accepted that electricity markets require an independent system operator (ISO). Although an accepted principle the role of the ISO is a hotly debated topic. Some market structures require a large role for the ISO while others require a minimalist approach. This operator should be fair and non-discriminatory to all participants of the market hence the term *independent* system operator. In general the ISO will be responsible for tasks such as coordination of the day-ahead scheduling, real-time balancing of load for all users and ensuring compliance with all regional operating and reliability standards. The ISO will operate the transmission facilities to provide non-discriminatory, open access to the transmission grid to all users and management of transmission congestion and constraints on a network basis [25]. To promote competition and to maintain system reliability the ISO should also play a role in the planning of transmission system enhancements. The ISO may also procure ancillary services on a competitive and unbundled basis and perform the settlement functions and provide transparent information flow.

The transmission grid has a number of aspects that need managing, ranging from connection policies, congestion management and the administration of energy losses. Connection policies are an important aspect of the ISO responsibilities. It sets the standards and charges that all participants must meet in order to connect to the grid and hence participate in the market. The transmission system is made up of a number of buses (nodes), where generators and consumers are located, and these buses are connected together by lines. These lines transport the electrical energy around the high-voltage transmission system and have limited capacities, which for security reasons should not be breached [26]. When a line is at its limit the system is congested and typically this will limit power injections at every node in the system. Relieving this congestion requires generators and/or consumers to alter their quantities. Therefore congestion puts a constraint on the energy markets and in many instances may render them non-competitive [27]. Losses on an electricity grid are unavoidable and can be substantial. A market participant's physical location on the transmission system, i.e. what buses it is connected to, is an important factor in wholesale electricity markets. The book by Schweppe *et al.* [28] gives the fundamentals of establishing the instantaneous locational price (spot price) of electricity. For example, a generator that is injecting power into the system at one location at one instant in time can cause substantially different losses and congestion than a similar injection at another location and/or time. The cost of these losses and the usage of the transmission system need to be allocated in some manner to the participants in the electricity market and this is not a trivial task [29]. The revenues collected by the ISO from the generators and loads for these transmission services (connection, usage, congestion and losses) need to pay for the transmission system in the short and long term [30].

In the VIU environment the least cost objective typically referred only to the cost of energy [31]. The ancillary services such as reserve and voltage control were treated as constraints in the optimisation process and their cost may not have been explicitly represented. The provision of these ancillary services is costly, and the quantification of these costs is difficult, e.g. ramping and reserve are services that generating units provide to the system and they have significant costs associated with them [32]. Generators will not provide these services unless they are adequately compensated [33]. In some cases, however, generators may be obliged to provide these services in order to be allowed to participate in the energy market. Ancillary services can be self-provided by the energy market participants and the ISO is responsible for acquiring the balance. Physically self-provision of these services may be inefficient and energy market participants may purchase these services from others. Therefore in competitive wholesale electricity markets there is a need for the establishment of suitable ancillary services markets [6,34].

Power system reliability is a difficult engineering task in situations where the system is run as a VIU [13]. With the advent of competitive markets for electric energy and the drive towards competitive ancillary services markets this task is changing [35,36]. When a generating unit is forced out, the system dynamics are aggravated which, may cause load shedding or put the system in a state where load shedding is more likely to occur [37,38]. Also, the replacement energy and reserve that must be procured by the ISO are costly. Sheblé [39] recommended penalising users who aggravate the dynamic behaviour of the system and rewarding those who improve it. If generator-forced outage probabilities are considered in the market process, then the expected impact of contingencies can be lowered thus enhancing the security of the system [40]. Generator reliability is improved through regular maintenance and good operating practice. In competitive electricity markets, generation owners are responsible for maintaining their units. This can mean that generators are not maintained often enough because of cost and lost revenue. A system whereby reliable generators are rewarded can be used to compensate generators for maintaining units. The use of fines when units are forced out can provide a further incentive to maintain units [41]. A strong argument can be made that these rewards and fines are not necessary, as pure market forces will encourage unit reliability. A unit that is unreliable will miss out on profitable opportunities to participate in the market. Reliability comes at a cost and in the competitive environment if this cost is deemed greater than its benefit [42] then in the event of a shortfall in generation the ISO may interrupt customers rather than pay for reserve [43,44].

3.3 Wholesale Electricity Market Characteristics

In a wholesale electricity market multiple products that may not be very distinctive from one another are being traded over multiple time periods using several different mechanisms. Here electricity market characteristics are described to give a sense of the number of choices that are available in designing an electricity market. A small power system model is used to illustrate the basic characteristics where appropriate. Some references to existing wholesale electricity markets are used here to illustrate these characteristics but it should be noted that in general these existing markets have peculiarities which do not allow them to be characterised in a simple manner. It should also be noted that even at the time of this writing many of these markets are changing. Where available the relevant web sites are given which will provide up-to-date information. Furthermore the quantity and quality of information that is easily available varies greatly and the reader should bear this in mind.

3.3.1 Small Test System

Consider the following simple test system consisting of a supply side with two generators, a demand side with two loads and a simple three-bus network.

Supply Side
It is assumed that the two generators have quadratic production cost curves [12] and minimum and maximum generating constraints given by

$$Cost_{gen,1} = 0.01P_1^2 + 12P_1 + 300 \ \$/h, \qquad 50 \leq P_1 \leq 500 \qquad (3.1)$$

$$Cost_{gen,2} = 0.015P_2^2 + 6P_2 + 400 \ \$/h, \qquad 100 \leq P_2 \leq 600 \qquad (3.2)$$

where P_1 and P_2 are the power outputs in MW of generator #1 and generator #2, respectively.

Demand Side
The two loads are assumed to have quadratic utility curves and minimum and maximum requirements given by

$$Utility_{load,1} = -0.016L_1^2 + 35L_1 \ \$/h \qquad 0 \leq L_1 \leq 900 \qquad (3.3)$$

$$Utility_{load,2} = 40L_2 \ \$/h \qquad 0 \leq L_2 \leq 200 \qquad (3.4)$$

where L_1 and L_2 are the power consumed in MW by load #1 and load #2 respectively.

Transmission system
A simple three-bus, three-line transmission is assumed (Figure 3.1) with generator #1 at bus A, generator #2 at bus B and both loads at bus C. Lines AC and BC have equal reactance and line AB has a reactance of twice their value. Lines AB, AC and BC have 100 MW, 200 MW and 400 MW line limits respectively. Using a DC load flow model (lossless, rated voltages and small angles) [12] then the following are the transmission congestion constraints:

Line AB $\quad -100 \le 0.25P_1 - 0.25P_2 \le 100$ (3.5)

Line AC $\quad -200 \le 0.75P_1 + 0.25P_2 \le 200$ (3.6)

Line BC $\quad -400 \le 0.25P_1 + 0.75P_2 \le 400$ (3.7)

The coefficients of P_1 and P_2 are the line sensitivities of the respective lines to injections at the A and B buses respectively [27].

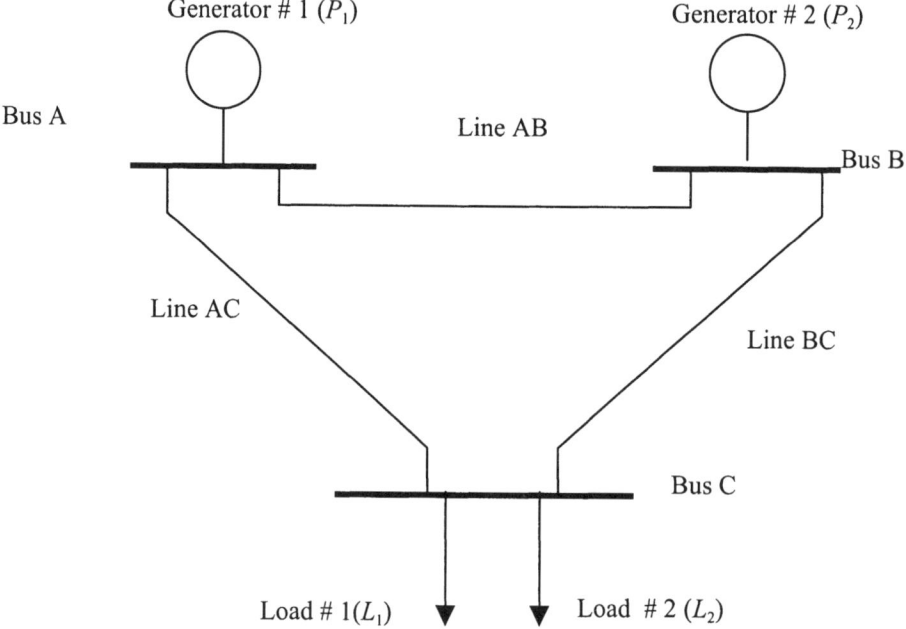

Figure 3.1 Test system

3.3.2   Central Auction

There are two distinct trading mechanisms, the central auction and bilateral trading [45]. In a centralised auction suppliers and customers both submit bids to a central pool or exchange and from these bids the market clears, i.e. determines quantities and prices at equilibrium [46]. In their simplest forms these centralised auctions are identical in function to a simple merit order economic dispatch algorithm [12]. The California Power Exchange (CalPX) is a simple centralised auction for energy [54]. The England and Wales power pool [47] is a highly complex auction mechanism, which is based on a unit commitment algorithm.

3.3.3 Bidding

Bidding into a simple central auction is similar to the process of each generator submitting cost data and each load submitting utility (willingness-to-pay) data to the VIU. This data is used by the VIU to dispatch the system. In an ideal world with a truly competitive electricity market the bid data should be the same as the production cost (utility) data or opportunity cost, whichever is greater. The opportunity cost is the revenue that a participant would expect to get by selling in a different market. This price-taking assumption in a competitive market is an optimal strategy for a market participant [48]. The pricing mechanism is an important factor in this price-taking assumption and the interested reader is referred to the seminal paper by Vickrey [49]. The fixed costs are not relevant to set the price and quantity, i.e. clearing the market. The incremental costs (utilities) are all that are needed to clear the market. Here it will be assumed that there are no opportunity costs and that all market participants bid at incremental cost (utility). The case where bids vary from incremental cost (utility) is dealt with later in the section on gaming (Section 3.3.9). The cost (utility) curves and the incremental cost (utility) curves for the small test system are given in Figures 3.2 and 3.3 respectively. The quadratic cost (utility) curves result in linear incremental cost (utility) curves.

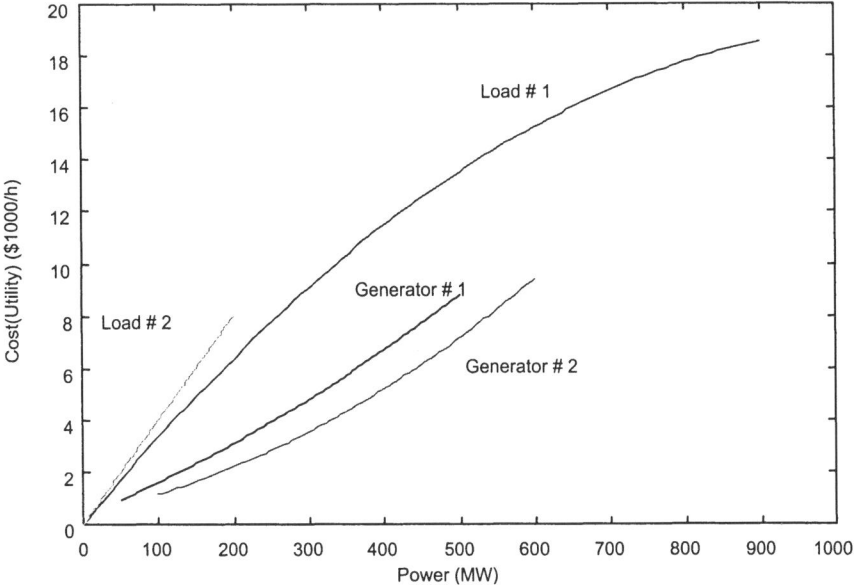

Figure 3.2 Cost (utility) curves for the small test system

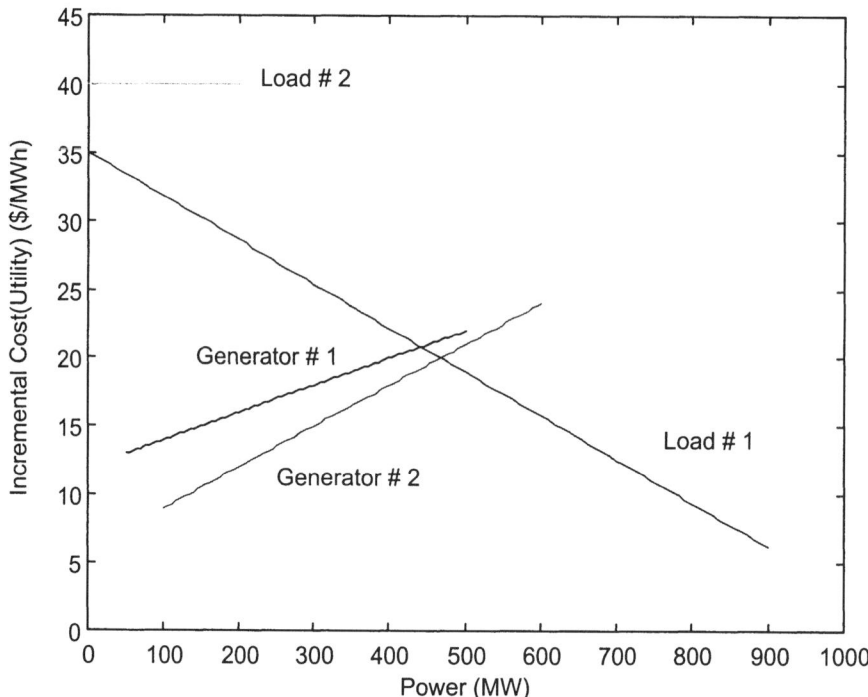

Figure 3.3 Incremental cost (utility) curves for the small test system

3.3.4 *Market Clearing and Pricing*

Supply-demand equilibrium will set the market price and quantities. This equilibrium point is the point where the social welfare, subject to constraints, is optimal. Social welfare is the difference between utility and cost [50]. Graphically this equilibrium point can be described as the point where the incremental aggregate cost curve and the incremental aggregate utility curve cross. At this point any additional trade between the supply and demand side will reduce the social welfare. Figure 3.4 graphically illustrates the test system market equilibrium for a simple auction process without transmission constraints. Aggregating the individual incremental (marginal) cost bids of the generators forms the supply curve and the demand curve is formed by aggregating the incremental (marginal) utility of the loads (Figure 3.3).

In a central auction where multiple generators and loads bid then the price can be set in a number of ways. The most common method is the uniform price method where the price is set at the incremental cost at the point of market equilibrium. This is illustrated graphically in Figure 3.4 for a market with no transmission constraints. This price corresponds to the spot price as described in Schweppe *et al.* [28]. Every generator receives this uniform price for each MWh and every load pays it. For a particular price each generator (load) will produce (consume) a quantity, subject to limits, given by its incremental cost (utility) curve (Figure 3.3). This type of auction concept is implemented in the CalPX forward energy markets and in the Norwegian and Swedish energy markets.

Profits for the generators are calculated by taking the difference between the revenue (price * quantity) and the cost. The cost in these calculations is taken to be the production cost (3.1), (3.2) which ignores other fixed costs such as capital costs etc. The surplus for the load is the difference between the utility and the payment (price * quantity). An alternative pricing mechanism is pay as you bid where participants pay (or are paid) their bid price. It is proposed that this type of discriminatory pricing will be used in the new balancing market in the UK [51].

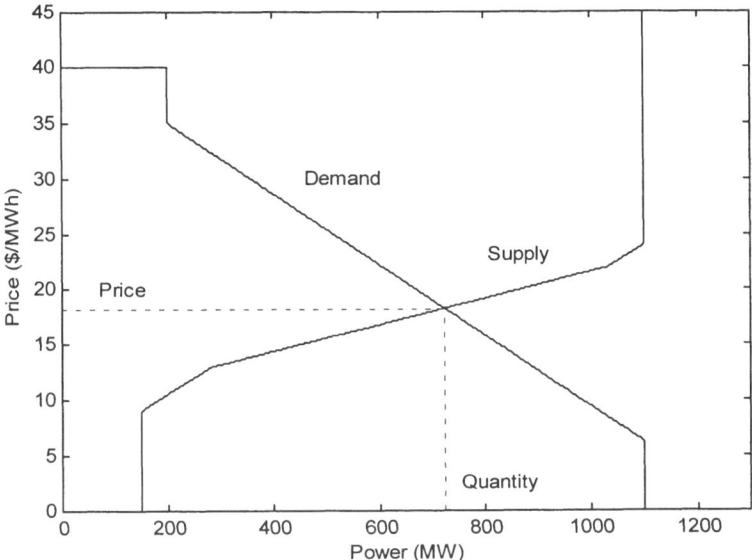

Figure 3.4. Market equilibrium, transmission unconstrained

The market clearing in Figure 3.4 can be expressed mathematically as the maximisation of social welfare

$$\max_{P_1,P_2,L_1,L_2} (Utility_{load,1} + Utility_{load,2} - Cost_{gen,1} - Cost_{gen,2}) \quad (3.8)$$

subject to the unit constraints (3.1), (3.2), (3.3) and (3.4) and the load balance constraint

$$P_1 + P_2 - L_1 - L_2 = 0 \quad (3.9)$$

In this optimisation process (a quadratic programming problem, see Wood and Wollenberg, [12] for details of solution) the no-load and fixed costs are irrelevant and from a bidding perspective incremental cost (utility) is all that is required (Figure 3.3). In order to clear a market in this manner without ambiguity the incremental costs (utilities) should be monotonically increasing (decreasing), i.e. the cost (utility) curves are convex (concave). In this unconstrained market Table 3.1 summarises the resulting quantities, prices and profits for the participants. Because of the uniform price, the load balance constraint (3.9) and the assumption of a lossless system, the pool (central auction) is revenue neutral, i.e. what is paid in by the loads is paid out to the generators.

Table 3.1 Market clearing, transmission unconstrained

Generator/Load	Quantity (MW)	Price ($/MWh)	Profit (Surplus) ($/h)
Generator #1	313.6	18.3	683.7
Generator #2	409.1	18.3	2110.3
Load #1	522.7	18.3	4371.9
Load #2	200.00	18.3	4345.5

3.3.5 Market Timing

Owing to the stochastic nature of the demand [4] and the need to schedule generation resources in advance, electricity markets can be characterised by timing. Forward electricity markets are run in advance of the delivery time. This enables suppliers to schedule generation to meet the demand and for the ISO to coordinate transmission and ancillary service needs. The forward markets also perform a very important financial service for participants by locking in prices and quantities in advance, which avoids the volatility of the real-time (spot) markets and is a prudent business practice [52]. In power systems with large thermal plants that are inflexible because of unit constraints these forward markets are particularly useful. There may be a multitude of forward markets at different time scales: year ahead, month ahead and day ahead. In California the power exchange (PX) runs three different types of forward markets [53,54]. The day-ahead market establishes prices and quantity of electricity for delivery during each hour of the following day. The day-of/hour-ahead market operates similar to the day-ahead market, but offers trading closer to the delivery hour. In the block forwards trading market participants can buy and sell energy months in advance. Markets that are operating in the order of minutes in advance of delivery are deemed to be real-time. As the time of delivery approaches real-time markets are needed to ensure supply and demand balance and to adapt to unforeseen circumstances. These real-time markets are invariably operated by the ISO. In California and Norway the respective ISOs operate an imbalance (regulating) market for real-time adjustments [55, 10]. The power pool of Alberta operates only on a real-time basis but this is set to change with the introduction of a binding day-ahead market [56].

3.3.6 Sequential and Simultaneous Markets

The core product being sold in electricity markets is energy. Ultimately the coordination of units (scheduling) and of the transmission and ancillary services enables its delivery. A sequential electricity market structure is one in which the energy traded is done so independently of the transmission and ancillary services. The provision of the consequential transmission and ancillary services needs follows the energy trading in a sequential manner. In California forward energy markets are conducted independently of real-time energy, congestion management and ancillary services markets [57]. Spain takes a similar approach to California [58]. There is a strong physical coupling between energy, ancillary services and congestion management and this is reflected in some markets where they are traded simultaneously. A simultaneous electricity market structure is one in which the energy traded is done simultaneously with the transmission and ancillary service markets [59]. The PJM (Pennsylvania, New Jersey, Maryland, USA) Interconnection [60] and New Zealand [61] both exhibit this simultaneous characteristic. The interim arrangement in the New England power pool is to use a hybrid sequential/simultaneous approach [62]. The proposed new market structure in Alberta may also be a hybrid as the intention is to co-optimise between the energy market and the ancillary services markets as opposed to running them simultaneously [56]. In the unconstrained market-clearing example above (Figure 3.4, Table 3.1) the transmission line power flows are given in Table 3.2.

Table 3.2 Power flow, market clearing, transmission unconstrained

Line	Line flows (MW)	Line limits (MW)
Line AB (P_{AB})	-23.9	-100
Line AC (P_{AC})	337.5	200
Line BC (P_{BC})	385.2	400

The line connecting bus A to bus C is overloaded by 137.5 MW. An elegant option for dealing with this congestion is to clear the energy and transmission market simultaneously. Although the example that will be given here will deal with energy and transmission simultaneously, it can be generalised to deal with energy, ancillary services and transmission [59,31]. By including the transmission constraints in the market clearing the quantities will not result in an overloaded line. This can be achieved by maximising the social welfare (3.8) subject to unit constraints (3.1), (3.2), (3.3) and (3.4), the load balance constraint (3.9) and the transmission constraint (3.5), (3.6) and (3.7) [12]. There is no simple graphic that illustrates this market-clearing mechanism. With the transmission constraints enforced Table 3.3 gives the quantities, prices and profits and Table 3.4 gives the power flows.

Table 3.3 Market clearing with transmission constraints

Generator/Load	Quantity (MW)	Price ($/MWh)	Profit (Surplus) ($/h)
Generator #1	111.0	14.2	-176.8
Generator #2	467.0	20.0	2871.1
Load #1	378.0	22.9	2286.0
Load #2	200.00	22.9	3419.1

Table 3.4 Power flow, market clearing with transmission constraints

Line	Line flows (MW)	Line limits (MW)
Line AB (P_{AB})	-89.0	-100
Line AC (P_{AC})	200.0	200
Line BC (P_{BC})	378.0	400

Notice that in Table 3.3 the price at each bus is different hence the introduction of the term locational marginal pricing or nodal pricing [28]. The price at each bus is the marginal cost of the next megawatt of power supplied at that bus. If a transmission constraint is active then typically the price at each bus is different. With this pricing mechanism market participants at different buses receive (pay) a different price and this is justified on the basis that the incremental cost is different at different locations. Locational pricing sends participants the appropriate price signals regarding their location. Generator #1 is poorly located in comparison with generator #2 as it is more dependent (0.75 sensitivity as opposed to 0.25 (3.6)) on the congested line to get its power to the load; hence its price is lower ($14.2 as opposed to $20.0). If these price differentials persist over long periods of time this should encourage generation to locate at bus C where the price is highest ($22.9). The price signals may also encourage load to locate at bus A where prices are lowest. The price differential between buses is the marginal cost of transmission between the two buses and hence it makes it an optimal price for transmission congestion management [28,7]. Losses, which have been ignored here, will also result in locational prices if the losses are included in the optimisation process. With locational pricing there is a net revenue gain as the revenue for the generators will be less than the revenue paid by the loads [9]. The net revenue gain in the example above is 2315.8 $/h. This excess revenue, known as congestion rent, can be used to invest in the transmission system to relieve congestion, to compensate the ISO which manages the transmission system, or can be distributed in some other manner [9,63]. PJM and New Zealand both manage congestion using locational marginal prices.

Although elegant and economically pure [28] this type of transmission management results in a price at every bus which makes it difficult to set up complementary financial markets with enough liquidity to hedge risk. Another concern with some participants is that to manage congestion the ISO can rearrange the result of the energy market. Contrasting Table 3.1 and Table 3.3 the transmission management process has had a large impact on the energy market. This interference from the ISO in the energy market is seen by some as unacceptable [57,25]. There are other approaches to congestion management. One involves ignoring the locational aspect and for the ISO simply to purchase increments (adjustment up) and pay for decrements (adjustment down) from the market participants. The cost of this process to the ISO is passed on in the form of a charge to the users of the transmission system. This approach lacks any locational pricing signal and can be severely abused in some market models [27]. A compromise between these two techniques is to set up zones. These zones are defined to be groups of buses within which (intra-zonal) there is little or no congestion expected. The congestion is expected to happen between zones (inter-zonal). If inter-zonal congestion occurs then separate energy markets are conducted in each zone. Intra-zonal congestion is managed with incremental and decremental bids. In Norway this zonal congestion management approach has worked well. In California, however, a similar zonal methodology exists but there have been serious gaming problems with the intra-zonal aspects [64].

3.3.7 Bilateral Trading

A bilateral market is one in which trades (quantity and price) are determined directly between suppliers and customers. This trading approach has the positive attribute that loads will vigorously seek out the cheapest generators. Bilateral markets exist in California, Norway and PJM but are not currently permitted in the England and Wales market where virtually all energy must be traded through the central auction process, the pool. This mandatory pool participation is set to change in England and Wales, allowing participants to trade bilaterally [51].

With bilateral trading there may be net injections which may congest the system or may relieve congestion. In any congestion management process these bilateral trades need to be accounted for [65]. Consider two bilateral trades: bilateral #1 is an injection of 10 MW at bus A and bilateral #2 is an injection of 20 MW at bus B. The loads associated with both these bilateral trades are located at bus C. If these transactions are to be allowed then effectively the line ratings for the system have been changed as the power flow from these bilaterals must be accommodated. These bilateral trades result in the following transmission constraints for the central auction:

Line AB $\quad -100 \leq 0.25(P_1 + 10) - 0.25(P_2 + 20) \leq 100$ (3.10)

Line AC $\quad -200 \leq 0.75(P_1 + 10) + 0.25(P_2 + 20) \leq 200$ (3.11)

Line BC $\quad -400 \leq 0.25(P_1 + 10) + 0.75(P_2 + 20) \leq 400$ (3.12)

The changes are asymmetric effectively increasing the capacity in one direction while decreasing it in the other. Depending on the result of the central auction some bilaterals can be beneficial, helping to relieve congestion, and some can make congestion worse. The market-clearing results are given in Table 3.5 and the power flows are given in Table 3.6. In this particular example, the bilaterals have made the congestion worse as the line between bus A and bus C has an additional 12.5 MW (0.75*10 MW+0.25*20 MW) on it, which leaves only 187.5 MW capacity for the other participants.

Table 3.5 Market clearing with transmission constraints and bilateral transactions

Generator/Load	Quantity (MW)	Price ($/MWh)	Profit (Surplus) ($/h)
Generator #1	92.6	13.9	-214.3
Generator #2	472.2	20.2	2945.3
Load #1	364.8	23.3	2129.6
Load #2	200.0	23.3	3334.9

Table 3.6 Power flow, market clearing with transmission constraints and bilaterals

Line	Line flows (MW)	Line limits (MW)
Line AB (P_{AB})	-97.4	-100
Line AC (P_{Ac})	200.0	200
Line BC (P_{BC})	394.8	400

The bilateral trades have altered the central market result. In order for the bilaterals to be allowed they need to pay for the transmission service. The transmission charge for bilateral #1 is the product of the quantity (10 MW) by the incremental cost of transmission between bus A and bus C ((23.3-13.9) $/MWh), i.e. 94 $/h. The transmission charge for bilateral #2 is the product of the quantity (20 MW) by the incremental cost of transmission between bus B and bus C ((23.3-20.2) $/MWh), i.e. 62 $/h. If the bilateral trades were in a direction that relieved congestion the price differentials would be negative and the transmission charge would be negative, i.e. the bilateral trade would be rewarded for relieving congestion.

Without congestion the bilateral trades can be conducted independently of the central auction. When the system becomes congested then there are locational prices and in order to pay the transmission charge these bilateral trades have been effectively forced into the central auction. This concept is recognised in Norway where zonal pricing is in use and for congestion management purposes bilateral trades between zones must participate in the central auction [20]. PJM with locational (nodal) pricing and California with zonal pricing are effectively the same. This mandatory participation is seen by many as interference in the energy market.

3.3.8 Scheduling

In Table 3.2 generator #1 is bidding at incremental cost and yet is losing money. This is because the market-clearing process is only concerned with incremental cost. At low power outputs units can be getting paid their incremental cost for all their output but because of the no-load costs (300 $/h for generator #1 (3.1) and 400 $/h for generator #2 (3.2)) this is unprofitable. Generator #1 needs to operate above 173.2 MW to make a profit and generator # 2 needs to operate above 163.3 MW to make a profit. However, the market-clearing process described above is assuming only a single time period. Considering a day-ahead market this time period will be typically 1 hour and therefore there will be 24 individual hourly markets. Generator #1 will be interested in its profits over the whole 24 hour period. Turning off in unprofitable hours may not be the most optimal strategy over a 24 hour period because of the cost of starting up again later. There are also intertemporal unit constraints (e.g. minimum up and down time constraints) which should be accounted for in the participant's bids [66]. In some wholesale electricity markets the resulting hourly quantities may be so variable that units' ramping abilities may not be able to match the pattern [67]. It is recognised that some participants may find the resulting schedule from a central auction process financially unattractive or technically impossible to implement. Attempts to deal with these difficulties range from being able to state minimum income and load requirements in Spain [68] to a proposed iterative bidding scheme in the California PX [54]. It is interesting to note that this iterative bidding scheme proposed for California proved impractical and has not been implemented.

In the VIU environment generators were typically scheduled centrally using a unit commitment algorithm. This UC algorithm uses cost data including incremental cost, no load cost, and startup costs and accounts for the intertemporal constraints such as minimum up and down times and ramping rates [15]. In some electricity market these constraints need to be internalised in the bids of the participants [66]. The participants must predict the prices in advance [69] and bid so that the expected quantities are both feasible and profitable. This self-scheduling approach is in existence in the California PX [54], New Zealand [61] and Norway [70]. Bilateral trades are by their nature self-scheduling. For reliability and security reasons, self-scheduling may be subject to approval by the ISO [71]. Bidding into a central auction process can also involve a firm that owns multiple units submitting portfolio bids. These bids represent an aggregate offer. After market clearing the firm can then decide how it will schedule its own units to supply the quantities. The CalPX allows portfolio bids.

An alternative to self-scheduling is centralised scheduling where a UC-type algorithm is used to clear the market [41]. Bidding information in this auction mechanism is very detailed, including all cost data and appropriate technical constraints. In the market-clearing examples above the optimisation problem variables were the quantities which are continuous. In a centrally scheduled system the objective is the optimisation of social welfare, subject to constraints, but the variables are both continuous (quantities) and discrete (turn a generator on or off) [72]. In PJM some units can choose to be centrally scheduled while others with bilateral contracts can self-schedule. In the UK system the energy market is a centrally optimised UC process but this is set to change [51].

3.3.9 Gaming

In perfectly competitive electricity markets the most profitable strategy for a market participant is to act as a price taker and bid at incremental cost [48]. This price-taking assumption assumes an infinite number of competitors so the bidding behaviour of any one player cannot affect the markets, i.e. influence prices. In the real world, however, there are only a finite number of market participants and each participant has to some degree market power, i.e. they can bid strategically to increase their own profits. There is a multitude of possibilities for this type of gaming behaviour in electricity markets [73,74]. Deviations of bids from incremental cost can also occur because a participant wants to ensure a feasible schedule [66] or it knows it has another opportunity in another market and it bids its incremental opportunity cost.

As an example consider generator #1 in the constrained market above. By bidding incremental cost generator #1 is making a loss (Table 3.3). In this case generator #1 can alter its bidding strategy so as to avoid this loss. Table 3.7 gives the result for one strategy where generator #1 increases the linear part of its bid (3.1) from 12 $/MWh to 14 $/MWh.

Table 3.7 Market clearing with transmission constraints and generator #1 bidding strategically, i.e. the linear coefficient is changed from 12 $/MWh to 14 $/MWh

Generator/Load	Quantity (MW)	Price ($/MWh)	Profit (Surplus) ($/h)
Generator #1	106.3	16.1	25.2
Generator #2	481.3	20.4	3075.1
Load #1	387.6	22.6	2403.2
Load #2	200.0	22.6	3480.4

Generator #1 by altering its bid away from incremental cost has turned a loss of 176.8 $/h (Table 3.3) into a profit of 25.2 $/h (Table 3.7). Generator # 2 has also benefited, increasing its profits from 2871.1 $/h (Table 3.3) to 3075.1 $/h (Table 3.7). Load # 1 and load # 2 have both also gained as their surpluses have increased. By bidding above its incremental cost generator # 1 has increased its price and reduced its quantity and most strategies are a balance between increasing price and reducing quantity. With inelastic load there is more scope for driving up prices without excessively reducing quantities [75]. Elastic loads can also bid strategically. The price differentials between bus A and B and bus C has reduced (Table 3.3, Table 3.7) and hence the loser in this game is the ISO whose revenue from the congestion management has reduced from 2315.8 $/h to 1726.3 $/h. There are many other gaming possibilities and these exist in situations where a participant has market power and/or where participants are in collusion [76,73]. Generally the larger the participant the larger the market power. However, transmission congestion can effectively fragment the market into small pockets with very few participants. In this situation strategically located participants with little market power globally can have very significant local market power [77]. Transmission systems that are prone to congestion make the implementation of a competitive electricity market difficult. Services such as reactive power that are highly localised in their nature [26] pose similar difficulties [78]. In California generators have been identified that must run for reliability reasons and these reliability must run (RMR) generators are compensated outside the market process. These RMR contracts have given some participants market power in the California ISO (CAISO) ancillary services markets [79]. Network operators can also act strategically to manipulate market results [80].

In the example above it should be noted that the strategy of generator # 1 is not optimal. Optimal strategies can be found, however [81,82]. Successful gaming behaviour requires participants to have good information about other participants' bids and to consider the stochastic nature of the problem [83].

Constraints and gaming behaviour act to reduce social welfare. Table 3.8 below summarises the social welfare for some of the examples we have considered above. With no transmission constraints and all participants bidding at incremental cost (utility) the social welfare is a maximum. With the transmission constraints more expensive power is dispatched in place of cheaper power and the social welfare reduces. When participants are bidding away from incremental cost (utility) then the social welfare is further reduced.

Table 3.8 Social welfare

Market	Social welfare ($/h)
No transmission constraints (Table 3.1)	11,511
Transmission constraints (Table 3.3)	10,715
Gaming and transmission constraints (Table 3.7)	10,711

3.3.10 Ancillary Services

Ancillary services are required for the reliable operation of the power system [84]. A standard definition of these services is not globally accepted. AGC, reserve (spinning and standby), load following, voltage control and black-start capability would be some of the more commonly recognised services. The generators typically provide these ancillary services although the load can also provide some. New Zealand is an example of a system where a large part of the reserve is provided by the load [61]. Competitive markets for these services are not widespread and this is an active research topic [85]. Many of these services are obtained by long-term contracts. Voltage control in New Zealand and California is acquired by long-term contract. Some of these services can be self-provided by participants in the energy market. In PJM, AGC is self-provided by participants and if they are unwilling or unable then the ISO operates a market to acquire it on their behalf [60]. Reserve in PJM is acquired by constraining the centralised scheduling and dispatching process. If in the process of acquiring this reserve a participant loses profits then it is compensated for its opportunity cost. In California some ancillary services are acquired by specific ancillary services markets, which are run sequentially following the energy and transmission congestion management markets.

Consider the simple test system. Assume that the ISO has determined that 200 MW of spinning reserve is required for system reliability and that this needs to be scheduled in advance by acquiring it in a forward market, e.g. day ahead. Spinning reserve is the ability of an on-line generator (load) to increase (decrease) its output (consumption) in a short period of time. The time period will be determined by the system but for smaller systems the time period is generally smaller in order to avoid large frequency deviations [38,86]. Assume that generator # 1 and generator # 2 can ramp up by 25% and 50% respectively of their unused capacities in the spinning reserve time period. It is also assumed that the two loads are incapable of providing spinning reserve. Therefore the reserve constraint is

$$0.25(500 - P_1) + 0.5(600 - P_2) \geq 200 \qquad (3.13)$$

Table 3.9 shows the results of clearing the market with the above constraint (3.13) and ignoring the transmission constraints (3.5), (3.6) and (3.7).

Table 3.9 Market clearing, reserve constraint, transmission unconstrained

Generator/Load	Quantity (MW)	Price ($/MWh)	Profit (Surplus) ($/h)
Generator #1	316.9	21.9	1842.4
Generator #2	291.5	21.9	2969.2
Load #1	408.4	21.9	2669.3
Load #2	200.00	21.9	3614.1

The first thing to notice about Table 3.9 is that in comparison with Table 3.1 the quantities have altered substantially. In order to meet the reserve constraint (3.13) generator #2 has had its quantity reduced and load #1 has also been reduced. Generator #1 is largely unchanged and load #2 is unchanged. Although generator #2 has had a large reduction in quantity it is more profitable than the unconstrained case (Table 3.1). The reason for this is that the price has increased. Although generator #2 has lost quantity it cannot complain about its profits. The biggest gainer out of this situation is generator #1 whose profits have more than tripled. This highlights the possibility of strategic bidding with technical parameters, i.e. generator #1 has profited by offering a poor ramping ability. It should be noted that if both generator #1 and #2 had the ability to ramp up to maximum output within the spinning reserve time period then the market would clear at the same price and quantity as in Table 3.1, i.e. the reserve constraint (3.13) will not be binding. Here the binding reserve constraint has caused the social welfare to reduce to 11095 $/h from 11511 $/h in the unconstrained case (Table 3.8). It should also be noted that in the event of this reserve being used then generators #1 and #2 would be paid the real-time price for their energy. This scenario, where both generators are better off because of the ancillary services, is not always the case and therefore if a constraint causes a reduction in profits a participant should be compensated for its opportunity cost [60]. The hybrid approach in the New England Pool requires the calculation of this opportunity cost for some reserve [62].

An alternative approach for the provision of ancillary services is to set up markets. In a competitive environment the bid curves for reserve and other ancillary services should reflect a participant's expected opportunity cost. Expected opportunity cost will require forecasting of the energy spot price [69]. In California the ancillary services markets follow in sequence after the energy and congestion management markets. In this way capacity is progressively assigned to the various tasks [55]. In New Zealand the reserve market is cleared simultaneously with the energy and transmission markets. With transmission and reserve constraints there may be a need to account for the interaction between the two, i.e. in the event that reserve is needed it will require transmission [87].

3.3.11 Physical and Financial Markets

Markets can be physical or financial. If the market is physical then the quantities are to be physically delivered in contrast to a financial market where no physical delivery is required. In advance of physical delivery the ISO may well receive information that is indicative of the physical deliveries. However, at some point in time the ISO needs to be informed of the binding physical commitments so it can coordinate the system to ensure security and reliability. Deviations from these binding commitments are typically dealt with by buying or selling the differences at the real-time price. In California, participants submit binding schedules and any imbalances are adjusted in the real-time market that is operated by the CAISO.

Because of price volatility many participants in a central auction process may wish to acquire financial contracts which hedge their position. In Alberta 85% of the market is effectively hedged against the pool price. Alberta currently has only a real-time market and this has been possible because of the large-scale hedging which protected the participants from price volatility. This situation is set to change with forward markets being introduced in the near future [56].

Bilateral trading is one mechanism that can be used to hedge the volatility in a central auction. A generator and a load that are participating in a central auction can have a contract to supply Q MW at a price P_{rc}. If the pool market has a uniform price of P_{rm} then the generator gets paid $Q * P_{rm}$ and the load pays the same amount. This can be achieved by the generator bidding Q MW at zero price and the load requesting Q MW and indicating it will pay any price for it. If the market price P_{rm} is higher than the contracted price P_{rc} then the generator pays $Q * (P_{rm} - P_{rc})$ to the load. If the market price P_{rm} is lower than the contracted price P_{rc} then the load pays $Q * (P_{rc} - P_{rm})$ to the generator. The net effect is that the generator and load have traded Q MW at a price P_{rc}. The two parties to the bilateral contract are perfectly hedged because of the uniform price. The simple form of hedge is known as a Contract for Difference (CFD).

However, this hedging mechanism is undermined in a system that is regularly congested and has a locational congestion management system. If the load and generator are at the same bus then the hedge is still perfect. If, however, the load and generator are at different buses they will have to pay for transferring Q MW from the generator bus to the load bus. This payment could be revenue depending on the price differential. How the load and the generator split this payment (revenue) between them is their own business. There is now a third party involved: the ISO which collects the charges for congestion management. The locational prices can be very variable and hence the price of transmission can be highly volatile. A solution to this difficulty is the concept of financial transmission rights (FTRs) where participants can in advance purchase from the ISO the right to collect the transmission charge for Q MW between two buses [63]. In this way the load and generator are again hedged. If these transmission rights are competitively traded then their price should reflect the expected price differential between the load and generator buses. The ISO must also ensure that these transmission rights are feasible, i.e. its transmission congestion income in the physical market covers the payments due on these rights. Transmission rights may also be subject to gaming behaviour [88]. With the existence of multiple trading opportunities (bilateral, spot market, forward, FTRs, etc.) in the wholesale electricity market participants will endeavour to optimise their portfolios [89,90].

3.4 Market models

From the electricity market characteristics described above it is evident that there is a multitude of design choices for electricity markets. With the level of complexity and the small number of existing markets it is difficult to make strict classifications of market models. In the literature there are two dominant models, the poolco model and the bilateral model [45]. In the poolco model all energy and related transmission and ancillary services are traded in a central auction mechanism in a coordinated manner. In the bilateral model all energy and related transmission and ancillary services are traded bilaterally. In practice, however, most wholesale electricity markets have characteristics of both models. PJM has simultaneously got both pool and bilateral aspects with participants choosing between Network Integration Service (pool model) and Point to Point Transmission Service (bilateral model).

The level of involvement of the ISO may be a more practical criterion for the classification of electricity markets [91]. Where the ISO operates all aspects of the physical electricity market, i.e. runs a centralised auction mechanisms in a coordinated manner for energy, transmission and ancillary services over multiple time frames, then it can be described as a maximalist ISO. In the minimalist ISO model the ISO only operates those aspects of the market that are necessary to facilitate the trading of electricity by the participants in any manner they choose. The maximalist ISO is obviously closely related to the poolco model while the minimalist ISO is related to the bilateral model; however, these classifications are less ambiguous than the poolco and bilateral classifications. Other terminology that would be appropriate here would be centralised (maximalist ISO) and decentralised (minimalist ISO).

3.4.1 Maximalist ISO

In a maximalist ISO model, resources are pooled together to achieve efficiency. Here the ISO operates all physical markets forward and in real-time. This model consists of a centrally optimised pool with all aspects dealt with simultaneously and it is commonly seen as a natural progression of the VIU mentality where scheduling algorithms become market-clearing algorithms and cost (utility) information is replaced with bid information. In this type of pool market very detailed bidding information is required, including incremental bid, no-load bid, startup bid and technical information such as minimum up and down times, ramping rates, etc. A very complex scheduling algorithm that simultaneously solves the unit commitment and OPF problems is used to clear all the markets [91]. Because of the intertemporal constraints there can be very tight linkages between these markets and the simple supply-demand equilibrium as in Figure 3.4 does not exist. This type of pool will compensate the participants for their startup costs and their no-load costs and the schedules will be feasible. The market-clearing process produces shadow prices which may not be directly related to the bids of any particular unit but do reflect the true marginal cost of supplying the next megawatt hour at each individual bus. The prices for energy are specific to each bus (locational marginal prices, LMP) and therefore transmission congestion is explicitly managed. Ancillary services are also explicitly priced by shadow prices [59]. Examples of markets that exhibit some of these characteristics include PJM,

New Zealand, England and Wales. It should be noted that none of these markets exhibit all aspects of a maximalist ISO; for example, in New Zealand there is no unit commitment and participants self-schedule.

This model offers all the benefits of comprehensive coordination of all the power system resources. Mandatory participation in the centralised scheduling and dispatch is a criticism of this model. Another valid criticism of this type of market relates to the fact that the unit commitment solution is typically not unique and the market equilibrium point is very sensitive to algorithm parameters which could lead to disputes [92]. This is a consequence of the integer nature of the optimisation process [72]. In addition to this difficulty the prices are set by a very complex algorithm which is not easy to understand by market participants [41]. Also, with different prices at every node any supporting financial markets will have very few participants and consequently these markets will be illiquid. This liquidity difficulty can be addressed by combining buses into hubs [60].

3.4.2   Minimalist ISO Model

In a minimalist ISO model, generators and loads will act largely independently of one another, though pools can be formed if the participants so choose. Here the ISO is given a minimalist role in the electricity market. Its role is only to facilitate the market by ensuring the power system is operated in a reliable manner. This typically requires that it is responsible for running the real-time energy market and any aspect of the ancillary services and congestion management that have not been resolved by the rest of the electricity market. The ISO congestion management approach should minimise any impact on the forward energy markets. Other than the ISO's role and the responsibilities of participants to provide information and data to the ISO all other aspects of this type of market are largely unstructured. The bulk of the energy is traded bilaterally between suppliers and customers in advance. The prices that are set are confidential to the two sides of the transaction. These energy trades are then reported to the ISO and the ISO runs markets to manage transmission congestion and to procure the required amount of ancillary services. The real-time market and operation follow this. Markets that fall into this classification include California, Norway and Spain.

This model offers participants the flexibility to determine their own prices, quantities and to self-schedule to meet their obligations. The difficulty with this structure is that because of the lack of coordination between the markets there is no guarantee that the energy trades can be accommodated operationally [71]. This could lead to transmission congestion and/or reliability difficulties that cannot be resolved by the transmission and ancillary services markets. In these situations the ISO may ironically have to revert to a maximalist ISO taking control of the entire market to ensure reliability and security.

3.5 Challenges

Electricity markets are highly complex systems that consist of a number of interrelated markets for different commodities (energy, transmission and ancillary services) and different time frames (real-time, hour ahead and day ahead). There are still many unresolved problems in the design and operation of electricity markets. Practical difficulties arise when the pure economic theory is applied to a power system with operational constraints. The economists want the electricity markets to embrace the laws of supply and demand and with simply ideal examples they can show the benefits of such a competitive environment. The real-time nature, physical constraints and reliability issue all act to make the development of an ideal market impossible. It should be noted that it is well accepted that all markets, even those for simple commodities, are not ideal. Therefore the goal should be to develop a market that is a *best fit* to the ideal.

Several wholesale electricity markets have been established around the world and most of these are in a continuous process of change. This evolutionary process is being driven by the need to address some of the outstanding issues in the design and implementation of these markets. Here some of these challenges are outlined.

3.5.1   Market Power Evaluation and Mitigation

Evaluation of market models can have many different viewpoints. The market must function in a reliable, efficient and fair manner. The generators will want to maximise their profits through the markets. The consumers will seek the best value for the service they receive which may conflict with the aims of the generators [78]. This will necessitate analysing the social benefit that the market offers and the prices that are charged. It will also be prudent to ensure that market power and gaming do not exist and that markets are not overly volatile.

For market evaluation there are some available simulation and analytic tools. Otero-Novas [68] proposed a simulation model that considers the market structure and estimates the expected bid prices and quantities. Kumar and Sheblé [93] have developed an auction market simulator. Hobbs *et al.* [81] have developed a framework to investigate supply curve market equilibrium when all participants are maximising their own position. Green and Newbery [94] investigated the UK market using the supply curve approach.

There is little doubt that market power is being exercised regularly in many electricity markets [95,96]. This practice is characterised by prices, which are well above competitive levels. The result is typically very profitable for the generator and ultimately costly for consumers. This power can be exercised in many ways. Generators with global market power can manipulate the marginal (spot) price as in the England and Wales power pool [96]. Transmission congestion can give participants local market power and they can manipulate the locational marginal prices [97]. Some possible solutions to this problem include the following [76]:

- Better market design. Some markets have experienced difficulties, which could be resolved by better design [24]. The congestion management process in California has a gaming problem and the Federal Energy Regulatory Commission (FERC) appears to be encouraging the adoption of locational marginal pricing as a solution [3,64].
- Breaking up the large generating companies into smaller competitive units. This is a regulatory/political issue, which may not fully solve the problem. In the UK market it is perceived that the two dominant generation companies exercised their market power to raise prices above competitive levels [96].
- Building more transmission so as to avoid creating opportunities for local market power. Over-building transmission may seem wasteful but with this transmission capacity in place local market power can be removed and generators may act more competitively [8]. This additional transmission will also increase the reliability of the system. There is, however, significant environmental concerns related to building more transmission lines.
- Making the load more responsive to price. In the examples given here the load is responsive (3.3); however, this masks the reality where in most wholesale electricity markets the load is largely inelastic. Any generator hoping to push up the price may find that a responsive load will reduce its quantity and reduce the potential for profit [75]. For domestic customers this may be very difficult to implement although some large customers may be capable of installing equipment that can respond to the market price.
- In the long run new technologies may make distributed generation (e.g. fuel cells) more prevalent and this will reduce the need for further investment in transmission [98]. It will also combat market power, in particular, if this type of generation is owned by groups of consumers (i.e. if the market price is too high they will generate themselves). If this does happen then the electricity market will become part of a larger energy market.
- In some markets if the price rises above certain levels the prices are capped; however, this distorts the price signal and may have long-term negative consequences. Price capping has been used at one time or another in most wholesale electricity markets. For example, the California ancillary services markets have had price caps imposed [79].

3.5.2 System Capacity

The issue of planning in generation and transmission must be addressed with a view to maintenance and enhancements to meet increasing demand. On the generation side these functions are generally left to the market, the assumption being that energy prices will signal the best times to maintain units and when to build new plant. The energy price spikes in the Mid West (USA) in June 1998 highlight this issue. A market for generating capacity over a longer time frame (more than one year) may provide the necessary market signals to ensure that the system will expand according to the needs of the consumers [23]. The concept of marginal cost pricing [99,100] for electricity is based on fundamental microeconomic principles [50]. In an ideal market bidding at incremental cost is an optimal strategy [48]. However, the resulting schedule may be unprofitable because of costs such as no-load costs, startup costs and fixed costs (Table 3.3). In the VIU environment with spot pricing Schweppe *et al.* [28] introduced the concept of revenue reconciliation where marginal pricing may not be sufficient to cover all costs and give a reasonable profit. In competitive markets revenue reconciliation should be redundant as it can be argued that marginal cost pricing will in the long run resolve this issue. In the long run if the revenue participants receive is not sufficient to cover their fixed costs plus expected profits then they should not be in business. However, this issue is still a matter of debate [101]. In PJM Interconnection and in the England and Wales power pool there are capacity payments which participants receive in addition to the market spot price. In California no such capacity payments exist in the energy market but there is a capacity component paid for ancillary services [55].

Virtually all the current market structures would appear to fail in encouraging investment in new transmission [8,102]. This may be a premature statement as transmission investment is a long-term issue [30] and the markets have only been recently introduced. Also it could be argued that the transmission system was over-built in the past and the excess capacity is only being utilised recently. In addition environmental concerns are also a factor in the lack of investment in the transmission system. However, if investment in transmission does not keep pace with the increasing demands for electricity there will be long-term economic and reliability problems. The pure market concept dictates that when transmission capacity is needed the market will build it. However, because of construction delays etc. this could lead to periods of unreliability and inefficiencies.

3.5.3 Reliability

While it is desirable to encourage competition in the electricity market to reduce the costs and improve the service quality for consumers, it is also vitally important to maintain the system reliability [103]. In an operational environment, an important reliability measure is system security. System security refers to a system's ability to withstand likely disturbances. A system is said to be in a secure state if it is able to meet the load demand without violating the operating constraints in case of a likely contingency, such as a line or generator outage [104]. In other words, security is defined with respect to a set of next contingencies that are likely to occur. Catastrophic failures of power systems are often caused by cascaded events that are combinations of natural calamities (e.g., weather-related causes), equipment malfunctions, design flaws and/or human errors [105]. The goal of security assessment is to reduce the likelihood of catastrophic failures.

Much effort in the past decades has been devoted to the development of computational tools for system security assessment. These tools include state estimation, contingency selection, contingency evaluation, external network equivalents and load forecast. As the power industry evolves into a competitive environment, system security continues to be an important function. In this new environment, the primary responsible party for system security is the ISO or a similar entity. Since the environment is market driven, however, there are new technical challenges. For example, the level of uncertainty in the generation pattern has increased significantly. This is due to the fact that the market decides the generation patterns and the market outcome may not be easily predictable. Consequently, a system engineer at the ISO who studies system security may find it difficult to predict the future generation and load conditions for evaluation of system security.

Available transfer capability (ATC) is a term defined for the electricity market environment. The North American Electric Reliability Council (NERC) defines ATC as 'the Total Transfer Capability (TTC), less the Transmission Reliability Margin (TRM), less the sum of existing transmission commitments (which includes retail customer service) and the Capacity Benefit Margin (CBM)' [106]. Note that ATC is defined for a fictitious path, say, from node X to node Y. TTC represents the amount of power that can be transferred from X to Y without violating the transmission system constraints, such as line flow limits. TTC depends on the operating condition of a power system; the system constraints that need to be considered include thermal, voltage and stability limits. TRM is used to represent the effect of various uncertainties in system conditions on ATC. CBM accounts for the transmission transfer capability reserved by load-serving entities to ensure their access to generations from interconnections to meet the system reliability requirements.

An example of the ATC determination based on power flow calculations can be found in Bergen and Vittal [107]. To determine the ATC for a path from X to Y, one can inject an amount of power at node X and remove the same amount of power at Y and calculate the power flows. When the injected/removed amount is increased to a level that causes a transmission line to reach its capacity, the amount can no longer be increased. This level of power transfer is the TTC. When a given line contingency is taken into account, the power flows of the post-contingency operating conditions also need to meet the transmission line constraints. Consequently, the ATC may not be as high as the case where the contingency condition is not considered. The steady-state power flow method can be extended to include system dynamics. Time domain simulations can be performed for various levels of power transfer to evaluate system stability including voltage stability and synchronism of the generator rotors. When dynamic security is considered in addition to the steady-state operating constraints, the resulting ATC may further be reduced. Note that the availability of ancillary services such as reactive power sources can also affect the ATC of a power system.

One needs to be aware of the limitations of the path-based ATC concept [108]. The existence of the multiple transactions is a reality in the market environment. When the ATC of a path from X to Y is being evaluated, one needs to consider other transactions that have to be accommodated. For the power flow method, other transactions are represented by power injected into and removed from other areas of the system. These transactions need to be taken into account simultaneously when the ATC for path X to Y is calculated. Clearly, different multiple transaction patterns will lead to different values of the ATC for path X to Y. Figure 3.5 illustrates the concept of multi-dimensional security regions where P_{T1}, P_{T2} and P_{T3} are the power transfers over tie lines 1, 2 and 3 respectively. Note that the projection of the three-dimensional region on the P_{T1}-P_{T2} plane resembles the 'nomograms' used by the industry that describe the secure power transfer levels under certain assumptions. An operating point inside the three-dimensional region indicates a power transfer pattern that does not violate the security constraints. The projection on other planes can be interpreted in a similar manner. Now suppose the power transfer P_{T3} is zero. When the power transfer level P_{T1} is at the value of P_{T11}, the maximally allowable level for P_{T2} is P_{T21}. As the tie line flow P_{T1} decreases to the value of P_{T12}, the maximally allowable transfer level for P_{T2} increases to P_{T22}. Now it is not difficult to see the changes as the power transfer level P_{T3} increases from the zero level. To summarise, the path-based ATC depends on other parameters of the operating conditions, including the other transactions. The ATC of a path, represented by a tie line in Figure 3.5, depends on the levels of other transactions represented by other tie lines.

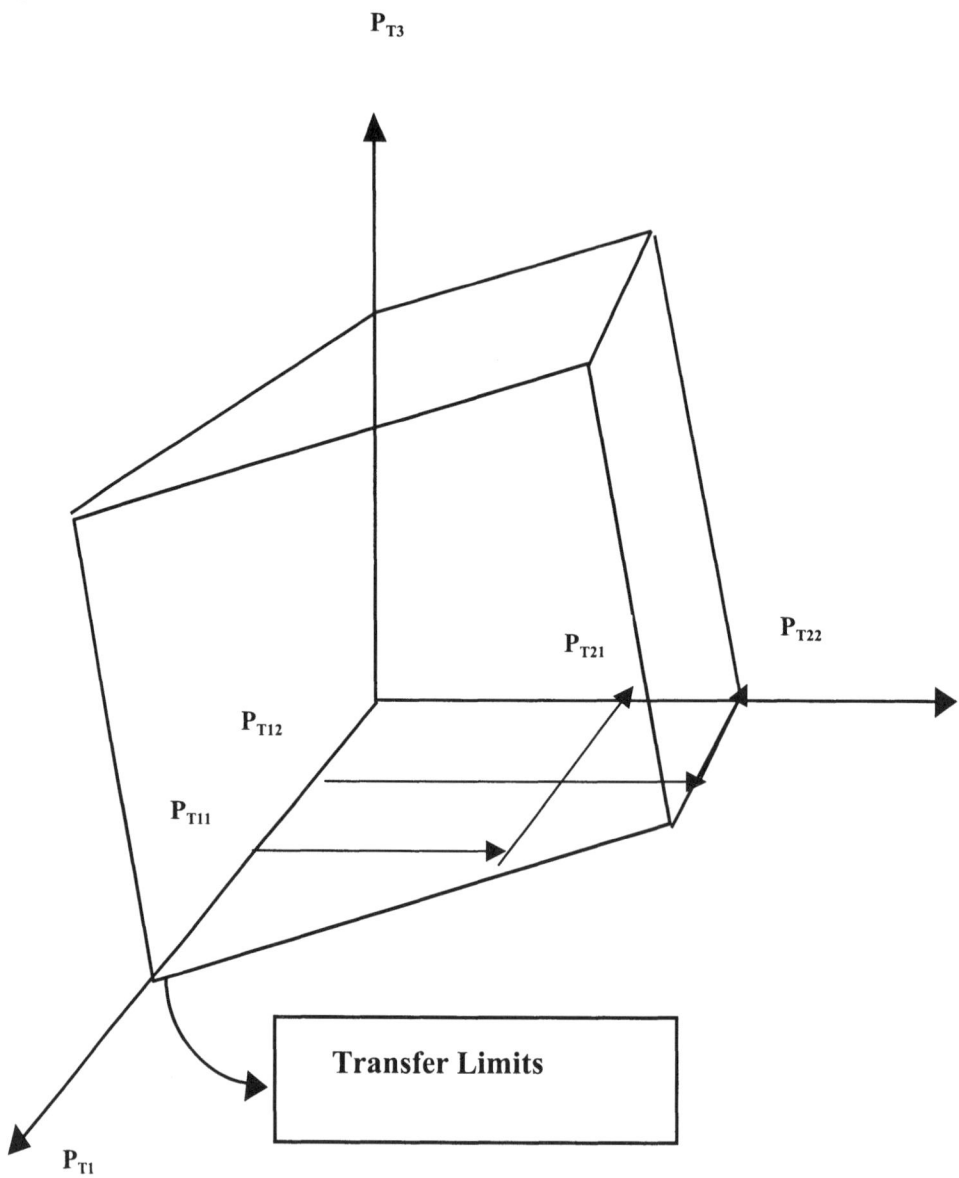

Figure 3.5 Illustration of a power system security region

3.5.4 Technical Issues

Regardless of wholesale electricity markets power system planning and operation has many technical challenges. With the advent of wholesale electricity markets new and different technical challenges may arise which need to be addressed. The computational aspects of the electricity markets are one obvious area of interest [109]. There are also interesting technical challenges related to the management of a large number of transactions [110]. The OPF algorithm which is at the heart of the marginal cost pricing paradigm [28] and of power system security analysis will have to meet ever-increasing challenges [111].

In the minimalist ISO model with self-scheduling the UC algorithm is being implicitly solved in a distributed manner by the market participants [112], which may or may not produce results which are as good as conventional UC algorithms. In the interest of efficiency these decentralised UC approaches need to be analysed. In the maximalist ISO model a centralised UC/OPF-type algorithm is required [113]. Although security-constrained UC algorithms exist [16,17] a UC algorithm with a full OPF formulation for a practical-size power system is still a significant computational challenge. The UC algorithm itself is still a very active research area with many issues unresolved [114,14]. In particular, solutions are invariably suboptimal and not robust [92].

In the short-term, regulators, system operators and market participants will have to face the challenges described above. However, any actions need to allow market forces to *push* the industry towards possible long-term competitive solutions.

3.6 Acknowledgements

The authors would like to thank ESB National Grid, UCD President Research Awards and Fulbright for their financial support. This work is partially supported by US National Science Foundation through Grant ECS-9612636 with matching funds from Alstom ESCA Corp. The authors would also like to thank Prof. Richard Christie, University of Washington, and Mr John Kennedy, ESB National Grid, for their useful comments and insights.

3.7 References

[1] S. Hunt and G. Shuttleworth, *Competition and choice in Electricity*, John Wiley & Sons, 1996.

[2] EU, European Union Council Directive 96/92/EC, 1996.

[3] FERC, Regional Transmission Organisations, Docket No. RM99-2-000, available at www.ferc.fed.us, 2000.

[4] D.W. Bunn, 'Forecasting loads and prices in competitive power markets', *IEEE Proceedings*, Vol.88, pp.163-169.

[5] T. Gjengedal, J.O. Gjerde and R. Flolo, 'Ancillary services in deregulated power systems: What are they: Who need them and who pays', *Proceedings of the Power System Technology Conference*, 1, 1998, pp.704-709.

[6] H. Singh and A. Papalexopoulos, 'Competitive procurement of ancillary services by an independent system operator', *IEEE Transactions on Power Systems*, Paper No. PE-427-PWRS-0- 06-1998, 2000.

[7] T.W. Gedra, 'On transmission congestion and pricing', *IEEE Transactions on Power Systems*, Vol.14, 1999, pp.241-248.

[8] L. Hyman, 'Transmission, congestion, pricing, and incentives', *IEEE Power Engineering Review*, August 1999, pp.4-10.

[9] R.D. Christie, B.F. Wollenberg and I. Wangensteen, 'Transmission management in the deregulated environment', *IEEE Proceedings*, Vol.88, 2000, pp.170-195.

[10] R.D. Christie and I. Wangensteen, 'The energy market in Norway and Sweden: Introduction', *IEEE Power Engineering Review*, February 1998, pp.44-45.

[11] G. Strbac, E.D. Farmer and B.J. Cory, 'Framework for the incorporation of demand-side in a competitive electricity market', *IEE Proceedings - Generation, Transmission and Distribution*, Vol.143, 1996, pp.232-237.

[12] A.J. Wood and B.F. Wollenberg, *Power Generation, Operation and Control*, Wiley Interscience, New York, 1996.

[13] R. Billinton and R.N. Allen, *Reliability Evaluation of Power Systems*, Longman/Plenum Press, 1996.

[14] R. Baldick, 'The generalized unit commitment problem', *IEEE Transactions on Power Systems*, Vol.10, 1995, pp.465-475.

[15] M. Walsh and M.J. O'Malley, 'Augmented Hopfield network for constrained generator scheduling', *IEEE Transactions on Power Systems*, Vol.14, 1999, pp.765 – 771.

[16] J. Batut and A. Renaud, 'Daily generation scheduling optimisation with transmission constraints: a new class of algorithms', *IEEE Transactions on Power Systems*, Vol.7, 1992, pp.982-989.

[17] J.J. Shaw, 'A direct method for security constrained unit commitment', *IEEE Transaction on Power Systems*, Vol.10, 1995, pp.1329-1340.

[18] L.F. Escudero, J. Salmeron, I. Paradinas and M. Sanchez, 'SEGEM: A simulation approach for electric generation management', *IEEE Transactions on Power Systems*, Vol.13, 1998, pp.738-747.

[19] R.J. Green, 'Competition in generation: the economic foundation', *IEEE Proceedings*, Vol.88, 2000, pp.128-139.

[20] R.D. Christie and I. Wangensteen, 'The energy market in Norway and Sweden: Congestion management', *IEEE Power Engineering Review*, May 1998, pp.61-63.

[21] R.D. Christie and I. Wangensteen, 'The energy market in Norway and Sweden: The spot and futures markets', *IEEE Power Engineering Review*, March 1998, pp.55-56.

[22] J. Ancona, 'A bid solicitation and selection method for developing a competitive spot priced electric market', *IEEE Transactions on Power Systems*, Vol.12, 1997, pp.743-748.

[23] N.S. Rau, 'The need for capacity markets in the deregulated electrical industry - a review', *Proceedings of the IEEE Power Engineering Society, Winter Meeting*, New York, 1, 1999, pp.411-415.

[24] J.M. Jacobs, 'Artificial power markets and unintended consequences', *IEEE Transactions Power Systems*, Vol.12, 1997, pp.968-972.

[25] H. Singh, S. Hao and A. Papalexopoulos, 'Transmission congestion management in competitive electricity markets', *IEEE Transactions on Power Systems*, Vol.13, 1998, pp.672-680.

[26] C. Taylor, *Power System Voltage Stability*, McGraw Hill, 1994.
[27] K. Seeley, J. Lawarree, J. and C.-C. Liu, 'Analysis of electricity market rules and their effect on strategic behavior in a non congested grid', *IEEE Transactions on Power Systems*, Vol.15, 2000, pp.157-162.
[28] F.C. Schweppe, M.C. Caraminis, R.D. Tabors and R.E. Bohn, *Spot Pricing of Electricity*, Kluwer Academic Publishers, 1998.
[29] G. Strbac, D. Kirschen and S. Ahmed, 'Allocating transmission usage on the basis of traceable contributions of generators and loads to flows', *IEEE Transactions on Power Systems*, Vol.13, 1998, pp.527-533.
[30] J.W. Marangon Lima and E.J. de Oliveira, 'The long-term impact of transmission pricing', *IEEE Transactions on Power Systems*, Vol.13, 1998, pp.1514-1520.
[31] N.S. Rau, 'Optimal system dispatch of a system based on offers and bids - a mixed integer LP formulation', *IEEE Transactions on Power Systems*, Vol.14, 1999, pp.274-279.
[32] B.H. Bakken, Technical and Economic Aspects of Operation of Thermal and Hydro Power Systems, PhD thesis, Norwegian University of Science and Technology, 1997.
[33] M. Flynn, M.. Walsh and M.J. O'Malley, 'Efficient use of generator resources in emerging electricity markets', *IEEE Transactions on Power Systems*, Vol.15, 2000, pp.241-249.
[34] J. Kumar and G.B. Sheblé, 'Framework for energy brokerage system with reserve margin and transmission losses', *IEEE Transactions on Power Systems*, Vol.11, 1996, pp.1763-1769.
[35] N.S. Rau, 'Assignment capability obligation to entities in competitive markets – the concept of reliability equity', *IEEE Transactions on Power Systems*, Vol.14, 1999, pp.884-889.
[36] R. Billinton, L. Salvaderi, J.D. McCalley, H. Chao, T. Seitz, R.N. Allan, J. Odom and C. Fallon, 'Reliability issues in today's power utility environment', *IEEE Transactions on Power Systems*, Vol.12, 1997, pp.1706-1714.
[37] C. Concordia, L.H. Fink and G. Poulilkas, 'Load shedding on an isolated system', *IEEE Transactions Power Systems*, Vol.10, 1995, pp.1467-1472.
[38] J.W. O'Sullivan and M.J. O'Malley, 'Economic dispatch of a small utility with a frequency based reserve policy', *IEEE Transactions on Power Systems*, Vol.11, 1996, pp.1648-1653.
[39] G.B. Sheblé, 'Priced based operation in an auction market structure', *IEEE Transactions on Power Systems*, Vol.11, 1996, pp.1770-1777.
[40] H.B. Gooi, D.P. Mendes, K.R.W. Bell, and D.S. Kirschen, 'Optimal scheduling of spinning reserve', *IEEE Transactions on Power Systems*, Vol.14, 1999, pp.1485-1492.
[41] M. Flynn, P. Sheridan, J. Dillon and M.J. O'Malley, 'Reliability and reserve in competitive electricity market scheduling', *IEEE Transactions on Power Systems*, in press, PE-012PRS (11-2000).
[42] M.J. Sullivan, B.N. Suddeth, T. Vardeli and A. Vojdani, 'Interruption costs, customer satisfaction and expectations for service reliability', *IEEE Transactions on Power Systems*, Vol.11, 1996, pp.989-995.
[43] J.W. O'Sullivan and M.J. O'Malley, 'A new methodology for the provision of reserve in an isolated power system', *IEEE Transactions on Power Systems*, Vol.14, 1999, pp.174-183.
[44] T.W. Gedra and P. Varaiya, 'Markets and pricing for interruptible electric power', *IEEE Transactions on Power Systems*, Vol.8, 1993, pp.122-128.
[45] H. Rudnick, R. Varela and W. Hogan, 'Evaluation of alternatives for power system coordination and pooling in a competitive environment', *IEEE Transactions on Power Systems*, Vol.12, 1997, pp.605-613.

[46] R.P. McAfee and J. McMillan, 'Auctions and bidding', *Journal of Economic Literature*, June 1987, pp.699-738.
[47] R.D. Tabors, 'Lessons form the UK and Norway', *IEEE Spectrum*, August 1996, pp.45-49.
[48] G. Gross, D. Finlay G. and Deltas, 'Strategic bidding in electricity generation supply markets', *Proceedings of the IEEE Power Engineering Society, Winter Meeting*, New York, 1, 1999, pp.309-315.
[49] W. Vickrey, 'Counterspeculation, auctions, and competitive sealed tenders', *Journal of Finance*, March 1961, pp.8-37.
[50] H.R. Varian, *Intermediate Microeconomics: A modern approach*, 5th Edition, Norton, New York, 1999.
[51] OFFER (2000) Office of Electricity Regulation, United Kingdom. www.ofgas.gov.uk.
[52] R.B. Bjorgan, C.-C. Liu and J. Lawarrée, 'Financial risk management in a competitive electricity market', *IEEE Transactions on Power Systems*, Vol.14, 1999, pp.1285-1291.
[53] R.J. Kaye, H.R. Outhred and C.H. Bannister, 'Forward contracts for the operation of an electricity industry under spot pricing', *IEEE Transactions Power Systems*, Vol.5, 1990, pp.46-52.
[54] CalPX, California Power Exchange, www.calpx.com, 2000.
[55] CAISO, California Independent System Operator, www.caiso.com, 2000.
[56] Power Pool of Alberta, www.powerpool.ab.ca, 2000.
[57] P.R. Gribik, G.A. Angelidis and R.R. Kovacs, 'Transmission access and pricing with multiple separate energy forward markets', *IEEE Transactions Power Systems*, Vol.14, 1999, pp. 865-876.
[58] J.J. Gonzalez and P. Basagotti, 'Spanish power exchange market and information system design concepts, and operating experience', *Proceedings of the 21st Power Industry Computer Applications Conference*, 1999, pp.245-252.
[59] X. Ma, D. Sun and K. Cheung, 'Energy and reserve dispatch in a multi-zone electricity market', *IEEE Transactions on Power Systems*, Vol.14, 1999, pp.913-919.
[60] PJM, Pennsylvania, New-Jersey and Maryland Interconnection, www.pjm.com, 2000.
[61] T. Alvey, D. Goodwin, X. Ma, D. Streiffert, and D. Sun, 'A security-constrained bid-clearing system for the New Zealand wholesale electricity market', *IEEE Transactions on Power Systems*, Vol.13, 1998, pp.340-346.
[62] K.W. Cheung, P. Shamsollahi, D. Sun, J. Milligan and M. Potishnak, 'Energy and ancillary service dispatch for the interim ISO New England electricity market', *Proceedings of the 21st Power Industry Computer Applications Conference*, 1999, pp.47-53.
[63] W.W. Hogan, 'Contract networks for electric power transmission', *Journal of Regulatory Economics*, Vol.4, 1992, pp.211-242.
[64] FERC, Order accepting for filing in part and rejecting in part proposed tariff amendment and direction revaluation of approach to addressing intrazonal congestion, California ISO, Docket No. ER00-555-00, available at www.ferc.fed.us, 2000.
[65] M.E. Baran, V. Banunarayanan and K.E.A. Garren, 'Transaction assessment method for allocation of transmission services', *IEEE Transactions on Power Systems*, Vol.14, 1999, pp.920-928.
[66] C. Li, A.J. Svoboda, X. Guan H. and Singh, 'Revenue adequate bidding strategies in competitive electricity markets', *IEEE Transactions on Power Systems*, Vol.14, 1999, pp.492-497.

[67] IEEE Current Operating Problems subcommittee, 'Operational aspects of generation cycling', *IEEE Transactions Power Systems*, Vol.5, 1990, pp.1194-1203.

[68] I. Otero-Novas, C. Meseguer, C. Batlle and J.J. Alba, 'A simulation model for a competitive generation market', *IEEE Transactions on Power Systems*, Vol.15, 2000, pp.250-256.

[69] B.R. Szkuta, L.A. Sanabria and T.S. Dillon, 'Electricity price short-term forecasting using artificial neural networks', *IEEE Transactions on Power Systems*, Vol.14, 1999, pp.851-857.

[70] O.B. Fosso, A. Gjelsvik, A. Haugstad, B. Mo and I. Wangensteen, 'Generation scheduling in a deregulated system: The Norwegian case', *IEEE Transactions on Power Systems*, Vol.14, 1999, pp.75-81.

[71] J.W.M. Cheng, F.D. Galiana and D.T. McGillis, 'Studies of bilateral contracts with respect to steady state security in a deregulated environment', *IEEE Transactions Power Systems*, Vol.13, 1998, pp.1020-1025.

[72] Z. Yu, 'Mixed integer social welfare maximization (MI-SWM) and implications in optimal electricity pricing', *IEEE Power Engineering Review*, July 1999, pp.53-54.

[73] H. Singh, *IEEE tutorial on game theory applications in electric power markets*, IEEE Catalog No. 99TP136-0, 1999.

[74] J.W. Lamont and S. Rajan, 'Strategic bidding in an energy brokerage', *IEEE Transactions on Power Systems*, Vol.12, 1997, pp.1729-1733.

[75] P. Sheridan, M. Flynn and M.J. O'Malley, 'Generator scheduling with demand bids', *Proceedings of the IEEE Power Engineering Society, Summer Meeting*, Seattle, Paper No. 2000SM-012, 2000.

[76] D.A. Lusan, Z.Yu and F.T. Sparrow, 'Market gaming and market power mitigation in deregulated electricity markets', *Proceedings of the IEEE Power Engineering Society, Winter Meeting*, New York, 2, 1999, pp.839-843.

[77] Z. Younnes and M. Ilic, 'Generation strategies for gaming transmission constraints: Will the deregulated electric power market be an oligopoly', *Proceedings of the 31st Annual Hawaii International Conference on System Sciences*, 3, 1998, pp.112-121.

[78] S. Hao and A. Papalexopoulos, 'Reactive power pricing and management', *IEEE Transactions Power Systems*, Vol.12, 1997, pp.95-104.

[79] F. Wolak, R. Nordhaus and C. Shapiro, Preliminary report on the operation of ancillary services markets of the California Independent System Operator, California Independent System Operator. www.caiso.com, 1998.

[80] R. Baldick and E. Kahn, 'Contract paths, phase shifters, and efficient electricity trade', *IEEE Transactions on Power Systems*, Vol.12, 1997, pp.749-755.

[81] B..F. Hobbs, C.B. Metzler and J.S. Pang, 'Strategic gaming analysis for electric power systems: an MPEC approach', *IEEE Transactions on Power Systems*, Vol.15, 2000, pp.638-645.

[82] C.W. Richter and G.B. Sheblé, 'Genetic algorithm evolution of utility bidding strategies for the competitive marketplace', *IEEE Transactions on Power Systems*, Vol.13, 1998, pp.256-261.

[83] H. Song, C.-C. Liu, J. Lawarrée and R. Dahlgren, 'Optimal electricity supply bidding by Markov decision process', *IEEE Transactions on Power Systems*, Vol.15, 2000, pp.618-624.

[84] M. Gibescu and C.-C. Liu, 'Optimization of ancillary services for system security', *Proceedings Bulk Power System Dynamics and Control IV - Restructuring, Symposiums*, 1998, pp.351-358.

[85] E.H. Allen and D. Ilic, 'Reserve markets for power systems reliability', *IEEE Transactions on Power Systems*, Vol.15, 2000, pp.228-233.

[86] D.R. Bobo, D.M. Mauzy and F.J. Trefny, 'Economic generation dispatch with responsive spinning reserve constraints', *IEEE Transactions on Power Systems*, Vol.9, 1994, pp.555-559.

[87] M. Aganagic, K.H. Abdul-Rahman and J.G. Waight, 'Spot pricing of capacities for generation and transmission of reserve in an extended poolco model', *IEEE Transactions on Power Systems*, Vol.13, 1998, pp.1128-1135.

[88] J.B. Cardell, C.C. Hitt and W.W. Hogan, 'Market power and strategic interaction in electricity networks', *Resource and Energy Economics*, Vol.19, 1997, pp109-137.

[89] T.W. Gedra, 'Optional forward contracts for electric power markets', *IEEE Transactions on Power Systems*, Vol.9, 1994, pp.1766-1773.

[90] R. Bjorgan, R., H. Song, C.-C. Liu and R. Dahlgren, 'Pricing flexible electricity contracts', *IEEE Transactions on Power Systems*, Vol.15, 2000, pp.477-482.

[91] R.J. Thomas and T.R. Schneider, 'Underlying technical issues in electricity deregulation', *Proceedings of the 30th annual Hawaii International Conference on System Sciences*, 5, 1997, pp.561-570.

[92] S.S. Oren, A.J. Svoboda and R.B. Johnson, 'Volatility of unit commitment in competitive electricity markets', *Proceedings of the 30th Annual Hawaii International Conference on System Sciences*, 5, 1997, pp.594-601.

[93] J. Kumar and G.B. Sheblé, 'Auction market simulator for priced based operation', *IEEE Transactions on Power Systems*, Vol.13, 1998, pp.250-255.

[94] R.J. Green and D.M. Newbery, 'Competition in the British electricity spot market', *Journal of Political Economy*, 100, 1992, pp.929-953.

[95] S. Borenstein, J. Bushnell and F. Wolak, Diagnosing market power in California's deregulated wholesale electricity market, University of California Energy Institute, 1999.

[96] C.D. Wolfram, 'Strategic bidding in a multiunit auction: An empirical analysis of bids to supply electricity in England and Wales', *Rand Journal of Economics*, 29, 1998, pp.703-725.

[97] K. Miyakawa, Analysis of strategic behaviors in an open access transmission grid. MSc Thesis, University of Washington, 2000.

[98] J.B. O'Sullivan, 'Fuel cells in distributed generation', *Proceedings of the IEEE Power Engineering Society, Summer Meeting*, Edmonton, 1, 1999, pp.568-572.

[99] M.L. Baughman, S.N. Siddiqi and J.W. Zarnikau, 'Advanced pricing in electrical systems. I. Theory', *IEEE Transactions on Power Systems*, Vol.12, 1997, pp.489-495.

[100] M.L. Baughman, S.N. Siddiqi and J.W. Zarnikau, 'Advanced pricing in electrical systems. II. Implementation', *IEEE Transactions on Power Systems*, Vol.12, 1997, pp.496-502.

[101] B.H. Kim and M.L. Baughman, 'The economic efficiency impacts of alternatives for revenue reconciliation', *IEEE Transactions on Power Systems*, Vol.12, 1997, pp.1129-1135.

[102] I.J. Perez-Arriaga, F.J. Rubio, J.F. Puerta, J. Arceluz, and J. Marin, 'Marginal pricing of transmission services: An analysis of cost recovery', *IEEE Transactions on Power Systems*, Vol.10, 1995, pp.546-553.

[103] DOE, Maintaining reliability in a competitive U.S. electricity industry, Final report of the task force on electric system reliability, United States Department of Energy, 1998.

[104] C.-C. Liu and R. Christie, Energy Management System, in *Encyclopaedia of Computer Science and Technology* (ed. A. Kent, J.G. Williams, C.M. Hall and R. Kent), 26, Supplement 11, 1992, pp.177-194.

[105] WSCC, Disturbance report for the power system outage that occurred on the Western Interconnection on August 10th, 1996 at 1548 PAST, Western System Coordinating Council, www.wscc.com, 1996.

[106] NERC, Available Transfer Capability Definitions and Determination, North American Electric Reliability Council, 1996.
[107] A.R. Bergen and V. Vittal, *Power System Analysis*, 2nd Edition, Prentice Hall, 2000, pp.436-440.
[108] M.D. Ilic, T.Y. Yong and A. Zobian, 'Available transfer capacity (ATC) and its value under open access', *IEEE Transactions on Power Systems*, Vol. 12, 1997, pp.636-645.
[109] G.B. Sheblé, *Computational Auction Mechanisms for Restructured Power Industry Operation*, Kluwer Academic Publishers, 1999.
[110] G.W. Rosenwald and C.-C. Liu, 'Consistency evaluation in an operational environment involving many transactions', *IEEE Transactions on Power Systems*, Vol.11, 1996, pp.1757-1762.
[111] J.A. Momoh, R.J. Kosselar, M.S. Bond, B. Stott, D. Sun, Papalexpoulos and P. Ristanovic, 'Challenges to optimal power flow', *IEEE Transactions on Power Systems*, Vol.12, 1997, pp.444-454.
[112] S. Hao, G.A. Angelidis, H. Singh, and A. Papalexopoulos, 'Consumer payment minimization in power pool auctions', *IEEE Transactions on Power Systems*, Vol.13, 1998, pp.986-991.
[113] D. Chattopadhyay, 'Daily generation scheduling: Quest for new models', *IEEE Transactions on Power Systems*, Vol.13, 1998, pp.624-629.
[114] M. Walsh and M.J. O'Malley, 'Augmented Hopfield network for unit commitment and economic dispatch', *IEEE Transactions on Power Systems*, Vol.12, 1997, pp.1765-1775.

4

Distribution in a Deregulated Market

Cliff Walton
London Electricity plc,
UK

Robert Friel
London Electricity plc,
UK

Dr Loi Lei Lai
City University, London
UK

4.1   Introduction to the UK Environment

Transmission and distribution are still regarded as the natural monopoly elements in the restructured UK energy market. Since privatisation in 1990 there have been a number of changes in the structure of the industry which have impacted on the distribution businesses and the introduction of the Utilities Bill heralds a further change in the relationship with government, the regulatory body and consumers.

One of the main objectives of privatisation was to promote competition. This has focused on the supply (i.e. the retailing) of electricity and gas and has encompassed the associated aspects of metering.

A competitive framework was developed for new connections to distribution networks defining certain elements as contestable work, but to date this area has not seen the wide-scale competitive activity expected and the regulator has indicated his intention to review competition in the gas and electricity connection markets by March 2001.

The biggest effects on the distribution businesses have resulted from the price control mechanism. Distribution businesses in the UK are price regulated, a part of which is to allow a return on the assets purchased at vesting and the investments made in the subsequent years. The latest price review, which came into effect in April 2000, saw the regulator propose reductions in distribution business income similar to those following the last review. None of the UK distribution companies have challenged the outcome of the latest review, which implies that the companies believe that they can achieve these savings. The only structural change at the time of writing has been the announcement by London

Electricity and Eastern Electricity of a joint venture to operate their networks, with the asset ownership remaining with the parent companies.

Since the latest price control the UK regulator has implemented a programme to look at the information provided by the regulated distribution companies and method of providing incentives to introduce an element of competition. The details of any such scheme are still to be decided, but it sends a clear signal that after 10 years there remains much scope for developments in regulatory practice and the electricity distribution industry.

4.2 The Development of Competition

4.2.1 Competition in Supply

The development of competition in the energy supply market in the UK has led to the development of two distinct activities in the UK public electricity suppliers (PESs), supply and distribution. The supply businesses are responsible for the sale of electricity and gas whilst the distribution businesses manage the cables, lines, transformers and switchgear which form the power supply networks between the EHV grid system and the end users.

The development of competition in supply as part of the process of deregulation is a subject in its own right. Certain aspects of the process to deliver a competitive supply market have had a significant impact on the PESs' distribution businesses.

The most visible aspect of this has been the moves both physically and financially to separate the retail and distribution business, and in some cases the sale of retail businesses to third parties. This process has involved the rebranding of the separate businesses. It was intended that the PES distribution businesses be rebranded and this will largely be the case, although those generators who have acquired retail interests (National Power/npower and PowerGen) have essentially rebranded their retail arms (Midlands and East Midlands respectively).

4.2.2 The Responsibilities of Retail and Distribution

Retail
The retail business bulk purchases both electricity and gas and supplies them to their customers over the electricity and gas networks. A great deal of work has been required to put in place the necessary systems to effect a competitive market.

As part of this marketplace the retail businesses have taken responsibility for meter reading, with the intent that this service be procured from meter-reading service providers on a competitive basis. The development of competition in the metering sector will eventually see the provision of meters to suppliers by meter asset managers and operators.

Distribution
The distribution companies manage and maintain the electricity distribution network. This involves both the technical asset management and planning services and the billing of use of system charges from suppliers and the management and maintenance of the existing meter assets.

The meter operation services organisation in London has been established as a separate entity from the network asset manager, with an agreed scope and level of service between them.

4.2.3 Why Separate Distribution and Supply?

The drive for separation of businesses is a requirement of establishing a competitive market in the supply of energy [1]. In essence the incumbent distributor may try to protect the position of the dominant host supplier to the detriment of customers and potential competitors. Five potential means of achieving this have been identified:

1. A PES will operate a combined distribution and supply business to maximise shareholder returns that could disadvantage competing suppliers. For example, to use distribution use of system charges to support the suppliers' retail tariffs.
2. The supply business potentially has access to information that other suppliers will not have. For example, the names of customers supplied by a second-tier supplier.
3. Access to future intentions of a distribution business. For example, advanced warning with regard to the size and nature of changes to use of system prices.
4. Cross-subsidisation by a disproportionate allocation of costs such as IT and corporate overheads to the distribution business. A small reallocation will have a relatively large impact in a retail industry that has very small margins. The regulator attempted to address this issue in the 1999 price control review by reallocating monies from distribution to supply before assessing the relative efficiencies of the distribution businesses.
5. A distribution business will in some way downgrade the service to a customer that has a second-tier supplier. For example, the response to power outages.

4.2.4 Key Issues for Distribution Businesses

The key issues for distribution businesses resulting from competition in supply and separation of the two business areas are:

- Information management
- Distribution use of system (DuoS) billing
- Customer service
- Metering.

The impact on the UK distribution businesses of the implementation of the New Electricity Trading Arrangements (NETA) was yet to be fully established at the time of writing. The most significant area of likely impact appeared to be in resolving system-balancing-related issues due to outages constraining suppliers' sales for embedded generators' energy sales.

4.2.5 Information Management

Information management is a key issue for any business and this is particularly true with a large distributed asset base. Most of the technical data management issues have resulted from the need to improve data handling to cope with a changing skill set and reduced workforce. Many businesses have adopted graphical information systems to manage their records. Some still transfer these to paper records for use by operational staff in the field but an increasing number are issuing electronic records to field staff, affectionately known as 'plans in vans'. The use of IT systems and in particular hand-held devices will soon enable a further improvement in the process of gathering data and maintaining information systems.

One of the most significant impacts of the separation of the distribution businesses related to the management of customer data. It was decided that there was no need for distribution businesses to maintain records of the customers connected to the distribution system; a record of the physical connection points and/or metering points would suffice. In the relatively rare instances where customer information may be required the supplier would then provide it. In most PES businesses this required segregation of information within their combined customer information and meter point registration systems.

A great deal of time has been spent in assessing the impact upon customer information and the management of customer enquiries, as the distribution business will no longer have sufficient information to provide a personal service to individual customers. This requires careful management so that the customers still receive the level of service they expect.

4.2.6 Use of System Billing

In a competitive market the use of system (DUoS) charge is distinct and separate from purchase and sales costs of electricity. An energy supplier has energy purchase costs, transmission use of system costs, distribution use of system costs, metering costs, its own costs and profit in its standard charges. Energy purchase costs are directly with the generators and the pool (or the balancing market, post the implementation of NETA in the UK). National Grid Company (NGC) charges the suppliers for use of its system and the distribution companies will charge suppliers for use of their own systems.

The point of contact for the distribution business is therefore with suppliers and not the end customers. This arrangement is controlled via the use of system agreement that each supplier must sign before being able to approach customers on the distributor's network.

Two distinct types of billing exist.

1. Site specific (based upon half-hourly consumption data) for larger greater than 100 kW customers.
2. Profiled (time-adjusted estimates) for smaller (mainly quarterly read customers).

Charges to suppliers consist of a fixed portion or standing charge and a variable per-unit charge. There is also an availability charge to larger customers. In order to maintain correct billing the distributors are able to use pool data.

4.2.7 Customer Service

The management of customer relations is another area where competition in supply creates a number of options.

In New Zealand, the initial approach routed all customer contact through the supply businesses. This simplifies the contact issues for the customer, but necessitates careful management of the interface between suppliers and distribution businesses to ensure that the correct information is available to inform the customer.

In the UK, the distribution businesses have kept an interface with customers in relation to supply outages. Hence customers have two points of contact. Although this simplifies the management of information flows on outages between distribution businesses and suppliers (there is none) the management of the routing of calls to the wrong number needs care. The future solution to some of these issues is already apparent in automated trouble call systems. These are already being installed to provide information on outages and are of particular use in the extreme circumstances of wide-scale power outages when call centres become overwhelmed.

Internet technology will soon provide accurate supply of information on outages and likely restoration times to both customers and suppliers - the integration of fault reporting and the telephone network. It is possible to generate specific voice-activated messages according to postcodes or dialling code information. It may even go as far as proactive message information (i.e. ring the customer). London Electricity has already been detailing current power outages on its internal web site for some time and is planning to make this facility available via the Internet once suitable security safeguards have been proven.

4.2.8 Competition in Metering

Competition has existed in the 1 MW and 100 kW markets since 1991 and 1994 respectively. Third parties have been able to own and operate meter assets in both these areas since 1995. In conjunction with the full competition in energy supply and the physical and financial separation of the distribution and supply businesses, the principles of competitive metering are being extended to the remaining small business and domestic market sectors. This has created two distinct business streams, meter reading and meter asset management (otherwise known as meter operations). Meter-reading companies will provide data retrieval and data processing (DR and DP as they are known) services. Meter providers will provide meter asset provision and management (known as MAP and MAM).

The distribution businesses no longer have any part in the reading of meters, which is handled by the energy retail business via a contractual arrangement with a meter reader. This is a natural progression in that meter readers could provide services for reading any utility service meter. Long-term developments are likely to result in meters that are read remotely.

The existing meter assets are presently owned by the distribution businesses. The energy supplier, via a meter asset management company, can provide new meters either for new customers or as replacements for existing meters. In the medium term competition will develop in the provision of meters and the management of these assets.

To ensure the availability of these services, the PES energy retail businesses will continue to provide a meter-reading service of last resort and the distribution companies will provide a meter operations service of last resort.

4.2.9   Scope for Demand-side Management

Demand-side management aims to reduce the peak demand either by reducing the level of demand through the promotion of more efficient usage or by moving elements of the load at time of peak demand, thereby improving the system load factor.

Demand-side management can provide real alternatives to the need for generation, transmission or distribution reinforcement. In general terms these benefits are greatest at the transmission and generation level where the cumulative effect is greatest.

Demand-side management has been encouraged by the use of tariffs, e.g. lower off-peak tariffs for storage heating. In this instance the move away from electric bar and convection heating to off-peak storage will impact on both generation and transmission reducing the installed capacity required. However, at a distribution level the widespread use of storage heating has resulted in peak demand occurring at night, promoting the need to reinforce.

In a deregulated market the promotion of energy efficiency and demand-side management is less clear. The generation capacity available becomes a function of market. Contracts exist between energy suppliers or retailers and various generators, with the system operator responsible for balancing the system's requirements and therefore maintaining system voltage, frequency and security. Large consumers can play a part in this market by making themselves available for disconnection as a 'negative generator'. However, the administrative complexities of participating in such a market probably limit this to a few very large businesses.

A distributor still has recourse to modifying use of system tariffs to promote alternative usage patterns. However, the energy supplier may not be obliged to reflect these to the end user, which could blur any resulting incentives. Additionally in the UK, only the supply businesses will have customer contact and knowledge of their particulars and so it would be difficult for the distribution system operator to manage the system load actively via the end customer.

The promotion of embedded generation to offset the need for reinforcement has become a major subject of debate. The principal difficulty is that promoting and ensuring the installation of suitable generation in advance of the reinforcement requirement is far from a simple task. Not only this but the appropriate commercial arrangements must be in place for the risk to be minimised sufficiently for the generation to represent an equitable alternative to traditional reinforcement.

Therefore it can be seen that in the marketplace active demand-side management would require a time of day price signal to be sent to customers via tariff arrangements at generation, transmission and distribution levels designed to reflect the local delivery costs.

However, it is unclear how significant this would be in reducing the need for system reinforcement, which must be proactive if system security is to be maintained. Any reduction in system capacity margins to allow for the effects of demand-side management would require careful risk assessment. Variances between actual and predicted demand would need to be reflected in variable use of system charges in order to manage the demand. It is not the intent to discuss here whether variable use of system charges would be desirable or practicable as these would have to be passed on to customers through variable tariffs or greater margins.

4.3 Maintaining Distribution Planning

This section will consider the planning of distribution system investment and associated network development issues. Since deregulation this has been dominated in the UK by the two major price reviews conducted by the industry regulator. During these reviews the privatised distribution companies have had to defend their past investment record and proposed detailed investment proposals for the subsequent review period. The regulator and his consultants then considered these proposals along with the other companies' submissions. The regulator's view considers the financial cost to the consumers against the overall level of investment in asset replacement, system reinforcement and quality of supply expenditure.

The aim has been to reduce the cost to the consumer whilst improving service. Increasingly the focus is on the relationship between investment and ongoing costs. The experiences of regulators in other utility services such as the rail industry have highlighted the need to assess and encourage long-term investment to preserve the integrity of the distribution systems. At the time of writing a project is commencing that will consider in more detail the relationships between ongoing investment, operating costs and the direct and indirect incentives of the regulatory regime.

4.3.1   Regulatory Incentives

Throughout the 1990s the regulator has driven the privatised electricity companies towards ever greater cost savings. During the last price control review in 1999, the retail and distribution businesses were considered separately. This approach was targeted at preventing cross-subsidisation of activities in the competitive retail market by activities in the distribution market.

The regulatory focus has remained one of achieving cost savings, primarily through reductions in operating costs. Much time was spent during the 1999 price control review assessing the distributors' cost efficiency through an attempt to identify a benchmark cost level and through regression analysis. The regression analysis created an efficiency frontier based on the lowest score results based on a Cobb-Douglas-type model which considered ratios of numbers of customers, units and network length as proxies. The model used was the subject of extensive debate, particularly surrounding the adjustments to base data, choice of weights and the resulting fit. At the time of writing, the regulator had just initiated further work to develop improved models before the next review in 2005 in order to attempt to reduce the perceived risk from regulatory price controls.

Capital investment has been aimed at meeting network reinforcement needs, replacing or refurbishing major assets, e.g. 11 kV panel switchgear, and improving quality of supply. There has been a significant incentive to achieve the minimum requirements at the least cost. Not all distribution companies have adopted such an approach, with many seeking to deliver the most effective investment within the allowed revenue for capital expenditure.

Distribution in a Deregulated Market 117

4.3.2   Technical Issues

The restructuring of the electricity industry in the UK has not altered the technical fundamentals of distribution network design. Voltage tolerances, fault ratings and power quality requirements remain as they were before the privatisation of the regional electricity companies (except for changes to harmonise within the European Union). That said, the harmonics standards are being updated to reflect EMI standards and customers are becoming more aware of these and other issues such as transient voltage dips associated with faults. This has not been due to the restructuring of the industry, but as a result of the increasingly sensitive load connected and the unwillingness of customers to accept low standards from the new 'for-profit' distribution companies.

What has changed through regulation is quality of supply as measured in terms of the number and duration of interruptions and customer service. The development of quality of supply drivers is discussed in the following section.

The following discussion covers the drivers affecting planning, the tools available to planners and the basic concepts of planning asset replacement and quality of supply improvements.

4.3.3   Planning Drivers

Any distribution system design must meet the following requirements:

- It must be able to supply the system demand whilst meeting the appropriate technical standards.
- It must meet the requisite security of supply standards.

In the modern UK systems the networks must also:

- Deliver the appropriate level of quality of supply.
- Be designed to manage the risk of major incidents.
- Reduce the long term operating costs of the network.

Security of Supply
In the UK, the security of supply requirements for distribution networks has been defined by Engineering Recommendation P2/5. This is a deterministic standard that was based on a great deal of economic and risk assessment. Whilst much has changed in the economy of the UK since the standard was written in the 1970s, the standard still remains reasonably robust although it only addresses the scale and duration of a loss of supply and not the frequency of which such incidents can be expected to occur.

In recent times the revision of this document has been considered to reflect present quality of supply targets and performance, which are beginning to drive network design beyond the security levels given in the standard, particularly in respect of the frequency of interruptions. The difficulty here has been to find a common approach that meets the differing performance levels that result from the different environments the various networks operate in. The most recent driver for change has come from the political desire to make the best use of embedded generation connected to the distribution networks at 132 kV and below. This generation has not been considered in system security to date. This is due to:

- Simple methods of connection of most generation.
- The lack of an appropriate commercial framework.
- The difficulties of ensuring the presence of the necessary generation before systems need reinforcing.

These issues still remained to be solved at the time of writing but it is likely that the security of supply standard will be amended with a means of assessing the contribution of embedded generation.

Quality of Supply

Quality of supply to customers has been measured for some time by the following measures:

- Availability: System average interruption duration index (SAIDI) or customer minutes lost (CML) as it is often known.
- Frequency: System average interruption frequency index (SAIFI) or its equivalent in interruptions per 100 connected customers.
- Restoration in 3 hours.

These measures represent the average performance of the system and so do not accurately reflect what the individual customer may experience [2,3]. In the UK any interruption of duration greater than 1 minute is counted towards these statistics although in Europe 3 minutes is the norm to allow for the benefits of system automation.

Experience indicates that customers prefer not to be interrupted, but in the event of an interruption speed of restoration and accurate information on likely outage times become very important. The provision of such information is possible both through call centre technology and via Internet services.

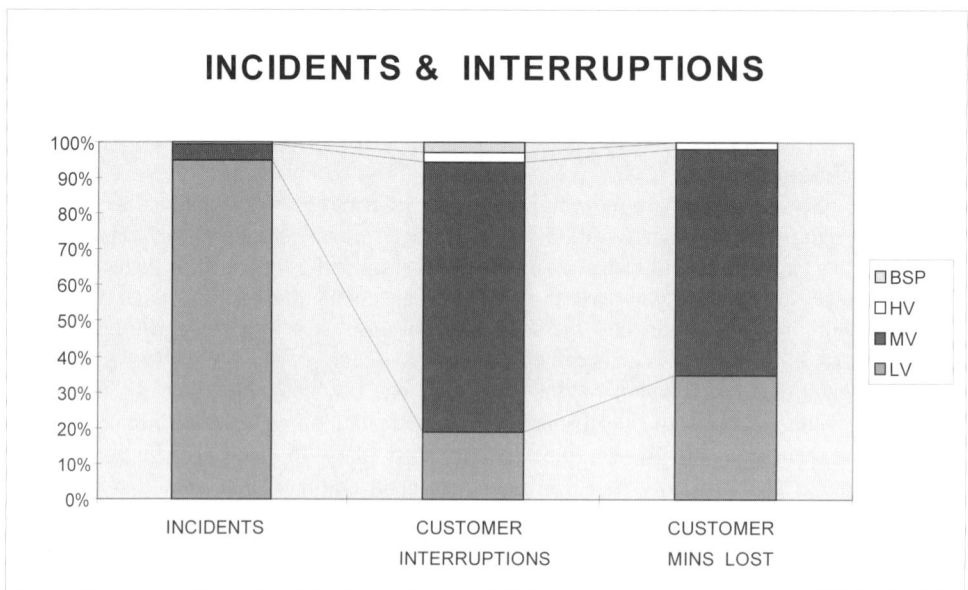

Figure 4.1 Incidents and interruptions

Figure 4.1 demonstrates the relationship between incidents, interruptions and customer minutes lost. This clearly shows that whilst by far the most incidents occur at low voltage, the medium-voltage (MV) 11 kV and 6.6 kV systems cause by far the most customer disruption. It is also the easiest area in which to improve network performance as there are cost-effective solutions available. It comes as no surprise therefore that investment has centred on reducing the impact of MV system incidents and disruption.

The most troublesome impact on customers is that resulting from frequent or multiple interruptions. The primary measures discussed above have therefore recently been supplemented by a measure of multiple interruptions, which will determine the percentage of customers who experience more than a given number of interruptions per annum.

During the second post-privatisation price control the distribution companies set targets to be achieved by the end of the second review period. During the third review which took place in 1998/9, the electricity industry regulator decided to set targets for the electricity companies rather than to allow the companies to set their own. In addition to this a system of incentives based around performance against some of these measures is to be implemented in 2002 [4]. Much discussion remained at the time of writing as to the emphasis to be given to which measures and therefore it is difficult to resolve how these measures will further affect the development of distribution planning. The design of networks to meet quality of supply requirements targets the reduction of the impact of network failures at two main areas:

- The prevention of interruptions
- The restoration of supplies

Since deregulation, the UK's electricity regulator has increased the focus on improving this aspect of customer service, amongst others. Different approaches have been adopted by different companies driven by their particular regional and network problems. The significant areas of investment have been in insulated or semi-insulated conductors and auto reclosers on overhead lines and network remote control and SCADA systems.

Insulated overhead conductors have been used to reduce significantly transient interruptions due to trees touching lines, which can lead to more prolonged interruptions. Pole-mounted auto reclosers have been introduced in conjunction to clear transient interruptions automatically and are therefore sometimes considered as a form of network automation.

Network secondary system (i.e. MV systems) remote control systems have been introduced in varying degrees by distributors in the UK. The most significant investments have been by London Electricity, Eastern Electricity and Southern Electricity. These systems allow much faster restoration of supplies following faults on cable networks or on systems relatively distant from operational centres. They also reduce the need for manual switching, reducing the amount of time an engineer needs to spend switching and also reducing the risks associated with this activity. The case study at the end of this section considers the scheme implemented by London Electricity, the benefits realised and the long-term potential foreseen for the system.

These projects have as part of their implementation significantly contributed to the replacement or upgrading of problematic network apparatus, be it overhead line or switchgear.

Supplier Requirements

A possible outcome of the competitive market in energy retail could ultimately be different quality and security of supply requirements being set by different energy suppliers.

Specific local suppliers in a given area, e.g. the City of London or a major business park where the networks are owned and operated by the incumbent public system operator, would most likely drive such changes. The incumbent public system operator is unlikely to be able to discriminate in such a way in more general networks, unless different networks with different performance levels were installed to operate in parallel or additional customer-specific control or automation facilities are provided. The former solution would take a great deal of expense over a considerable period of time and would create a two-tier system, whilst the second could be possible with differential charging.

Operational Drivers

Probably the most difficult driver affecting the designer of distribution systems in the restructured UK industry is the balance between capital and operating expenditure.

Capital expenditure has been managed separately from operational cost until recently. The large savings made since privatisation were largely through process review and multi-skilling, enabling the elimination of inefficiencies. Future savings will increasingly need to stem from investments made to reduce the inherent day-to-day costs of maintaining and operating the networks.

Load Forecasting - Capacity Planning

Until the 1970s, before energy efficiency gained public attention, electricity load forecasting was relatively simple. It was a direct reflection of the customers' immediate demand that was not restricted or controlled. Load forecasting, at that time, could simply mean a matter of projection based upon weather and type of day information with concern mainly on narrowly defined objectives such as justification for future plant expansion. However, owing to the increasing societal consciousness of clean environment and energy efficiency, it can involve treatment on least cost analysis, reliability, flexibility, social, cultural and environmental acceptability. Inevitably, load demand nowadays has to deal with more sophisticated considerations for which demand-side planning includes factors such as customer energy efficiency, whilst supply-side planning includes views on reduced transmission loss, for instance.

Today, many electric utilities and their consumers in many countries have used or will use the real-time pricing (RTP) system instead of the traditional time of use (TOU) system because of the proven benefits under the RTP system. The electricity prices under the traditional TOU system are fixed in a few different periods of a day, and the load forecast under this condition is more important for utilities than for customers. Because under the TOU system the electricity price is fixed in a relatively long period in a day, there is no price change for customers to get benefits in this period. For utilities, it is still necessary for them to find out how much energy will be needed in this period, i.e. to forecast the amount of electric energy that will be supplied to customers.

Under the RTP system, the condition is different. Because the electricity price under the RTP system is changing hourly, and the cost savings that customers can get under this system are based on the changes of the electricity price, the price changes bring some benefits for customers. Therefore, if customers know more clearly their energy needs in the near future, they can optimise their electricity application scheduling to get the maximum cost savings under the RTP system.

In a deregulated, competitive power market, utilities tend to maintain their generation reserve close to the minimum required by an independent system operator. This creates a need for an accurate instantaneous-load forecast for the next several dozen minutes. The load-frequency control and economic dispatch functions of the EMS require load forecasts within shorter time leads, from single minutes to several dozen minutes. They are referred to as very short-time load forecasts (VSTLFs). These forecasts, integrated with the information about scheduled wheeling transactions, transmission availability, generation cost, spot market energy pricing, and spinning reserve requirements imposed by an independent system operator, are used to determine the best strategy for the utility resources. Very short-term load forecasting is important in today's deregulated power industry. The instantaneous-load predictions for the next several dozen minutes are also required to operate the power system reliably and economically.

Many methods have been developed for short-term load forecasting. They can be divided into three broad categories: parametric, non-parametric, and artificial-intelligence-based methods. The parametric methods formulate a mathematical or statistical model of load by examining qualitative relationships between the load and load-affecting factors. Some examples of the models used are explicit time functions, polynomial functions, auto-regressive moving average (ARMA) models, Fourier series and multiple linear regression [5]. Non-parametric methods allow a load forecast to be calculated directly from historical data. For example, using non-parametric regression, a load forecast can be calculated as a local average of observed past loads with the size of the local neighbourhood and the specific weights on the loads defined by a multivariate product kernel [6]. Artificial-intelligence-based methods use artificial neural networks (ANNs) as load models. The main advantage of using neural networks lies in their abilities to team the mentioned dependencies directly from the historical data without the necessity of selecting an appropriate model. There are a few types of neural networks that have been applied for load forecasting. A multi-layer feedforward network with one hidden neuron layer is most commonly used [7].

Very short-term load forecasting requires a different approach. Instead of modelling relationships between load, time, weather conditions and other load-affecting factors, it would require to focus on extrapolating the recently observed load pattern to the nearest future. Methods for very short-term load forecasting are not so numerous. Some reported techniques include first or second order polynomial extrapolation, auto-regressive (AR) and ARMA models, ANNs [8].

Reference [9] reports the use of ANNs to model load dynamics to predict very short-term load forecasting. This shifts neural networks' task from forecasting actual loads to forecasting relative increments in load and leads to a better accuracy. Time of day and relative increments in load from the recent past are used as input variables. The proposed algorithm is more robust compared with the approach when actual loads are forecasted and used as input variables. It is less sensitive to the requirement of having the training data representative of the entire spectrum of possible load and weather conditions. Its superiority is especially evident when the current weather conditions are different from those used for training. Traditional neural-network-based forecasters may produce erroneous forecasts in such circumstances.

The proposed method for a VSTLF has been successfully implemented in a power utility in the USA, and is used by dispatchers for on-line load forecasting. The developed forecasting system predicts eight values of load for the time leads from 20 to 90 minutes in 10 minute increments. To provide dispatchers with the information about expected forecast errors, mean absolute percentage errors (MAPEs) are calculated based on the recent forecasts for which load information has become available. For the 20, 30, .., 60 minute forecasts, the mean absolute percentage error lies in a range of 0.4-1.1% [9]. Load data is obtained from the automatic generation control (AGC) system every 4 seconds. The upcoming data is converted into 1minute integrated loads which are considered as 'instantaneous' loads. These loads are used as input information for computing load predictions and they are also stored for training. The forecaster's neural networks are automatically retrained once a day.

It is not the intent to discuss in detail load forecasting and its disaggregation to a level where the installed capacity in the distribution networks can be established.

The advent of the increased demand for telecoms data and Internet services has also begun to add significant loads to the distribution systems particularly for the associated switching centres. The magnitude of such loads (10-40 MW) and time scales (12-18 months) for such developments are such that the distribution company has to be in a position to respond creatively and flexibly if it is to avoid losing the potential business either to another company in a different location where supply can be afforded quicker or cheaper, or to a competitor who is prepared to establish a separate distribution or private network. The large urban centres such as London are not seeing the predicted reductions in system maximum demand due to more efficient loads as these are being offset by these IT-related increases.

Planning Horizon
Most of the privatised distribution businesses have adopted a five-year planning horizon, largely due to the five-yearly review period and the difficulties of predicting demand beyond this. However, the construction times related to major system changes, changes in technology and large capital substation projects make the five-year period relatively short and a longer term view of the technologies and structure of the networks must be developed. Methodologies to achieve this end are discussed in more detail in the following section on long-term planning.

Asset Replacement
The challenge of combining asset management and planning is to integrate as far as possible the replacement of poorly performing or high-risk asset types with other network changes. Planning departments will need to play an increasingly larger part in integrating asset replacement programmes with network reinforcement and major new connections works. The planning of asset replacement is discussed later.

4.3.4   Long-term Planning

The essential point of long-term planning is to determine how external influences, including growth of new business and changes in the regulatory environment, will affect the development of the network and the levels of investment that will be required.

Changes in Demand

The impact of load growth and new business can be determined by the economic study of demand growth. This is a significant subject in its own right and no attempt will be made to discuss the building of a demand model here.

Any such model should address sufficient growth scenarios to allow the planner to assess the sensitivity of their proposals. It is normal to plan against a nominal base case load forecast. The base case forecast would define the most likely growth scenario, but being a long-term prediction it is unlikely to be very accurate in the short (3-5 year) term.

Two alternative load growth scenarios should also be considered. These are normally a high-growth scenario and a low-growth scenario. This is illustrated in Figure 4.2. Actual load would be expected to vary between these two predictions, hopefully showing a linear trend similar to the base case forecast.

Developing a long-term plan against such forecasts allows the planner to refine its long-term developments to maximise the flexibility of the network. As discussed, the current regulatory regime has encouraged a five-year planning cycle. The problem with the five-year plan is that construction of the latter stages of a plan has to begin very shortly after the plan has been defined, limiting the flexibility of the planner to change the proposal to meet changes in demands on the system.

Whilst it is often considered that planning more than five years ahead is of little benefit because of changes in load growth, longer term planning against the three forecasts previously described allows a better view to be taken as to when capacity limits may be reached. This will give an improved view of the risk of large capital investments needing to be brought forward or deferred. The planner can therefore continually adapt the long-term development proposals to the changing environment.

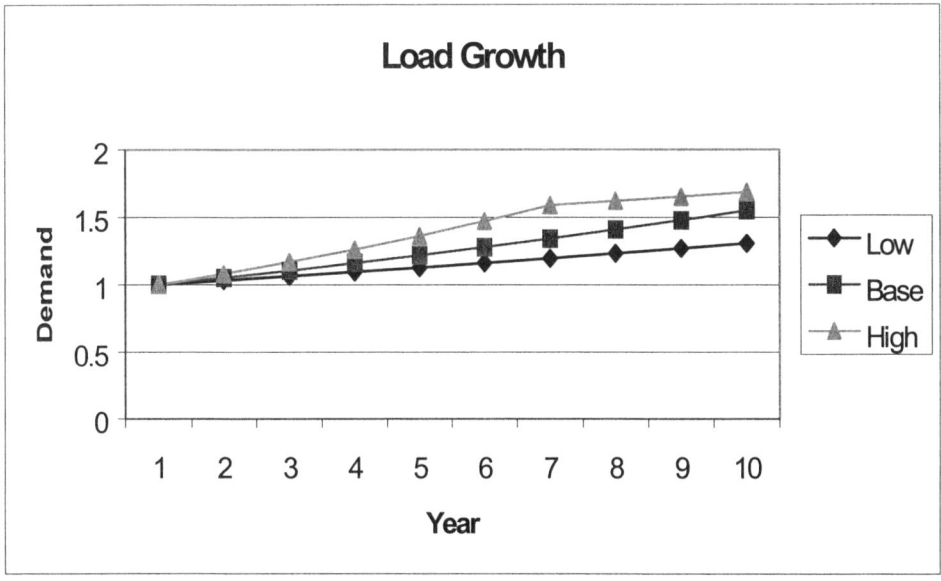

Figure 4.2 Load growth scenarios

In time adopting such a methodology should lead to an improved match between system developments and the demands they must supply. Such techniques have been successfully applied in a number of projects worldwide and are typically standard practice when developing plans for international financing agencies.

Scenario Planning

The previous section discussed the use of a set of load growth scenarios for network planning purposes. A similar technique using three alternative scenarios can be applied to all aspects of business planning. Three views of all aspects affecting the businesses' operating environment and investment drivers are normally developed being:

1. A 'steady as she goes' view: a stable environment based on existing economic conditions encompassing a reasonable view of the effects of known developments and trends.
2. A 'rose-tinted spectacles' view: a positive view of the development of the economy and how this will impact on the demands on the business.
3. A 'dark and gloomy' view: a more negative view considering the impact of a contracting economic environment and how this would impact on the business.

Against each scenario one would look at a range of business factors, determine a likely impact and an appropriate strategy or strategies. A tabulation of the strategies against the scenarios and business factors allows the planner to identify key strategies that appear to be robust against more than one scenario, as illustrated in Table 4.1.

Table 4.1 Scenario mapping

	Business Factor 1	Business Factor 2	Business Factor 3	Business Factor 4
Scenario 1	Strategy 1A	Strategy 2A	Strategy 3A	Strategy 4A
Scenario 2	Strategy 1B	Strategy 2A	Strategy 3B	Strategy 4B
Scenario 3	Strategy 1C	Strategy 2C	Strategy 3B	Strategy 4C

It can be seen that Strategies 2A and 3B are robust to more than one scenario, from which it could be concluded that these are the more robust strategies and should be adopted. Such techniques are widely used in high-level business planning but can also be of great use in identifying key long-term technical developments.

4.3.5 *Network Planning Tools*

Many different software tools exist to aid the design of power systems. Most of these concentrate on analytical load flow and fault level types studies and a few offer network reliability calculation tools. However, these typically use a fault rate per kilometre approach which does not, in the authors' experience, assist very precisely targeted asset replacement owing to the difficulty in determining appropriate fault rates. The other problem with a fault rate approach is that it cannot, through its historical basis and its lack of consideration of the underlying causation mechanisms, predict future fault rates.

4.3.6 Asset Replacement Planning

The asset management discipline and network planning's most significant interface is in the planning of asset replacement. The development of asset management policies is covered in detail elsewhere. The planner's role is to convert these policies into actual work programmes. In doing this the planner must consider the asset management policy objectives and the condition and performance of the networks where the work is to be carried out.

4.3.7 Risk Assessment

Risk assessment methodologies are useful in any business and distribution businesses are no exception. Risk assessment is applied at two levels, the business level and for individual asset assessments as part of the asset replacement planning process.

Business risk analysis considers all areas, including network performance, finance, commercial (e.g. use of system income), contractual and regulation. Potential risks in each area are identified and probabilities and consequences determined. Finally mitigation measures and appropriate actions to control the risks identified are established.

In deregulated distribution businesses, particularly where there is fully developed competition in supply, the largest risks are often associated with the management of income streams owing to the complexities of the data acquisition and aggregation process and the number of different parties involved.

However, network risks must not be ignored. Historic control measures exist through planning and construction standards such as the UK's Engineering Recommendation P2/5. In planning individual investments, risk assessments are normally conducted to identify likely common mode failures etc. Major network failures such as that which occurred in Auckland, New Zealand, and the recent weather-related incidents in Canada and France have prompted further debate on the appropriateness of existing design standards and the cost and benefit of changing these.

4.3.8 Skills and Resources

Since the privatisation of the distributors in the UK there has been ongoing pressure to reduce costs. This has inevitably resulted in a very significant reduction in the companies' staffing levels.

Whilst companies have in general set up asset management organisations these have a much wider remit than simply the planning of the networks. It is evident from the discussion of planning issues that there are two principal activities, new connections and infrastructure planning. In London these have been separated, the practical impact of which has already been discussed. In terms of the staffing of these organisations, the new connections activity requires a different set of competencies than infrastructure planning. Much work has been undertaken in attempting to benchmark best practice in these areas, and in particular comparing the skill levels of various companies' planning staff.

Both activities have suffered from the downsizing undertaken during the 1990's with significant skills and experience being lost. Whilst one can no longer expect to staff these activities with the most highly skilled engineers, a mix of experienced planners with less highly qualified but able technicians has been established. The planning skills of the more experienced staff are gradually being transferred to the less experienced team members and the lost competencies being replaced.

Whilst it can be argued that these skills have been retained by some of those utilities who purport to best practice, the skills gap is being gradually redressed. With the advent of the promoting of competition in connections design and provision, the area of greatest need will be to maintain and develop the infrastructure planning skills necessary to review the overall network successfully and determine where action will be required to maintain and improve existing levels of service. This is probably one of the areas most under pressure at the present time, especially with the increasing complexity of systems such as remote control and automation being introduced and the increasing asset-management-derived workload. The use of expert systems to capture experience and make it widely available has not yet been widely adopted but is an obvious opportunity to supplement process charts in more complex and/or less routine operations.

4.3.9   Network Design

There are three elements to network planning that need to be considered, these being the connection of new load, the reinforcement of the system and improvements to meet quality of supply targets.

Connections

The design of new connections is driven by the regulatory requirement to offer the lowest cost connection and the need to meet larger customer's needs. The former of these has given rise to a conflict with some aspects of network design to meet quality of supply targets.

For example, the simplest design to connect a voltage load of less than 1 MW to the MV system is to create a new substation and connect it to the existing network via a tee off an existing circuit. If a large number of customers are supplied from this single source, such as a large housing development, then ideally the substation should be connected so as to be looped into the existing circuit, as shown in Figure 4.3, in order to lessen the risk of a repair time outage.

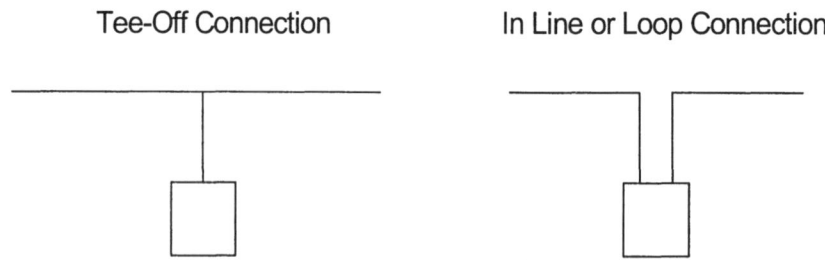

Figure 4.3 Tee vs. loop connections

The decision as to whether to fund this out of quality of supply monies will depend on the number of customers, the distance from the main circuit and the additional energy losses incurred. The advent of competition in connection services would further complicate this issue. The network manager will have either to pay the contractor to install the additional cable at the same time or retrofit the additional cable at a later date. The customer will not be expected to pay for the additional costs related to the quality of supply as part of his connection charge.

This conundrum will also apply to the installation of spurs to feed a number of customers at low voltage (LV) where no alternative back-feed arrangements from other networks are available or where the installation of remote terminal units for SCADA or remote control may be desirable. Evidently this becomes easier to manage as the size of the load increases and the number of connection requests decreases.

Reinforcement
General load growth and the connection of new load drive the need for network reinforcement. Typically the impact affects the thermal ratings of the network apparatus, security of supply or the voltage performance of the networks, but recently greater thought is having to be given to managing power quality issues, particularly harmonics.

In most established networks the general growth of load is relatively low. In certain areas, particularly highly urbanised areas, redevelopment has seen prospective loads increase owing to new office developments and the associated IT-related loads. At the time of writing this would seem to be a developing trend, the forecasting of which represents a significant challenge.

The management of reinforcement with the connection of new load has become the most significant challenge. The management of the new connections process is becoming progressively more detached and this is likely to lead to an increased need for the asset manager to monitor connections activity and identify reinforcement requirements and implement them in an appropriate time scale. The failure of this process will ultimately impact upon a distributor's ability to meet a customer's connection requirements within its schedule. In large urban areas this may have not simply a financial impact on the distributor but also an economic one.

The present regulatory process in the UK which involves five-yearly reviews to fix income for the following five-year period increases the risk of increased reinforcement expenditure affecting other capital programmes.

4.3.10   Distribution Automation

Introduction
As discussed in the previous section distribution automation in its simplest form has begun to be used to improve quality of supply. At this level the automation installed consists of auto reclosers and auto change-over devices.

The installation of remote terminal units provides the basis for a distributed communications system that could be used to implement some degree of automation. For the convenience of the readers, an appendix is included below to detail distribution automation and communication systems under a competitive environment.

Levels of Automation
Two levels of automation exist:

- Via central control systems
- Via embedded systems.

Embedded Automation
In their simplest form, embedded automation systems such as auto reclosers and auto change-over systems are probably the easiest distribution automation systems to implement. No high-speed communication systems are required as the devices function independently. Some form of communication to inform the central control room or system of their operation is desirable but not essential to realising the benefits.

However, embedded systems have greater possibilities if high-speed local communication systems can be implemented. Possibilities could include the automatic sectionalisation of faulted circuits allowing meshed network operation without the need for costly traditional directional protection schemes or unit protection. The high-speed communications required could be achieved by a number of means including:

- Power line carrier
- Low power radio
- Fixed line communications.

Centralised Automation
Centralised automation could produce many of the same benefits, but it would be unlikely to be able to operate as quickly. The communications path and central control system would have to cope with the information traffic from the entire network and may have to process several scenarios simultaneously. However, such systems may have advantages where the rate of development or changes in network configuration are high as the connectivity and logical sequence can be maintained directly from the centralised control system.

Benefits of Automation
Automation could deliver the following benefits:

- Short-duration interruptions (less than 3 minutes) to customers rather than extended interruptions.
- More precise fault location to assist emergency response teams.
- Reduced workload on control engineers.
- Reduced operational manual switching workload.
- Increased utilisation of assets.

Remote control and indication can deliver much reduced interruption times without the complexities of introducing automation. However, it cannot reduce switching times to less than the 3 minutes allowed under the European standard short duration or the 1 minute interruption benchmark used in the UK.

A noticeable impact of the introduction of a remote control system is the increase in the workload on the control engineer. This is manifest both in the volume of switching operations undertaken by control staff that were previously carried out in the field and in the volumes of data and information that can be presented to the control engineer. An important facet of the development of any automation system must be the handling of the data and the presentation of useful information to the control engineer as to the actions that have been taken by the system. Improvements in the speed of restoration or securing of supplies following a network fault may in future allow increased asset utilisation by permitting higher short-time loading levels as the duration could be managed automatically. This will of course depend upon the network configuration, but it increases the potential benefits of moving towards meshed network operation in the medium to long term. A reduction in the need to carry out manual switching has a significant safety benefit, particularly with ageing oil-filled switchgear, as an operative does not need to be in the immediate vicinity. This is enhanced by the possibilities for local condition monitoring and remote indication of alarms relating to possible hazardous conditions.

4.3.11 Automation Case Study - Remote Control in London Electricity

Introduction
This case study considers the planning issues concerned with the long-term development of London Electricity's networks, particularly the changes made to improve quality of supply.

The introduction of a remote control and data acquisition system has been central to London Electricity's development plans for its secondary (MV and LV) networks over the past five years.

This case study considers the philosophy behind the programme, the equipment deployed, the advantages London believes these offer over alternative systems and the performance delivered to date.

Philosophy
The concepts developed from a review of secondary system performance in 1992. The impetus to implement the programme was brought to the fore during the second price control review in 1994/5 when the regulator's focus on quality of supply improvement increased.

The remote operation of network switches allows switching to occur in the time before an engineer can reach the affected area, often in excess of an hour in London, significantly reducing the interruption time seen by a large number of affected customers. The programme was therefore targeted to reduce customer minutes lost from around 53 (1995) to a target level of 40 in 2000.

The overall asset management plan required that the remote terminal units (RTU's) selected had to deliver data acquisition and intelligent monitoring facilities in addition to the basic control necessities. This approach has proved well founded in view of the long-term development of the system, as will be discussed later.

In order to achieve the target performance improvement a three-stage approach was adopted aimed at deploying up to 6000 RTUs across the 6.6 kV and 11 kV networks.

Planning
The three-stage approach aimed to deliver as much benefit as possible in the initial phase. The first stage was aimed at targeting one in every four ring main units in the worst performing networks. The MV networks in London Electricity's system are configured as groups of about four circuits. These groups are run radially with a number of open points between them, effectively creating an open four-feeder ring. However, these can be divided into two further categories depending on whether the associated LV systems are

operated radially or interconnected, i.e. operated as a mesh. The meshed LV systems are typical of the centre of London and assist in coping with the high load density in Westminster and the City of London.

The greatest initial benefit in quality of supply performance was to be gained from targeting the areas where the LV networks are operated radially (the radial areas) and offer no support in the event of an MV fault, which is a characteristic of the interconnected LV system. The feeder groups supplying the radial areas were ranked in order of their performance over the pervious years, bearing in mind any major asset replacement undertaken to correct high fault rates.

The RTU installation programme was then targeted in these networks at open points and every fourth ring main unit which offered a suitable switching point. In order to achieve the switching functionality in the existing ring main units a programme of retrofit actuator solutions was developed to mitigate the need to replace switchgear. Initially this was targeted at modern SF_6 ring main units and some of the more modern oil-filled units which were deemed suitable. This resulted in remote switching being available at an open point and at the approximate mid point of each circuit. A fault passage indicator with provision for remote indication was fitted with each installation. This would allow approximately 50% of each feeder to be restored by remote switching.

The second stage of the programme extended the provision of remote control facilities to approximately one in two ring main units, again with the initial concentration being on the worst-performing feeder groups. In turn this would allow up to 75% of customers to be restored by remote switching.

The third stage extended these facilities to those networks operating with interconnected LV systems. This is a more complex task as each MV/LV substation is equipped with an LV air circuit breaker (ACB). This is installed to prevent network collapse in the event of an MV feeder fault, due to either a fault infeed from the LV system or resulting network overloading (it being preferable for the ACB to open than a number of network fuses which then have to be identified). It is operationally desirable for the ACB to be controlled to reduce the number of site visits by engineers in the event of a loss of supply. It is also necessary to know the status of the ACB while attempting to secure supplies following a fault, so remote indication had to be provided.

A one-in-two strategy was adopted as this was felt to be the minimum necessary to provide the degree of control required to secure supplies remotely without the need for an engineer to be present in the field.

Performance
By the end of 1999, London Electricity had equipped 3000 MV/LV substations with remote control facilities as part of stages 1 and 2 of the programme described above. Stage 3 was initiated in late 1999 and would begin to take effect in the least well-performing interconnected networks during 2000. The majority of these, 2000 of them, were installed during 1999. The performance of the programme has been excellent with customer minutes lost visibly reducing with the numbers of units in commission. Supplies are now restored to all customers within 1 hour for over 50% of all MV network faults. Most pleasing of all has been the number of routine switching operations that were soon being carried out using the system.

Figure 4.4 shows the theoretical performance estimates made when the project was conceived. The top curve shows the predicted performance against the bottom curve, which identifies the theoretical best performance. 45 CML is the target performance for the year 2000. It can be seen that approximately 3500 RTUs would be needed to achieve the target performance.

Figure 4.5 shows the performance achieved by the system at the time of writing. It can be seen that the targeted programme has been successful in achieving the 40 CML target with only 3000 RTU's installed. The improvement over expected performance is a direct result of the improvement in supply restoration times for MV faults. This is shown in Figure 4.6.

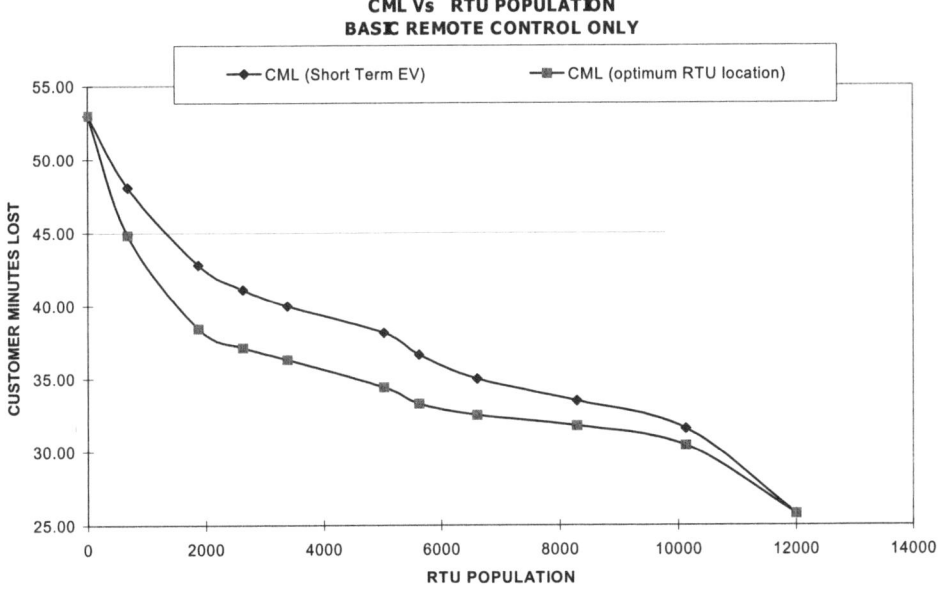

Figure 4.4 Theoretical benefits of remote control

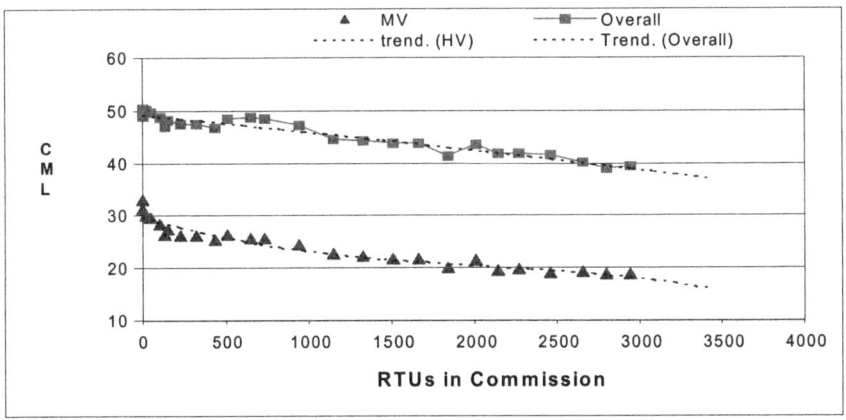

Figure 4.5 Actual performance

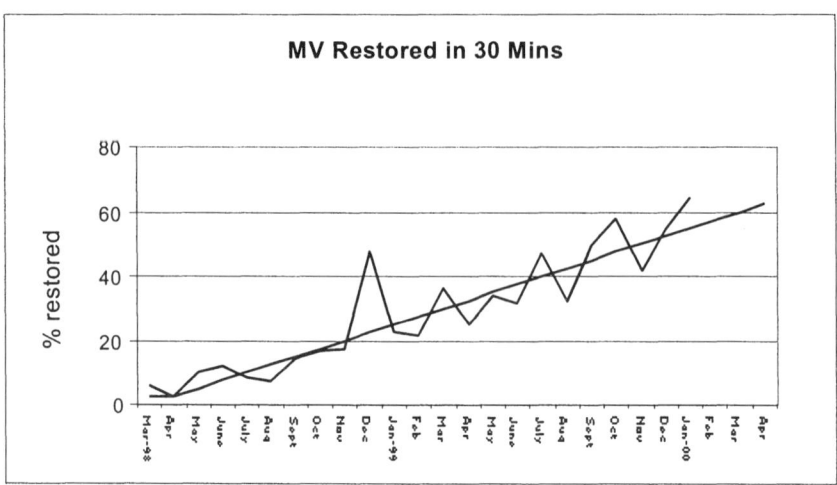

Figure 4.6 Restoration performance

4.4 Future Development

The original vision for the development of the initial remote control system was to create an integrated network management system. The advanced, expandable and to a limited extent intelligent RTUs were chosen to facilitate this development. The initial elements of the future plans are already being implemented. An auto change-over mechanism has been implemented in the RTU's logic to allow supplies at open points to be restored in less than 1 minute following an MV network fault. This delivers an improvement in the customer interruptions statistic as well as the customer minutes lost.

Such a concept can be expanded to deliver automated restoration of the majority of customers following fault. There are still many issues to be overcome including the fault location logic and communications. The need to communicate to a number of sites more than once in less than a minute will pose a significant challenge.

There are, however, other aspects of network management beyond the control of the MV system which the RTU was specified to cope with. These include condition monitoring of the MV system and monitoring and control of the LV system via additional monitoring technology designed to interface with the RTU. The condition monitoring of the MV system could be developed to include discharge levels in cables and various indicators of the condition of switchgear and transformers. The system already allows the monitoring of the LV load on a single-phase and three-phase basis and can also measure the distortion of the voltage waveform and the harmonic components of that distortion. The monitoring of the LV could also be extended to encompass operation of fuses and perhaps even the location of faults.

Whilst many of these developments are designed to assist in the operation of the system, the data will be invaluable to the planner in maximising the use of the existing assets and optimising future developments of the networks. However, the planner must also understand these technologies, their benefits and consider their application when designing the system.

4.5 Appendix: Distribution Automation in a Deregulated Environment

4.5.1 Introduction

Deregulation is one of the main reasons for utilities to advance automation of distribution which gives benefits on customer service. Deregulation is evolving to establish some customer satisfaction indexes in order to penalise utilities in case of low-quality service. In deregulation, the importance of efficiency will grow, and high peak prices for spot power could happen. Broadcast load control will help manage the risk associated with peak demands in a 'customer-friendly' way.

Distribution automation provides the ability to communicate vital information. It also gives the ability to monitor and control that information from a central location. In short, the system could tell whose electricity is out - before anyone calls to complain. And it could save money by restoring power sooner than ever before. Distribution automation allows continuous monitoring to provide the real-time knowledge needed to optimise power system operation and control. And it can maximise customer satisfaction with improved reliability, consistent quality and high-tech services.

Distribution automation refers to the automation of repetitive tasks on an electric utility's distribution system. For the electric distribution system, these repeated tasks include reading kilowatt hour and kilowatt-demand meters, reading temperatures, taking load checks at distribution substations and along the feeders, opening and closing feeder circuit-switching devices, operating capacitor banks, and raising and lowering taps on voltage regulators. These tasks involve communicating and turning things on/off, counting things and measuring things. Most electric utilities already have automation at certain levels. This automation may range from time clocks for turning street lights on and off, to a computerised energy management system for economic dispatch of the generation and transmission of bulk power. When we speak of distribution automation, we are talking about a fully integrated system that includes all of the functional data flow and control involved in the distribution system. The distribution automation system (DAS) will be a natural extension and enhancement of the level of automation that exists in the individual utility today. Automation provides timely control and data acquisition through communications with remote devices.

The DAS should be viewed as a tool that can be used to address corporate needs to do repetitive tasks in a systematic way. The DAS provides flexibility and the power of choice to the utility and to the customers it serves. The technology required to implement the DAS is available today. The challenge for the utility is to identify and evaluate potential distribution automation functions and determine those appropriate for implementation. It is important to remember that although all electric utilities are similar, no two are alike. Each utility has developed under different geographic and regulatory conditions and has unique customer bases and prevailing economic conditions. Once desired functions have been identified by the utility, the integration of these functions and tasks into one system requires a combination of specialised hardware and software.

The scope of the distribution automation concept has certain implications. The evolution of distribution automation in a specific utility will be based on that individual utility's organisation and requirements. In recent times, distribution automation has become a watchword for utilities wanting to improve their commercial position -

improving revenues and reducing costs. Distribution automation is a complex subject comprising the following major components:

- remote terminal units (RTUs);
- SCADA master station (open systems design, full graphics);
- software functionality (feeder sectionalisation, cold load pickup, topology processor, voltage/var control, graphic feeder tracing, switching order preparation, special reports, automatic meter reading, etc.);
- operations and maintenance procedures (safety, tagging, permits, clearances, work orders, preventative, routine and restorative practices, spare parts, service agreements);
- system integration, design and management; and
- communication system (e.g. cellular phone, radio, power line carrier, ripple).

4.5.2 Remote Terminal Units

Constantly involved with improvements in RTU technology are things such as the development of ladder/sequence logic/PID algorithm capabilities, multiple serial interfaces to accommodate smart meters and relays, peer-to-peer protocols and direct WAN communicating RTUs.

For distribution automation purposes, small, low-power, weather-proof, compact RTUs are available. These come in a variety of enclosure packages, with specifically constrained points counts, direct CT/VT inputs, AC analyser modules to give a variety of calculated information, and more. The units can perform data logging to minimise the need for constant polling via the communications system. In some applications, peer-to-peer communications have been utilised to facilitate independent islands of automation (e.g. for voltage/var control) that do not rely on the master station for decision making and issuing control actions.

RTU-initiated report-by-exception protocols are being utilised by some vendors to keep the RTU's power draw (from constant communications with the master station) to a minimum. Solar-powered units are in common usage. Compact, low-maintenance battery backup power is also available. With the advent of two-way communication on distribution power circuits, a great future lies ahead for distribution automation and demand-side management applications at the customer level for functions such as remote meter reading, selective control of customer loads, surveillance of customer installations, and customer choice of electricity rates.

4.5.3 SCADA Master Station

The design of modern 'open systems' SCADA configurations makes use of considerable communications technology to spread the risk that a single failure will wipe out or disable a mission-critical system. Modern LAN/WAN technology permits highly distributed processing and achievement of graceful degradation upon failure. High-powered workstations and personal computers give users full graphics depictions of system assets, often in proper geographic orientation, and provide a very flexible window into the systems they are controlling. The dissemination of computing elements and the inherent flexibility of full graphics interfaces have increased the burden upon the system designer and user to manage these numerous features. The control system hierarchy for the whole of the power system must be flexible, with its topology adaptable to meet the changing environment. Distributed databases need to be kept synchronised so that all users view coincident data. Displays must be constructed in a manner that makes operator navigation and problem recognition simple under the most stressful situations. These items, if overlooked, can lead to poor operator acceptance of the new tool.

SCADA system architecture can now be dynamic, allowing change and enhancement over time as both user needs and technology change. Relational database management systems have facilitated easier, more functional interchange of data with other corporate systems (accounting, customer records, maintenance management, work dispatch, etc.). Compact, high-density storage media have simplified the tasks of regular backup, keeping historical data and managing archives. Disk shadowing provides hot standby data requirements for key operational areas.

Distribution automation functionality in the SCADA must work with the actions and reactions of the distribution system protection equipment. Actions taken by the logic or control algorithms of both systems (e.g. protection and SCADA) must not compete with one another to cause additional problems. Those applying distribution automation to existing power systems must ensure that all protection schemes and systems are thoroughly understood and catered for. One needs to have considerable experience with protection schemes involving:

- smart relays which feature multiple settings and communications interface capabilities;
- relay, recloser and tap/transformer fuse coordination; and
- provisions for feeders in parallel operations between buses or substations.

Distribution Management System (DMS) Functionality
Typical functionality contained in a DMS, additional to standard power system SCADA functions, may be as follows:

- feeder load shedding - coordinated manual, rotational and under frequency schemes;
- connectivity processor - provides up to the moment topology and energisation status;
- switching procedure generation and management;
- distribution power flow;
- cold-load pickup estimation;
- transformer load management;
- supply interruption reporting and outage management;
- fault location;
- transfer optimisation (load transfer and reconfiguration capability);
- automated feeder reconfiguration and service restoration;
- voltage/var control - transformer tap and switched-capacitor management;
- distribution short-circuit analysis;
- demand-side management - load management and time of use strategies;
- computer graphics capability to provide schematic display, switch position marking (manual or telemetered), feeder connectivity status, and energisation status information in real-time on the operators' VDUs; and
- training simulators.

4.5.4 Software Functionality

Whilst the hardware of a SCADA system is of great importance, the key investment is primarily in system and applications software. The SCADA system software dictates the system's functionality, supportability and maintainability – e.g. its usefulness over time. Modern systems built to internationally accepted open-system standards utilise:

- UNIX or Windows NT with real-time extensions;
- X-Window - the OSI Windows environment;
- C++ language for ease of programming and portability;
- application program interfaces (APIs) - for almost any language to permit diverse applications to be easily integrated with the SCADA system;
- parallel processing - use of dedicated servers to handle time-critical and computing-intensive functions, without the necessity for them to compete for computing resources;
- ODBC - open database connection to allow easy passing of real-time and historical data among diverse systems;
- object-oriented database and system topology - assists with ease of system building and display building, and makes topology problems much easier; and
- SQL - structured query language to enable simple requests for information to be made within the systems and by those users connected via the LAN. It is vitally important to specify and select systems that are based on tried and proven software. Developments in this area of the system can have significant consequences on delivery schedules and cost. Rigorous design checking, verification procedures, and factory and site acceptance testing are recommended.

4.5.5 Operations and Maintenance (O&M)

A communication system for distribution automation is a complex combination of transmitters, receivers and data links. The system should be designed so that operation and maintenance will be as easy as possible. Personnel will have to be trained for the new skills involved and new tools will need to be purchased (the cost of this should be included in the present-worth analysis of a potential system). Selection of rugged, easy-to-use communication equipment will significantly improve the transition from a manual to an automated system. The use of standardised components and communication protocols would not only allow better compatibility with existing communications equipment but also include the likelihood that the system will remain compatible with communications and automation equipment developed in the future. This will help to reduce operation and maintenance costs to the utility.

Utilities need to develop appropriate O&M philosophies and practices to meet their particular needs. Risk management analysis is a valuable tool for sorting out how O&M matters should be approached by each utility.

There are several aspects of O&M for distribution automation to be considered. These include:

- Operations philosophy with distribution automation installed (who does what, why, when, how).
- Person/machine interface procedures to interact with and carry out the operations philosophy (authorisations, modes of operation).
- Distribution system operations procedures that take into account the distribution automation/SCADA capabilities (e.g. tagging and permits, work order dispatching).
- Maintenance procedure safety considerations for the distribution system given the potential for automatic restoration, feeder paralleling, etc.
- Maintenance of the distribution automation equipment.
- Equipment condition monitoring.
- Maintenance of the master station equipment, software, database and displays (configuration management, spare parts holding, service contracts).
- Training to support the accepted philosophies of operations and maintenance - both classroom style and using simulator scenarios.

4.5.6 System Integration, Design and Management

The traditional 'point-to-point' wiring between the equipment controller and the RTU was expensive, difficult to maintain and impossible to migrate into future technologies. In addition, an inexpensive SCADA solution is needed that would allow:

- Remote monitoring and control.
- The ability to address multiple communications ports with different drivers/protocols.
- On-line configuration.
- Handling a large number of points while maintaining real-time performance The right combination of SCADA software, field equipment, system integration, communications and feeder design to install distribution automation has to be found.
- Introduction of an integrated system with both high-voltage sub-transmission networks and MV distribution feeders. This would help to optimise operations requiring the coordination of field crew working at different voltage levels.
- Integration of power network diagrams, plant data and telemetry network data regardless of voltage level or mode of control (remote or manual). This aims to reduce the overhead of data entry into multiple databases and the problems due to inconsistencies between them.
- Integration of power analysis tools for planning and optimisation studies.
- Provision of facilities for interfaces to fault management and customer information systems.
- Provision of facilities for interfaces to asset and map management systems.
- Provision of reports for regulatory purposes.

Systems integration is critical to the success of any telecommunications and networking initiative. Since most systems are not developed in a vacuum, integration of existing or legacy systems and networks must be addressed. This integration must be addressed at two levels: (1) the network or communications level, which ensures that existing equipment and protocols are capable of working in the new system to ensure integration, and (2) the applications level, which ensures that information generated and passed from one application can be accessed by another application. Both levels are critical to the success of the system and the organisation's operations. Access and availability of information in a timely, accurate and user-friendly manner are necessary for the system to be a success. The development and implementation of any telecommunication system will affect organisational operations. The success of any project is a direct result of the attention to detail given to system specification, design and installation. This includes verification that what was specified and procured has been delivered, testing of system components and applications, and ensuring that the system satisfies implementation requirements and regulatory guidelines.

Management information systems (MIS) are becoming an increasingly important tool in the daily operations of electric utilities. The information system is more than just a collector, repository and transport mechanism for information. A well-designed information system is a combination of hardware, software and communications capability that forms the foundation of efficient operations and decision making.

A modern integrated network management system is used to control HV substations remotely and to supervise manual operations on MV distribution equipment. The system automatically processes topology and highlights de-energised feeders when devices change state after telemetry input or manual dressing. System Alterations and switching schedules are prepared in advance and operations can be automatically checked against utility-defined safety rules. Power analysis functions can analyse the HV sub-transmission network or individual distribution feeders. One of the major benefits is that SCADA, world-map schematic diagrams, plant parameters and network connectivity are stored in one consistent system. Data is held at a variety of levels of detail to suit both high-level analysis and detailed device operation.

Network operation functions are 'those functions which enable control and supervision of the distribution network facilities' and include control, monitoring, fault management and operating statistics. Operational planning functions are 'facilities to define, prepare and optimise the sequence of operations required for carrying out maintenance work on the system' and include network simulation and switch action scheduling [10,11].

The primary purpose of a network management system for network operations is to enable the dispatch of field crew (people) to maintain and repair the network, safely and promptly. This differs from an energy management system (EMS) whose primary purpose is the dispatch of power (MW and MVAr). The modern competitive market emphasises that utilities need to monitor and improve levels of customer satisfaction as well as optimising network operations and controlling operational costs.

Deregulation creates a new wave of electronic brokering as electricity is bought and sold in a commodities market. Tracking of these transactions within a given utility should be manageable; however, most of these transactions will span multiple companies. In order to achieve interoperability, implementation of a common information model is recommended. The common information model has a data structure that is common to almost all EMS proprietary systems.

Most information networks will be connected to the EMS to provide accurate real-time data to support the available transfer capability (ATC) calculations. Hypertext markup language (HTML) will be used to present information to customers. HTML forms will be provided by the transmission services information networks for customers to use to request purchases from a provider. The Hypertext transport protocol (HTTP) is used for data access, including both data downloads and uploads. The request is usually made through an HTML form which will in turn will create the HTTP arguments through a Common Gateway Interface (CGI) program to query or select data for browsing or downloading. Similarly, a program can be executed by the World Wide Web (web) browser to upload data through the HTTP, which would then be processed by the provider's server through a CGI program.

Looking into the near future, operation of the intelligent electronic devices (IEDs), the network, host computers and operator interfaces will be totally separated. In other words, failure of the host computers would have no effect on the inherent operation of the system. All IEDs on the network will respond to requests from any other IED on the network. Many next-generation IEDs will be able to synchronise their data capture and measurements to the global positioning satellite (GPS) system. Such synchronisation will allow measurements that give an instantaneous snapshot of the system state. Given the observability of the state, optimal control theory indicates that there is a good chance of being able to control the power system. For example, if it was able to detect a potentially unstable power swing in progress on the power grid, control actions could be taken in tenths of a second to bring the power system back to a stable state. The network architecture (both LAN/WAN) provides the foundation to effect the high-speed measurements and subsequent control actions required to implement this functionality.

System Security
With increased energy use, the electric power system is experiencing much greater levels of regional transfers. These new requirements push the system to its limits for maximum economic benefit, while maintaining adequate security margins requires on-line real-time network analysis. An interconnected system can collapse owing to a number of different limits being exceeded, e.g. thermal limits, voltage stability, transient stability and dynamic stability.

The transaction management system (TMS) allows marketers to enter transmission reservations, energy schedules and ancillary service reservations, and to receive notices of curtailments and reservation confirmation [12]. TMS also allows the tracking and approval of these reservations and schedules. The reservations and schedules are needed for the security applications to determine next-hour and future security constraints on the electrical grid. Stability limits for the operation of electric power systems are necessarily conservative because off-line simulation studies used to compute them assume worst-case conditions. The situation is no longer adequate in increasingly competitive energy markets. This calculation can be improved considerably for use in today's competitive energy market. Dynamic security analysis (DSA) will be an on-line function in an EMS that is capable of calculating stability limits using state estimated data from on-line system conditions. DSA will provide measures of dynamic security to allow the system operator to take timely preventive actions. The deliverable is an on-line EMS software program capable of assessing 30-40 dynamic contingencies in a time frame that is useful to system operation (i.e. 20 minutes). DSA will identify critical power system interfaces and calculate stability limits for these interfaces, as well as other necessary system and unit operating limits.

DSA uses a contingency screening tool to determine any contingencies that may affect the stability of the program and then uses parallel processing, unique stopping criteria and a very fast stability algorithm to determine the transfer limits of a critical interface. On-line

DSA will allow stability-constrained utilities safely to increase the loading of transmission facilities and can also improve system reliability. On-line DSA will also allow timely assessment of the security impact of transactions that occur as result of open access to the transmission system.

Reference [13] reported that a preliminary demonstration using a simplified, partially functional prototype of DSA has resulted in a potential savings of $1.4 million over 3 years for one utility. A typical benefit of the production version might be a 5 % increase in transmission capacity across a constrained 1900 MW interface. Assuming 30 % of this increased capacity could be used on average over a year and a 10 $/MWh price differential across the interface, this could result in an annual financial benefit of $1.3 million.

On-line voltage security analysis (VSA) will continuously monitor the voltage stability of the grid using state estimated data from the EMS. Online VSA will complete an assessment of voltage security of the current system state within 20 minutes for large systems, providing various security indexes for the operator and a list of contingencies that could lead to instability. It will also identify measures to mitigate an evolving voltage collapse, such as load curtailment, redispatching, emergency loading of VAR sources, etc. The tool will be capable of performing on-line studies of various operating decision strategies, such as options for power transfers and scheduling planned outages for maintenance. For voltage stability constrained utilities, VSA will allow increased utilisation of grid assets while maintaining the reliability of power delivery. This is achieved through the calculation of limits dynamically using actual system conditions. A typical benefit might be a 15 % increase in transmission capacity across a constrained 1000 MW interface. Assuming 30 % of this increased capacity could be used on average over a year and a 10 $/MWh price differential across the interface, this could result in an annual financial benefit of $3.9 million.

4.5.7 Communication Systems

Electric utility deregulation is moving ahead at full speed throughout the world. As a result, the integration, consolidation and dissemination of information both inter- and intra-utility are important. Information traditionally used only within a given utility now becomes desired by many players. The general trend in the industry has been towards the use of the Internet for the transfer of data for operation and scheduling.

Beyond data sharing, new mandates are being placed on distribution utilities to minimise outage times, provide rate alternatives (e.g. through demand-side management), maintain operating data archives, and, in general, push more power through existing power lines. On the financial side, deregulation introduces the need for the sharing of accounting data among utilities, international standards organisation (ISO), metering firms, billing firms and independent power producers (IPPs). Inter-utility billing must be correct and standardised.

Another outcome from deregulation is the merger and consolidation of many of the existing utilities. Mergers will require the establishment of intra-company communication and the integration of data from companies' control centres, power plants and substations. Implementing this integration with different data models and communication protocols will add considerable time and money to the process. The possibility of the need to perform this integration process should drive all utilities towards the standardisation of data models and communication protocols.

Telecommunications play an increasingly important role in the daily operation of electric utilities. More than just a means of providing connectivity between one person and another, well-designed and engineered telecommunication systems are the cornerstone of efficient operations, enabling better communication, not only between employees inside the organisation, but with citizens and customers as well. With the onset of restructuring within the utility industry and recent changes in the telecommunications industry, business opportunities such as cable television, security and Internet connectivity pose attractive new revenue streams and the means to establish and maintain strong customer relationships.

Communication requires transmission channels, which may consist of copper or optical fiber, or free space lasers. While there are many companies today that sell only point-to-point communications channels, a communication infrastructure generally means a network. A network implies a set of channels linked by some form of switching that enables any two parties connected to the network to send signals between them. Success in business requires low-cost production. That in turn requires investment in state-of-the-art technology for production, i.e. transmission facilities.

Prior to 1980, most companies as well as most homes were wired for communications only with twisted copper pairs. If a computer terminal was needed in the office, special wiring would be pulled on an *ad hoc* basis. Between 1980 and 1990, many companies went from less than 20 % of white-collar workers with terminals to more than 70%. As a consequence, they began to think of data communications wiring as part of the building infrastructure that should be installed once and managed as infrastructure. Accordingly, the industry developed a number of standard data wiring schemes based on coaxial cable, shielded twisted pair cable or unshielded data-grade twisted pair, with some use of fiber as well. Today most new office buildings are provided with a data wiring infrastructure just as they include electric wiring. With advances in technology, copper pair wires have been made to support data rates up to 100 Mbps for short distances from the desk to a wiring closet. At the wiring closet sophisticated wiring 'hubs' provide the first level of traffic concentration, justifying the use of more expensive optical fiber to link these concentration points in campus-wide networks. The ability to carry up to 100 Mbps on copper seems likely to forestall widespread introduction of fiber to the desktop for another 5 to 10 years.

Leasing full-period channels is costly, however, when traffic has a high peak requirement, but relatively low average throughput. This creates demand for switched data services from the carriers who can take advantage of traffic statistics to provide high peak capacity to each user, while charging only for the average throughput actually consumed. The first carrier services meeting these objectives were public switched packet networks (PSDNs) which carried data in packets of 128 characters at speeds up to 48 kbps. These served primarily terminal to mainframe computer traffic. The introduction of desktop computers created a demand for high-speed switching of bursty traffic between machines. The idea of distributing the switching function among all the attached machines rather than having a central switch led to the development of LAN technology. In early LANs, the transmission medium is configured as a bus or ring and its capacity is shared by all of the users as they transmit their data in high-speed bursts. Each node on the LAN is responsible for assuring that its transmissions do not interfere with any others. Very quickly many companies found themselves with campus-wide LANs capable of efficiently handling bursty traffic at speeds of 4 Mbps or more. By comparison, the data services offered by the public carriers were slow and not well suited to computer-to-computer - as opposed to terminal-to-computer - traffic. Corporate users again had to rely on leased lines and premises-based switches ('routers') to link LANs at their various locations.

In the early 1990s, carriers began to introduce new, higher speed switched services suitable for linking the high-speed LANs. Frame relay is a stripped-down version of traditional packet switching. Relying on the lower error rates of fiber-based transmission, and the increasing intelligence of data terminating equipment, frame relay nets dispense with the error correction service offered by traditional PSDNs and in return realise high speeds, lower delay and lower costs. Responsibility for error correction is left with the equipment at each end, much as is done in LANs.

For medium to large business customers, however, fiber has clearly proved its worth. Large business users, because they concentrate traffic from many offices, can make use of high-capacity links from their premises to the carrier's central office. Business users need the capacity of fiber. Fiber not only provides more capacity than copper today, but also promises easy expansion of capacity simply by installing more capable electronics in the future. The demand by businesses for fiber-based access has induced a number of new companies to construct metropolitan fiber networks to compete with the local exchange carriers. By providing quick service, competitive prices, and an alternate path for reliability, these competitive access providers (CAPs) have gained many satisfied customers. In a cellular telephone system subscriber terminals are linked to a base station by radio signals rather than wires. These base stations are then linked by wire or microwave to a switching node. As mobile subscribers move from the area or 'cell' served by one base station to a cell served by another, the wireless access link is automatically switched to the new base station.

As new-technology developments increase the capacity of wireless systems and reduce their cost, it is likely that existing communications networks based on fixed wires for the access link will find themselves in direct competition with new technology based on wireless local access. Using wireless access, these important customers can be served quickly. In some urban areas, carriers are looking at wireless loops as less susceptible to vandalism. The ultimate radio access link could be satellite based. However, new research on wireless access at frequencies around 30 GHz may make possible wideband wireless access links. These higher frequencies are currently more costly to exploit, however, and more susceptible to interference due to inclement weather.

By 1992, several LAN vendors had begun offering asynchronous transfer mode (ATM) switches as successors to LAN hubs for linking increasingly powerful workstations at speeds above 100 Mbps. While there are some who still question whether ATM switching will prove to be the optimum solution for integrated broadband networks, the concept has such momentum within the international telecommunications community that it is virtually certain to be widely deployed. The flexibility of service offerings over the network is greatly affected by the mechanisms implemented for control of the switching functions. These include rules for routing information, load sharing, simplified addressing, and sophisticated accounting and billing. The local exchange carriers have begun to introduce common channel signalling in their own networks to provide such services as call return and repeat call. Common channel signalling is also a pre-requisite for full deployment of integrated services digital networks which bring common channel signalling right to the end user's terminal.

Ethernet was chosen as the data link layer because of its predominance in the marketplace and the subsequent availability of low-cost implementations and associated network hardware (such as bridges and routers). In addition, the scalability of Ethernet is well defined with 100 Mb implementations being fairly common and 1Gb Ethernet well on its way. Processors are available today with multiple 10 Mb Ethernet ports integrated into the chip, and next-generation designs are planned with 100 Mb ports.

As it was deemed desirable to be able to access data from any device, two solutions were adopted for the network layer: Transmission Control Protocol/Internet Protocol (TCP/IP) and the ISO Open System Interconnect (OSI). TCP/IP is a network layer that operates over the Internet. For international standardisation [14,15], it was decided to include the ISO network layers. These layers have robust flow control and buffering capabilities that are very useful on a busy substation LAN. Both network layers support the concept of broadcasting a message for all devices on the bus to hear. This feature is very desirable for functions such as data capture triggering, time synchronisation, and even control messages to multiple devices.

To facilitate data organisation, object models [16,17] are used because they can easily be shared with others. For example, a relay makes measurements of voltage, current and power on a monitored transmission line. The measurements made by the relay can be organised in a measurement model containing all the elements mentioned above. If additional measurements such as power quality and power factor are added at a later date, the original model is easily expanded to accommodate this data.

Fully interactive communication systems provide a full range of functions including voice/data transfer, remote access and control, entertainment and education. Added communications, network capability and services may also have an impact on the staffing and organisational needs of the organisation. An evaluation of the existing staff, roles and responsibilities is necessary to determine if the required capabilities exist or if new staff and training of existing staff are required.

With the constantly changing telecommunications environment and the ability of almost any organisation to offer telecommunications capabilities and services commercially, the legal issues related to a system must be evaluated at the local and national level so that the organisation can address these early in the development process.

Automation of power distribution systems requires the use of an effective communication system to transmit control and data signals between control centres and a large number of remotely located devices. Since there are a wide range of available communication technologies capable of performing this task, selecting the appropriate communication system requires a thorough understanding of the strengths and weaknesses of each communication technology. Presently, no single communication technology has been demonstrated as being best suited for all distribution automation needs. Each distribution automation scheme has unique communication requirements, and therefore a communications technique for distribution automation must be chosen based on those unique requirements.

Communication Requirements
The communication requirements for distribution automation depend on the size, complexity and degree of automation of the distribution system. In general, it is desirable

that a communication system for distribution automation has the following characteristics [18,19]:
- Communication reliability.
- Cost-effectiveness.
- Ability to meet present and future data rate requirements.
- Two-way capability.
- Ability to communicate with power outage areas/faults.
- Ease of operation and maintenance.
- Conforming to the architecture of data flow.

For distribution applications, we could employ combinations of commercial paging network, utility-owned digital communication, power line carrier, telephone (e.g. ISDN), pilot circuits, dedicated or trunked VHF/UHF radio (including spread spectrum), fiber optic cables or cores in overhead earth wires, cellular phone, and ripple control signals (one way only). Radio is by far the most popular medium.

Coupling this communication flexibility with an extensive set of load control strategies, there is a platform for controlling residential and small commercial loads, distributed generation and other advanced applications.

Power Line Carrier

Power line carrier (PLC) was first introduced in the 1920s. Since then, PLC technology has evolved into a mature and reliable communications technique for power transmission systems. Today it is used primarily for protective relaying, SCADA and voice data on transmission systems. Utilities have looked favourably upon PLC because of the positive experience they have had with it in transmission applications.

PLC utilises a carrier frequency to transmit information over existing transmission lines [20]. For transmission line applications the carrier frequency is usually between 20 kHz and 300 kHz. Information is encoded on the carrier through the use of amplitude modulation (AM), single side band (SSB), frequency modulation (FM) or frequency shift keying (FSK). At the sending end, the modulated carrier is injected onto a transmission line by means of a coupling capacitor and tuner. The modulated signal propagates down the transmission line to the receiving end. At the receiving end, a coupling capacitor and tuner separates the PLC signal from the power frequency voltage and a demodulator extracts the information encoded in the signal. Line traps at either end of the line prevent the carrier from travelling down undesired paths.

Growing competitive pressures resulting from the advancing privatisation of the international energy market combined with the liberalisation of the telecommunications market have sparked energy suppliers' interest in power line communication, a new technology that sends data through existing electric cables alongside electric current and is set to turn the largest existing network in the world, the electricity distribution grid, into a data transmission network [21]. Since the development of complex frequency modulation processes such as OFDM (orthogonal frequency division multiplexing), together with highly integrated and inexpensive semiconductor chips, data transmission rates of 1 Mbps and more have been possible (1 Mbps is 1 million bits per seconds - 16 times faster than ISDN (integrated service digital network)). It will make power line communication possible both to phone and surf the Internet over power lines. Long-distance monitoring of alarms and air-conditioning systems, comfortable control of intelligent household appliances and off-site readings of electricity meters will all become feasible simply via the power grid. There is no inherent physical barrier to the further development, improvement and commercialisation of power line communications. It can be expected that it will be driven by business needs (specially for new operators) and that if it is seen as viable, then higher performance systems will surely appear. For this reason, power line communication technology has the potential to be a serious competitor to traditional network operators.

Automatic Metering
The Internet and the explosive growth in the number of its users are also helping to make it both feasible and inexpensive to provide not only electricity but also high-speed data services – an important edge on the competition for just about every local power company. Because it is a PLC technology, the lines of communication already go to every meter. Direct control (load management) is built into the communication technology, and is often integrated with the automatic metering and aggregation (AMR). In other words, we not only read the meter, but can perform scheduled control tasks at the same time. This will be a valuable function in the future days of 'time-based' rates, because a 'metering company' can offer to control loads for aggregated groups who want to make use of the lowest rates offered by energy service providers (ESPs). We can communicate digitally to electronic meters using the fixed network. We are talking about reading the actual revenue registers from the meter, resetting the meter remotely, and synchronising the time - instant, on-line communication that takes only seconds per read, for polyphase revenue meters [22]. Other features may also be implemented, such as:

- The area containing the fault may be automatically isolated and power from a parallel feeder automatically connected to bypass the isolated area.
- The power factor on the three-phase feeder may be determined and compared established set points; control commands may be sent to associated capacitor banks to manage the power factor intelligently.
- Switches may be opened and tagged by line crews while they work on a line section; the tagged status prevents automatic or manual operation of the switch, and the tagged/not-tagged status is reported to load control.
- The SCADA system will monitor the security of all remote sites in the system. Unauthorised entry, high- or low-temperature, high-humidity, etc., conditions may be reported as alarms requiring human intervention.
- Communications may occur directly among the multiple sites in the SCADA system and with the central computer on the same radio system utilised for voice communications.

The AMR scheme can be realised partly by PLC technology (from consumer to substation) and partly by telephone or fiber optic communications (from substation to the central station). Customer interaction can be achieved by selling and installing the meter-reading software in telemetering or by having customer display units for AMR. Depending upon broadcast messages, load shedding becomes obsolete because all premises can be rapidly contacted on an individual basis, with instructions for selective load shedding. Equally each of the premises can then report back to confirm that the instructions have been received and acted upon. Customers can select the most suitable tariff for their needs from a wide range, and, with a little help from the supplier, optimise electricity consumption to minimise costs. At the same time, suppliers can achieve vastly improved control over demand.

AMR technology provides near-instantaneous information on supply failure, down to the level of a single household. It supports rapid, report network reconfiguration to restore lost supplies, and to optimise network loading for reduced energy losses - in effect, providing SCADA facilities for the lower levels of the distribution network. And because AMR technology can monitor deviations from established usage patterns, it provides a warning of possible tampering. Variable rates can be remotely programmed, schedule recovery of debts (through flexible prepayment) can be easily arranged, and in the case of an empty building, remote disconnection can take place with complete certainty.

Further new tariffs can be quickly and easily programmed into any customer's meter down the wires whenever required. The data can be transferred interactively through the electricity distribution system to the distributor's own headquarters, with benefits to both supplier and customer. Financial applications, such as real-time credit card transactions between stores and finance organisations, can become simple and inexpensive.

AMR technology could use the existing electricity distribution infrastructure for data transfer between user and supplier. As a result, no costs are incurred in the provision of separate carrier media. Two-way data transmission takes place over existing cables - and not only on LV networks, but also on their HV equivalents.

Fiber Optic Communication

Around 1910 it was shown [23] that a circularly symmetric transverse magnetic mode can be guided by a dielectric cylinder in free space. And although there were demonstrations of the effect from time to time, by the early 1960s there had been no practical applications, since the losses in transmission in what was then considered good optical glass were in the order of 1000 dB/km.

In the mid 1960s, Kao and Hockam [24], then at Standard Telecommunications Laboratories in England, realised that the losses were not due to an inherent property of the glass, but were in fact caused by impurities. They wrote a paper that examined the economics of optical communication taking into account repeater costs and fiber losses [25]. This set the stage for the commercial development of fiber communication systems, and indicated to the telephone companies in particular the way to achieve the greatest benefit/cost ratio. For a telephone application, the economics are most advantageous under two conditions:

1. The distance between repeaters should be maximised. Repeaters are expensive pieces of hardware, and the fewer the better.
2. The bandwidth of the channel should be maximised. In this way the maximum number of calls can be routed on a given channel, and the cost shared among many subscribers.

Because of these factors, the first widespread use of fiber optics for communications was on the long-distance trunk lines of the telephone companies. The same driving forces have led to improvements in the performance of the fiber, to the extent that today repeaterless fibers are usable over large distances. At the same time, the cost of the fiber cable has been reduced to the point where it is comparable, on a length-for-length basis, with copper conductor. The implication of these developments is that for the phone company it is clear that fiber optics can be used to replace copper trunk cables, and provide greater capability; and submarine optical cables could carry more traffic to Europe than satellites.

Two-way Capability

Two-way communication is required for most distribution automation functions. Load control is an example of a function suited to one-way communication. A control signal is sent from the control centre that informs the loads to toggle on or off. More advanced load control systems can have addressing information transmitted along with the load control information so that individual loads or groups of loads can be controlled. Real-time verification of the load status with a return signal from the load is not necessary for controlling the devices. Some ways of determining the health of the load control devices is needed. Two-way capability would help determine which load control devices are not working.

Fault isolation and service restoration is an example of an automatic function requiring two-way communications. In this case fault detectors must communicate with the utility control centre so that the fault location can be determined, then signals must be sent from the control centre to sectionalising switches to isolate the faulted section. Two-way communication is a requirement for any automated distribution system with advanced functions.

Communication Reliability

A communication system for distribution automation will be exposed to the severe environment outdoors. This means constant exposure to adverse weather conditions, such as rain, snow, hail, severe wind and electrical storms. In addition, long-term exposure to ultraviolet light from the sun can lead to deterioration of some materials. The communication system must be designed so that it can withstand these rigours with only routine maintenance.

The communication system will be exposed to electromagnetic interference (EMI), which can seriously affect its data transmission reliability. EMI can occur in the form of radio frequencies (caused by gap noise, discharges, the corona or other radio sources), or the 50/60 Hz fields associated with normal operation of the distribution system. Temporary bursts of high-intensity EMI can occur during lightning flashes, faults or switching surges. The degree to which EMI problems will be tolerated is dependent on the automated function being performed. As an example, remote meter reading does not require a communication system that is immune to transient bursts of EMI (lightning, faults, etc.), since this operation can be performed under steady-state system conditions. On the other hand, an automation function such as fault isolation and service restoration requires the operation of the communication system during periods of system disturbances; therefore the communication system should be hardened against transient EMI. The ability to communicate through faults and to areas without power is another factor seriously affecting communication reliability. Fiber optic communications in distribution automation are immune to interference.

A generalised fiber optic communication system for distribution automation can be described, based on a few assumptions. The communication system is assumed to be collocated with the distribution system, both in the substation and along the feeders and laterals.

While most power systems are operated radially they are customarily built as a series of open loops. There is almost always an alternative way of bringing power to any given feeder, if the right switching operations are done in the system. The fiber can be arranged to cross an open power switch. This means that the fiber optic communication network is arranged not as a conventional ring, star or bus system, but as a series of interconnected loops.

The data rate needed for distribution automation is very small compared with the capability of a fiber-based network. Data acquisition and control may be concentrated at the distribution substation. If there is a device, e.g. a switch, a transformer or a voltage control capacitor, there is a monitoring or control point on the fiber. The speed on the optical system is about 1 Mb/s, and the transceivers are intelligent. There is no practical limit to the number of RTUs that could be served on one loop. The two ends of the loop can be at different substations, which means that the fiber could pick up all the RTUs between two substations.

Unlike copper wires, fibers cannot be tapped arbitrarily. When a wire is tapped, more power is drawn from the source. When a fiber is tapped, the available optical energy has to be split. After a few such taps, the optical energy at downstream locations is much less than that at upstream locations. The design of such a network is very difficult, and the approach does not lend itself to modification. Rather than reduce the optical energy by tapping the fiber, it is conventional to use a repeater to produce copies of incoming messages for each outgoing fiber.

Comparison of Fiber with other Media

The capability of the communication system has limited the application of distribution automation. Some media are unsuited to access a large number of points and some are capable of only low-rate signalling.

From the point of view of the utility considering distribution automation, communication systems can be divided into four categories: those that are under utility control and use existing power lines for the signal path, those that are under external control and must be leased, broad coverage systems using radio, and finally, systems requiring installation of a signal path. Table 4.2 summarises the communications options available for distribution automation.

No single communication technology has been demonstrated as being best for all distribution automation needs. Each automation scheme has unique communication requirements, and therefore a communication system must be engineered on an individual basis, using a combination of media [18]. The combination selected will depend on a number of factors, including the functions to be implemented, and the nature of the utility. A hilly environment might rule out the use of radio, for example, and a mixed overhead and underground distribution system might make PLC difficult.

Table 4.2 Categorisation of communication techniques and some limitations

Communication Method	Signal Path	Utility Control	Limitations
Power Line Carrier	Carrier Distribution Line	Yes	Too slow, high error rate
Ripple Control	Distribution Line	Yes	
Zero Crossing Technique	Distribution Line	Yes	
Telephone (all kinds)	Telephone Line	External	Access delays
Cable TV	CATV Network	External	
Radio: Broadcast	Free Space	External	One way
VHF/UHF	Free Space	Yes	Limited coverage

Satellite	Free Space	External
Microwave	Free Space	Yes
Fiber Optics	Optical Fiber	Yes

Data Rate Requirements

The nature of applications will normally establish the requirements for the communication system. Meter reading, for example, requires one-way communication with a large number of locations, but relatively little data from each, and that only rarely. Power system monitoring requires access to fewer points, but data must be sent both ways, and promptly. Reconfiguration may require communication with locations in a part of the power system that has become isolated. In general, these various requirements can be used to derive a data rate and a network topology, and to choose a communication protocol.

Every communication system has a bandwidth limit. The smaller the bandwidth the lower the maximum possible data rate. A communication system for distribution automation must not only meet its present data rate requirements but also have sufficient bandwidth margin to allow for future expansion of the system. Many distribution automation functions can be performed with communications systems having a 300 bps data rate or less. For large systems, higher capacity links may be desired at the top of the communication hierarchy. Functions such as load control can be performed with very low data rates (less than 10 bps). Two main application areas for data transmission on LV electric distribution networks are between MV/LV transformer and electricity meter, and between electricity meter and in-house equipment. For the latter application, products are now commercially available for low data transmission (typically 100kbps).

Prior to selection of a communication system, a data rate audit of a distribution automation scheme must be performed [26,27]. The audit will analyse each automatic function and make a determination of the bit rate required to perform the function. Worst-case scenarios should be considered. The communication system should have at least enough bandwidth along each of its respective signal paths to meet the data rate requirements determined in the audit. A large margin will allow for future growth of the system and increased system flexibility. In cases where the data rate is low, distributed intelligence in conjunction with low data rate can be a viable option. For example, a satisfactory feeder automation system can be made based on the use of UHF radio. Low-power UHF radio, operating at about 950 MHz, can provide two-way communications for line-of-sight locations. The bandwidth available is not as impressive as the frequency might imply, but signalling rates up to about 10 kbps can be achieved. However, there is likely to be a good deal of customising required to make such a system work well in any particular location.

This case-by-case approach would generate a costly proposition for the utility, and there are still likely to be problems. For example, in the case of UHF radio systems, spectrum limitations can lead to interference. Some relief is found by the use of frequency hopping, or spread spectrum, However, even with this technique, the throughput is reduced as the number of users increases. Continued growth of the number of users - and remember that in addition to utilities, there are pagers and other unlicensed low-power transmitters - will increase system noise, and further reduce throughput. Some intelligence will generally be used to reduce the communication traffic.

The scheme uses local intelligence to examine local data to see if there is anything worth reporting to the polling station. These methods usually break down when there is a widespread problem, because each local set of data seems to be worth reporting. Automated functions such as feeder deployment switching and automatic sectionalising require the use of a communication system that will operate in areas with power outages. Communication systems that rely on the power line as a signal path may have difficulty communicating with outage areas. Distribution line carrier, ripple control and zero crossing techniques all use the power line as the signal path. The effects of faults or open circuits on these systems must be considered. Another concern is terminal equipment in areas of a power outage. Remotely located communications equipment in outage areas may require backup power from batteries or some other sources during the outage. Further work is required to improve this approach.

Cost Comparison
Since the cost of the communication system is significant, selection of the best combination of real cost and overall performance could yield substantial savings to the utility. If the proper communication system is not selected, its high cost may offset the benefits of distribution automation. Both first cost and life time operation and maintenance costs must be evaluated.

For any communication network, capital costs consist of two components, the one-off cost of the central unit, and the costs of the remote nodes and the channel to them, which vary as a function of their number and location. It may be estimated that the cost of the optical central unit, a low-power device, is comparable with or less than the central unit of any other medium, e.g. a UHF radio transceiver. It may in fact be much less expensive, since no antenna structure is required. The same is true for the remote nodes: the received signal is of such high quality that, in quantity, the hardware may actually be simpler and cheaper than for nodes using other media.

The difference in overall cost between a fiber system and the 'conventional' media is mainly the channel cost, which is zero for radio systems and PLC, but not for telephone-based systems or fiber optics. Fiber cable can be installed on a typical distribution circuit for less than $3000 per km.

A fiber-based network has the advantage of being all dielectric. In the power distribution system, where tree and ice damage may be common, and a copious supply of energy is very close, there is much to be said for maintaining dielectric isolation. For a similar cost, the fiber has a large advantage.

It is important to note that the fiber optic system becomes more attractive as the number of applications increases. This is because the channel cost can be considered to be zero once the first application is installed. This is not the case with the other media. If an additional function is required, it is usually necessary to add another communication channel, starting from scratch, For example, suppose a feeder monitoring system is installed, with a node (RTU) every 500 m. To estimate the cost of communications, to the cost of each RTU must be added the cost of 500 m of cable. Now, suppose load management is to be added. The central unit and the channel are already in place. All that is needed is the software, and the nodes for load management.

If yet more functions are added, the cost advantage of fiber optics becomes greater still. The conclusion is that a fiber optics communication system can be competitive in cost. One of the goals of using fiber optics to implement a communication system for distribution automation must be to remove any obstacles that might be presented by the conventional media. Fiber allows the communication engineer to design a system that will meet all the worst-case requirements, that can access as many locations as necessary and can handle the highest data rate likely to be required by any application. It is not difficult to expect that some full-capability distribution automation will happen in the near future.

Further innovations in the control of switching are being developed by the local exchange carriers under the heading 'Advanced Intelligent Network' (AIN) [28]. The goal of AIN is to make it easier for carriers to offer advanced call control features on a customised basis. In a period of great technological ferment, competition is seen as more likely to ensure that technological opportunities are exploited. Thus, the way to ensure that the communications infrastructure is based on the most cost-effective technologies is to encourage competition at all levels.

In contrast to the rapid progress of recent years in developing and deploying new transmission technologies or new switching techniques, we are just beginning to learn what a services infrastructure might consist of, and what will be required to put it into place. Given connectivity among millions of students, business persons, professionals and researchers who clearly have a need for information sharing, to make this information sharing simple and available to the average user will still need lots of effort.

4.6 References

[1] Separation of businesses: proposals and consultation. Office of Gas and Electricity Markets (Ofgem) May 1999.

[2] K.K. Kariuki and R.N. Allan, 'Assessment of customer outage costs due to electric serviceinterruptions: residential sector', *IEE Proceedings - Generation, Transmission and Distribution*, Vol.143, 1996, pp.163-170.

[3] K.K. Kariuki and R.N. Allan, 'Evaluation of reliability worth and value of lost load', *IEE Proceedings - Generation, Transmission and Distribution*, Vol.143, 1996, pp.171-180.

[4] 'Information and Incentives Project: Defining output measures and incentive regimes for PES distribution businesses Update', Office of Gas and Electricity Markets (Ofgem) March 2000.

[5] I. Moghram and S. Rahman, 'Analysis and evaluation of five short-term load forecasting techniques', *IEEE Transactions on Power Systems*, Vol.4, No.4, October 1989, pp.1484-1491.

[6] W. Charytoniuk, M.S. Chen and P. Van Olinda, 'Nonparametric regression based short-term load forecasting', *IEEE Transactions on Power Systems,* Vol.13, No.3, August 1998, pp.725-730.

[7] A. Khotanzad, R. Afkhami-Rohani. T.L. Lu, A. Abaye, M. Davis, and D.J. Maratukulam, 'ANNSTLF: A neural-network-based electric load forecasting system', *IEEE Transactions on Neural Networks*, Vol.8, No.4, July 1997, pp.835-846.

[8] R.L. King and R. Luck, 'Very short term load forecasting algorithms', *EPRI Electric Utility Forecasting in an Era of Deregulation Conference*, Dallas, TX, November 1996.

[9] Wiktor Charytoniuk and Mo-Shing Chen, 'Very short-term load forecasting using artificial neural networks', *IEEE Transactions on Power Systems*, Vol.15, No.1, February 2000, pp.263-268.

[10] CIRED Ad-Hoc Working Group 2, 'Distribution Automation: functions and data', CIRED 95, Session 4.

[11] Draft IEC 61968 System Interfaces for Distribution Management – Part 1: Interface Architecture and General Requirements, IEC 1999.
[12] Gerry Cauley, Peter Hirsch, Ali Vojdani, Terry Saxton, and Frances Cleveland, 'Information network supports open access', *IEEE Computer Applications in Power*, 1996, pp.12-19.
[13] Peter Hirsch and Stephen Lee, 'Security applications and architecture for an open market', *IEEE Computer Applications in Power*, July 1999, pp.26-31.
[14] Mark Adamiak and William Premerlani, 'Data communications in a deregulated environment', *IEEE Computer Applications in Power*, July 1999, pp.36-39.
[15] ISO/IEC DISP 11188-3: 1994 Information technology - International Standardized Profile - Common Upper Layer Requirements - Part 3: Minimal OSI upper layer facilities.
[16] M. Blaha and W. Premerlani, *Object Oriented Modeling and Design for Data Base Applications*, Prentice Hall, 1998.
[17] 'Test Methodologies, Setup, and Result Documentation', EPRI Sponsored Benchmark of Ethernet for Protection Control, Version 1.0, May 1997.
[18] IEEE Working Group on Distribution Automation (Edited), Distribution Automation, IEEE Tutorial Course, 88EH0280-8-PWR, 1988.
[19] J. Bunch, Guidelines for Evaluating Distribution Automation (EPRI EL-3728, Research Project 2021-1, Final Report), 1984.
[20] 643-1980 IEEE Guide for Power-Line Carrier Applications.
[21] G. Griepentrog, R. Maier and A. Meusling, 'Sparkling new solutions in industrial communications', Research and Innovation, *Siemens' Science and Technology Magazine '99*.
[22] ISO/IEC 10039: 1991, Information Technology - Telecommunication and Information Exchange Between Systems- Local and Metropolitan Area Networks- Common Specification- Part 1: Medium Access Control (MAC) Service Definition.
[23] D. Hondros and I. Debye, Electromagnetische Wellen an Dielektrischen Drahten. Ann. Physik. Vol.82, 1910, pp.465-470.
[24] K.C. Kao and G.A. Hockam, 'Dielectric fiber surface waveguides for optical frequencies', *IEEE Proceedings*, Vol.113, 1966, pp.1151-1158.
[25] R.J. Landman and B. Louie, 'A multidrop fiberoptic communications system for supervisory control and data acquisition in underground networks', *Proceedings of Power Engineering Society Transmission and Distribution Conference*, Dallas, 1991, pp.407-412.
[26] H. Kirkham, A.R. Johnston and G. Allen, 'Design considerations for a fiberoptic, communications network for power systems', *IEEE Transactions on Power Delivery*, Vol.9, No.1, 1994, pp.510-518.
[27] 1138-1994 IEEE Standard Construction of Composite Fiber Optic Ground Wire (OPGW) for Use on Electric Utility Power Lines
[28] Tutorial Seminar, Intelligent Networks: Advanced services and their Management, IEE, London, May 1994.

5

Transmission Expansion in the New Environment

Mr Yong T. Yoon
Massachusetts Institute of Technology
Cambridge, USA

Prof. Marija D. Ilic
Massachusetts Institute of Technology
Cambridge, USA

5.1　　Introduction

At the initial stage of electricity restructuring in the early 1990s there were various reports estimating the expected improvements in efficiency with the introduction of competition. They ranged from the short-term effects, i.e. savings of $24 billion to $80 billion per year, or 10 % to 40 % off the average electric bill, to the long-term consequences, i.e. technological innovations and increase in reliability.

Indeed the experience from the deregulation of the telecommunications industry gave every indication that similar benefits would be capitalised by simply dividing vertically integrated utilities into generation, transmission and distribution sectors and allowing competition to take place in generation sectors through divesture.

However, the reality is that the electricity restructuring process has achieved only a few successes, far below expectations, as well as with many more difficulties than that of the telecommunications industry. Did people just expect too much? In order to answer this question, we must look at the assumptions that often follow the introduction of competition.

The competition forces market participants to be more aware of their own profits. In simple economic terms, the profit consists of two parts: revenue and cost. From the supplier point of view, an increase in profit can be achieved either by decreasing costs or by increasing revenues. A decrease in costs is possible when the supplier can achieve higher efficiency from its existing plants, thus reducing the associated O&M costs. An increase in revenues is possible when the supplier can expand its customer basis (we make an implied assumption here that no supplier has the market power to raise its price to increase its revenue). From the consumer point of view, an increase in profit is directly related to finding a supplier who can offer the same quality goods at lower prices.

In the electricity industry the suppliers are the generators. Their costs consist of various parts depending on the particular technology used to produce (electric) power; running a nuclear plant, for example, requires the incursion of (plutonium) fuel costs, O&M costs, fixed costs, etc. Their revenues are the product of (electric) energy produced and corresponding electricity prices. The consumers, on the other hand, consist of distribution companies, electric cooperatives, market aggregators and in some instances large industrial users. Their costs are the electricity prices at which their loads are served.

In many parts of the USA the energy market is structured in such a way that there is no direct access between suppliers and consumers. As far as suppliers are concerned, their only customer is the transmission provider (TP), and for consumers, the TP plays the role of the sole supplier. This is due to the peculiar nature of electricity. Because there is no good practical means of storing electricity, the supply and the demand must be balanced continually. Also, unlike the telecommunications industry where a failure to execute a transaction results in a 'busy' signal, a failure to balance the system can result in a system-wide blackout which can amount to astronomical figures in terms of losses. Therefore, the TP who is also the system operator must lead the coordinating effort in meeting the supply and demand with the scarce transmission capacity at times, and the easiest possible way to do this is by being in the middle and acting as the sole purchaser to suppliers and the sole seller to consumers.

Unfortunately in this market setup, the competition is always in a confined situation. In the short run, without the direct access which allows an active interaction between suppliers and consumers, there is a limit to how much suppliers are willing to lower prices in order to expand their customer base. More importantly, however, in the long run no direct access means no customer choices, which is often the key to technological innovations. To make matters worse, the market is structured so that in connecting suppliers and consumers, the TP does not assume any financial involvements due to its monopolistic stance.

In order to overcome this dire situation, the current electricity market must undergo a little evolutionary step so that there is a proliferation of direct access in the form of bilateral contracts.

Bilateral contracts are financial contracts written on the physical basis of energy transfer involving only a subset of suppliers and consumers without the TP. Actually, as a financial contract the bilateral contract need not be limited to physical transfer. However, for simplicity and without loss of generality we consider in this chapter only those contracts associated with physical transfer. As with other financial contracts, there are a number of risks associated with bilateral contracts. The two major ones are the risks associated with future electricity prices and with transmission capacity. Because the participants enter into the contracts in advance, they are exposed to the risk of future energy prices set by the TP on which the strike price is determined. This is, however, well understood in the world of finance, and there are many financial tools to deal with such risk. When the transmission capacity is scarce owing to a high level of demand, energy transfers from certain parts of the transmission system to certain other parts are simply not possible or extremely uneconomical. Owing to the high level of complexity in mapping financial bilateral deals to physical transfers, this risk is extremely hard to measure and there are relatively few financial tools that can be of help.

With the presence of bilateral contracts (and various other financial deals on the transfer of electricity), the TP faces not only an increase in operational difficulties with added complexity, but also a conundrum in planning as the market need changes far more rapidly than the transmission system can evolve. This has serious consequences in reliability as evidenced by recent system-wide blackouts. In the subsequent sections below, we present a particular market structure that equips the TP with market-based solutions to conducting as energy market with a large number of bilateral transactions. This market structure also permits the TP to become actively involved in the market process despite the monopolistic stance. By allowing the TP to pursue profit, it is shown that the transmission expansion problem can also be solved in an efficient way as intended with the introduction of competition.

5.2 Role of the TP

The electric transmission system is one of the most complex constructed systems. Owing to the externality stemming from the operation of the transmission system, implementing the market mechanism to the industry requires a fair level of understanding of not only the economic, financial and regulatory aspects but also the engineering consequences of restructuring.

Figure 5.1 shows the evolution of the role of the TP in the industry (as at the time of writing).

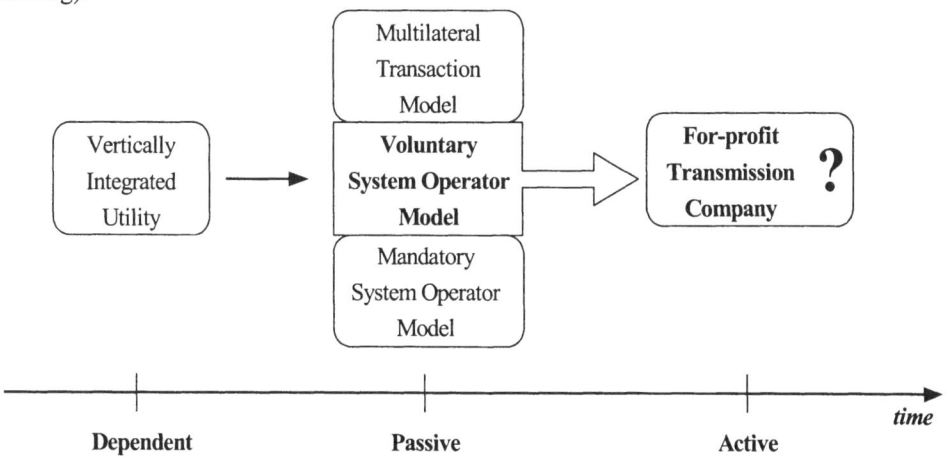

Figure 5.1 Evolution of the role of the TP

In the *dependent* phase the TP functions as a part of the vertically integrated utility. In the *passive* phase the TP stands alone and oversees overall market activities. The market participants are required to submit their intended use of the system to the TP and based on that information the TP allocates transmission capacities following strict rules set by the regulators. The TP assumes no financial responsibilities and has minimal interactions with market participants. As shown in Figure 5.1 there are three different structures of the TP under this phase.

In the *active* phase the TP participates in every phase of market activities. The functions of the TP in this phase can be categorised as that of market maker and that of service provider. Of these two only the function of market maker is under strict regulation. As a service provider the TP assumes full financial liability but is under no regulation.

We will discuss the role of the TP in each phase in detail in the following sections.

5.2.1 Vertically Integrated Utility

In the *dependent* phase the TP exists only as a part of a vertically integrated utility. The vertically integrated utility owns and operates a considerable amount of physical assets, including all the generating plants, transmission system and distribution networks over a sizeable geographical area. The consumers in the area are the captive customers often referred to as 'native load' and the utility is obligated to serve them under a strict regulation. In return, the utility is guaranteed to recover the cost of its prudent investment.

The operation and planning of the system by the TP, therefore, can be viewed as a combined optimisation problem of short-term generation scheduling and investment in new generation and transmission to balance load demand deviations ranging from hourly through seasonal and long term and to do this at the lowest cost. A possible mathematical formulation of this problem is given in [1]:

$$\min_{P_{i,a}^G, I_{i,a}^G, I_l^T} E \left\{ \sum_{i,a} \int_{t_0}^{T} e^{-rt} \left[c_{i,a}\left(P_{i,a}(t), t\right) + C_{i,a}^G\left(K_{i,a}^G(t), I_{i,a}^G(t), t\right) \right] dt \right. \tag{5.1}$$

$$\left. + \sum_{l} \int_{t_0}^{T} e^{-rt} C_l^T\left(K_l^T(t), I_l^T(t), t\right) dt \right\}$$

subject to

$$\frac{dK_{i,a}^G}{dt} = I_{i,a}^G(t) ; \quad K_{i,a}^G(t_0) = K_{i,a;t_0}^G \tag{5.2}$$

$$\frac{dK_l^T}{dt} = I_l^T(t) ; \quad K_l^T(t_0) = K_{l;t_0}^T \tag{5.3}$$

$$I_{i,a}^G(t) \geq 0 , \quad I_l^T(t) \geq 0 \tag{5.4}$$

$$F_l\left(P_G(t), P_L(t)\right) \leq F_l^{\max}\left(K_l^T(t), t\right): \mu_l(t) \tag{5.5}$$

$$P_{i,a}(t) \leq K_{i,a}^G(t): \eta_i(t) \tag{5.6}$$

$$\sum_{i,a} P_{i,a}(t) = \sum_{j} P_{L_j}(t): \lambda(t) \tag{5.7}$$

where

$K_{i,a}^G$: the amount of installed generation capacity at node i and technology a

K_l^T : the amount of installed transmission capacity for line l

$I_{i,a}^G$: the rate of investment in generation capacity using technology a

$C_{i,a}^G(K_{i,a}^G(t), I_{i,a}^G(t), t)$: the cost of investment using technology a at node i

$C_l^T(K_l^T(t), I_l^T(t), t)$: the cost investment in line l

$P_{i,a}(t)$: the production using technology a at node i, at time t;

$$P_G(t) = [P_{1,a_1}(t) \ldots P_{n,a_n}(t)]$$

$c_{i,a}(t)$: the cost of generation using technology a at node i, excluding capacity cost.

$P_{L_j}(t)$: the uncertain (uncontrolled) load at node j, at time t;

$$P_L(t) = [P_1(t) \ldots P_n(t)]$$

$F_l(P_G(t), P_L(t))$: the flow on line l as a function of system generation and demand

$F_l^{max}(K_l)$: the maximum allowable flow on line l as a function of the amount of installed transmission capacity; owing to secure constraints, $F_l^{max} \ll K_l$

r: the discount rate of risk-free investment

$\mu_l(t)$, $\eta_i(t)$, λ: Lagrangian multipliers for corresponding constraints.

The optimisation period, T in problem (5.1), is the longer of two time intervals over which the generation or transmission investments are valued. As the system operator/planner decides the level of production and the rate of investment on generation and transmission, $P_{i,a}(t)$, $I_{i,a}^G$ and I_l^T serve as control variables in this formulation. The state variables of the system are $\mu_l(t)$, $\eta_i(t)$, $\lambda(t)$, $K_{i,a}^G$ and K_l^T for the status of the system operation can be accurately appreciated by examining these variables.

This formulation captures many well-known trade-offs relevant for the efficiency of the power industry: the relationship between the investment timing and the balance of the costs and benefits over time, the value of different technologies at different locations used to produce power, and complementarily of generation capacity and transmission capacity.

There are two noticeable features considering the operation and planning of the system by the TP (as a part of the vertically integrated utility) as the combined optimisation problem: the apparent complexity of the problem (5.1) and the implied assumption of return on investment based on costs $C_{i,a}^G$, C_l^T and $c_{i,a}$. Owing to the complexity, the solution to the problem is not readily available, and thus the actual operation and planning of the system are performed suboptimally in many cases. Further, since the rate of return on investment is determined based on costs, the optimality condition of the formulation is limited to concern $I_{i,a}^G$, I_l^T and $P_{i,a}^G(t)$. Nevertheless, problem (5.1) is a valuable benchmark in studying the efficiency of the industry as the restructuring takes place.

5.2.2 Three Models of the Electricity Market

In the *passive* phase the TP exists as the final authority in administering the market activities separate from the generation and distribution sectors and indifferent from the financial consequences in a market environment. A newly created entity, called the system operator (or the grid operator), manages the system in order to ensure the independence of the TP. The specific functions carried out by the TP are tailored to the market structure of the region the TP serves.

The structure of markets, both existing and developing, is highly non-uniform. Depending on particular regional characteristics some markets admit centralised day-ahead and hour-ahead markets for wholesale trading and a real-time energy market for balancing, while others only offer one or two centralised markets and still others offer only bilateral contracts among market participants with no centralised markets. Most of the markets in various regions within the USA can be represented by one of three simplified market models: the multilateral transaction model, the mandatory system operator model and voluntary system operator model as shown in Figure 5.1.

The multilateral transaction model is based on bilateral transactions among market participants. For example, the proposed structure of the Midwest ISO is closely related to the multilateral transaction model. The model consists of three stages in completing transactions. Firstly, individual buyers and sellers make bilateral trades with one another without disclosing the price and propose the agreed trades to the TP for physical implementation. The TP, upon receiving the proposed transactions, makes decisions whether or not to allow the transactions based on an analysis of transmission network constraints. If the proposed transactions do not violate any constraints, then they are accepted without any modifications. This is the most desired case. If the proposed transactions result in violation of constraints, then the TP accepts none or a part of the proposed transactions and suggests necessary modifications to the transactions in the form of public information called 'loading vector' [2]. Based on this information, the market participants make a new set of trades to satisfy the unmet demand while observing system limits. Figure 5.2 shows the interaction among various market participants for the model. In this model, the function of the TP is limited to verifying whether proposed transactions will result in violation of system limits.

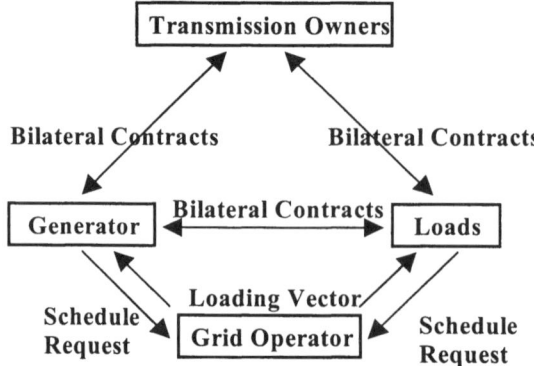

Figure 5.2 Multilateral transaction model

The mandatory system operator model is developed based on the existing practices of tight power pools. The structure of the PJM ISO resembles the mandatory system operator model. In this model the TP becomes the sole centralised market maker for overseeing economically and functionally bundled energy and transmission trades. The spot market refers to a place where this type of centralised market-based trades takes place. Figure 5.3 shows the relation among market participants.

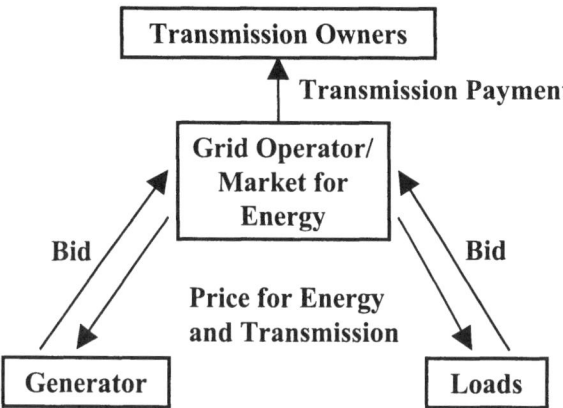

Figure 5.3 Mandatory system operator model

Initially, market participants bid supply curves to the TP, although a generalisation can be made to include elastic demand in the formulation; for the rest of the chapter we assume the consumers' demand is inelastic since not much is lost in terms of the main purpose of the chapter. The TP then simultaneously dispatches generators and allocates transmission capacity using an optimal power flow program, which determines the most economical mix of generations for given load. The voluntary system operator model supports a multi-tiered structure that minimises the TP's influence on profits by market participants while achieving acceptable levels of reliability. Figure 5.4 shows the basic schematic of the model.

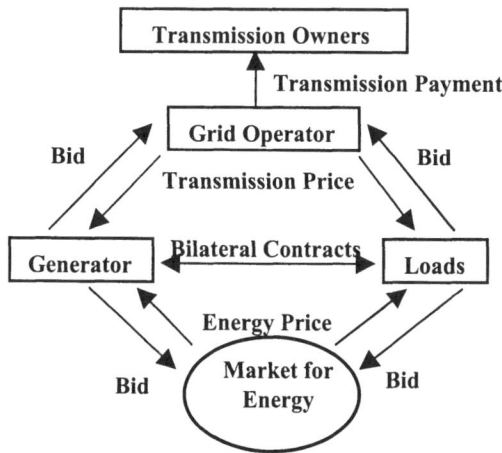

Figure 5.4 Voluntary system operator model

In this model both bilateral and centralised market-based trades are allowed. The presence of spot market transactions is desired because of the requirement for continual balance of instantaneous supply with uncertain demand, particular to the electricity industry, while direct access and customer choice are achieved via bilateral trades.

The level of efficiency can be compared for each of the above three market structures by studying the operation and planning of the system under each structure. Under the perfect market assumption with complete information, both the multilateral transaction model and the mandatory system operator model lead to an equilibrium solution of the following optimisation problem:

$$\min_{P_{i,a}^G} E\left\{ \sum_i \int_{t_0}^T e^{-rt} c_{i,a}\left(P_{i,a}(t),t\right) dt \right\} \quad (5.8)$$

subject to

$$F_l\left(P_G(t), P_L(t)\right) \leq F_l^{\max}\left(K_l^T(t), t\right): \quad \mu_l(t) \quad (5.9)$$

$$P_{i,a}(t) \leq K_{i,a}^G(t): \quad \eta_i(t) \quad (5.10)$$

$$\sum_{i,a} P_{i,a}(t) = \sum_i P_l(t): \quad \lambda(t) \quad (5.11)$$

The optimisation defined in (5.8) is known as the (short-term) optimal power flow (OPF) problem. The result of solving the OPF problem yields

$$p_i^*(t) = \frac{dc_{i,a}}{dP_{i,a}} + \eta_i(t) = \lambda(t) - \sum_l \mu_l \left.\frac{\partial F_l}{\partial P_{i,a}}\right|_{P_G^*} \quad (5.12)$$

We make a couple of observations when comparing the optimisation problems in (5.1) and (5.8). Firstly, the formulation given in (5.8) is much more manageable than that in (5.1) owing to the reduced complexity. This is the strength of the market mechanism; by allowing decentralised decision making, the overall system performance improves, as the seemingly unattainable solution to the optimisation problem becomes reachable.

Secondly, the implicit assumption in the formulation in (5.8) is that $I_{i,a}^G$ and I_l^T are determined by individual suppliers and the TP respectively without considering the interaction between the two. This is a possible weakness of the market mechanism; the expansion of the system could end up at a suboptimal state due to the lack of coordination between generation investment and transmission investment. It is very difficult (if not impossible) to formulate the planning aspect of the problem with (5.12) as the only coupling relation. As it stands, therefore, using either the multilateral transaction model or the mandatory system operator model and solving the optimisation problem in (5.8) leads to a suboptimal solution different from that of (5.1) in most of the cases.

On the other hand, the flexibility inherent in the voluntary system operator model allows the formulation of another optimisation problem needed to create another coupling relation between operation and planning. This problem is posed by defining the interaction of dynamics between bilateral transactions and the spot market. However, the TP must evolve into an active economic entity in the energy market before such a problem can be posed.

5.2.3 For-profit TP

Before discussing the additional restructuring steps needed in order to allow the TP to participate actively in the market process, it is important to understand the underlying reasoning behind many regulatory rules ensuring the independence of the TP from the generation sector and the distribution sector so that any regulatory change complies with the minimum requirement for independence based on the reasons given.

Unlike generation assets, the efficient operation of the transmission system requires a single grid configuration rather than multiple grids serving customers in the same geographical region. This is due to the high degree of the economies of scale and the economies of scope related to transmission.

By having a transmission system covering a wide area of the region, a large number of generators can be connected to a large number of customers. Under this configuration, the TP can serve the suppliers and consumers with a total transmission capacity smaller than the absolute sum of capacity demanded by individual transactions between suppliers and consumers. For example, this point is simple to illustrate through an example involving various transactions which commonly create counterflows on a single grid. Therefore, the TP as well as system users benefit from the economies of scale. In addition, by serving customers whose generation and consumption patterns differ from one another, a significant amount of system-wide savings can be achieved.

For instance, it is easy to show that the total generation capacity required to serve various loads is far smaller than the sum of peak consumption levels of each load. Thus, the TP enjoys the economies of scope.

Given this monopolistic nature, a strict regulatory oversight is necessary when the TP takes on the role of a market maker. As a market maker, the TP allocates the transmission capacity by setting the price of energy and deciding which generating units are to be dispatched. This is an important function of the TP, especially at the time of scarcity in transmission capacity since this allocation process directly affects the profits of each market participant. It is not an overstatement to say that the success of deregulation in the industry depends on how well this process works so that the transmission capacity is distributed without resulting in distorted price signals and the overall electricity market achieves the highest efficiency level possible.

Because the TP as a market maker exerts a considerable influence on each participant's market activities, much time and effort has been spent in establishing the pricing for the allocation of transmission capacity which will promote the efficient use of transmission and subsequently generation and consumption. The three market structures presented in the previous section all attempt to achieve this objective. The function of the TP as a market maker is, therefore, under strict regulation and will remain so regardless of any future regulatory changes.

The voluntary system operator model differs significantly from the other two models because the function of the TP is no longer limited to that of a market maker. In the multilateral transaction model the TP operates as a market maker by supplying the information necessary in accommodating bilateral trades. No explicit price setting by the

TP takes place in this model. Therefore, the transmission revenue is strictly equal to the level designated through the rate of return regulation. In the mandatory system operator model, the TP sets the bundled energy and transmission price that minimises the overall cost of satisfying the system load at each given instant. The transmission revenue in this model has two parts. The first cut is specified by computing the difference between the marginal prices assigned to consumers and to suppliers. The difference in the level of allowed investment recovery and the computed first cut is then compensated through approved access fees and usage charges assigned to consumers.

In the voluntary system operator model the TP is in a unique position to assign an explicit price for using transmission capacity as a result of the dual functions required for furnishing explicit bilateral transactions requested by a subset of market participants and for operating the spot market for the rest. The TP still assumes the role of a market maker in operating the spot market and is subject to strict regulation as in the other two models. However, in allocating the necessary transmission capacity for bilateral trades the TP functions as a *service provider*. As a service provider, the TP designs the appropriate transmission rates to be charged to each bilateral trade implemented *without the regulatory oversight*.

In implementing bilateral trades the TP conducts business just as any other *for-profit* entities; for given variable and fixed costs the TP functions in order to maximise profit. We refer to the for-profit TP as an independent transmission company (ITC) [3].

An increase in profit can be achieved by increasing revenues and/or by decreasing costs. An increase in revenues is possible when the ITC can either raise the price service or expand the customer base. Since the ITC can only set the transmission rates to be charged to bilateral trades while the spot market is operated under strict regulation, there is a clear ceiling to how much the price can be raised despite the monopolistic nature. On the contrary, the ITC has every incentive to lower the rate in order to expand the customer base. Further, related to the expansion of the customer base, the system reliability is expected to improve in order to attract more customers.

A decrease in costs is possible when the ITC can achieve higher efficiency from the existing system. This is directly related to intelligent handling of the limit F_l^{max} and flow $F_l(P_G(t), P_L(t))$ on each transmission line in (5.5). For example, by improving the real-time system coordination through better control design the ITC can effectively manage $F_l^{max}(t)$ as the operating condition changes since the system is no longer required to operate at the conservative static F_l^{max} limits. It is important to notice that $F_l^{max}(t)$ is a function of time as it is no longer static under the presence of the profit-maximising ITC. New technology such as the flexible AC transmission system (FACTS) also becomes attractive for such devices allow direct control of flows on individual transmission lines, $F_l(P_G(t), P_L(t))$.

The investment on transmission system expansion is expected to become more prudent as well. Before committing to any major transmission project the ITC is expected to spend a considerable amount of time and effort in analysing the interaction between the investment and the decentralised decisions by participants on investment and production. Whether done analytically or through trial and error, the resulting effect of this type of pre-project activity is the understanding of the complex relationship among I_l^T, $I_{i,a}^G$ and $P_{i,a}$ in the optimisation problem (5.1).

Therefore, it is essential to transform the TP into an ITC by extending the function to include both a market maker and a service provider for achieving the short-run objectives of efficient system operation and the long-run objectives of prudent investment and technological innovations in introducing competition. An initial analysis shows that not only will the market forces eventually solve the optimisation problem posed in (5.1) despite its apparent complexity, but also that the overall system can possibly outperform the benchmark by incorporating $F_l^{max}(t)$ and $F_l(P_G(t), P_L(t))$ as other control variables. The significance of this is that before restructuring there is no incentive for the TP (or the vertically integrated utility) to include $F_l^{max}(t)$ and $F_l(P_G(t), P_L(t))$ as possible other control variables in the optimization problem (5.1) since the return on investment is determined from only costs $C_{i,a}^G$, C_l^T and $c_{i,a}$ which are direct functions of I_l^T, $I_{i,a}^G$ and $P_{i,a}$ respectively. In the subsequent sections, we discuss the regulatory changes needed for transforming the TP into an ITC.

5.3 New Market Organisation

From the perspective of overall market development the transformation of the TP into an ITC is a consequential effect of an evolutionary process of changing the role of the TP from a passive phase to an active one. As we will see in the subsequent sections the system being operated by the ITC can admit a highly sophisticated market. The ITC is assumed to be derived from the voluntary system operator model of the passive phase as illustrated in Figure 5.1. In the active phase of system operation the competition in the energy market is supported and promoted by the well-functioning transmission ownership and operating structure of the ITC.

A few modifications to the regulatory policies become necessary so that the newly created ITC can serve as both a market maker and a service provider. It is important to recognise that any modifications to the policies must ensure that the primary objective of achieving greater market efficiency remains intact, by meeting the following minimum criteria:

1. Assuring open access: the ITC should continue operating as a common carrier.
2. Maintaining adequate reliability: the ITC should ensure the quality of transmission services to be acceptable by market participants.
3. Planning necessary transmission system expansion: the ITC should engage in transmission projects essential for meeting the changing needs of the market.

In accordance with the spirit of the free market, the market participants in all the generation, transmission and distribution sectors should participate in proposing the regulatory changes detailing the rights/incentives and responsibilities/penalties for the formation of the ITC. The role of the regulator should be limited to verifying whether the proposed changes meet the minimum criteria above while refraining from explicitly defining the functional requirements of the ITC.

The regulatory changes are directly related to the rate design applied to the ITC in allocating transmission capacity. A properly designed rate structure yields appropriate incentives for investment by the ITC. Equivalently, the ITC enters into the transmission project that reduces transmission system congestion only when the associated increase in social welfare equals the marginal cost of the investment. As mentioned earlier, the investment by the ITC is not restricted to the increase in transmission capacity, $K_l^T(t)$, but includes the improvement in the security limit, F_l^{max}, and control of flow, $F_l(P_G(t), P_L(t))$. It is evident that the ITC must be prevented from benefiting from congestion under the rate structure.

Avoiding customer bypass is another direct benefit of a properly designed rate structure. Customer bypass in the energy market is related to the uneconomical usage of the transmission system. If the rate is set too low, a number of users are excluded from utilising the system because of the inefficient allocation of transmission capacity. If the rate is set too high, some users will choose to avoid using the system entirely. This customer bypass is a well-understood phenomenon from the era of the vertically integrated utility. Under top-down pricing the transmission tariff is set such that the avoided cost of generation is less than the total cost of distributed generation.

The rate design problem for the ITC can be approached by studying the development of rate structures in the era of the passive TP. In terms of formulating problem the only difference between the passive TP and the ITC lies in the inclusion of the long-run activities of transmission system investment in the short-run function of efficient operation. The coupling of two separate time-scale functions becomes feasible by considering the service provider aspect of the ITC in implementing bilateral trades. The incentives for the ITC to accommodate bilateral trades are created as part of the rate design problem so that the trades are implemented as long as the system configuration allows for a given operating condition. Prima facie the rate design incorporates the incentives for supporting the operating condition that grants a high degree of well-balanced economic efficiency and technical reliability in the short-run spot market just as required by the passive TP. Thus, the rate design problem is discussed by considering only the operation of the spot market or equivalently the strict regulation required on the ITC in carrying out the role of a market maker.

5.3.1 Incentive Rate Design – Price-cap Regulation

The marginal pricing of transmission has been subject to extensive studies for the rate design assuming the passive TC. The main idea of the marginal pricing is the geographical differentiation in spot prices when a subset of transmission constraints is reached for a desired operating condition [4]. Under marginal pricing all significant network effects in operating the system are internalised in the resulting spot prices. Equivalently the resulting spot prices provide a short-term signal for the economic utilisation of the system. This operating condition is also regarded as the most economical with respect to the system-wide variable cost of generation.

The transmission rate design based on marginal pricing, however, is insufficient to generate enough revenues for the adequate recovery of investment in the transmission system in supporting competitive energy markets. This is due to the inherent assumption in marginal pricing that the revenue is computed excluding the usage 'rent' on uncongested transmission lines; the higher revenue results only with the higher level of congestion.

With the introduction of the ITC, the rate design based on marginal pricing is likely to lead to inefficient utilisation of the transmission system. For example, by restricting transmission flows and thus raising the usage charges the ITC collects a higher net revenue. With this type of perverse incentive the ITC is expected to use all three means of restricting flows: manipulating generation dispatch, reducing transmission capacity through poor maintenance and delaying system expansion.

With the transmission revenue from marginal pricing ranging typically around 25 % of the total cost, the investment recovery problem requires a stipulation of a complementary charge in the rate design. The complementary charge is meant to recover the fixed cost element of transmission investment without distorting short-run prices. This turns out to be a hard problem under general setting since no simple approach can meet the requirements of efficiency, objectivity and implementability. Some simplifying assumptions can be made if consumers are restricted to having only inelastic demand.

Under one approach to the ITC rate design the transmission charges have two components: market-based usage charges with some relation to marginal pricing and access fees with the level agreed upon by the ITC and the regulator. The proper level of access fees should yield the ITC's revenue for recovering both fixed and variable costs with 'appropriate' profits designed to foster economic efficiency. The rate design thus includes a periodic review by the regulator of the ITC's profit and adjustment of access fees. In order to create explicit incentives for promoting short- and long-run economic efficiency, price-cap regulation is suggested for the design of the ITC rate comprising market-based usage charges and access fees. For price-cap regulation the rate that can be charged by the ITC is capped by the price index. Although there is little experience with this type of incentive regulation for electric transmission in the USA, the regulator can draw from the knowledge gained in the gas industry.

In the gas industry it was found that by incorporating either the x-factor or profit sharing the overall system efficiency improves despite the effect of uncertainty and the regulator's imperfect information. Both the x-factor and profit sharing are a mechanism that passes a portion of the benefits gained from cost reduction to consumers, by defining the change in the price index over a given period. In implementing the x-factor the regulator and the ITC agree on the part of profit to be distributed back to consumers on an *ex ante* basis, whereas applying profit sharing determines the portion on an *ex post* basis. When there is a high degree of uncertainty in an expected improvement in productivity, the ITC may prefer profit sharing than the x-factor since the former reduces the responsibility of the ITC. On the other hand, the flexibility bestowed by the x-factor may convince the ITC for a higher increase in efficiency.

The practical application of price cap regulation requires a thoughtful consideration on the following issues:

1. Setting the initial rates: the initial rates are set by comparing the rate level to the total cost of the ITC including transmission losses, O&M and fixed costs. The rate level refers to the expected revenue from the designed rate structure.
2. Revising the ongoing rates: the ongoing rates are periodically revised under regulatory review in order to ensure the proper performance incentives for each review period. It is important to recognise that the operating costs are not relevant during the period of the price cap since pricing is independent of cost once the rate level is determined.
3. Deciding the level of flexibility in rate structure: the rate structure is usually determined by the ITC in order to carry out necessary operational adjustments for improvement. The rate structure here refers to the size of each component in computing the price index. However, this may lead to a significant volatility in pricing. The regulator may require to limit the volatility by placing restrictions on the cross-subsidy under the price cap.

It is worthwhile at this point to investigate the practices used for computing the market-based usage charges and the access fees related to the specific rate design, especially setting the initial rates.

The market-based usage charges are commonly referred to as congestion charges. The nodal pricing and zonal pricing methods are two widely used methods in computing the congestion charges. The nodal pricing method computes the transmission rates by solving the OPF given in (5.8). For a given time instant t, the problem can be solved by constructing a Lagrangian function of the form

$$L = \sum_i c_{i,a}(P_{i,a}) + \lambda \left(\sum_j P_{L_j} - \sum_i P_{i,a} \right) \quad (5.13)$$

$$+ \sum_l \mu_l \left(\sum_i H_{li} P_{i,a} + \sum_j H_{lL_j} P_{L_j} - F_l^{max} \right) + \sum_i \eta_i \left(P_{i,a} - K_i^G \right)$$

where $\mu_l \neq 0$ if and only if $F_l(P_G, P_L) = F_l^{max}$. For simplicity, we use DC power flow in computing the flows on each line H_{li} in the system. The DC power flow equations in matrix notation are written as

$$B\delta = P_G - P_L \quad (5.14)$$

where δ is the voltage angle vector. Taking the first derivative of L with respect to $P_{i,a}$ and setting it equal to zero yields

$$\frac{dc_{i,a}}{dP_{i,a}} + \eta_i(t) = \lambda(t) + \sum_l \mu_l H_{li} \quad (5.15)$$

Suppose the generation cost of supplier i, $c_{i,a}$, is a quadratic function of the output given by

$$c_{i,a}(P_{i,a}) = a_i P_{i,a}^2 \quad (5.16)$$

Then, under the perfectly competitive market condition, the optimal supply bid by supplier i, $b_{i,a}$, is the marginal cost bid given by

$$b_{i,a} = \frac{dc_{i,a}}{dP_{i,a}} \quad (5.17)$$
$$= 2a_i P_{i,a}$$

Matching the solution in (5.15) and the supply bid in (5.17) the system operator can set the price at node i, ρ_i and the dispatch amount, $P_{i,a}$, as

$$\rho_i = \lambda(t) + \sum_l \mu_l H_{li} \quad (5.18)$$

Finally, the transmission rate is set by the difference in the ρ_i, i.e. $\rho_{ij} = \rho_i - \rho_j$.

The zonal pricing method consists of two steps: (1) aggregation of individual nodes into zones and (2) computation of zonal prices. The system is first divided into a number of smaller markets by aggregating individual nodes into zones whenever there is little expectation of congestion within each market. The transmission rate is then computed by solving a similar optimisation problem as given in (5.8); the cost $c_{i,a}(P_{i,a})$ now represents the average cost of generation in zone i. The line flow constraints are now the congestion interface flow limit constraints, i.e. the power flow on any line l along only the congestion *interfaces* is within the maximum rating of the line. The transmission rate is $\rho_{ij} = \rho_i - \rho_j$ where *i* and *j* now represent zones rather than nodes.

Although higher sophistication may be required in order to implement the zonal pricing method from the ITC perspective, a significant reduction in computation complexity can be achieved in the rate design under price-cap regulation since only a small number of zonal prices are needed to be considered rather than many nodal prices as is the case in nodal pricing. Further, there is a greater advantage to be gained in implementing zonal pricing in accommodating bilateral trades as is illustrated in the subsequent section.

The access fees are intended to recover the fixed part of the ITC's costs and are thus independent of actual usage. However, usage-independent charging for the access fees is impractical and may result in improper incentives for the ITC. In order to stipulate a meaningful charging mechanism, some measure of base-load capacity needs to be given. A practical approach is to compute the access charges based on a coincidental peak consumption of loads. The '12-CP' method [3] is one such approach. The portion of individual access fees is computed as

$$S_i(t) = \frac{L_i(t)}{\sum_i L_i(t)} \qquad (5.19)$$

where $S_i(t)$ is the load *i*'s share of system coincident peak, and $L_i(t)$ is the average of the load in month *t* at peak loading condition of each day. As the total revenue from this charge is equal to the product of access charge and the coincidental peak of each load, the approach provides the ITC with incentives to increase individual base-load capacity.

Therefore, price-cap regulation and the rate design consisting of the market-based transmission usage charges and regulator-approved access fees offer the ITC an opportunity to recover the investment with some incentives for improvement in efficiency. However, the resulting rate structure does not immediately yield proper incentives for transmission expanding. In the subsequent section, a market mechanism called the priority insurance service is discussed in terms of complementing price-cap regulation in order to provide the right set of incentives to enhance the transmission system.

5.3.2 Priority Insurance Scheme

The driving forces of deregulation aim to establish a more competitive market in order to achieve lower rates for consumers and higher efficiency for suppliers. Through bilateral trades, consumers can establish various service contracts with any supplier in order to obtain the lowest rate and most desirable service. Bilateral contracts specifying the amount of power, the time and duration of the service and the associated rate and possible compensation are negotiated and agreed upon between the suppliers and consumers. Since the proliferation of competition is directly related to the bilateral trades which allow direct access and customer choice, the success of the market is dependent on the ITC's ability to administer the bilateral trades.

Since the transmission grid is a physical system, the ITC is able to honour and execute these bilateral contracts as far as the system design and operating conditions permit. Unlike in the spot market, the ITC is not allowed to participate directly in re-dispatching resources when executing these bilateral trades. Thus, the ITC relieves transmission system congestion by curtailing balanced bilateral trades or by creating counter flows in the spot market by systematically adjusting the rate structure. All bilateral contracts are required to specify the replacement resources in case of interruptions die to either the congestion curtailment or generator-related contingencies.

Without loss of generality all bilateral contracts consist of the following specification:

Q_{ij} : the quantity of energy transfer

i, z_i : the injection point and the corresponding zone

j, z_j : the withdrawal point and the corresponding zone

c_i^G : penalty payment by generator i for generator-related contingencies determined *ex ante*.

In the case of curtailment it is assumed that the load is satisfied through the replacement generation purchased in the spot market of zone z_j at zonal price $\rho_{zj}(t)$ paid by the load j. Otherwise, generator i pays the transmission charge $\rho_{zi\,zj}(t)$ to the ITC.

By entering into a bilateral contract the supplier and the consumer face a few uncertainties related to transmission provision. Suppose the bilateral contract is signed over the period $t \in T = [t_1, t_N]$. Over the period T, any time the transaction is curtailed owing to system congestion, the load is responsible for paying $\rho_{zj}(t)$ which is determined *ex post*. Otherwise, the generator is obligated to pay $\rho_{zi\,zj}(t)$ which is also determined *ex post*. Thus, it is implicitly required that each time participants enter into bilateral contracts, they must estimate possibly highly volatile prices, $\rho_{zj}(t)$ and $\rho_{zi\,zj}(t)$, and the probability of being curtailed, R_{ij}, over the period T. This puts an extra burden on participants.

The priority insurance service offered by the ITC is designed to take away this extra burden. When the interested parties in the bilateral contract request an allocation of transmission capacity for physical implementation of the trade, the ITC offers an alternative transmission pricing where they can agree on the transmission charge $\rho_{zi\,zj}(t)$ determined *ex ante* for each time the trade is executed and the insurance payment paid to load j by the ITC, $I_{zi\,zj}(t) = \rho_{zj}(t) - c_i^T$ where c_i^T is equivalent to a deductible payment. In addition, the system operator insures that the bilateral trade is interrupted no more than x_{ij} times over the period T. By purchasing this service, the interested parties are completely protected from the volatility of $\rho_{ij}(t)$ and $\rho_{zj}(t)$.

In deciding to purchase the priority insurance, the interested parties must consider several factors, including:

- the opportunity cost of a curtailed transaction, and
- the probability of being curtailed with the upper bound given as x_{ij}/N.

When the ITC provides priority insurance services to some interested parties, the transmission charges are determined through the bottom-up approach since in essence the marginal valuation of the transaction is reflected in $\rho_{zi\,zj}(t)$.

In real-time operation, the ITC determines the curtailment of bilateral trades needed to relieve transmission congestion along with computing spot market prices for each zone by optimising the profit from both accommodating bilateral trades and conducting spot markets. Equivalently the priority insurance is implemented by solving an instantaneous minimum compensation problem over the entire network. Since there is a strict regulation on the ITC's role as a market maker, the profit from conducting spot markets is restricted. However, by imposing no regulation on priority insurance, the ITC may increase the profit as long as the market can take it. This creates an attractive incentive for expanding the transmission system as a substantial effort by the ITC is expected in order to increase the customer base for priority insurance service. The advantage of this method is that because of the presence of the spot market, which is under strict regulation, the market's willingness to take the ITC's profit is well capped. Over time, the shortage (or excess) of customers in either the spot market or bilateral trade is expected to level out depending on the ITC's ability to meet the changing needs of the market. The ITC with the system having better ability to adapt to market evolution is likely to have a relatively higher level of customers subscribing to priority insurance service and to enjoy profits from business with no regulation.

In the subsequent section we discuss the effect of reduced regulatory uncertainty in revenue by providing priority insurance services on transmission expansion.

5.3.3 Transmission Expansion

The new market organisation described at the beginning of this section provides a fundamental setting for systematic transmission expansion. This task becomes an inherent part of the proposed ITC structure. A forward-looking transmission provider actively learns the needs of its customers based on frequency and magnitude of congestion within its transmission system. Careful analysis of this data, together with the development of tools necessary for enhancing the system, provides a basis for successful transmission enhancement and expansion. The timing of these enhancements and expansions in response to energy market needs becomes a very important aspect of the overall successful ITC business.

It is worthwhile observing that most of the immediate opportunities are likely to be for transmission enhancements by means of new control and information technologies. The prospects for new rights-of-way for new transmission lines are not promising.

Another important aspect of dynamic transmission investments is the interaction with the regulators. The proposed ITC structure is dependent on performance base rate (PBR) regulation. It is not easy to design such regulation for networks, and it becomes necessary to develop the right dialogue between the ITC, considering the details of the technical challenges, on one side, and the regulators, concerning the details of meaningful price-cap design, on the other side, which is sufficiently flexible to take into consideration locational and temporal aspects of electric power delivery. Only by having a sufficient understanding of these details could one begin to harvest the potential benefits which come from using control, communications and information at the right location and at the right time.

Possibly the most challenging part of transmission pricing for robust design is the need for price design and rates which are sufficiently simple and transparent to users. The proposed market structure provides simplicity through effective zonal aggregation. This aggregation helps the users appreciate the need for new transmission investments, internalise their value and, at the same time, actively develop a secondary transmission market by exchanging transmission rights in near real-time. The liquidity of the transmission market is an essential part of successful implementation of ITCs. Therefore, the transparency provided at the zonal level could serve as a fundamental catalyst for restructuring the industry.

5.4 Conclusions

The development of new market tools for operating the transmission system becomes essential as the ITC moves into the active phase of management. In this phase the ITC is required to make complex business decisions over a wide range of time scales: long-term, short-term and near real-time.

The long-term decisions deal with expansion of the transmission system. A fundamental question is related to computing the impact of future demand on the system constraints and making system reinforcements in order to meet this demand. It is shown in the proposed transmission rate design that the investment cost is not directly affected by the congestion rent on the spot market owing to the high fixed cost element. Thus, no market tool for investment decision is required for the ITC based on spot market activities. Even when there is significant congestion sustained over a long period of the time, the investment needed for relieving this congestion is a decision to be made by the regulator upon reviewing the performance of the ITC, since the authority to modify the transmission rate lies with the regulator and not the ITC. The activities in implementing bilateral trades and the implicated priority insurance services, on the other hand, are an immediate concern of the ITC in making investment decisions. Typically the bilateral trades take place over an extended period and thus provide adequate revenue sources for recovering the investment. The new market tools in the long-term project the demand in bilateral trades and in priority insurance services. The new market tools should make this projection based on the historical patterns of users subscribing to bilateral transactions and sometimes supported by priority insurance services, as well as the expected changes in the customer base. An investment in a new efficient generator by a participant is likely to be followed by a request for implementing bilateral trades since such investment requires a steady flow of revenue. The better the projection that the new market tools can produce, the more prudent investment the ITC makes and subsequently the higher the earnings.

The short-term decisions deal with pricing priority insurance services. This is perhaps the most difficult task for the ITC since the success of the ITC as an independent market entity depends on the ITC's ability to function as an insurance service provider.

There are three aspects to consider in the pricing. The first is refining the projection of bilateral trades from the long-term market decision tools. Although only the aggregate volume is important in making investment decisions, the short-term decision requires an accurate projection of the locational and temporal patterns and the opportunity costs for each bilateral transaction. Over time, the market tools in this time scale can discover the patterns and the costs by extrapolating from the previous seasonal behaviour of the participants.

The second is the valuation of insurance services given the specifications of a bilateral contract as described in Section 5.3. The ITC has a menu of prices defined for different levels of reliability. In this formulation, the reliability is explicitly given in terms of the maximum number of interruptions x_{ij} by the ITC over the contract period, T. This problem is similar to the option valuation and may be solved using similar market tools.

Finally, the third aspect of pricing is relating the decisions in providing the insurance services to spot market activities. Because the amount of compensation depends on the deductible c_i^T as well as the prices at the spot market, there is a high correlation between accepting bilateral trades and operating the spot market. The ITC is required, therefore, to solve for the optimal balance between bilateral trades and spot market transactions in terms of the profit. For instance, if the ITC deviates from this optimal path and leans too much on the bilateral trades, there is an expected deterioration in the short-term efficiency for which the ITC is responsible through the strict rate design. If the ITC, on the other hand, relies heavily on the spot market while neglecting the bilateral trades, the ITC may not be able to function as an active market entity. There are very few market tools available for solving this type of problem in other financial markets, but some active studies are under way.

The near-real-time decisions involve computing a combined optimisation problem for minimising insurance compensation to bilateral trades while maximising the spot market throughput. These two are conflicting objectives and thus require of the definition of some offsetting weights when solving the combined optimisation problem. The ITC can expand the conventional OPF tools as the new market tools needed to tackle the problem.

As the industry moves into the more mature stage of deregulation, the role of the TP becomes more important. The new market tools described above are only the minimal changes required in the way the TP conducts business as an active market participant, i.e. the ITC. It is, therefore, critical to build the tools that are consistent with the way they function over different time scales as well as with the other new business-oriented tools that are used by the participants.

5.5 References

[1] C-N Yu, J-P Leotard and M.D. Ilic, 'Dynamics of Transmission Provision in a Competitive Power Industry', *Journal Discrete Even Dynamic Systems: Theory and Applications*, Kluwer Publishing Company, 1999, pp.351-388.

[2] F.F. Wu and P. Varaiya, 'Coordinated Multilateral Trades for Operation in a Competitive Open Access Environment', *Electricity Journal*, 1996.

[3] S. Awerbuch, L. Hyman and A. Vesey, *Independent Transmission Companies: Unlocking the Benefits of Electricity Restructuring*, (Edited), Public Utilities Reports Inc., 1999.

[4] F.C. Schweppe, M.C. Caramanis, R.D. Tabors and R.E. Bohn, *Spot Pricing of Electricity*, Kluwer Academic Publishers, 1988.

6

Transmission Open Access

Prof. A.K. David
Hong Kong Polytechnic University
Hong Kong, China

Dr Fushuan Wen
Hong Kong Polytechnic University
Hong Kong, China

6.1 Introduction

6.1.1 The Traditional Power Industry

The electricity supply industry in nearly every country for about the last hundred years has been a natural monopoly and as a monopoly attracted regulation by government. Without exception, the industry has been operated as a vertically integrated regulated monopoly that owned the generation, transmission and distribution facilities. It was also a local monopoly, in the sense that in any area one company or government agency sold electric power and services to all customers. In many countries, especially developing countries, the electric utility was owned by the state or local government, and in other countries, such as the USA, by an investor-owned private entity. The traditional power industry had several characteristics [1]:

(1) Monopoly franchise:
Only the national or local electric utility was permitted to produce, transmit, distribute and sell commercial electric power within its service territory.

(2) Obligation to serve:
The utility had to provide electricity for the needs of all consumers in its service area, not just those that were profitable.

(3) Regulatory oversight:
The utility's business and operating practices had to conform to guidelines and rules set down by government regulators.

(4) Regulated rates:
The electric utility's rates were either set or regulated in accordance with government regulatory rules and guidelines.

(5) Guaranteed rate of return:
The government guaranteed that regulated rates would provide the electric utility with a 'reasonable' or 'fair' profit margin above its cost.

(6) Least cost operation:
The electric utility was required to operate in a manner that minimised overall revenue requirements.

In less developed countries the obligation to serve was often not enforced while the guaranteed rate of return concept was usually replaced by government ownership, subsidies or other economic arrangements.

6.1.2   Motivations for Restructuring the Power Industry

Since the 1980s, the electricity supply industry has been undergoing rapid and irreversible change reshaping an industry that for a long time has been remarkably stable and had served the public well. A significant feature of these changes is to allow for competition among generators and to create market conditions in the industry, which are seen as necessary to reduce the costs of energy production and distribution, eliminate certain inefficiencies, shed labour and increase customer choice. This transition towards a competitive power market is commonly referred to as *electricity supply industry restructuring* or *deregulation*. A modified form of restructuring is occurring in the countries of Asia [2], driven by a need for rapid expansion of capacity in all three branches of generation, transmission and distribution. Hence a great variety of organisational forms is emerging.

Restructuring started in the 1980s in Chile and the UK and spread to the Latin American countries, such as Argentina, and accelerated in the 1990s in diverse forms in the USA, Australia, the Nordic and a number of Asian countries. Many factors such as technology advances, changes in political and ideological attitudes, regulatory failures, high tariffs, managerial inadequacy, global financial drives, the rise of environmentalism, and the shortage of public resources for investment in developing countries, contributed to the world wide trend towards restructuring. Some specific issues are as follows [1].

The main drive for electricity industry restructuring in the classic case of the UK came from the government's belief that the advantages of competition among energy suppliers, and wide choice for electricity consumers, outweighed the benefits of the long-established arrangement. Although restructuring of the power industry inevitably results in some new problems, governments and consumers in many countries believed that the benefits of the restructuring would outweigh potential problems.

The change in generation economies of scale that occurred throughout the 1980s was an important stimulus to industry restructuring. Advances in gas turbine technology led to more efficient small turbines and generators. As a result, smaller generators could nearly match the efficiency of very large units, particularly if run on natural gas rather than coal. The price of natural gas declined and the prohibition on gas burning for electricity generation was removed in this period.

Restructuring of the government-owned electricity industry encouraged privatisation, although privatisation does not have to be part of a restructuring effort. In the 1980s and early 1990s several Western governments were of the view that private organisations could do a better job of running the power industry, and that higher operating efficiencies and reduction in labour could be achieved by privatisation. Private utilities also refuse to subsidise rates and have a greater interest in eliminating power thefts and managerial or workplace inefficiencies. In other countries either ownership or responsibility for various functions was transferred to cooperative or to private organisations, or to new types of public corporations or quasi-governmental entities that could act like market-sensitive economic agents. Ownership and functional restructuring have therefore taken many different forms. Incremental involvement of private capital, which has played an important role in Asia, can be considered as private sector participation rather than privatisation.

Price is expected to drop but become more volatile. Service may improve as a result of the restructuring, but there is also a serious concern in many countries about falling maintenance standards. Competition breeds innovation, efficiency, and lowers costs but also leads to short-termism. A competitive power industry will provide rewards to risk takers and encourage the use of new technologies and business approaches. The regulated monopoly scheme was unable to provide incentives for innovation since the utility had little motivation to use new ideas and technologies to lower costs under a regulated rate of return framework. Lack of competition also gave electric utilities little incentive to improve service, and in countries such as India and China, the standards of service were generally very low. A more commercial ethos could be helpful in improving standards of service to customers.

6.1.3 *Unbundling Generation, Transmission and Distribution*

Notwithstanding the variations discussed above the experiences of restructuring have reached similar patterns in certain respects: (1) The generation subsystem has a high degree of freedom in the selection of energy sources, ranging from capital-intensive low-operation cost resources to others with low capital and high operation costs. Hence, significant economies of scale or natural monopoly features, except in the case of large-scale hydro-electric potential, are not an impediment to competition in generation. (2) There are no clear economies of scale, but there is a geographic monopoly, in the distribution subsystem, and therefore some form of regulation is needed. The distribution business has, however, been further unbundled into (a) a wires business, which maintains the distribution network and provides facilities for electricity delivery, and (b) a retail or supply business, which provides electric energy to end consumers. (3). The transmission subsystem is a natural monopoly in the economic, the geographic and the technical (control) sense, and therefore must continue to function as an integrated and regulated entity. However, to implement competition in the generation and retail sides, it is necessary to unbundle these two from the transmission system and ensure that the latter offers open access on an equitable basis to all power suppliers and consumers. The transmission system thus becomes the focus of attention in organising competition and must act as a 'level playing field', and the rules for managing access by all participants must be transparent and non-discriminatory. This chapter focuses on this critical subject.

6.2 Components of Restructured Systems

The structural components representing various segments of the electricity market are generation companies (Gencos), distribution companies (Discos), scheduling coordinators (SCs), transmission owners (TOs), an independent system operator (ISO), and a power exchange (PX). Depending on the structure and the regulatory framework, some of these components may be consolidated together, or may be further unbundled. In Asian countries various regional, state, provincial or independent generators coexist. In these cases the financial and technical interrelationships are murky and are in a process of rapid evolution.

6.2.1 Gencos

Gencos are responsible for operating and maintaining generating plant in the generation sector and in most of cases are the owners of the plant. Open transmission access allows Gencos to access the transmission network without distinction and to compete.

6.2.2 BOT Plant Operators and Contracted IPPs

Build, operate and transfer BOT; (or build, operate and own) plant or IPPs who have long-term contracts with surrounding, usually national, utilities play an important role in providing additional generation in many fast-growing systems. Take-or-pay power purchase agreements are often in force as an economic incentive to investors.

6.2.3 Discos and Retailers

Discos assume the same responsibility on the distribution side as in a traditional supply utility. However, a trend in deregulation is that Discos may now be restricted to maintaining the distribution network and providing facilities for electricity delivery while retailers are separated from Discos and provide electric energy sales to end consumers. Another trend in developing countries is to sell to an investor, or to corporatise, portions of the distribution system so that investment for reinforcement can be raised and better operating practices implemented.

6.2.4 Transmission Owners (TOs)

Where the transmission network was state owned before restructuring, obviously this integrity will be retained and a distinction between owner and operator is redundant. In cases like the USA, former electric utilities may sell off their other assets and become regulated T&D companies. A basic premise of open transmission access is that transmission operators treat all users on a non-discriminatory basis in respect of access and use of services. This requirement cannot be ensured if transmission owners have financial interests in energy generation or supply. A requirement, therefore, is to designate an independent system operator to operate the transmission system.

6.2.5 Independent System Operator (ISO)

The ISO is the supreme entity in the control of the transmission system. The basic requirement of an ISO is disassociation from all market participants and absence from any financial interest in the generation and distribution business. However, there is no requirement, in the context of open access, to separate transmission ownership and operation. For example, the National Grid Company (NGC) in England and Wales (E&W) is both the transmission owner and the operator. The roles and responsibilities of ISOs vary widely and this issue will be discussed later in the next section. In countries such as India and China where regional grids are owned by regional or state governments, and system interconnection is only now growing and growing rapidly towards national grid status, the protocols of future ownership and operation are still being evolved.

6.2.6 Power Exchange (PX)

The PX handles the electric power pool, which provides a forum to match electric energy supply and demand based on bid prices. The time horizon of the pool market may range from half an hour to a week or longer. The most usual is the day-ahead market to facilitate energy trading one day before each operating day. An hour-ahead market is also useful since it provides additional opportunities for energy trading to redress short-term imbalance. In the E&W system the ISO and PX functions both exist within the NGC. In the evolving Asian systems the arrangements for the future are still under intense discussion, e.g. in China, India, Thailand, Indonesia and also in most of Africa.

6.2.7 Scheduling Coordinators (SCs)

SCs aggregate participants in the energy trade and are free to use protocols that may differ from pool rules. In other words, market participants may enter an SC's market under the SC's rules and this could give rise to different market strategies. In some markets such as E&W, SCs are not allowed to operate. In many new situations such as California, SCs are an integral component of the market.

6.3 PX and ISO: Functions and Responsibilities

6.3.1 PX Functions and Responsibilities

A PX of some form is essential for efficient trading in electricity. The PX establishes an environment in which generators and consumers bid to sell and buy energy. Parties to bilateral contracts can operate their own separate energy trades and schedule their transactions outside the PX's market. The primary function of the PX is to provide a forum to match electric energy supply and demand in the current and forward energy markets. As mentioned before, the market horizon may range from a half-hour to a few months but the most usual situation is a day-ahead market. Depending on the market design, the day-ahead market may be preceded by a longer term market and supplemented by an hour-ahead market. The so-called 'hour-ahead market' provides energy trading opportunities up to 1 or 2 hours before the operating hour.

In its simplest form a PX provides a bulletin board type of environment for energy suppliers and energy customers to engage in bilateral forward contracts. However, the more usual function is to act as a pool for energy supply and demand bids, and to establish a market-clearing price (MCP). The MCP is the basis for the settlement of forward market commitments. Regardless of asking prices all selected bidders are paid the MCP. This approach encourages bidders in a competitive market to price energy close to their marginal production costs.

Depending on the market design and activity rules the energy bids may include several price components (multi-part) or a single price component (single-part) [3,4]. A multi-part bid may include separate prices for unit start up, no-load operation and energy. A single-part bid is an energy price inclusive of fixed and variable costs. In either case the energy bid may include several energy price segments depending on the amount of energy; that is, the bid may take the form of a separate $/MWh quotation for each block of energy from the same unit or a portfolio of units.

The market design, bidding protocols and bid selection (scheduling) process, impacts on the computer applications needed to support the PX. In the case where the market design is based on single-part bids a simple market-clearing process based on the intersection of supply and demand bid curves is sufficient to determine the winning bids, MCP and schedules for each hour. However, if the market design is based on multi-part bids, unit commitment software, possibly with enhancements to take into account security constraints, may be needed.

The complexity of the bidding infrastructure and system is also dictated by the market design and the protocols used. If iterative bidding is allowed the bidding infrastructure and supporting software may have stringent performance requirements. Since multi-part bid systems do not require bidding iterations performance requirements may be less stringent than for single-part bids.

Basically, the working process of the PX is: (1) receive bids from power producers and customers; (2) match the bids, decide the MCP prepare scheduling plan; (3) provide schedules to the ISO or transmission system operators; (4) adjust the scheduling plan when the transmission system is congested.

6.3.2 *California Power Exchange*

The California power exchange (CalPX) is a widely cited example of PX structuring, and is briefly described here. CalPX, which was put into operation on 31 March 1998, is a non-profit independent organisation with a mandate to establish an efficient market for forward electricity transactions in the deregulated California system. Its initial design was based on the E&W power pool concept; however, CalPX is a specific SC with the following functional features:

1. Run a forward market in which parties can bid to buy and sell energy.
2. Develop a preferred schedule for the forward market.
3. Develop the MCP energy transactions.
4. Submit the preferred schedule to the ISO and work with the latter to adjust schedules when necessary.

CalPX markets are essentially short-term forward markets in which generators bid for the right to serve load and loads bid for the opportunity to have their demands satisfied. The CalPX conducts two distinct markets: day ahead and hour ahead. The CalPX day-ahead market consists of 24 separate double auctions, one for each hour, without considering transmission costs, constraint management or congestion management issues. Supply-side bidders may bid as a single generator or as a portfolio of generators; however, a unit may bid in only one portfolio though a bidder may submit multiple portfolio bids.

When there is congestion the CalPX sets a maximum price and buyers may submit price-insensitive demand bids for non-curtailable loads; this actually represents a willingness-to-pay up to this price.

The CalPX hour-ahead market auction conducts a single iteration auction similar to the day-ahead market. This occasion provides a competitive mechanism to allow loads and generators to adjust their day-ahead schedules in the light of new information such as short-term load forecasts and unit status.

6.3.3 ISO Functions and Responsibilities

The system operator plays a critical role in both the traditional utility environment and the emerging unbundled systems, although some activities and responsibilities have changed considerably. In the traditional utility environment the system is vertically integrated and the operator, as the top manager, takes over the entire business so far as operating the physical system is concerned. In vertically integrated traditional utilities the range of operator responsibilities (which encompasses operational aspects of corporate economics) as well as the ownership of the system are maximised in one corporate entity.

In the new market structures there are a variety of arrangements for the system operator and since the operator must be disassociated from all participants the name independent system operator (ISO) is a natural choice. The ISO has three objectives: security maintenance, service quality assurance and promotion of economic efficiency and equity [5].

To achieve these objectives the ISO performs one or more of the following functions:

(1) Power system operations function:

This fundamental function includes the operation-planning function and real-time control. The operation-planning function includes:

- Perform power system scheduling.
- Co-ordinate with energy markets.
- Perform power system dispatch.
- Determine available transfer capabilities (ATCs).
- Determine real-time ATCs.
- Pre-calculate short-run costs and prices for transmission-related services.
- Calculate hourly prices for transmission-related services.

Real-time control includes:

- Monitor power system operation status.
- Monitor system security.
- Conduct physical network operations and network switching.
- Deal with outages and emergencies.
- Coordinate real-time system operation.

(2) Power market administration function:

There are two types of energy markets: the pool market and the contract (bilateral and multilateral transactions) market. The former could be run by the PX or an ISO-PX combine while the latter may be coordinated by one or more SCs. The pool market includes:

- Run a power pool where parties can bid to buy and sell energy.
- Develop a preferred schedule for the pool.
- Submit the supply and load schedule to the ISO according to pre-specified protocols (this is for the case when the PX is separated from the ISO).

The contract market includes:

- Manage bilateral and multilateral transactions.
- Manage and coordinate submissions from SCs.
- Submit preferred schedules to the ISO according to pre-specified protocols.

(3) Ancillary services provision function:

- Own certain ancillary services for satisfactory grid operation.
- Purchase ancillary services transactions from market participants according to pre-specified protocols.
- Provide ancillary services to transmission users.
- Allocate costs of ancillary services among all users.

(4) Transmission facilities provision function:

- Maintain the transmission network.
- Provide transmission facilities for all supplies and loads.
- Plan transmission, reactive power and FACTs expansion and ensure that resources for future investment are generated.
- Plan and commission own ancillary services.

6.3.4 Classification of ISO types

There can and will be many forms in which an ISO is set up depending on the ownership structure of the reformed and unbundled system. The debates on the ISO issue have given rise to two ISO models called the MinISO in the USA (hereafter called MicroISO in this chapter) and the MaxISO, which refer to the two ends of the scale [5,6].

At a minimum the responsibility of the MicroISO must include ensuring security in system operation. Such a MicroISO is not involved in the energy markets and its role in generation (or transaction) scheduling will be limited to ensuring that the submitted schedules are feasible. This MicroISO does not perform real-time control of power system facilities, which is done by separate system control centres that are hierarchically dependent on it as shown in Figure 6.1. It does, however, monitor system operation to ensure adequacy of available reserves, and other pertinent ancillary services. It will co-ordinate measures to alleviate transmission congestion and will perform contingency analysis to ensure system security against credible contingencies. The term MinISO may be used for the case where the ISO is distinct from the PX but is fully responsible for real-time system control and network switching; this case is shown in Figure 6.2 (the figures in this section are developed using the approach in reference [7]).

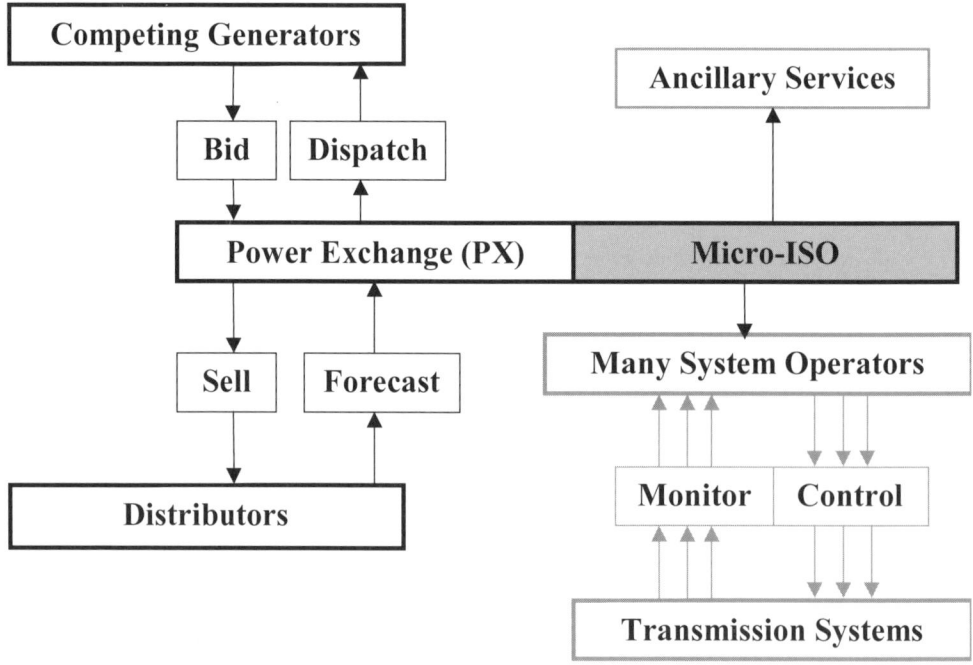

Figure 6.1 The MicroISO

At the other end of the scale, the MaxISO has a wide range of responsibilities and authority, and combines the MinISO and PX functions – Figure 6.3. The MaxISO performs generation scheduling (possibly including unit commitment), and scheduling of ancillary services (possibly simultaneously with energy/power scheduling). It would perform scheduling and pricing of transmission facilities. It would dispatch generation for imbalance, ancillary services and congestion management and would perform real-time control of generation and transmission. Standard features would include state estimation and security analysis, possibly on a real-time basis.

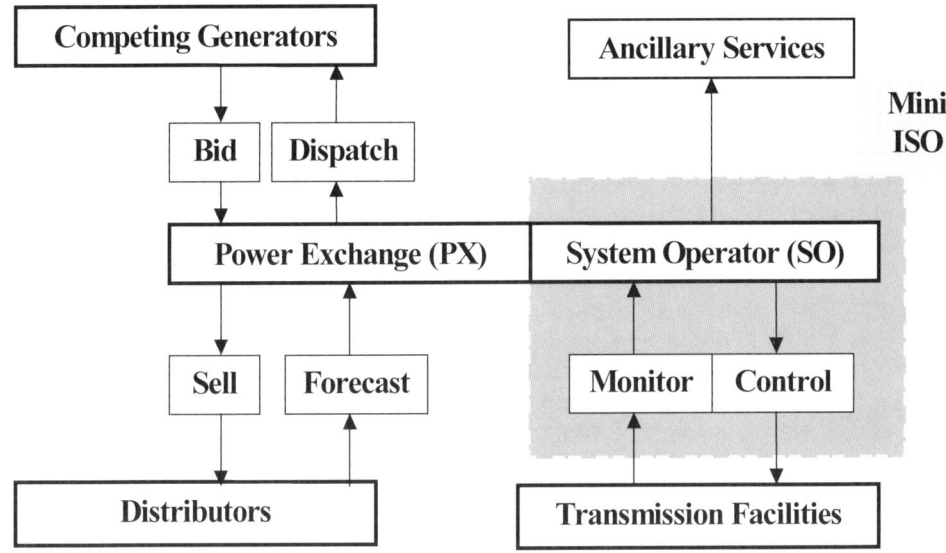

Figure 6.2 The MiniISO

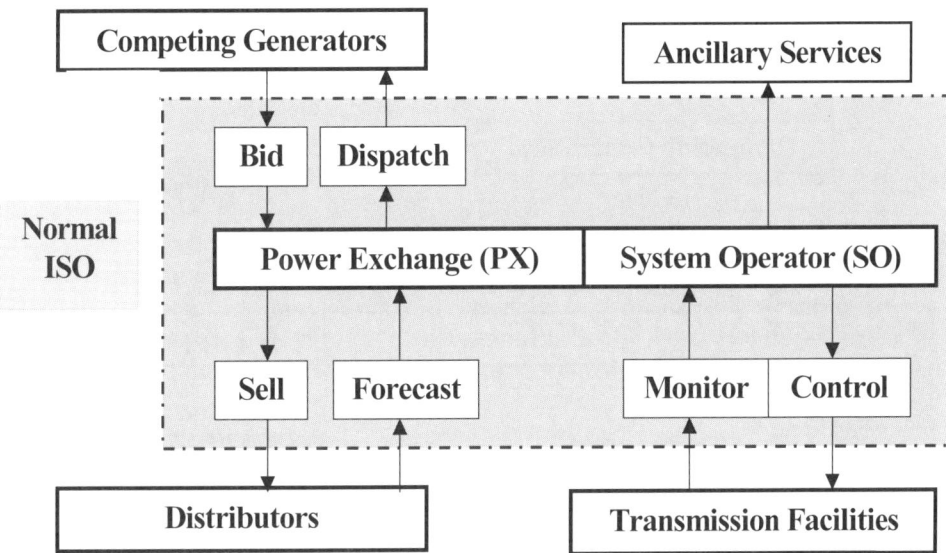

Figure 6.3 Typical MaxISO incorporating the PX

The NGC in the E&W system is an example of a MaxISO, which is also the owner of transmission assets as shown in Figure 6.4. The ISOs in Chile, Argentina and East Australia fall close to the MaxISO category with individual modifications. The California ISO [8] shown in Figure 6.5 falls somewhere between the Mini- and MicroISO structures and is not a transmission owner.

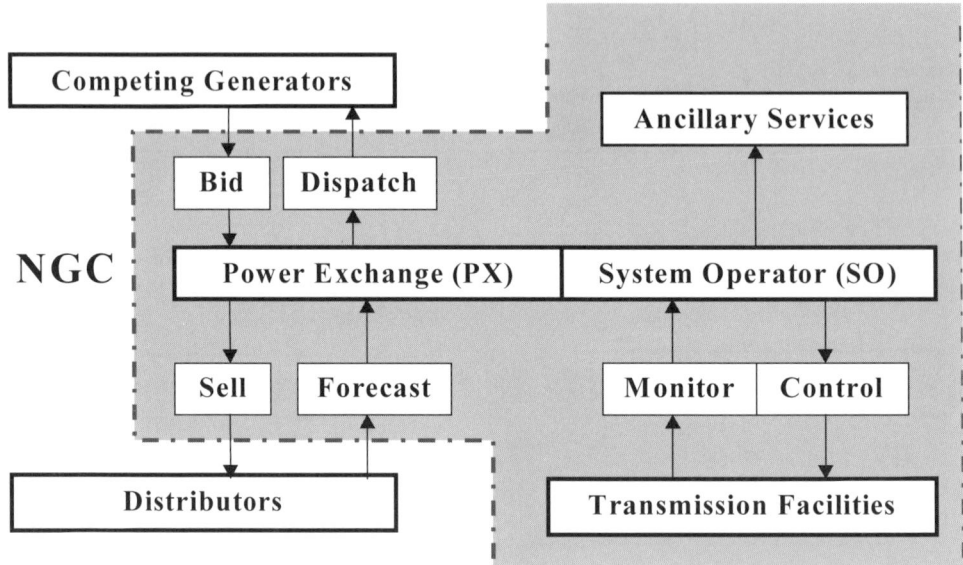

Figure 6.4 The NGC of the England & Wales power pool

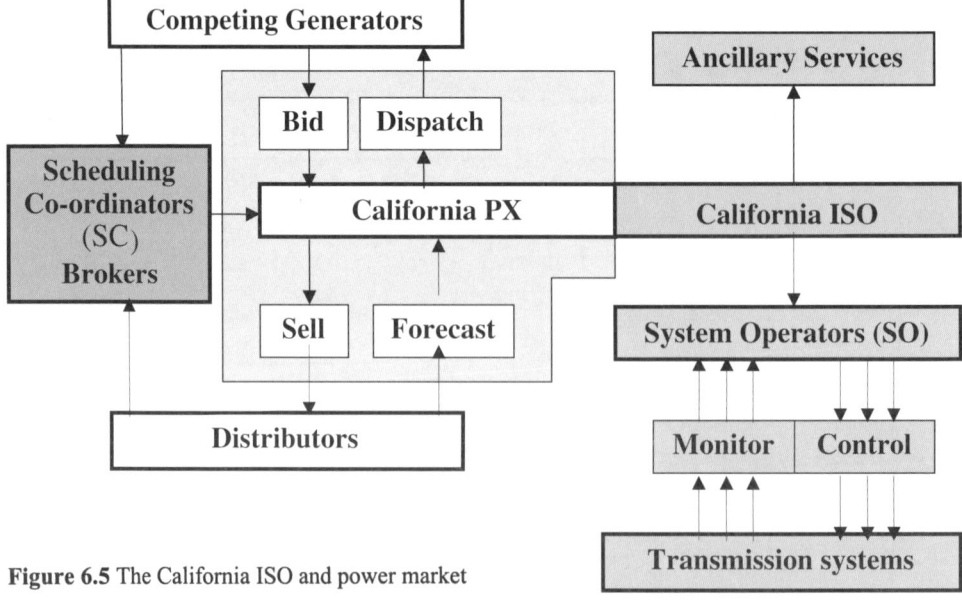

Figure 6.5 The California ISO and power market

6.4 Trading Arrangements

It is possible to conceptualise and model the new trading structures into a few alternative categories as discussed below.

6.4.1 The Pool

In the pool model shown in Figure 6.6, competition is initiated in the generation business by creating more than one Genco and is gradually brought to the distribution side where retailers could be separated from Discos and where consumers could be allowed to phase in a choice of retail supply. The transmission system is centrally controlled by a combination of an independent system operator and a power exchange (ISO+PX) which is disassociated from all market participants and ensures open access. The ISO+PX operates the electricity pool to perform a price-based dispatch and provides a forum for setting the system prices and handling electricity trades. Hedging contracts become a major option and are popular in the E&W system under the name 'Contracts for Difference'. The restructuring models in Chile, Argentina and East Australia also fall into this category with some modifications to the basic structure of Figure 6.6.

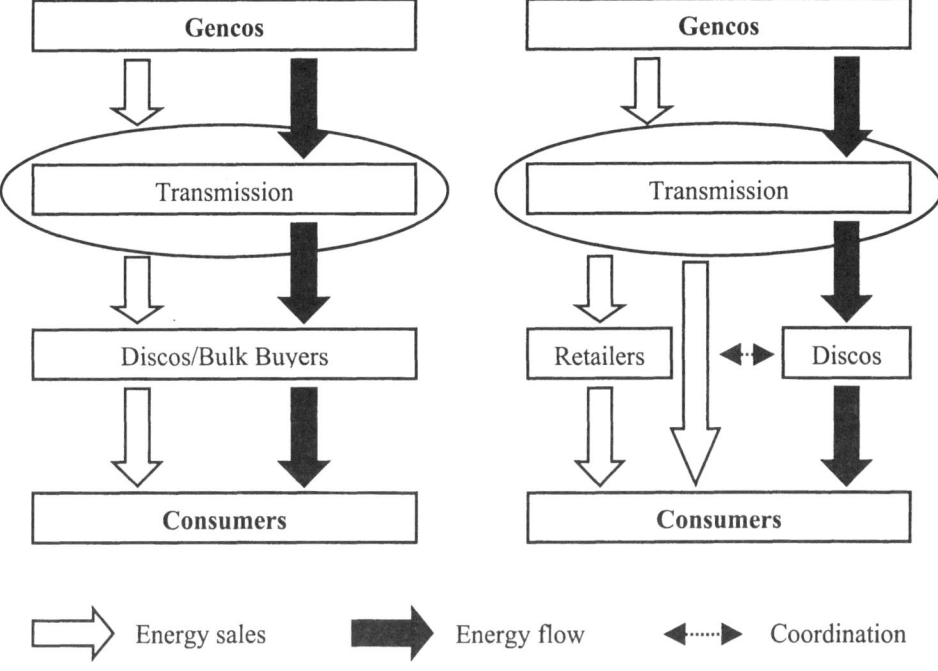

Figure 6.6 Trading in the power pool

6.4.2 Pool and Bilateral Trades

The most likely arrangement, which will emerge in practical systems in the future, is that a pool will exist simultaneously with bilateral and multilateral transactions. The significant difference between this model and the pool model is that the transmission sector is unbundled into a 'market' sector and a 'security' sector (see Figure 6.7). In the market sector, there are multiple separate energy markets, containing a pool market taken care of by the PX and bilateral contracts established by the SCs. The ISO is responsible for system operation and guarantees system security and in operational matters holds a superior position over the PX and SCs. The existence of a power pool is not mandatory in this model but will invariably be the case. Market participants may not only bid into the pool but also make bilateral contracts with each other. Therefore, this model provides more flexible options for transmission access. The California model is a representative of this category. The Nordic model and the New Zealand model also fall into this category with some modifications. Other models such as the New York power pool (NYPP) and the PJM model fall somewhere in between these three categories.

The power system in the new environment could be further disaggregated and reference [9] has proposed a fully disaggregated competitive electricity market. The model envisages what it calls 'co-ordinated multilateral trading' and completely dispenses with the PX in favour of a multitude of individual market-driven transactions.

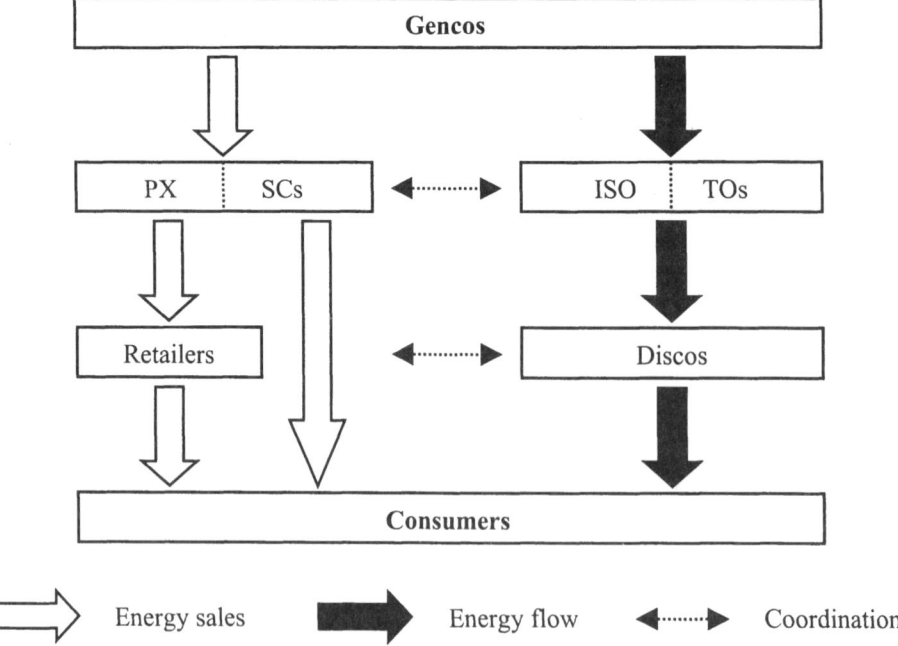

Figure 6.7 Trading with pool and bilateral contracts

6.4.3 Multilateral Trades

Multilateral trades are a generalisation of bilateral transactions where an SC or power broker puts together a group of energy producers and buyers to form a balanced transaction. In practice, multilateral and bilateral transactions will coexist with a power pool. Conceptually the extreme case is where the concepts of pool and the PX disappear into this multi-market structure as illustrated in Figure 6.8. Each market is managed by an SC or a broker under its individual rules. Differences in the rules of the different markets could give rise to different strategies for participants. In this model, the ISO is a MicroISO whose objectives are restricted to operation and security. The contracts in the energy markets will be respected by the ISO without discrimination. Only when system security is threatened will the ISO interfere in managing contracted dispatches. Changes in the electricity supply business bring opportunities for new participants. Many of the new entrants will be intermediarie; e.g. SCs are aggregators for better deals. Marketers, who buy and then resell electricity supply contracts, and brokers who arrange transactions between buyers and sellers, will enter the markets. These intermediaries will have a constructive role in promoting competition but there is also the danger of price volatility and market instability and many concerns have been voiced. Furthermore, some companies may play more than one role in this scenario. For instance, a Genco could also be a broker or a marketer and many business acquisitions have already taken place.

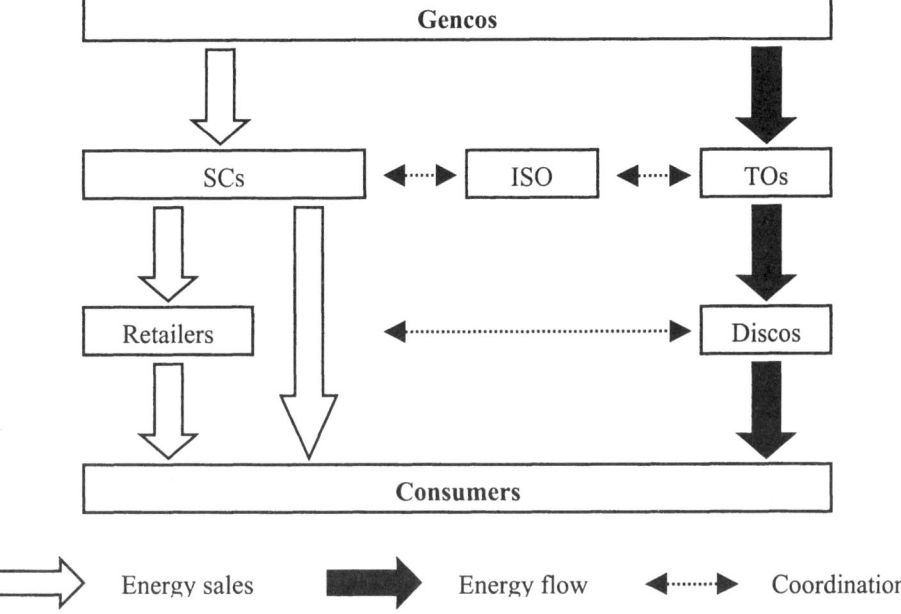

Figure 6.8 Structure with bilateral/multilateral trades only

Changes in the industry also bring opportunities to manage risk associated with a competitive environment. In order to hedge against price volatility due to market competition, fuel availability and load fluctuations, financial instruments, such as Contracts for Differences and other forward market variations, are used. Futures contracts, in which the right to take delivery of electricity can be bought or sold, are another tool for trading off price and risk. For example, futures markets have been set up in the Nord Pool and the USA. The distinction between a forward market and a futures market is that the former regulates specific transactions and is a technical and economic instrument while the latter is a purely financial instrument tradable in a futures exchange.

6.5 Transmission Pricing in Open-access Systems

6.5.1 Introduction

A key feature of the open transmission system in an unbundled environment is the need to charge all customers on a non-discriminatory basis for transmission services. Derivation of charges for the different kinds of transmission services should be simple and transparent and the signals these charges provide should be stable. 'Correct' pricing of transmission services is useful in providing economic signals for efficient short-run operations, recovery of costs, long-term capital investments and fair allocation of costs among participants.

There are five types of costs associated with transmission services: operating costs, capital or embedded costs, opportunity costs, reliability costs and system expansion costs. These are described briefly below.

Operating costs include transmission losses, costs incurred by generation redispatch due to operating constraints such as transmission and bus voltage limits, costs incurred in the provision of ancillary services and management and maintenance expenses.

Recovery of capital costs of the transmission system over the lifetime of the facilities is likely to be larger than the operating costs of the grid system. Individual transmission facilities are variably loaded as system conditions change with time, and the extent to which any particular supplier-to-consumer transmission transaction utilises any particular circuit is difficult to discern.

Opportunity costs are the benefits that the grid company forgoes as a consequence of operating constraints that arise owing to a specific transaction. Each transmission transaction may change the service reliability level and this affects expected outage cost and hence results in *reliability costs*. Reliability costs are very difficult to assess because they are related to many factors such as the timing, the duration, the extent of the service outage, the backup power available to the customer and customer location. These costs have been ignored in the past in designing transmission service rates.

System expansion costs include investment to accommodate new users and the extremely important issue of long-term transmission expansion in developing countries.

Transmission pricing is one of the most complicated issues in restructuring electricity supply because of the physical laws that govern power flow in the transmission network, and the need to balance supply and demand at all times. Since generators and customers are all connected to the same network, actions by one participant can have significant consequences on others making it difficult to investigate the cost each participant is responsible for. In addition to operating costs, embedded costs and financing of future expansion should be reflected in the tariff structure. Since embedded costs in the transmission system are very large compared with operating expenses, it is especially important to design a reasonable pricing structure for its recovery from all participants on a reasonable basis. As to expansion, it is largely driven by policy objectives.

There are three main pricing paradigms: rolled-in methods, incremental methods and composite embedded/incremental (marginal) methods [10]. A brief introduction is provided below followed by a concise description of the transmission pricing method used in the NGC, UK.

6.5.2 Rolled-in Pricing Methods

The simplest method of charging for transmission services is the so-called postage stamp method, which depends only on the amount of power moved and the duration of use, irrespective of supply and delivery points, distance of transmission usage or the distribution of loading imposed on different transmission circuits by a specific transaction. The main disadvantage of this method is the ignorance of the impact of any particular transaction on actual system operations. As a result, it is likely to send incorrect economic signals to users. A participant who uses the transmission system lightly, i.e. at a short electrical distance, actually subsidises others who use the system heavily. Obviously, this method is not equitable among users.

The MW-Mile method is an attempt to compensate for these shortcomings. The basic concept is that the loading of each transmission line due to each transaction is to be obtained separately; this is multiplied by the line length and then summed over all lines in the grid to obtain a measure of how much each transaction uses the grid. Different transactions are then charged in proportion to their utilisation of the grid. Neglecting temporal considerations this may be expressed mathematically by setting C_{T_i}, the charge levied on transaction T_i, where

$$C_{T_i} = \sum_j \left[\frac{P_{j;T_i} F_j}{\sum_i P_{j;T_i}} \right] \quad (6.1)$$

and $P_{j;T_i}$ is the loading of line j due to transaction T_i, F_j is the revenue required from this line. $P_{j;T_i}$ has to be computed using a sensitivity method or some linearised approach.

The MW-Mile method is a more equitable approach for transmission pricing but continues to suffer from the defects of lumping operating and embedded costs together and from a failure to distinguish between the relative importance of different lines for the secure operation of the system as a whole and to the reliability of each transmission transaction.

6.5.3 Incremental (Marginal) Pricing Methods

Nodal Pricing
The most complicated but accurate pricing method is nodal pricing derived from marginal cost theory. Nodal pricing can be short-run marginal (incremental) cost based or long-run marginal (incremental) cost based, depending on whether the recovery of operating costs, capital costs or expansion costs is desired. Short-run marginal pricing is discussed next; similar procedure can be applied to the other cases.

This method determines prices for power at each bus of the system accounting for all costs and transmission constraints. The nodal prices are typically calculated as dual variables or Lagrange multipliers of an optimal power flow (OPF) calculation [11]. Marginal transmission prices, or marginal wheeling rates as they are sometimes called, are derived directly from nodal prices. If MP_i and MP_j are the marginal nodal prices of electricity at buses i and j, the marginal transmission price is $MP_i - MP_j$ and is a measure of what it costs the grid to accept an additional unit of power at i and to deliver it at j.

Short-run nodal prices change dynamically as a function of load distributions, system structure and generation output patterns. Therefore, they must be recomputed on a periodic basis. This pricing scheme reflects only operating costs, so that capital recovery and management cost must be added. A major advantage of this method is that the right operational pricing signals are revealed. The major disadvantage is the need for a separate revenue reconciliation exercise.

This pricing scheme has three further disadvantages. The first is the intense real-time computational effort required. The second is the incentive problem; this method produces a perverse incentive for the monopoly transmission owner to cause, or not to relieve, constraints. Another disadvantage is the dramatic volatility of the price. Up to now, there is no fully operative example of nodal pricing in the industry.

The nodal pricing method can be applied to the pricing of real as well as reactive power [12]. The pricing of real power can be carried out, approximately, solely by using a DC load flow model. However, the pricing of reactive power has to be done using an AC load flow model simultaneously with real power pricing and also incorporating operating constraints, especially those concerning reactive power (such as reactive power balances at all buses and bus voltage limits).

Zonal Pricing

Notwithstanding its compelling theoretical basis the nodal pricing method has been deemed to be too complicated for applications, at least in the foreseeable future. The zonal pricing scheme has been proposed and used as an alternative. This pricing method represents a combination of the postage stamp method and the nodal pricing method and attempts to simplify the pricing process and at the same time to reflect the varying costs of supplying power to different areas.

Like the nodal pricing method the zonal pricing method can be short-run marginal (incremental) cost based or long-run marginal (incremental) cost based, depending on the objective as mentioned before.

Generally, the nodal pricing method is first utilised to obtain the prices in all buses, and then a weighed average value of all nodal prices within a zone is set as the zonal price. This is done periodically to account for the change of system conditions. The nodal prices at all buses in a zone should be reasonably close to each other for the method to be meaningful. Otherwise zonal boundaries should be adjusted. The Norwegian electricity market uses this approach with four or five zones. A long-run marginal-cost-based zonal pricing method is employed in the NGC, UK, for recovery of capital costs [13,14].

Another important purpose of zonal pricing is congestion management. Congestion occurs when the dispatch of all pool and contracted transactions in full would result in the violation of operational constraints. For example, in California, there are 24 adjustable price zones. Under normal circumstances without congestion, all zones have the same postage stamp rate, and there are no inter-zonal charges if power is transported across zone boundaries. The ISO raises the price in the congested zone when congestion occurs thereby providing a price signal to encourage users to sell or buy power in their own zone. Since the congested lines change with time the zones are adjustable.

The determination of zonal prices is somewhat subjective and this represents a major disadvantage, but the method is simple and easy to implement.

6.5.4 Embedded Cost Recovery

In general, the price of the transmission service should depend on the recovery of embedded costs as well as the incremental (operating) costs discussed above. This section discusses the former. The primary function of a transmission line is to transmit power from source point(s) to loading point(s). However, major lines also play an important role in security. Sometimes a particular transaction cannot take place during some contingency if some other line, which appears to be unimportant and lightly loaded during normal operation, was not present in the system. Moreover a specific line may be more important for the security of a particular transaction than it is for others. Clearly security/reliability is a relevant consideration in equitable embedded cost allocations.

A method for transmission line embedded cost allocation between transmission transactions accounting for both line capacity use and reliability benefits [5] is introduced here. The capacity use of a line, by a transaction, is the amount of power transmitted over that line because of that transaction, while the line reliability benefit for a particular transaction is calculated as the increment in the total probability of system failure, with the line out of service, compared with the failure probability when the line is in service. Capacity use is determined using a full AC power flow and hence the effects of reactive power can also be investigated.

Allocation Based on Capacity Use only
Since conditions are constantly changing the flow pattern in a network will depend on the values of generation, load and wheeling transactions at various times. Therefore, the host utility must set aside adequate capacity on each line to meet the maximum flow that may reasonably be expected to occur. Hence $P_{j;T_i}$, the load of line j due to transaction T_i, in (6.1) should be interpreted as follows:

$$P_{j;T_i} = \underset{\delta}{\text{Max}} (P_{j;T_i})$$

where δ indicates the combination of all transaction and transmission system conditions that could reasonably be expected to occur during the pricing time interval. The C_{T_i} value in (6.1) now provides a price allocation based on capacity use only.

Allocation Based on Reliability Benefit only
It is important to recognise that some lines may usually be lightly loaded but serve an extremely important purpose from a reliability point of view because they contribute much to system security during contingencies and emergencies. A charge can reasonably be made for this reliability benefit.

For each transaction, the reliability of transmission with all lines in service, and with all but line j in service, are first calculated. The reliability benefit $R_{j;T_i}$ derived from line j by transaction T_i is calculated as the increment in transaction failure probability caused by absence of the line. Similar probabilities are calculated in respect of each transaction with respect to line j. The embedded cost of the transmission line j, allocated to transaction T_i, based on reliability benefits only is now defined as:

$$C_{j';T_i} = \left(\frac{P_{T_i} R_{j;T_i}}{\sum_i P_{T_i} R_{j;T_i}} \right) \cdot F_j$$

where P_{T_i} is the magnitude of transaction T_i and F_j is the same as in (6.1). The specific method used to compute probability of transaction failure is in itself not central to this discussion. For example, the failure probability of a transaction may simply be taken as the probability that a path will not exist from the transaction sending bus to the transaction receiving bus. This approach to failure calculation is readily applicable to simple configurations but becomes complicated and virtually unworkable in large interconnected structures. For networked circuits, a cut set method [15] or a conditional probability approach [16] are suitable for calculating this availability.

Composites Based on Capacity Use and Reliability Benefits
Embedded costs are allocated to capacity use and reliability benefits in a ratio which has to be specified exogenously and depends on the judgement of the transmission planner. Hence, the composite charge is set at

$$C_{j;T_i} = \left[a \left(\frac{P_{j;T_i}}{\sum_i P_{j;T_i}} \right) + b \left(\frac{P_{T_i} R_{j;T_i}}{\sum_i P_{T_i} R_{j;T_i}} \right) \right] F_j$$

where $a+b=1$.

It should be noted that embedded transmission costs have to include facilities such as transformers, substation equipment, switchgear and shunt/series capacitors in addition to transmission lines and cables.

There is no single 'best' pricing mechanism and different methods may be appropriate for different situations. The various unbundled portions of the transmission service may also be priced using different methods; hence, an integrated approach needs to be used to price transmission services, as is the case in some electricity markets. The different tariff elements may be designed with time of use, geographical, quantitative and customer-specific variations in mind.

6.5.5 Transmission Pricing Method in the NGC, UK

A brief description of the transmission pricing in the NGC, UK, is provided [17] in this subsection.

The NGC is allowed to levy a charge to cover the costs of its assets but the costs of operating the system are passed through to consumers.

The 'System Marginal Price' (SMP) is the price quoted by the most expensive generator which is accepted for dispatch during each half-hourly time slot when transmission constraints are ignored – simple unconstrained dispatch.

The probability that demand will exceed capacity ('loss of load probability', or LOLP) is calculated by comparing expected demand with the capacity expected to be available. In order to encourage capacity offers from generators the pool purchase price (PPP) is obtained by augmenting SMP with this probability-weighted average. If an economic private generator is already selected in the unconstrained dispatch but prevented from generating owing to transmission constraints or other factors the generator would be entitled to a certain compensatory payment. This payment consists of the difference between PPP and the price at which the generator offered to supply power to the pool ('bid' price). Since the offer prices of such generators are lower than the PPP, the pool makes a loss on the deal. This additional expense is passed on to consumers. On the other hand, some uneconomic generators who are not selected in the day-ahead market but are called upon to generate owing to transmission constraints or other reasons, are paid their 'bid' price, which is higher than the prevailing PPP. This is, effectively, a payment for out-of-merit generator operation. Constrained-off costs, out-of-merit payments and several additional expenses such as transmission fixed charges, transmission losses, start up costs, and ancillary services charges are passed on to consumers in the *uplift*.

The customer side of the market is simpler: all energy is purchased at the pool selling price (PSP). All of the extra costs of energy above the PPP are simply lumped together in *'uplift'* and spread over all kWh taken by customers through the calculation of a single half-hourly consumer price, the PSP:

$$PSP = PPP + uplift$$

Strictly speaking, therefore, the PSP is fixed on a revenue reconciliation basis; that is, PSP multiplied by the electricity sold is made equal to payment to generators plus other costs incurred. Hence, the costs of transmission losses are also rolled into the uplift.

There are certain important implications to this. Although the NGC is a private company benefiting from a monopoly position, albeit regulated, the financial losses arising from its inability to provide adequate transmission access are passed on entirely to consumers. For this reason and since average allocations of *uplift* result in inequitable charges to different users, *uplift* has been the primary focus of contention since divestiture. Customers share the costs of operating the system and bear not those costs which they impose themselves but an average of the costs incurred by all users. The Office of the Electricity Regulator is re-examining these problems and recently incentives in the form of rewards and penalties for the NGC to improve its performance have been introduced.

6.6 Open Transmission System Operation

Technically, the greatest challenge facing the deregulated and unbundled electricity supply system is the operation of the grid. In the traditional structure the grid and the generating plant constituted a single business enterprise and the system control centre held a pre-eminent position in the hierarchy of control. The decision-making authority of the system control centre in respect of operational matters was unambiguous and the constraints imposed by the individual interests of numerous separate business entities were unknown. The gradual emergence of IPPs and BOT/BOO plants in the 1980s and 1990s did not change matters significantly. The experiences gained by different power pool operators in coordinating the operation of large interconnected systems was perhaps the only experience really relevant to the new circumstances.

There are and will be many variations in the deregulated scenarios for electric power systems around the world. Accordingly, there exist different schemes of opening access to transmission systems, and the differences may be due to historical, political, geographic or economic factors, specific to each country or area. However, each restructuring scheme must address the fundamental technical issues associated with properly operating an open transmission network.

This section will address some issues that the system operator will face in routine system operation, and how these issues can be resolved in a market environment. The normal (or uncongested) condition is when all pool demand and all bilateral and multilateral transactions are dispatched without violating operating constraints. All transactions will be serviced at their desired value and charged fairly and the ISO only needs to optimise pool dispatch and to manage ancillary services. In real world open access transmission as it is now evolving pool and contract dispatch will coexist. The issue for integrating them into a single dispatch strategy is one of the problems that has to be addressed.

6.6.1 Dispatch

Under normal operating conditions without any operating constraint violations, the major difference between the centrally dispatched utility and dispatch in the pool-type electricity market is the replacement of costs by bids. When demand-side bidding is permitted, a slightly more complicated dispatch procedure is required, and the objective, theoretically, will change to the maximisation of social welfare (in practice, just making an allowance for demand elasticity) rather than just bid cost minimisation. However, if bilateral and/or multilateral transactions are permitted and/or transmission congestion occurs the dispatch methods become more complex and are discussed in the next section of this chapter.

6.6.2 Transmission Loss Compensation

If the ISO owns generators, transmission losses can be supplied at charges which will, essentially, be set by the regulator. If the ISO does not own any generators loss compensation must be procured from power suppliers through pool purchases, auctions or contracts. An example is the California electricity market where the ISO broadcasts the loss distribution factors to all participants, the SCs (including the PX), in advance, and they are required to submit balanced schedules to the ISO with transmission losses made good. In other words, the participants provide the losses. However, losses cannot be fully compensated before the fact in this way since the system load cannot be accurately forecast and loss distribution factors are only approximate. Ultimately, the residual losses must be compensated through a real-time imbalance market which will be discussed in the next subsection.

6.6.3 System Control

The key responsibility of the ISO is ensuring system security while monitoring and controlling real-time system operation is its most routine task. During normal system operation the ISO will be responsible for maintaining system frequency and voltage within a tolerable range and ensuring that all operating constraints are respected.

Since electricity is delivered instantaneously, and supply and demand must be balanced at every instant, no matter which market structure is utilised, the system operator must rely on one or more generator outputs to follow load changes. An ISO which owns generators can provide this service, otherwise it must procure the service contractually from on-line generators. In fact, most generators have the ability to change generation outputs automatically to respond to the supply/demand imbalance which is reflected in small system frequency deviations; however, in a commercial market, this service will not be provided voluntarily. In any case, the ISO must tightly control generators which are contracted to supply this second-to-second balancing service. In E&W, the balancing function is provided at the day-ahead energy market stage. In California balancing power is priced separately at an *ex post* price that is determined in the real-time market based on an auction for power imbalance.

Automatic generation control (AGC), sometimes referred to as secondary frequency control, is another important control issue. AGC smoothes over large frequency deviations, usually the result of large load variation and/or generation outage. The ISO must ensure that there is a sufficient quantity of on-line unloaded capacity and/or quick-start generation capacity, and that this conforms to mandatory reliability criteria. These capacities must be provided or procured by the ISO.

Reactive power and voltage control are also of vital importance for secure operation. The ISO should first utilise the reactive power resources available in the transmission system; thereafter provision of extra reactive power by generators is necessary. Up to now, little effort has been directed at finding commercially efficient ways of providing the resources needed for voltage control. This issue will be discussed in the next subsection. When there is significant imbalance between reactive power supply and demand, installation of new reactive power resources is unavoidable. In this case, the ISO, the participants or a regulatory authority will need to decide who should undertake these tasks. Black start and restoration can be procured through long-term contracts or auctions.

6.6.4 Ancillary Service Provision

The term 'ancillary services' generally refers to power system services other than the provision of energy. Specifically, ancillary services are those functions performed by the equipment and people that generate, control, transmit and distribute electricity to support the basic task of transmission [18]. These functions include, but are not limited to, spinning reserve, non-spinning reserve (dispatchable load and generation), regulation, frequency control, AGC, reactive power and voltage control and black-start capability. In the USA FERC requires the ISOs to offer six ancillary services: scheduling, system control and dispatch; reactive power supply and voltage control; regulation and frequency response; energy imbalance; operating spinning reserve; operating reserve-supplemental reserve. In the past, these services were simply and naturally bundled into the main activities of generation and transmission and the costs associated with energy generation and with ancillary services provision were jointly recovered through the electricity tariff.

In the new competitive environment this natural or organic link is fractured giving rise to many debates about how these services should be procured. Two important considerations in procuring ancillary services are the costs of providing the services and the values of the services to the system. Depending on the organisational structures in different electricity markets, ancillary services may be provided by the system controller or be purchased.

Some ancillary services can be mandated; for example, all generators could be required to provide frequency control as a precondition of connecting to the network (unless the system operator wishes to encourage some generators to provide more than others). However, mandatory ancillary services provision is a little inconsistent with the objectives of unbundling. Similarly, charging for ancillary services using a bundled rate is not equitable to users since this does not reflect actual usage. Competitive market-based procuring and charging for unbundled ancillary services is required by FERC in the USA.

Generation-based ancillary services such as spinning reserves and AGC can be made competitive and separate from the energy market. On the transmission side, ancillary service provision, mainly reactive power, has to be priced differently from that on the generation side. However, reactive power support, which relies on both generators and requirements and installation of compensating devices, is probably better provided by a contribution of mandated equipment owned by the transmission provider. The location-specific nature of reactive power needs is sufficiently great that competitive markets may not be easy to develop; however, there is also a counter-view that to the extent that reactive supplies are not geographically restricted, it might be possible to create competitive markets for their acquisition locally [19]. Irrespective of these debates, restructuring will almost surely change the mechanisms that system operators use to acquire and deploy reactive power resources. In the USA, FERC differentiates between generation-based reactive power support and transmission-system-based resources, with the latter to be priced under the basic transmission tariff. The requirements and payment for generation-based reactive power support fall under FERC's voltage control service, and this service can be procured by the ISO following market management protocols. Transmission-based reactive power resources are addressed under the general transmission tariff based on embedded costs. Developing country system planners are certain to view FERC protocols as excessively market driven and avoid most of them.

Several different methods of procuring ancillary services have been practised in the existing electricity markets [20]. In the UK, the pool operator either provides ancillary services or procures them from generators and charges users through the 'uplift'.

Another approach to procuring and pricing ancillary services, such as reserves, is to procure reserves and energy simultaneously in a single combined auction and compensate the generators providing these reserves at an opportunity cost estimate derived from the energy auction. This approach can work in a market structure where the organisation responsible for procuring ancillary services is also responsible for operating the energy market. Examples of this include New Zealand and New England.

Procuring ancillary services, separate from the energy market, is better suited for market structures where the ISO is separate from the PX, as is the case in California. The ancillary services which the California ISO is responsible for procuring include spinning reserves, non-spinning reserves, AGC, replacement reserves, voltage support and black start. The first four services can be procured by the ISO through daily competitive auctions or be self-provided by users. However, self-provision is limited to mandatory or contractual arrangements. The use of all ancillary services, including those self-provided, is the exclusive responsibility of the ISO. Black-start capability and reactive power support must be provided or purchased by the ISO.

In some electricity markets, both the mandatory and market-based approaches are combined. For example, in Spain, the so-called 'primary reserves' service is mandatory for all generators. All plant must be equipped with a governor and there is no remuneration associated. This is intended to reduce the rate of frequency deviations during momentary power imbalances. If a generator cannot provide the required primary reserve it must buy it from other generators. On the other hand the procurement of AGC service is through a competitive (auction) market.

6.7 Congestion Management in Open-access Transmission Systems

Congestion [21] is not a new problem in power system operation and was a routine problem for the system operator in the traditional system. In electricity market environments, however, previously established practices for dealing with congestion can no longer be relied on since cooperation between market participants cannot be guaranteed. Any control measures adopted by the system operator to eliminate congestion must not only be technically justifiable, but also be fair to users and commercially transparent. In some electricity markets with bilateral and multilateral contract transactions this problem is more difficult to solve since these contract transactions introduce additional constraints on the system operator. For example, curtailment of a bilateral transaction requires simultaneous and equal reduction at the entry and exit points. All this makes congestion management a challenging problem and requires a combination of policy, pricing and operational responses. It is perhaps the thorniest issue in transmission operation.

6.7.1 Congestion Management in Normal Operation

Pool and contract models are separately addressed first and an approach to reconcile both models is then explored. Congestion management issues without consideration of contingency/security problems are discussed in this subsection. A fuller treatment of these topics can be found in [22].

Pool Model
The formulation given below assumes price-based dispatch built on spot pricing theory [11] and in its simplest terms, neglecting price elasticity effects and the significance of location, the dispatch algorithm may be stated as:

$$\operatorname*{Max}_{G_i, D_j} \sum_j B_j(D_j) - \sum_i C_i(G_i) \qquad (6.2)$$

subject to:

$$\sum_j D_j - \sum_i G_i + L = 0$$

$$G_i - G_{i.\max} \leq 0 \qquad : \forall i$$

$$Z_k \leq 0 \qquad : \forall k$$

where i and j are the set of producers and purchasers, G and D their respective generation and consumption, and producer offer (bid) price and purchaser benefit (utility) functions are given by C and B, respectively. The single load balance constraint will later be generalised to a set of augmented load flow equations. L is a transmission loss function, $G_{i.\max}$ generator i capacity and Z_k the kth operating constraint. Problem (6.2) leads to the solution and Kuhn-Tucker conditions:

$$\frac{\partial B_j}{\partial D_j} = p_j = \lambda\left(1 + \frac{\partial L}{\partial D_j}\right) + \sum_k \pi_k \frac{\partial Z_k}{\partial D_j} \qquad : \forall j$$

$$\frac{\partial C_i}{\partial G_i} = \lambda\left(1 - \frac{\partial L}{\partial G_i}\right) - \mu_i - \sum_k \pi_k \frac{\partial Z_k}{\partial G_i} \qquad : \forall i$$

$$\mu_i(G_i - G_{i.\max}) = 0 \quad \text{and} \quad \mu_i \geq 0 \qquad : \forall i$$

$$\pi_k Z_k = 0 \quad \text{and} \quad \pi_k \geq 0 \qquad : \forall k$$

where λ is the system incremental cost (dual multiplier on the equality constraint) and μ and π are the sets of Kuhn-Tucker dual variables on the capacity and operating constraints, respectively. The first statement above is the stipulation of a rational consumer who sets price p_j equal to marginal utility. The demand-price elasticity issue, which was incorporated as a part of pool dispatch in [23,24], is discussed in Section 6.8.

Contract Model
The contract model [25-27] is designated as power dispatch in a structure dominated by bilateral and multilateral transmission contracts. A bilateral transaction is made directly by a Genco-Disco pair while a multilateral transaction is an extension of this, which is supplemented by third parties, such as brokers or forward contractors, and involves several Gencos and Discos. These transactions are then brought to the ISO with a request that transmission be provided. If there is no congestion on the system the ISO simply dispatches all requested transactions and makes a fair charge for the service.

A separate notation appropriate for modelling bilateral and multilateral transactions is now introduced. Consider a power system with Genco bus bars and Discos bus bars. Position 1 is reserved for a location fictionally called the system regulation bus where the ISO purchases power to make good transmission losses. (There is no loss of generality since 1 can be merged with any one or more generators using the distributed slack bus concept [28].) Since any Genco and Disco may contract with several participants the total power at buses i and j is given by

$$P_i = \sum_{j \in J_D} P1_{ij} + \sum_{k \in K} P2_{ik} \qquad i \in I_G; i \neq 1$$

$$D_j = \sum_{i \in I_G; i \neq 1} D1_{ji} + \sum_{k \in K} D2_{jk} \qquad j \in J_D$$

where $P1_{ij}$ and $D1_{ji}$ are power injection at bus i and extraction at bus j, respectively, under the bilateral contract from i to j. $P2_{ik}$ and $D2_{jk}$ are power injection at bus i and extraction at bus j, respectively, under the kth multilateral contract. K is the number of multilateral contracts. There exist important modelling relationships pertaining to these contracts, namely

$$P1_{ij} = D1_{ji} \qquad i \in I_G; i \neq 1; j \in J_D \tag{6.3a}$$

$$\sum_{i \in I_G; i \neq 1} P2_{ik} = \sum_{j \in J_D} D2_{jk} \qquad k \in K \tag{6.3b}$$

where the first statement depicts the power balance equations in bilateral contracts and the second one is the group power balance equations in multilateral contracts. In the event of system congestion the group power balance requirement in (6.3b) needs to be supplemented by stipulations of how group loads will be balanced with group generation when the latter is curtailed for example, all loads may be curtailed proportionally, or the group may specify some other curtailment plan.

Congestion management, essentially, has to do with rationing of transmission access. Rationing has to follow a user-pay philosophy where willingness-to-pay to avoid curtailment is an indicator of the importance of unfettered dispatch for the participants in a transaction. Let u represent the vector of contracts mentioned in (6.3), whose elements can be either $P1_{ij}$ or $D1_{ji}$ (but not both) and a certain number of variables from the set $\{P2_{ik}, D2_{jk}\}$ selected with due regard to the permissible degrees of freedom in the power balance constraints (6.3b). The following optimisation problem is proposed to guide ISO operation under congested conditions:

$$\text{Min } f(u,x) = (u - u^\circ)^T \cdot w \cdot (u - u^\circ) \tag{6.4}$$

subject to:

$$L1(u,x) = 0$$

$$Z1(u,x) \leq 0$$

where u^o is the desired or target value of u while w is a diagonal matrix whose elements are 'willingness-to-pay' price premiums to avoid transmission curtailment and x is the set of dependent variables. The equality constraints in (6.4) are the system power flow equations augmented by a set of contracted transaction relationships, including (6.3a, 6.3b). The inequality constraints in (6.4) are an extension of the usual set of system operating constraints augmented by additional inequalities for the upper bounds u^o on u.

Detailed examples of the method with different curtailment strategies are provided in [27]. The proposed method may be extended to wheeling in interconnected systems, which may be viewed as a point-to-point transfer similar to a bilateral contract.

6.7.2 Integrated Transmission Dispatch Strategy

In real world open-access transmission, as it is now evolving, pool and contract dispatch will coexist and integrating them into a single dispatch strategy is an essential task of open-access congestion management [25,26,29]. An approach based on [29] is now discussed.

In a power system where pool and bilateral/multilateral transactions coexist, active power at generator bus i and load bus j can be respectively formulated as

$$P_i = P_{PL,i} + \sum_{k \in K} P_{T_k,i} + \sum_{k \in K} P_{LT_k,i} \quad i \in I_G \tag{6.5}$$

$$D_j = D_{PL,j} + \sum_{k \in K} D_{T_k,j} \quad j \in J_D$$

where K is the total number of bilateral/multilateral transactions and T_k the kth bilateral/multilateral transaction. $P_{PL,i}$ and $D_{PL,j}$ are bus i pool generation and bus j pool consumption, respectively. $P_{T_k,i}$ and $D_{T_k,j}$ are, respectively, power injection at bus i and power extraction at bus j under transaction T_k. $P_{LT_k,i}$ is the power supplied at bus i by bilateral/multilateral participants to make good transmission losses (see ancillary services, Section 6.6.4).

In actual operation of power systems the responsibility for transmission losses may be distributed between all dispatched transactions. Two different arrangements were addressed in [29] but for simplicity only the case in which transmission losses are supplied by pool generation is considered here. That is,

$$P_{LT_k,i} = 0 \quad i \in I_G; \quad k \in K$$

and the ISO will dispatch pool power to make good transmission losses, including the losses associated with the delivery of transactions T_k.

A two-stage dispatch process is now proposed. In the formulations below I, with various subscripts, is a row vector of ones of appropriate dimension used to denote a summation as a simple scalar product; for example,

$$I \cdot a = \sum_i a_i, \text{ where } a = [a_1,...,a_i...]^T$$

(1) Normal condition: The normal condition is when all pool demand and all bilateral and multilateral transactions are dispatched without system security violations. All these transactions will be serviced at their desired value and the ISO only needs to optimise pool dispatch and ancillary services. Hence a slightly modified traditional OPF follows:

$$\text{Min} u_{P_{PL}} \cdot \rho_{P_{PL}}^T \quad (6.6)$$

subject to:

$$L1(P_{PL}, D_{PL}^o, P_T^o, D_T^o, Q, V, \theta) = 0$$

$$Z1(P_{PL}, D_{PL}^o, P_T^o, D_T^o, Q, V, \theta) \leq 0$$

where $\rho_{P_{PL}}$ is a column vector with typical element $\rho_{P_{PL,i}}$, which is the product of $P_{PL,i}$ and bid price for this pool power; P_{PL} is a vector of pool powers with elements $P_{PL,i}$, which are the control variables of this problem; D_{PL}^o, P_T^o and D_T^o are vectors of desired values of pool consumption, and bilateral/multilateral injection and extraction with elements $D_{PL,j}$, $P_{T_k,i}$ and $D_{T_k,j}$, respectively; Q, V and θ are vectors of reactive nodal power, bus voltage magnitudes and angles, respectively. The first constraint in problem (6.6) is the conventional load flow equation set plus the set of nodal power expressions (6.5). The second constraint is a set of inequalities, including limits on pool power and system operating constraints such as bus voltage levels and line overloads.

(2) Congestion: This occurs when the dispatch of all pool and bilateral and multilateral transactions in full would result in the violation of operational constraints. The following dispatch problem is now formulated:

$$\text{Max}\ (I_{D_{PL}} \cdot \rho_{D_{PL}}^T - I_{P_{PL}} \cdot \rho_{P_{PL}}^T) - I_{D_{PL}} \cdot \gamma_{D_{PL}}^T + I_{P_T} \cdot \rho_{P_T}^T - I_{P_T} \cdot \gamma_{P_T}^T \quad (6.7)$$

subject to:

$$L1(P_{PL}, D_{PL}, P_T, D_T, Q, V, \theta) = 0$$

$$M1(P_T, D_T) = 0$$

$$Z1(P_{PL}, D_{PL}, P_T, D_T, Q, V, \theta) \leq 0$$

where the first term, within brackets, in the above objective represents the net pool welfare; ρ_{P_T} is a column vector of elements $\rho_{P_{T_k,i}}$, the transmission charge for delivering $P_{T_k,i}$; $\gamma_{D_{PL}}$ is a column vector of elements $(w_{D_{PL,j}} \cdot \Delta D_{PL,j}^2)$, where $w_{D_{PL,j}}$ is a willingness-to-pay factor and $\Delta D_{PL,j} = (D_{PL,j}^o - D_{PL,j})$, the pool customer shortfall, where $D_{PL,j}^o$ is the desired value of $D_{PL,j}$ and satisfies $D_{PL,j} \leq D_{PL,j}^o$; γ_{P_T} is a column vector of elements $(w_{P_{T_k,i}} \cdot \Delta P_{T_k,i}^2)$, where $w_{P_{T_k,i}}$ is also a willingness-to-pay factor and $\Delta P_{T_k,i} =$

$(P^o_{T_k,i} - P_{T_k,i})$, the bilateral/multilateral supply shortfall, where $P^o_{T_k,i}$ is the desired value of $P_{T_k,i}$ and satisfies $P_{T_k,i} \leq P^o_{T_k,i}$.

It is worth mentioning that willingness-to-pay factors, which have been introduced in the above contract model, accommodate the interests of both pool and bilateral/multilateral participants during congestion. That is, any participant may be willing to make extra payment to avoid curtailment. This arrangement has to be agreed with the ISO in advance and the ISO will determine magnitudes for the willingness-to-pay factors in order to ration transmission access accordingly.

The first constraint in problem (6.7) is similar to the first in (6.6), but with $\boldsymbol{D}_{PL}, \boldsymbol{P}_T$ and $\boldsymbol{D}_T$ as vectors of $D_{PL,j}$, $P_{T_k,i}$ and $D_{T_k,j}$, respectively. The second constraint in (6.7) is a set consisting of group power balance equations and group curtailment relations that have been mentioned in the contract model. These relations will be declared by bilateral/multilateral participants in advance. The inequality constraint in (6.7) is an extension of the inequality expression in (6.6), obtained by adding inequalities for the upper bounds on $D_{PL,j}$, $P_{T_k,i}$ and $D_{T_k,j}$. This completes the mathematical formulation of the congestion management problem in deregulated power systems which contain both pool and contract transactions.

6.7.3 Illustration Using a Small Power System

The five-bus system of Figure 6.9 is used here to illustrate the integrated transmission dispatch strategy discussed above. For simplicity only the impact of the willingness-to-pay for transmission factor is examined. System data including pool generator costs can be found in the Appendix.

The generators at buses 1 and 2 bid into the pool and the 80 MW load at bus 5 takes power from the pool at pool prices. The 200 MW load at bus 4 is divided into two equal parts, one-half takes power from the pool and the other enters into a bilateral contract with the independent generator at bus 3. The results of a power flow calculation show that if the pool demands at buses 4 and 5 are completely supplied and the bilateral transfer from 3 to 4 is transmitted in full, line 2-5 is overloaded. Therefore the ISO will execute curtailment by using the method of problem (6.7). The solutions, in which all line flows are brought within limits, are given in Table 6.1.

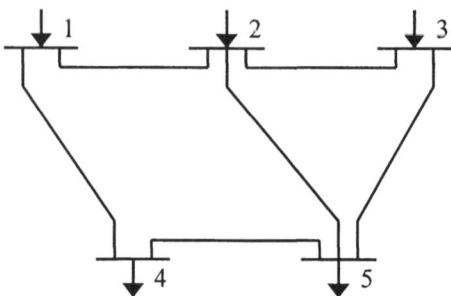

Figure 6.9 Example system

Case 1 assumes willingness-to-pay factors of the pool demands at buses 4 and 5 are the same as that of the bilateral transfer from 3 to 4, namely 20 $/MW^2h$. In case 2 willingness-to-pay factors of all pool demands are doubled while the bilateral transfer remains unchanged. As expected, the pool demands at buses 4 and 5 were curtailed less in case 2 than case 1 and the bilateral transfer from 3 to 4 was curtailed more since pool consumers were willing to pay more.

Table 6.1 Dispatch results with different willingness-to-pay

Transaction	Optimal Dispatch (MW)	
	Case 1	Case 2
Pool generation at bus 1	54.6	56.1
Pool generation at bus 2	119.1	119.5
Pool demand at bus 4	94.5	95.2 ↑
Pool demand at bus 5	73.8	75.2 ↑
Transfer from bus 3 to 4	96.8	94.3 ↓

The relation between curtailments and willingness-to-pay magnitudes as shown in Figure 6.10 is obtained as the willingness-to-pay of the bilateral transfer (from 3 to 4) is varied from 0.0 to 60.0 $/MW^2h$ in 10 $/MW^2h$ steps while other factors are retained the same as case 1.

The non-linear curve in Figure 6.10 shows that the more the willingness-to-pay, the less the curtailment and that when it becomes larger the bilateral transfer tends to the desired value (100 MW). It is important and interesting to emphasise that willingness-to-pay not only has a non-linear relation with the curtailment of its own transaction but also has an indirect influence on the curtailment of other loads and transactions.

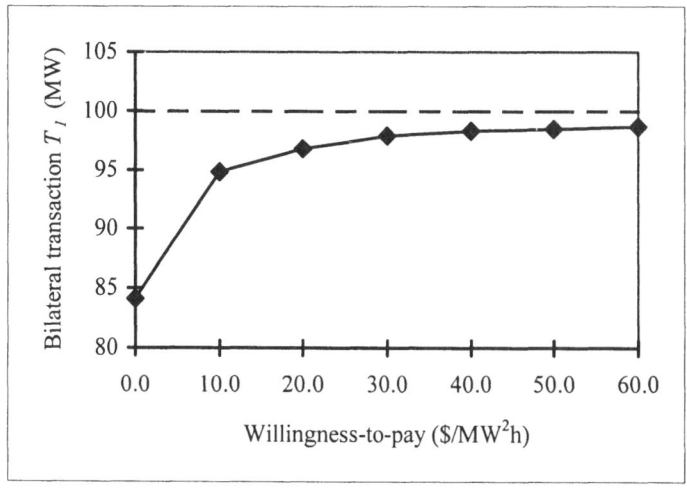

Figure 6.10 Impact of willingness-to-pay on curtailment

6.7.4 Static Security-constrained Rescheduling

A methodology [30] to reschedule pool and bilateral transactions taking account of system security, which may be helpful in providing an insight into the security challenges faced by an ISO in the context of system deregulation is presented next.

Conceptual Model
Security-induced rescheduling [31,32] is the preventive therapy of eliminating potentially dangerous operating conditions and bringing a vulnerable system into a secure operating state. This is an adjunct to on-line security monitoring and contingency evaluation and is implemented only when the system is found to be in a vulnerable state.

It is recognised from [33] that transaction post-contingency corrective capability will be helpful for eliminating constraint violation. Both post-contingency corrective control and preventive control are, therefore, taken into account here.

The aim of the method is to minimise deviations from transaction schedules while taking both preventive and post-contingency corrective measures to ensure system security. The methodology for the case of preventive control is based on the assumption that the current (un-rescheduled) transaction is the most economical. For post-contingency corrective control, the consideration is to reduce benefit loss of the previously scheduled transactions when a contingency occurs.

Mathematical Model
The discussion that follows is restricted to the case where the flow of power on some lines may be out of sustainable limits under line-outage conditions and deals with the rescheduling of pool generation and demand as well as bilateral transactions. A new notation is introduced here.

(1) Objective: The objective for the preventive aspect is formulated as

$$\text{Min } F_0 = \sum_{i \in NG} \rho_{Gi} |\Delta P_{Gi}| + \sum_{i \in ND} \rho_{Di} |\Delta P_{Di}| + \sum_{i \in NT} \rho_{Ti} |\Delta P_{Ti}|$$

where ΔP_{Gi}, ΔP_{Di} and ΔP_{Ti} are preventive changes in transaction P_{Gi} bid into the pool, pool to customer sales transaction P_{Di} and bilateral transaction P_{Ti}, respectively; NG, ND and NT are the total numbers of transactions of the type P_{Gi}, P_{Di} and P_{Ti}, respectively; ρ_{Gi} and ρ_{Di} are marginal prices of pool transactions P_{Gi} and P_{Di}, respectively, while ρ_{Ti} represents price information in respect of bilateral transaction P_{Ti} and is discussed later.

The objective function for the kth line-outage contingency is taken to be

$$\text{Min } F_k = \sum_{i \in NG} \rho_{Gi}^+ |\Delta P_{Gi}^k| + \sum_{i \in ND} \rho_{Di}^+ |\Delta P_{Di}^k| + \sum_{i \in NT} \rho_{Ti}^+ |\Delta P_{Ti}^k|$$

where ΔP_{Gi}^k, ΔP_{Di}^k and ΔP_{Ti}^k are post-contingency changes enforced on transactions P_{Gi}, P_{Di} and P_{Ti} in the event of contingency k; ρ_{Gi}^+, ρ_{Di}^+ and ρ_{Ti}^+ are alternative values for the prices ρ_{Gi}, ρ_{Di} and ρ_{Ti} under contingency. Market participants may (or may not) submit different prices for normal states and for contingent conditions depending on their aversion to occasional short-time interruptions or curtailments.

A multiple objective rescheduling problem is now formulated as follows:

$$\text{Min } F = w_0 \cdot F_0 + \sum_{k \in K} w_k \cdot F_k \quad (6.8)$$

where w_0 is a weight attached to F_0 and satisfies

$$w_0 = 1 - \sum_{k \in K} w_k$$

K is the total number of the postulated contingencies; w_k is a weight attached to the objective F_k, which is determined by the ISO using probability and duration expectations for line-outage contingency k; that is to say, the ISO uses these expectations to decide how much weightage to attach to different contingencies.

(2) Bilateral power balance equations: A bilateral transaction comprises a power injection at a generator bus and an extraction of the same quantity of power at a load bus. Therefore, preventive changes ΔP_{Ti} in bilateral transactions P_{Ti} require enforcement of the group power balance

$$\Delta P_{Ti,m} = -\Delta P_{Ti,n} = \Delta P_{Ti} \quad i \in NT \quad (6.9)$$

where $\Delta P_{Ti,m}$ and $\Delta P_{Ti,n}$ are specific preventive changes in bilateral transaction P_{Ti} at generator bus m and at load bus n, respectively.

Similarly post-contingency corrective changes ΔP_{Ti}^k require group power balances

$$\Delta P_{Ti,m}^k = -\Delta P_{Ti,n}^k = \Delta P_{Ti}^k \quad i \in NT; k \in K \quad (6.10)$$

where $\Delta P_{Ti,m}^k$ and $\Delta P_{Ti,n}^k$ are corrective changes in bilateral transaction P_{Ti} at buses m and n, respectively, when contingency k occurs.

From equations (6.9-6.10) it follows that if a bilateral transaction has to be curtailed a change is made simultaneously at two buses. In order to reflect benefit changes in the proposed objective F caused by the changes ΔP_{Ti}, the prices ρ_{Ti} should be the sum of marginal prices at the two buses of the bilateral transaction P_{Ti}, i.e.

$$\rho_{Ti} = \rho_{Ti,m} + \rho_{Ti,n} \quad i \in NT$$

where $\rho_{Ti,m}$ and $\rho_{Ti,n}$ are marginal prices at buses m and n, respectively, which are associated with bilateral transaction P_{Ti}.

However, for bilateral transactions, assuming short-term rational pricing, only transmission prices $(\rho_{Ti,m} - \rho_{Ti,n})$ are known to the ISO. This chapter assumes that the marginal prices $\rho_{Ti,m}$ and $\rho_{Ti,n}$ can be obtained as follows. 1) $\rho_{Ti,m}$ is set as pool marginal price at bus m (the calculation of pool price is outside the scope of this chapter, but see [23]); 2) $\rho_{Ti,n}$ is set equal to $\rho_{Ti,m}$ plus the transmission price of bilateral transaction P_{Ti}. The prices ρ_{Ti}^+ under contingency can be obtained in a similar way to ρ_{Ti}.

(3) Pool power balance equations: Preventive and post-contingency corrections must satisfy the conservation of power in the pool. Assuming that only transaction is set by the ISO to provide system regulation the associated generator bus can be chosen as the slack bus. That is to say, the slack bus power adjustment must balance the changes in generator and load power with due regard for transmission losses. A linear pool power balance equation in the normal state can then be written as

$$\Delta P_{G1} - \sum_{i \neq 1; i \in NG} r_{Gi} \cdot \Delta P_{Gi} - \sum_{i \in ND} r_{Di} \cdot \Delta P_{Di} - \sum_{i \in NT} r_{Ti} \cdot \Delta P_{Ti} = 0 \tag{6.11}$$

where r_{Gi}, r_{Di} and r_{Ti} are sensitivities of transactions $P_{Gi} (i \neq 1)$, P_{Di} and P_{Ti} with respect to slack bus power P_{G1} in the normal state.

If for any postulated contingency additional alterations to transmission losses are caused by post-contingency corrections the linear power balance equation under contingency can be written as

$$\Delta P_{G1}^k - \sum_{i \neq 1; i \in NG} r_{Gi}^k \cdot \Delta P_{Gi}^k - \sum_{i \in ND} r_{Di}^k \cdot \Delta P_{Di}^k - \sum_{i \in NT} r_{Ti}^k \cdot \Delta P_{Ti}^k = 0 \qquad k \in K \tag{6.12}$$

where r_{Gi}^k, r_{Di}^k and r_{Ti}^k are sensitivities of transactions $P_{Gi} (i \neq 1)$, P_{Di} and P_{Ti} with respect to slack bus power P_{G1} under contingency k.

(4) Limits on changes of line flow: The flow of power in a line can be denoted by the square of the current in the line and since current cannot be allowed to exceed certain limits,

$$0 \leq I_l^2 \leq (I_{l,\max})^2 \quad l \in L$$

The changes of line flow in the normal state must satisfy

$$\Delta(I_l^2) \geq -I_l^2 \quad l \in L \tag{6.13}$$

$$\Delta(I_l^2) \leq (I_{l,\max})^2 - I_l^2 \quad l \in L \tag{6.14}$$

where L is the total number of lines in the network and ΔI_l^2 can be expressed as a linear function in terms of a multitude of electricity transactions:

$$\Delta(I_l^2) = \sum_{i \neq 1; i \in NG} s_{Gi} \cdot \Delta P_{Gi} + \sum_{i \in ND} s_{Di} \cdot \Delta P_{Di} + \sum_{i \in NT} s_{Ti} \cdot \Delta P_{Ti}$$

where s_{Gi}, s_{Di} and s_{Ti} are sensitivities of the corresponding transactions to the square of the current in line l in the normal state.

Similarly, the changes of line flow under contingencies must satisfy

$$\Delta[(I_l^k)^2] \geq -(I_l^k)^2 \quad l \in L; k \in K; l \neq k \tag{6.15}$$

$$\Delta[(I_l^k)^2] \leq (I_{l,\max}^k)^2 - (I_l^k)^2 \quad l \in L; k \in K; l \neq k \tag{6.16}$$

where

$$\Delta[(I_l^k)^2] = \sum_{i \neq l; i \in NG} s_{Gi}^k \cdot (\Delta P_{Gi} + \Delta P_{Gi}^k) + \sum_{i \in ND} s_{Di}^k \cdot (\Delta P_{Di} + \Delta P_{Di}^k) + \sum_{i \in NT} s_{Ti}^k \cdot (\Delta P_{Ti} + \Delta P_{Ti}^k) \quad l \neq k$$

where s_{Gi}^k, s_{Di}^k and s_{Ti}^k are sensitivities of the corresponding transactions to the square of the current in line l under contingency k.

(5) Limits on the range of corrections: Whenever necessary upper and lower bounds, based especially on practical concerns, may need to be placed on all variables as follows:

$$\Delta P_{Gi,\min}^k \leq \Delta P_{Gi}^k \leq \Delta P_{Gi,\max}^k \quad i \in NG; k \in K \tag{6.17}$$

$$\Delta P_{Di,\min}^k \leq \Delta P_{Di}^k \leq \Delta P_{Di,\max}^k \quad i \in ND; k \in K \tag{6.18}$$

$$\Delta P_{Ti,\min}^k \leq \Delta P_{Ti}^k \leq \Delta P_{Ti,\max}^k \quad i \in NT; k \in K \tag{6.19}$$

$$\Delta P_{Gi,\min} \leq \Delta P_{Gi} + \Delta P_{Gi}^k \leq \Delta P_{Gi,\max} \quad i \in NG; k \in K \tag{6.20}$$

$$\Delta P_{Di,\min} \leq \Delta P_{Di} + \Delta P_{Di}^k \leq \Delta P_{Di,\max} \quad i \in ND; k \in K \tag{6.21}$$

$$\Delta P_{Ti,\min} \leq \Delta P_{Ti} + \Delta P_{Ti}^k \leq \Delta P_{Ti,\max} \quad i \in NT; k \in K \tag{6.22}$$

If the post-contingency rescheduling limits (with superscript k) are set to equal zero, the proposed problem becomes identical with the 'pure' preventive control, i.e. the conventional security-constrained dispatch, problem [31,32]. If the limits are set sufficiently large, contingency conditions become independent of the normal state and no preventive action needs to be taken.

The proposed multi-objective optimal rescheduling approach therefore consists of objective (6.8) and constraints (6.9-6.22), that can be easily transformed into a linear programming problem and then solved by any appropriate linear programming method.

6.7.5 Dynamic Security-constrained Rescheduling

The previous section has considered only the static aspect of security-constrained rescheduling in the deregulated power system where security is measured with respect to steady-state operation. However, there are many situations where dynamic constraints such as stability limits are reached before steady-state constraints. If a severe contingency, such as a line outage, could cause generators to lose synchronism, then the operator must take preventive action to modify the operating state. Hence, dynamic security-constrained rescheduling in the context of both pool and bilateral dispatch is also a very important issue.

An operating state can be modified in many different ways, and the operator must choose the action which will not only ensure system stability, but also achieve commercial equitability in an open-access environment. A transient energy function (TEF) approach is described in this section. The TEF is a Lyapunov-like function and the controlling unstable equilibrium point (UEP) for a particular fault is the most critical point on a boundary surface in the state space of generator angles. The transient energy margin (TEM), denoted by EM, is the difference between the transient energy of the system of generators at the instant of fault clearing and its value at the controlling UEP. The state space is that corresponding to the final post-disturbance system and topology. If EM is positive, i.e. if the transient energy is less than the potential energy corresponding to the controlling UEP, the system possesses transient stability for the fault in question.

The chief attraction of a TEM method is that it lends itself very conveniently to a sensitivity-based approach [34]. The sensitivities of greatest significance are the changes of EM with respect to individual generator power outputs. In the event that a particular dispatch configuration possesses dynamic security risks, the most immediate solution available to the ISO is to redispatch generator power outputs so as to make EM larger. Sensitivity information provides a clear signal of the most advantageous redispatch strategies. The numerical values of the sensitivities can be obtained either analytically or numerically by repeating time simulations of the transient disturbance, and the latter is utilised here. The approach described here is to use the sensitivity of EM with respect to the change in generation from critical generators to non-critical generators.

Several TEF-based methods are available, including the potential energy boundary surface (PEBS) method [35], the controlling unstable equilibrium point method [36], the extended equal area criterion (EEAC) method [37] and the hybrid method [38]. These methods can be used to compute the UEP, and thus EM for a given fault-clearing time t_{cl}, and the sensitivities can be obtained then. A method [39] to implement dynamic security-constrained rescheduling is presented next in which the hybrid method is used because it possesses many advantages [40].

Transient Energy Margin (TEM)

The equations of motion in the classical model [41] of n synchronous generators in the centre of inertia (COI) frame of reference are

$$M_i \dot{\tilde{\omega}}_i = P_i \{= P_{mi} - E_i^2 G_{ii}\} - P_{ei} \{= \sum_{\substack{j=1 \\ j \neq i}}^{n} [C_{ij} \sin(\theta_i - \theta_j) + D_{ij} \cos(\theta_i - \theta_j)]\} - \frac{M_i}{M_T} P_{COI}$$

$$\dot{\theta}_i = \tilde{\omega}_i$$

And, for generator i,
P_{mi} = mechanical power input
G_{ii} = driving point conductance
E_i = constant voltage behind direct axis transient reactance

where $M_T = \sum_{i=1}^{n} M_i$, $P_{COI} = \sum_{i=1}^{n}(P_i - P_{ei})$ and $C_{ij} = E_i E_j B_{ij}$, $D_{ij} = E_i E_j G_{ij}$

ω_i, θ_i = generator rotor speed and angle, respectively, w.r.t. COI frame of reference
M_i = moment of inertia
$B_{ij}(G_{ij})$ = transfer susceptance (conductance) in the reduced Y-matrix

For the above formulation, the TEM is given by

$$EM = -\frac{1}{2} M_{eq} \widetilde{\omega}_{eq}^{cl^2} - \sum_{i=1}^{n} P_i^{pf}(\theta_i^u - \theta_i^{cl}) - \sum_{i=1}^{n-1} \sum_{j=i+1}^{n} [C_{ij}^{pf}(\cos\theta_{ij}^u - \cos\theta_{ij}^{cl}) - \beta_{ij} D_{ij}^{pf}(\sin\theta_{ij}^u - \sin\theta_{ij}^{cl})] \quad (6.23)$$

where
β_{ij} = $[\theta_i^u + \theta_j^u - \theta_i^{cl} - \theta_j^{cl}]/(\theta_{ij}^u - \theta_{ij}^{cl})$
θ^{cl} = rotor angle positions at the end of the disturbance
θ^u = controlling UEP
M_{eq} = $M_{cr} M_{sys}/(M_{cr} + M_{sys})$
$\widetilde{\omega}_{eq}^{cl}$ = $(\widetilde{\omega}_{cr}^{cl} - \widetilde{\omega}_{sys}^{cl})$
M_{cr}, M_{sys} = inertia constants of the critical generators and the remaining generators, respectively
$\widetilde{\omega}_{cr}^{cl}, \widetilde{\omega}_{sys}^{cl}$ = speed of inertial centres of the critical generators and the remaining generators, respectively, at the instant when the fault is cleared.

Superscript ^{pf} stands for the values of variables in the final post-fault system configuration. The approximate controlling unstable equilibrium point (θ^u) is calculated using a hybrid method which is used as starting point for solving the post-fault system equations. Derivation of these results can be found in [42].

Sensitivity Analysis
For a given contingency and fault-clearing time, the initial energy margin, EM^0 is calculated. If EM^0 is positive, the system is stable and no further action is required in respect of that contingency. However, if EM^0 is negative then some corrective action must be taken to ensure system security. The sensitivity $\eta_{i \to j}$ is defined as the ratio of the change in system EM to a shift in real power generation from generator i to generator j while continuing to satisfy system demand in full. Therefore, it may be defined by

$$\eta_{i \to j} = \frac{\Delta EM}{\Delta P_{i \to j}}$$

where $\Delta EM = EM^{new} - EM^0$. The sign of $\eta_{i \to j}$ indicates the direction in which generation is to be shifted to enhance the EM. The magnitude of sensitivity corresponding to the change in the generation from the most advanced critical generator to the least advanced (non-critical) generators will be high and, hence, the best candidate for the rescheduling. The power generation of the critical and non-critical generators should be set as

$$P_i = P_i^0 + \frac{EM^0}{\eta_{i \to j}} \quad \text{and} \quad P_j = P_j^0 - \frac{EM^0}{\eta_{i \to j}}$$

It was observed from several simulation runs that the change in the EM with respect to the change in the generated power from a critical generator to a non-critical one is approximately linear for reasonable critical clearing time, t_{cl}, assumptions (practically acceptable value). This fact is also reported in some other studies [43]. There is a possibility that the required change in power from any one generator to another generator will be larger than operating restrictions on the machine will allow. That is to say, if it is found that, for generator i (critical) and j (non-critical), the following limitations occur

$$-\frac{EM^0}{\eta_{i \to j}} \geq P_i^0 - P_i^{min} \quad \text{or} \quad -\frac{EM^0}{\eta_{i \to j}} \leq P_j^{max} - P_j^0$$

(note that EM^0 is negative in the unstable case):

where P^0 is the generator base loading at which the system energy margin has been calculated; then the maximum or minimum power limit of the critical or non-critical generators P_i^{min} or P_j^{max} would limit the permissible rescheduling between these generators.

If a limit of this type is reached, a new energy margin (EM^n) can be calculated after utilising the largest possible change in power between i and j. Starting with this new EM, further rescheduling between other pairs of generators can be undertaken.

It has been observed that the change in the sensitivity with small change in generator loading is small. For a shift in generation from one critical to non-critical generator pair, the sensitivity for another pair of critical and non-critical generators is reduced slightly. The rescheduling with base case sensitivity will give a little over-correction but the system will still be in a transient stable mode.

The discussion up to this point has not made a distinction between a restructured open-access type power system and a conventional system. The difference arises in the way in which the restructuring is selected. Sensitivity is not the only criterion, the other is cost. Making the i-j pair which has the largest sensitivity multiplied by the cost differential between the two generators, the first choice for rescheduling, ensures that the system is rescheduled at minimal additional cost. This consideration too is applicable to the traditional system, but by supplementing the notion of cost by definitions such as the willingness-to-pay to avoid rescheduling, price information of bilateral transactions (ρ_{Ti} discussed in Section 6.7.4) or other yardsticks by which to weight market signals, a methodology for dynamic rescheduling in open-access systems becomes available.

Solution Methodology
The approximate UEP for each case is estimated using the hybrid method to calculate the exact UEP (θ^u). For a set of critical contingencies, EM is then calculated using (6.23). The algorithm provides a method for security evaluation and preventive control, and can be summarised as follows:

a. Choose a contingency from the given set.
b. Obtain the optimal dispatch using equation (6.2) (ensuring post-fault system static security is a part of this problem).
c. Compute θ^u and corresponding EM (equation (6.23)).
d. If EM is positive, then go to step a and select the next contingency. If EM is negative, go to step e.
e. Compute the numerical sensitivities, $\eta_{i \to j}$, for the set of system generators i and j as discussed above. For computation of new energy margin for change in real power generation from generator i to j, the base loading θ^u has been used as the starting point to get the new UEP corresponding to the change in the generation. It is found that 1-5 % perturbation in power gives the best result.
f. Weight the sensitivities by market price signals; compute the constraints on generators; redispatch the generators in pairs as described.
g. If all critical contingencies are tested, then stop. Otherwise go to step a.

Detailed numerical studies reveal that:

- The spot price of critical/non-critical generator buses is reduced/increased significantly after scheduling.
- Dynamic security increases spot prices at the buyer's bus but the increase is not uniform.
- The presence of bilateral contracts further increases the spot prices at the buyer's bus.

6.8 Open-access Coordination Strategies

In practice, congestion management when systems become more mature and more experience is gained in operating strategies will entail a more complex operating process than discussed so far. Hence, three additional issues in relation to congestion management are now presented.

6.8.1 Price Elasticity as a Means to Relieve Congestion

Some important points concerning the demand price elasticity issue [23,24,44], which may be used for congestion relief, are discussed next.

Demand at any time is treated as a function not only of contemporaneous price but also of prices at other times. Indeed, the inter temporal price dependence of demand lies at the heart of all dynamic tariff theory and is central to pool dispatch. The following definitions of elasticity are introduced:

$$e_{tt.j} = \frac{\partial D_{t.j}(p)}{\partial p_t} \qquad e_{tt'.j} = \frac{\partial D_{t.j}(p)}{\partial p_{t'}} \qquad (6.24)$$

where $e_{tt} \leq 0$ is the conventional price elasticity of demand and $e_{tt'} \geq 0, t \neq t'$, is the cross-time price elasticity of demand. The additional subscript j on e and D, the price elasticity and the pool demand, denotes each purchaser separately. p_t is an element of a time-dependent price vector p (for example 24 hourly pool prices).

The following Kuhn-Tucker conditions can be derived from the pool dispatch problem,

$$\frac{dC_i}{dG_{it}} = -\frac{\partial L_t}{\partial G_{ti}} \cdot \Lambda_t - \mu_{ti} - \frac{\partial Z_t}{\partial G_{ti}} \cdot \Pi_t \quad : \forall i, t \qquad (6.25)$$

$$e_t \cdot p = \sum_{\tau} e_{t\tau} \left(\frac{\partial L_\tau}{\partial D_t} \cdot \Lambda_\tau + \frac{\partial Z_\tau}{\partial D_t} \cdot \Pi_\tau \right) \quad : \forall t$$

where C and G are producer offer price and power output, respectively; Λ_t and Π_t are sets of matrices of dual variables, at time t, on the set of power flow equations L_t and the set of system operation constraints Z_t, respectively. $e_{t\tau}$ and e_t can be described, respectively, as

$$e_{t\tau} = [e_{t\tau.1}, e_{t\tau.2}, ..., e_{t\tau.J}]$$

$$e_t = [(\sum_j e_{1t.j}), (\sum_j e_{2t.j}), ..., (\sum_j e_{Tt.j})]$$

where J is the number of pool loads (load busbars).

The dispatch procedure would begin with a forecast of the day's prices, say at half-hourly intervals, and the corresponding expected set of demand vectors. The solution of the pool dispatch problem as well as (6.25) will provide an alternative set of prices corresponding to purchaser expectations. If these match the initially assumed price vector the dispatch problem is solved. If they do not, the expected customer demands are modified using the elasticities in (6.24) and the price deviations and the dispatch problem solved again. The procedure is repeated till convergence obtains. Detailed examples of the procedure are provided in [23].

6.8.2 Relieving Congestion by ISO Executed Price Signalling

Reference [44] presented an approach to use nodal congestion price signals to let power markets attain a socially optimum operating point in case of congestion. In a vertically integrated electric utility scenario, congestion is avoided by using an OPF tool to readjust power generation. This is based on the premise that the system operator knows exactly the marginal costs of production for each unit in the system.

However, in an open market, only bids, not costs, are known. The ISO has to estimate the true 'costs' by observing the actual or proposed behaviour of the various interacting parties and then send an extra cost signal to power producers having the character of costs to rearrange the power injections such that congestion is avoided. Consider a lossless network with I_G generators. Let the cost of generator i be

$$C_i = a_i + b_i P_i + \frac{1}{2} c_i (P_i)^2$$

Optimality under a normal condition is determined by defining a Lagrangian λ and finding its derivative with respect to all the variables:

$$b + C \cdot P = I \cdot \lambda$$
$$I^T \cdot P = P_D \qquad (6.26)$$

where P is a vector of generator powers, b and C are a vector and a diagonal matrix containing the cost coefficients, respectively, and P_D represents total system demand. I is a column vector of dimension I_G whose elements are all 1.

Similarly, optimality under congestion is determined as follows:

$$b + C \cdot P = I \cdot \lambda - s^T \cdot \mu \qquad (6.27)$$
$$I^T \cdot P = P_D$$
$$s \cdot P \leq P_{l,max} \qquad \mu \geq 0$$

where s is the matrix $(n_l \times n_b)$ of sensitivities of branch flows to bus injections, n_l is the number of binding congested branches, n_b is the number of buses in the system (excluding the reference bus), μ is a vector of the dual variables on the inequality constraints, and $P_{l,max}$ is a vector of maximum feasible flows in the binding lines.

Assuming that the cost coefficients remain constant, since changes in generation pattern ΔP can be observed in the power market, the authors of [44] suggest that the true costs can then be estimated based on the knowledge of a solution of (6.26) – (6.27). Suggested estimation procedures can be found in [44] and require at least one iteration between the suppliers' bids and the ISO proposed transmission charges. We call these estimates 'pseudo-costs'.

Once the pseudo-costs have been obtained, the ISO is able to induce the expected optimal congestion-relieving behaviour by issuing an additive price vector β^* obtained from a solution of the following equations:

$$b + C \cdot P^{new} = I \cdot \lambda - s^T \cdot \mu$$
$$I^T \cdot P^{new} = P_D$$
$$s \cdot P^{new} \leq P_{l,max} \qquad \mu \geq 0$$

and then setting $\beta^* = s^T \cdot \mu$. Both the solution P^{new} and the eventual market response to the additions β^* are identical to the optimal operating point P^*.

6.8.3 Coordination between Transactions

Coordination based on power flow sensitivities, such as loading sensitivities, is discussed here. The mechanism of coordination can be illustrated diagrammatically as follows.

Let the initial operating point be represented by point A in Figure 6.11, which is drawn in the state space of system voltages and power angles. In general one or more operating constraints (voltage limits, line loadings, etc.) will be binding and there will be feasible and infeasible directions in which the operating state can move if additional power injections and deliveries are undertaken. If the scalar product of the direction change vector and the normal vector to the subspace generated by the intersections of the surfaces generated by the set of binding constraints is negative the change vector is in a feasible direction.

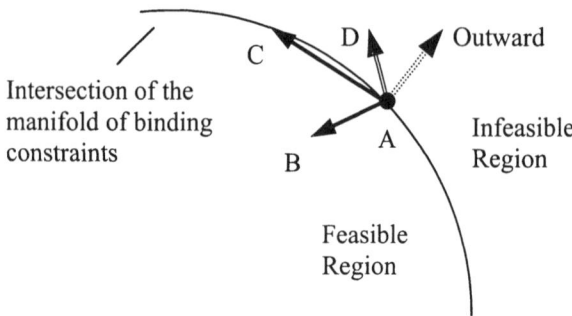

Figure 6.11 Feasible (B, C) and infeasible (D) power trades from an initial operating point A

A more compact notation is now introduced. Let $T = [T_1^T,...,T_K^T]$ be a K-vector of transactions, including pool, bilateral and multilateral transactions. The elements of the column vector $T_k, k \in K$, are real or reactive power injected or taken under contracts at buses of the power system and the formulation of nodal power, whose real part was given in (6.5), can be written as

$$T \cdot I^T = [P(A), Q(A)]^T \tag{6.28}$$

where I is a vector of ones. Argument A is an optimum operating state at some time interval. The elements of $Q(A)$ are assumed to be set by the ISO for secure transmission operation.

Let load flow equations and inequality constraints of the previous optimal dispatch problem (6.6) be rewritten as follows:

$$L(A) = [P(A), Q(A)]^T \tag{6.29}$$

$$Z(A) \leq 0 \tag{6.30}$$

It is now possible to obtain

$$\nabla_{V,\theta} L(A) \cdot [\Delta V, \Delta \theta]^T = [\Delta P, \Delta Q]^T \tag{6.31}$$

$$\nabla_{V,\theta} Z(A) \cdot [\Delta V, \Delta \theta]^T = \Delta Z(A) \tag{6.32}$$

and from (6.28), (6.31) and (6.32)

$$N_Z = \nabla_{V,\theta} Z(A) \cdot [\nabla_{V,\theta} L(A)]^{-1} \tag{6.33}$$

$$N_Z \cdot \Delta T \cdot I^T = \Delta Z(A) \tag{6.34}$$

where N_z is the matrix whose elements are sensitivity relations between P. and $Z(A)$ and when (6.30) remains true for a set of incremental transactions ΔT, this also satisfies

$$N_z \cdot \Delta T \cdot I^T \leq C^T \quad (6.35)$$

where the elements of vector C are constants, larger than or equal to zero, whose values depend on the state A. Inequality (6.35) is the condition that an additional transaction set ΔT proposed at state A, for the next interval, is in a feasible direction. Reference [25] provides a more detailed derivation and reference [9] the initial linear formulation on which the method is based.

Now let $\widetilde{Z}(A) = 0$ be the subset of $Z(A)$ of constraints in state A, which are on limit. If we focus attention on these problem areas, then the necessary condition is

$$N_{\widetilde{z}} \cdot \Delta T \cdot I^T \leq 0 \quad (6.36)$$

or

$$\sum_{k \in K} N_{\widetilde{z}m} \cdot \Delta T_k^T \leq 0 \ \forall m \quad (6.37)$$

where $N_{\widetilde{z}m}$ is the mth row vector of $N_{\widetilde{z}}$. Expression (6.37) provides the necessary coordination relations between various extra transactions $\Delta T_k, k \in K$, to avoid violating binding inequality constraint $\widetilde{Z}_m(A)$.

Now consider just one additional transaction $\Delta T_{k'} = [\Delta P_{k'1}, \cdots \Delta P_{k'N}, \Delta Q_{k'1}, \cdots, \Delta Q_{k'N}]^T$ where the $\Delta P, \Delta Q$ values refer only to the buses belonging to this group (k'). Suppose all others remain unchanged, namely $\Delta T_k = 0; k \in K, k \neq k'$, then relation (6.38) follows from (6.37),

$$N_{\widetilde{z}m} \cdot \Delta T_l^T \leq 0 \forall m \quad (6.38)$$

Inequality (6.38) reflects the coordination relations within a single transaction in the congested condition. That is, power inputs and outputs within a group can be coordinated to make $\Delta T_{k'}$ feasible. By extension of this argument coordination between different transactions is also possible.

Expressions (6.37) and (6.38) merely formulate the possibility of coordination from a mathematical viewpoint. In practice the transactions ΔT will be left to market participants and scheduling coordinators to negotiate while the responsibility of the ISO will be to broadcast the information in $N_{\widetilde{z}}$.

6.8.4 Illustration of Transaction Coordination

The example system considered here is the same as Figure 6.9 and system parameters remain unchanged except that line 2-3 now has its square of the current limit reduced to 0.25 pu. This system with the transaction set shown as Fixed+Additional (in regular type) of Table 6.2 is assumed to be initially in operation.

Assume now that an additional demand of 70 MW at bus 5 is required and that the Genco at bus 3, for commercial reasons, is responsible for supply. This extra request, represented by transaction T_A (70 MW initially from bus 3 to 5), will be submitted to the ISO for dispatch. How coordination between market participants and the ISO can significantly affect dispatch is now illustrated. The existing pool demands at buses 4 and 5 and the bilateral transaction from 3 to 4 are given priority in the following cases because they are assumed to have obtained prior commitments from the ISO.

Case 1: Bus 3 Genco supplies the additional transfer T_A alone and only bus 1 is marginally redispatched to make up the transmission losses caused by the delivery of T_A.

Case 2: Bus 3 Genco invites bus 2 Genco to join in the transaction T_A. Bus 1 is still responsible for making good extra losses.

Case 3: Bus 3 Genco is willing not only to coordinate with bus 2 Genco but also to coordinate with pool transactions through the ISO.

Optimal dispatch results of cases 1-3 are given in Table 6.2. Case 1 is the 'non-coordination' case and new transaction T_A, which is bilateral (from bus 3 to 5) in this case, is heavily curtailed. Case 2 shows coordination within transaction T_A, which now becomes multilateral since both bus 3 and 2 supply the extra load at bus 5. Since the loading sensitivity of line 2-3 with respect to the transfer from bus 2 to 5 is -0.194 p.u./p.u. while the sensitivity with respect to the transfer from 3 to 5 is 0.474 p.u./p.u., it is possible to introduce some coordination (equation (6.38)) where both transfers are simultaneously increased without causing violation of power flow on the problem line 2-3. Compared with case 1, not only is the transfer from 3 to 5 curtailed less in case 2 but also the additional request of 70 MW transfer is satisfied in full. That is, transaction T_A is completely dispatched in case 2.

A more complex type of coordination is illustrated in case 3. When the additional 70 MW transaction is included line 4-5 is found to be close to its capacity limit. However, increasing the transfers from 2 to 5 and from 3 to 5 ameliorates the loading of line 4-5. Therefore more expensive power at bus 1 can be shifted to bus 2 and the more the power shifted to bus 2, the less the loading of problem line 2-3 as well. The optimal strategy for bus 3 Genco is to provide a smaller share (21 MW) to bus 2 Genco than case 2 without causing curtailment to transaction T_A. Overall benefits are improved.

Table 6.2 Optimal results with/without coordination

Transaction (MW)		Fixed+ Additional	Case 1	Case 2	Case 3
Pool gen. at bus1		38.8	39.8 ↑	40.6 ↑	29.2 ↓
Pool gen. at bus2		75.0	75.0	75.0	86.8 ↑
Pool demand at bus4		100.0	100.0	100.0	100.0
Pool demand at bus5		10.0	10.0	10.0	10.0
Bilateral from 3 to 4		80.0	80.0	80.0	80.0
Multi-	from 3 to 5	70.0	38.1	47.3	49.0
Lateral	from 2 to 5	--	--	22.7	21.0

-- No data

Obviously different versions of coordination result in different outcomes. Coordinated approaches towards transmission system operation in power markets working in an open transmission scenario are necessary and developing techniques for power market administration, therefore, are very desirable.

6.8.5 Integrated Coordination Procedure

The demand price elasticity issue discussed in Section 6.8.1 is based on the assumption that market participants are perfectly rational and able to respond properly and accurately to price signals within appropriate time frames. However, demand elasticities of customers range from highly elastic to highly inelastic. Customers with high elasticities will be very sensitive to power prices while customers who are more inelastic will be inert to power prices and fail to react in time.

Transaction coordination as demonstrated above explores an alternative approach to ameliorate congestion and provides useful guidelines for market participants to negotiate power exchanges that avoid congestion and learn to depend on geographically diversified power supply (and demand) portfolios instead.

An integrated congestion management procedure [45] as shown in Figure 6.12, which incorporates the above two issues into the willingness-to-pay dispatch strategy, is briefly described.

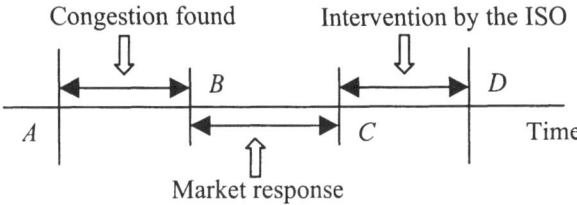

Figure 6.12 Integrated coordination procedure for congestion relief

Let AD represent a short time period, which is divided into three intervals, i.e. AB, BC and CD. (These three intervals need not be equal to each other.) Assume that in actual system operation congestion is found during interval AB; the ISO then broadcasts price information, which may be higher than usual, and information about the transmission operating status including $N_{\tilde{z}}$, and grants a time BC for market response. Some participants will adjust their transactions during BC responding to the price signals. Others may take advantage of the sensitivity signal to make additional contracts. If congestion is relieved owing to market reaction taken at BC, the ISO directly transmits all submitted transactions; however, if not, the ISO has to use the willingness-to-pay dispatch strategy (6.7) during interval CD to implement curtailment and relieve congestion.

In normal conditions (i.e. no AB), the procedure is simpler. The ISO continuously broadcasts information and then occasionally enhances prices when potential congestion is foreseen (BC). ISO interference (CD) is used when the response is inadequate. This procedure must iterate in the whole transmission operation time domain.

This integrated coordination procedure can be seen as a practical alternative to capacity rights, which are defined in a pairwise manner [46] or as rights on individual lines [47].

6.9 Conclusions

Restructuring of the power industry has received much attention around the world in the last two decades and up to now about 20 countries have restructured their systems and many others are actively pursuing similar paths. One of the most important issues in restructuring efforts is the transmission open access and this has been the theme of this chapter.

Firstly, this chapter describes characteristics of the traditional power industry, some motivations, which have led to marked changes in the institutional structures of the power industry, and the necessity of open transmission access to facilitate these changes. Secondly, a summary discussion of electricity market structures and associated structural components has been provided. The functions and responsibilities of the PX and ISO in open-market systems have been described and discussed. The economic issues associated with the transmission open access, i.e. costs of transmission services and the related pricing, were next presented briefly. Finally, this chapter provides a discussion of some important operational issues in the emerging market environment. Normal dispatch, congestion management and effects of security considerations on congestion management have all been discussed from the open-access viewpoint, and these sections represent the focuses of this chapter. Most of the discussion is based on current research work worldwide and some topics have been illustrated using simple examples.

6.10 Acknowledgements

The work of former PhD candidates at the Hong Kong Polytechnic University, Dr R.S. Fang (Section 6.7) and Dr C.W. Yu (Sections 6.5.2 and 6.5.4) and assistance from Dr S.N. Singh, Assistant Professor at Roorkee University, India, formerly research Fellow at the Hong Kong Polytechnic University, in preparing Section 6.7.5 are acknowledged. Financial support from the Research Grants Council (RGC), University Grants Committee (UGC), Hong Kong and the Hong Kong Polytechnic University for much of the research is gratefully acknowledged.

6.11 Appendix

System data used in the numerical example of subsection 6.7.3 are listed in Tables A-1 and A-2 below.

Table A-1 Pool and bilateral transactions

Bid generation	Rating MW	Bid price $/h *
$P_{PL,1}$	100.0	$6.0 P_{PL,1} + 0.06 P_{PL,1}^2$
$P_{PL,2}$	200.0	$3.0 P_{PL,2} + 0.03 P_{PL,2}^2$

Pool Load	Demand MW	Demand price $/h	Willingness(w) $/MW^{2}h
$D_{PL,4}$	100.0	$9.0 D_{PL,4}$	20.0
$D_{PL,5}$	80.0	$10.0 D_{PL,5}$	20.0

Bilateral Contract	Transfer MW	Delivery Price $/h	Willingness (w) $/MW^{2}h
T_1	100.0	$4.0 P_{T_1,3}$	80.0

* An arbitrary unit of currency.

Table A-2 Regulated operating conditions

Bus number	Voltage magnitude (pu)	Reactive power (MVAr)
1	1.02	--
2	1.04	--
3	1.05	--
4	**	20.0
5	**	20.0

** Voltages are kept within the range of 0.95-1.05.
-- No data.

6.12 References

[1] L. Philipson and H.L. Willis, *Understanding Electric Utilities and De-regulation*, Marcel Dekker Inc., 1998.

[2] A.K. David, 'Restructuring the electricity supply industry in Asia', *International Journal of Global Energy Issues*, Vol.10, Nos.2-4, 1998, pp.203-212.

[3] A.F. Vojdani and F.A. Rahimi, 'Electricity market structures', EPSOM'98, Zurich, September 1998.

[4] F.A. Rahimi and A.F. Vojdani, 'Meet the emerging transmission market segments', *IEEE Computer Applications in Power*, Vol.12, No.1, January 1999, pp.26-32.

[5] P. Varaiya and F.F. Wu, 'MinISO: A minimal independent system operator', *Proceedings of 30th Annual Hawaii International Conference on System Sciences*, January 1997, pp.7-10.

[6] R.J. Thomas and T.R. Schneider, 'Underlying technical issues in electricity deregulation', *Proceedings of 30th Annual Hawaii International Conference on System Sciences*, January 1997.

[7] D. Shirmohammadi, B. Wollenberg, A. Vojdani, P. Sandrin, M. Pereira, F. Rahimi, T. Schneider and B. Stott, 'Transmission dispatch and congestion management in the emerging energy market structures', *IEEE Transactions on Power Systems*, Vol.13, No.4, 1998, pp.1466-1474.

[8] F. Albuyeh and Z. Alaywan, 'California ISO formation and implementation', *IEEE Computer Applications in Power*, Vol.12, No.4, October 1999, pp.30-34.

[9] F.F. Wu and P. Varaiya, 'Coordinated Multilateral trades for electric power networks', *International Journal of Electrical Power and Energy Systems*, Vol.21, 1999, pp.75-102.

[10] D. Shirmohammadi, X.V. Filho, B. Gorenstin and M.V.P. Pereira, 'Some fundamental technical concepts about cost based transmission pricing', *IEEE Transactions on Power Systems*, Vol.11, No.2, 1996, pp.1002-1008.

[11] F.C. Schweppe, M.C. Caramanis, R.D. Tabors and R.E. Bohn, *Spot Pricing of Electricity*, Kluwer Academic Publishers, 1988.

[12] Y.Z. Li and A.K. David, 'Pricing reactive power conveyance', *IEE Proceedings-C*, Vol.140, No.3, 1993, pp.174-180.

[13] M.C. Calviou, R.M. Dunnett and P.H. Plumptre, 'Charging for use of a transmission system by marginal cost methods', *Proceedings of 11th Power Systems Computations Conference*, Avignon, France, 1993, pp.385-391.

[14] C.W. Yu and A.K. David, 'Pricing transmission services in the context of industry deregulation', *IEEE Transactions on Power Systems*, Vol.12, 1997, pp.503-510.

[15] J. Endrenyi, *Reliability Modelling of Electric Power Systems*, John Wiley & Sons, 1978.

[16] R. Billinton, 'Composite system reliability evaluation', *IEEE Transactions on Power Apparatus & Systems*, Vol.88, No.4, 1969, pp.276-281.

[17] S. Hunt and G. Shuttleworth, *Competition and Choice in Electricity*, John Wiley & Sons, Chichester, UK, 1996.

[18] B. Kirby and E. Hirst, 'Unbundling electricity ancillary services', *IEEE Power Engineering Review*, June 1996, pp.5-6.

[19] S. Hao and A. Papalexopoulos, 'Reactive power pricing and management', *IEEE Transactions on Power Systems*, Vol.12, No.1, 1997, pp.95-104.

[20] H. Singh, 'Auctions for ancillary services', *Decision Support Systems*, Vol.24, No.3/4, 1999, pp.183-191.

[21] R.D. Christie, B.F. Wollenberg and I. Wangensteen, 'Transmission management in the deregulated environment', *Proceedings of the IEEE*, Vol.88, No.2, 2000, pp.170-195.

[22] R.S. Fang, Transmission dispatch and congestion management in open market systems, PhD Thesis, Hong Kong Polytechnic University, 1999.

[23] A.K. David and Y.Z. Li, 'Electricity pricing with competitive supply conditions', *International Journal of Electrical Power & Energy Systems*, Vol.13, No.2, April 1991, pp.111-122.

[24] A.K. David and Y.Z. Li, 'Effect of inter-temporal factors on the real time pricing of electricity', *IEEE Transactions on Power Systems*, Vol.8, No.1, February 1993, pp.44-52.

[25] A.K. David, 'Dispatch methodologies for open access transmission systems', *IEEE Transactions on Power Systems*, Vol.13, No.1, February 1998, pp.46-53.

[26] A.K. David, 'Reconciling pool and contract dispatch in open access transmission operation', *IEE Proceedings - Generation, Transmission and Distribution*, Vol.145, No.4, July 1998, pp.468-472.

[27] R.S. Fang and A.K. David, 'Optimal dispatch under transmission contracts', *IEEE Transactions on Power Systems*, Vol.14, No.2, May 1999, pp.732-737.

[28] A. Zobian and M.D. Ilic, 'Unbundling of transmission and ancillary services Part I: Technical issues', *IEEE Transactions on Power Systems*, Vol.12, No.2, May 1997, pp.539-548.

[29] R.S. Fang and A.K. David, 'Transmission congestion management in an electricity market', *IEEE Transactions on Power Systems*, Vol.14, No.3, August 1999, pp.877-883.

[30] A.K. David and R.S. Fang, 'Security-based rescheduling of transactions in a deregulated power system', *IEE Proceedings - Generation, Transmission and Distribution*, Vol.146, No.1, January 1999, pp.13-18.

[31] J.C. Kaltenbach and L.P. Hajdu, 'Optimal corrective re-scheduling for power system security', *IEEE Transactions on Power Apparatus and Systems*, Vol.PAS-90, No.2, 1971, pp.843-851.

[32] A. Thanikachalam and J.R. Tudor, 'Optimal rescheduling of power for system reliability', *IEEE Transactions on Power Apparatus and Systems*, Vol.PAS 71, 1971, pp.2186-2192.

[33] A. Monticelli, M.V.F. Pereira and S. Granville, 'Security-constrained optimal power flow with post-contingency corrective rescheduling', *IEEE Transactions on Power Systems*, Vol.2, No.1, February 1987, pp.175-182.

[34] A.K. David and R.S. Fang, 'Significant operational issues in open access systems', *Australian Research Council Workshop*, paper presented at University of Western Australia, July 1998.

[35] N. Kakimoto, Y. Ohsawa and M. Hayashi, 'Transient stability analysis of electric power system via Lure-type Lyapunov function, Part I and II', IEE Japan, Vol.98, 1978, pp.62-79.

[36] T. Athay, R. Podmure and S. Virmani, 'A practical method of direct analysis of transient stability', *IEEE Trans. on Power Apparatus and Systems*, Vol.98, 1979, pp.573-584.

[37] Y. Xue, T.V. Cutsen and M. Pavella, 'Real time analytic sensitivity method for transient security assessment and preventive control', *IEE Proceeding,* Part C, Vol.135, No.2, 1988.

[38] H.D. Chiang, F.F. Wu and P.P. Varaiya, 'Foundations of direct methods of power system stability analysis', *IEEE Transactions on Circuits and Systems*, Vol.34, No.2, 1987.

[39] S.N. Singh and A.K. David, 'Dynamic security constrained congestion management in competitive electricity market', *Proceedings of the IEEE PES 2000 Winter Meeting*, Singapore, January 2000.

[40] D.Z. Fang, On-line Algorithms for Transient Stability Assessment and Security Control, PhD Thesis, Hong Kong Polytechnic University, 1995.

[41] P.M. Anderson and A.A. Fouad, *Power System Control and Stability*, Iowa State University Press, 1977.

[42] P.W. Sauer and M.A. Pai, *Power System Dynamic and Stability*, Prentice Hall, New Jersey, 1998.

[43] J. Sterling, M.A. Pai and P.W. Sauer, 'A methodology of secure and optimal operation of a power system for dynamic contingencies', *Electric Machines and Power Systems*, Vol.19, 1991, pp.639-655.

[44] H. Glavitsch and F. Alvarado, 'Management of multiple congested conditions in unbundled operation of a power system', *IEEE Transactions on Power Systems*, Vol.13, No.3, August 1998, pp.1013-1019.

[45] R.S. Fang and A.K. David, 'An integrated congestion management strategy for real-time system operation', *IEEE Power Engineering Review*, Vol.19, No.5, May 1999, pp.52-54.

[46] W.W. Hogan, 'Contract networks for electric power transmission', *Journal of Regulatory Economics*, Vol.4, 1992, pp.211-242.

[47] H.P. Chao and S.C. Peck, 'A market mechanism for electric power transmission', *Journal of Regulatory Economics*, Vol.10, 1996, pp.25-59.

7

Electric Power Industry Restructuring in China

Prof. Xifan Wang
X'ian JiaoTong University
China

Dr Loi Lei Lai
City University London
UK

7.1 Introduction

Since China initiated its first economic reforms in late 1978, energy consumption has grown only half as quickly as the gross domestic product. No other developing country has decoupled the relationship between energy consumption and economic growth so radically [1,2]. Surprisingly, higher energy prices have played a key role in containing energy consumption only in the past five years or so. Before that strict but effective regulations helped hold energy demand in check, and until recently, electricity shortages constrained consumption. Now that the market largely determines energy prices and the transition from central planing has matured, economic forces are likely to keep China's income elasticity of energy demand low. While the market can successfully promote energy efficiency in China, additional reforms are still needed.

In power, price reforms began in the mid 1980s, when the government changed the financing mechanism for new power plants. Traditionally, plant construction had been financed with a combination of grants and subsidised loans from the central government, with some funding from provincial and local utilities. In 1985, the grants were abolished and utilities ordered to set tariffs to recover investments on a cost-plus basis. Prices for electric power varied considerably across provinces and even within small regions. The government recently established new policies to create a 'small grid, same price'; that is, all plants connected to the same regional grid must charge the same price for electricity.

Another problem in the power sector during the late 1990s has been regional overcapacity. Power purchase agreements, which define how much power the utility guarantees to buy from the power plant, are not being honoured. Electricity from plants is thus often dispatched according to marginal cost; that is, old plants usually sell the most power since they have already paid off their capital costs and need to cover only fuel, operation and maintenance costs.

Recently, the development of China's electric power industry has gone through a very important period; that is, the power industry is gradually changing from a monopolised industry to one with market economy characteristics and the dominating position between power supply and demand is also changing from the sale side to the buying side. Especially as the reform of the electric power industry goes on, the decision-making structure in the electricity sector is changing a great deal and will have a great influence on the future development of China's electric power industry.

Some price controls (subsidies) still remain in effect and the environmental costs of energy use are not yet even partially accounted for. These factors must be included in pricing schemes to promote the sustainable use of energy [3-6].

Better transparency and legal recourse will ensure that contracts are honoured, accounting is standardised, and the decision-marking authority is clear. Also, the full economic cost of pollution needs to be considered so that true 'least cost' decisions are made. China's economic efficiency and environmental quality depend on these additional reforms.

The electric power industry in China has already gone through many changes over the past decade. At present, its transition from a planned economy to a market economy is under way. On 16th January 1997, the Ministry of Electric Power was succeeded by the semi-autonomous State Power Corporation, whose responsibilities were largely confined to management of the grids and to planning and research. While grid management is being unified and integrated, ownership of generation comes from many sources. The objective, according to the corporation's report on the power industry, is to create a market that will ensure creditable service and safeguard legal rights of investors.

The establishment of the State Power Corporation (SP) marked the reformation of China's electric power industry as it entered a new stage [7-16]. Pilot bidding systems are in place in Shanghai Municipality, Zhejiang Shandong provinces and, Liaoning, Jilin and Helongjiang provinces in 2000, and then nationwide by 2005. An even more competitive and orderly generating market will be fully established by 2010, when the completed Three Gorges hydro power plant will allow a wider grid [17].

The reform of China's electric power industry is at its initial stage. How to accelerate the pace of reform and a smooth make transition from existing conditions to a market-oriented electric power industry are the questions of the day. China has traditionally been ruled by direct administrative orders, and such a system is so deeply implanted, that even today administrative measures are still used to regulate and control the economy. Therefore fusing the electric power industry of China fully into the framework of a modern market economy presents problems that demand further exploration and confrontation; many issues remain to be satisfactorily resolved, from basic theory to concrete practical application.

This chapter will introduce the development and management system of China's electric power industry, and present its reformation plan. The problems and obstacles faced in restructuring are also discussed. The theories of electricity pricing and transmission pricing developed by the first author are presented in some detail and several case studies are used to demonstrate the usefulness of the proposed theories.

7.2 Development of Electric Power Industry in China

7.2.1 Successive Growth of Power Production and Installed Capacity

In order to cope with the increasing power demand of economic and social development, the annual total newly commissioned generating capacity exceeded 15 GW during the last six years. In 1998, total net growth of generating capacity amounted to 17.696 GW, of which large and medium-size units had a combined capacity of 12.862 GW, making the nation's total installed capacity of 277.289 GW by the end of 1999 [2].

During the period 1991-1998, electric power production also kept steady growth and the average yearly growth rate of nation's total electricity generation was nearly 9 %. The nation's total electricity generation reached 1157.7 TWh in 1998. The historical development of electricity generation and installed capacity is shown in Figure 7.1.

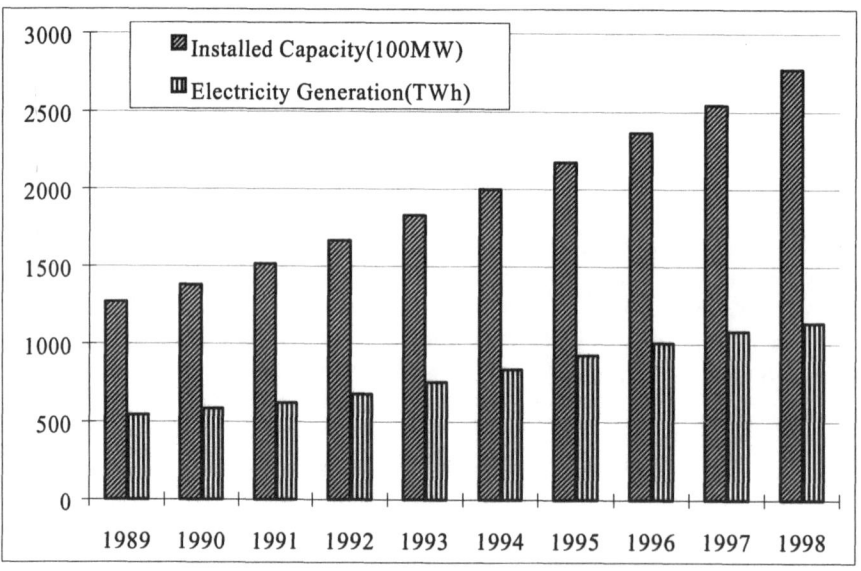

Figure 7.1 Recent records of total electricity generation and installed capacity

7.2.2 Further Expansion of Power Networks

With the steady growth of generating capacity, further expansion and interconnection of power networks has been made. Up to now, six inter-provincial power networks and five independent provincial power grids different size have been formed throughout the country. The aggregate installed capacity of these networks and grids amounted to 264.92 GW, representing 95.6% of the nation's total. By the end of 1998, the total length of 220 kV and above-level transmission lines reached 143029 km of which 500 kV lines cover over 20000 km.

The distribution of power network service areas and their installed generating capacities are shown in Table 7.1 and Figure 7.2. Actually, the first four inter-provincial regional networks shown in Table 7.1 have an installed capacity in excess of 30 GW each and the South China Interconnected Network with capacity of over 45 GW has been formed by interconnecting the Guangdong, Guangxi, Guizhou and Yunnan four provincial grids. HNPG is just below GXPG and has not been shown in Figure 7.2.

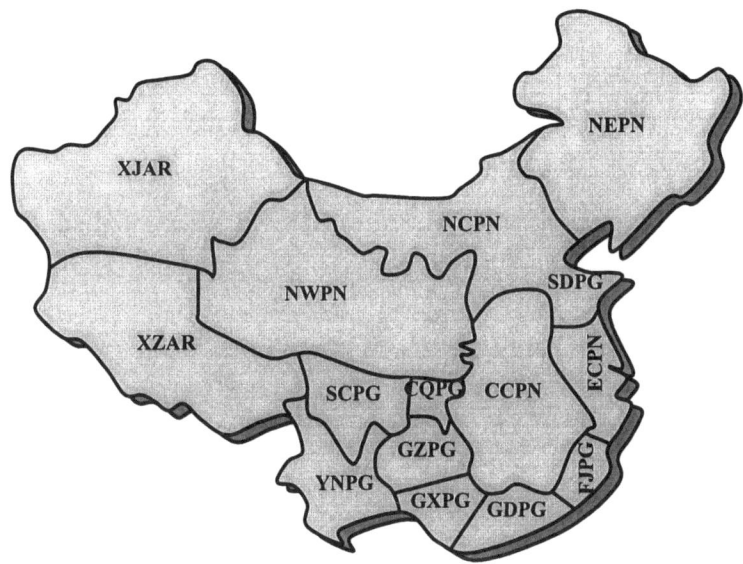

Figure 7.2 Distribution of power network service areas

Table 7.1 Installed capacity and electric generation in regions in 1998

Network & Region	Installed capacity		Electricity generation	
	Total (MW)	Hydro (%)	Total (TWh)	Hydro (%)
North China Power Net. (NCPN)	34312.1	15.96	141.15	5.62
Northeast Power Net. (NEPN)	37186.6	5.94	178.93	1.36
East China Power Net. (ECPN)	46121.0	9.62	211.45	5.40
Central China Power Net. (CCPN)	40749.3	30.60	160.37	28.70
Northwest Power Net. (NWPN)	19275.1	36.28	69.60	28.33
Shandong Provincial Grid (SDPG)	17380.1	0.27	84.06	0.09
Fujian Provincial Grid (FJPG)	8008.0	58.28	32.19	57.41
Guangdong Provincial Grid (GDPG)	29027.7	18.99	103.85	0.09
Guangxi ProvincialGrid (GXPG)	5645.0	58.32	22.78	56.94
Chongqing Power Grid (CQPG)	3157.0	9.93	12.64	9.37
Sichuan Power Grid (SCPG)	11942.3	57.14	44.37	49.94

Yunnan Provincial Grid (YNPG)	5995.6	61.91	24.14	61.26
Guizhou Provincial Grid (GZPG)	4578.2	37.64	23.58	29.30
Hainan Provincial Grid (HNPG)	1536.7	34.71	3.67	27.42
Xinjiang Autono. Region (XJAR)	2160.0	5.84	10.64	4.01
Xizang Autono. Region (XZAR)	159.0	68.68	0.28	61.92

7.2.3 Continuous Increase of Electricity Consumption

As a result of the progressive development of the electric power industry, electricity has become more and more significant for the progress of the national economy and the improvement of living standards. Total electricity consumption reached over 1134.73 TWh in 1998. The consumption structure and its variation are shown in Table 7.2 and Figure 7.3. From Figure 7.3 it is obvious that the residential consumption has the highest growth rate in various sectors and the share of residential use increased to 12.0% in 1998 from 6.0% in 1988.

Table 7.2 Electricity consumption structures in recent years

Year	Society total (TWh)	Share of industry (%)			Share of agriculture (%)	Share of transportation etc. (%)	Share of municipal commerce (%)	Share of urban & rural residential household (%)
		Whole	Heavy	Light				
1987	490.27	81.0	64.5	16.5	7.1	1.6	4.8	5.5
1988	535.87	80.3	64.1	16.2	7.0	1.6	5.1	6.0
1989	576.20	79.8	64.0	15.8	7.0	1.7	5.1	6.4
1990	612.60	78.7	62.6	16.1	6.8	1.7	5.3	7.5
1991	669.68	77.8	61.8	16.0	6.9	1.7	5.6	7.9
1992	745.54	77.1	61.2	15.9	6.8	1.8	5.8	8.5
1993	820.11	76.7	61.2	15.5	6.3	1.8	6.3	8.9
1994	904.65	75.4	60.3	15.1	6.3	1.9	6.8	9.7
1995	988.64	74.8	59.8	15.0	6.2	1.8	6.9	10.2
1996	1057.03	74.1	59.3	14.8	6.1	1.9	7.2	10.7
1997	1103.91	73.0	58.3	14.6	6.2	1.9	7.6	11.3
1998	1134.73	72.0	58.0	14.0	4.0	2.0	10.0	12.0

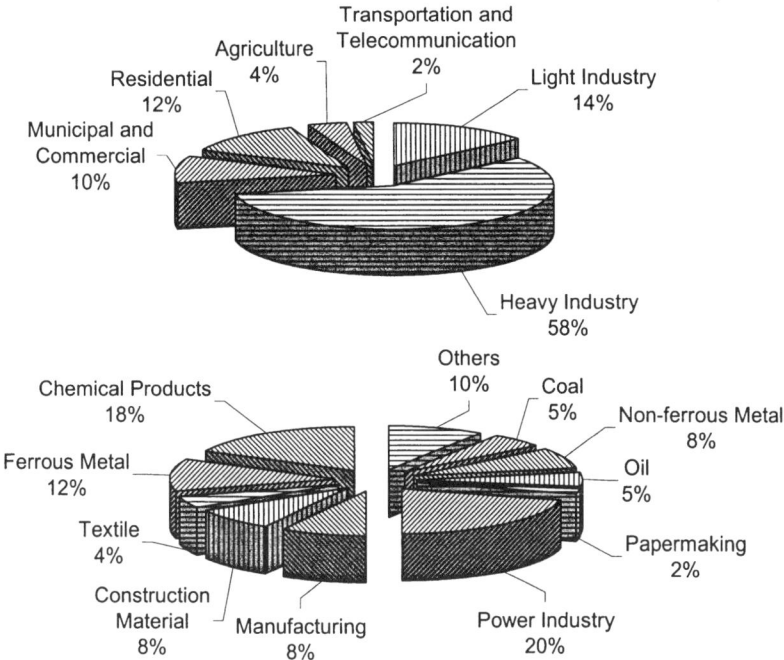

Figure 7.3 Electricity consumption structure in 1998

As an important index of the degree of electrification, the share of energy consumption for electricity generation in the total primary energy consumption rose to 32.6 % in 1998 from 22.4 % in 1987. With the increasing share of electricity in the total primary energy assumption and adopting various energy conservation measures, the energy intensity (energy consumption per unit output value of GDP) of 1998 decreased to 58 % of that in 1987.

7.3   Management System of Electric Power Industry in China

The development and reformation of the industry is according to the policy of

Separating administration from enterprise, making the provincial power companies into real entities, developing united networks and unified dispatching, raising funds from various channels, including from abroad, for power development.

7.3.1   The State Power Corporation

The State Power Corporation (SP), invested and established by the State Council, is the main body for investment and managing the state-owned asset authorised by the State Council. The SP is an economic entity running inter-regional power transmission and is an enterprise/legal entity undertaking unified management of the nation's power networks.

The SP, as an extremely large state-owned enterprise, has assets of 758.2 billion RMB assets. It plays a very important role in the future development of China's power industry whether running and managing its assets well or not. In two years' practice and exploration, the SP has set its development objective of creating a first-class enterprise in the world in terms of holding stock and group management, and this is a further development of the SP policy of corporatised restructuring, commercialised operation and legalised management. Thus, the SP has made a strategic decision-making plan in four steps for the future development of China's electric power industry [15]:

(1) The first step, from January 1997 to March 1998

Establishing the SP and the dismantling the Ministry of Electric Power, realising the transfer of government functions and professional management functions, preliminarily constructing a new decision-making system framework in compliance with a socialist market economy;

(2) The second step, from 1998 to 2000

Insisting on the policy of separating the government functions from the enterprise's functions and taking the provinces as entities, developing the generation market and completing the restructuring in the SP system;

(3) The third step, from 2001 to 2010

Realising the nationwide power network interconnection, fully realising the separation of generation from transmission and distribution, and complete competition in the generation market;

(4) The fourth step, after 2010

Gradually and fully opening the distribution and sale sectors and realising complete competition in the electric power market.

Upon the establishment of the SP, the functions of running the state-owned asset and enterprise management formerly undertaken by the Ministry of Electric Power were transferred to the SP. It possesses no governmental function, is subject to administrative management and supervision from related government sectors, and accepts sectional service from the China Electricity Council. The SP has a special listing in the state plan and its financial budget is allocated directly from national financial plan.

The organisation is shown in Figure 7.4 below. As defined by the State Council, seven power groups, namely Northeast, North China Power Group, East China Power Group, Central China Power Group and Northwest Electric Power Group, the China Huaneng Group and Gezhouba Engineering Group, as well as the China Power Grid Construction Company, are wholly owned subsidiaries of the SP. There are seven independent provincial power corporations, namely Shandong, Sichuan, Fujian, Yunnan, Guangxi, Guizhou provincial and Chongqing municiple power corporations, as well as the Southern Electric Power Joint-Venture Corporation, which are all exclusively owned subsidiaries under the SPC. These state-owned assets, held by the Armed Police Hydropower Construction Troops (also known as Anneng Corp.), belong to the SP's management. Other companies under the former Ministry of Electric Power are the SP's wholly owned subsidiaries, holding or jointly shared companies according to their property right structures. These corporations and institutions under the SP include the following:

Affiliated Enterprises

- China Huadian Power Plant Engineering General Corp.
- China Anneng Construction Corp.
- Longyuan Electric Power Group Corp.
- Zhongneng Power Tech. Development Co. Ltd
- China Electric Power Trust & Investment Co. Ltd
- China Electric Power Technology Import & Export Corp.
- China Fulin Wind Energy Development Corp.
- China Power Investment Co. Ltd
- China Power Investment Holding Corp.
- Zhongxing Electric Power Industry Development Corp.
- National Electric Power
- China Extra High Voltage Trans. & Substation Construction Corp.
- Zhongneng Electric Power Fuel Corp.
- Other enterprises under the management of the SP.

Affiliated Institutions

- Electric Power Planning & Engineering Institute
- Hydropower & Water Planning & Design General Institute
- National Power Control Centre of China
- China Electric Power Information Centre
- Electric Power Research Institute
- Thermal Power Research Institute
- Najing Automation Research Institute
- Wuhan High Voltage Research Institute
- North China Electric Power University
- China Electric Power News
- China Electric Power Press
- Other institutions under the management of the SP.

The operational and managerial functions of the SPC mainly include:

- Running the exclusively owned subsidiary companies and the holding or jointly shared companies and the state-owned stock rights in their affiliated units pursuant to state law, regulation, policy and strategy.
- Raising funds within the financing scope approved by the state to finance and invest in power projects and related enterprises; the income from investment and assets property transfer will be used for capital reinvestment pursuant to the regulation; taking charge of national power network interconnections.
- Running and managing the large power stations connected to regional networks or transmitting bulk power across regions and the necessary peaking and frequency regulating power stations.
- Planning and dispatching the national power network; supervising safe, stable, economic and high-quality operation on all power networks in the country.
- Exercising power network dispatching management on the national power network and the related generation, transmission and distribution enterprises based on the Regulation of Power System Dispatching.

The restructuring within the SP, separating generation from transmission and distribution, promoting the nationwide power network interconnection and speeding up rural power institutional reform are the current focuses of electric power industry reform and are listed as follows:

(1) To normalise the relationship among companies of different levels in the SP system.
In accordance with the requirements of setting up a modernised enterprise system, the SP is now constructing a multi-level enterprise/legal entity management system. The relationship between the SP and its subsidiary is an equal one in law among independent persons, and it is a capital link relationship in property between the investor and the enterprise invested. The pilot project was initiated by the Northeast Power Group Company (NPGC). The NPGC was reorganised as an affiliated entity of the SP, being an agency of the SP in Northeast China. The three provincial power companies in Liaoning, Jilm and Heilongjiang provinces, formerly affiliated to the NPGC were reorganised as companies with independent legal qualification, having complete self-running rights. The basic principle for reorganising is to set up one provincial power company for each province and its responsibility is to implement the unified planning and management of the provincial power network. The administrative functions are transferred to the synthesised economic management department of the local government.

(2) To promote separation of generation from transmission and distribution, introduce the competition mechanism and build a normalised power market.
The launch on the building power market was determined in December 1997. For establishing a normalised power market, a step-by-step method adopted. According to the policy, 'power plants can be run by multilaterals, power networks must be managed by the State'; the current objective is to separate generation from transmission and distribution and to build the generation-side power market. It has been determined to initiate pilot projects in five provinces and one city, i.e. Zhejiang, Shendong, Liaoning, Jilin, Heilongiiang provinces and Shanghai. Because of the complicated situations and various issues, the concrete approaches of these power companies are different. In accordance with the requirements of the SP, the following principles should be complied with:

- equal competition
- high transparency
- sharing benefit
- lowest cost
- operation by laws and regulations
- subject to supervision.

The concrete practice is to separate generation from transmission first, reorganise several generation group companies, and adopt a bid price mechanism in generation for the generating companies, but a few power plants, such as peak regulating units thermal units mainly used for supplying heat to the local area, are temporarily not included. For the sake of transition, the electricity generation could be divided into two categories: one is the basic part of electricity generation, the account of which is settled according to the current electricity generation price considering the repayment of principal with interest for newly built power plants; and the other is the competitive part of electricity generation, which is determined by the bidding price. As time goes on, the bidding part should be increased gradually. Finally, the principle of 'an equal electricity price for the same network and the same quality of electricit' should be carried out.

(3) To promote the implementation of the nationwide power network interconnection and realise the optimal disposition of resources. Owing to the distribution of energy sources and loads in China, implementing the nationwide power network interconnection and realising the optimal disposition of resources is an inevitable option. The construction of the extremely large Three Gorges hydro power station and its transmission system promotes the formation of the nationwide power network interconnection. It is planned that the interconnection between the Northeast and North China power networks will be accomplished in 2000, the interconnection between the Fujian provincial power network and the East China power network will be accomplished in 2001, and the interconnection between the Shantong provincial power network and the North China power network will be accomplished in 2003, the interconnection between the Sichuan provincial power network and the Northwest power network will be accomplished in 2004. Three cross-regional interconnected power networks in northern, middle and southern China will be basically formed around 2010. The unified interconnected power network of the whole country will be achieved between 2010 and 2020. The decisions for the above large engineering projects are all made on the basis of detailed preliminary feasibility studies of the benefits and effectiveness of interconnection. The formation of the nationwide power network interconnection will definitely accelerate the future development of China's power industry more economical and effective way.

(4) To speed up rural power management institutional reform.
Implementing rural power management institutional reform, technically renovating rural power networks and realising a unified electricity price for urban and rural areas in the same power network with the same quality of electricity are the current objectives of rural power system development. It will take three to five years. The task of this reform is mainly to simplify the management structure and to solve the chaos in rural electricity pricing, targeted at realising unified management, unified accounting, and a unified electricity price for urban and rural power networks. Technical renovation of the rural power network aim to reduce the losses of lines and transformers. The estimated investment is 180 billion RMB yuan. The line loss rate will be reduced to below 15 % from

more than 25 %, and the electricity price could be lowered by 0.1-0.2 RMB yuan per kWh after the renovation is completed. The realisation of the above-mentioned objectives will help to alleviate the peasants' burden, improve the peasants' living standards, develop a rural power market and promote rural economic prosperity.

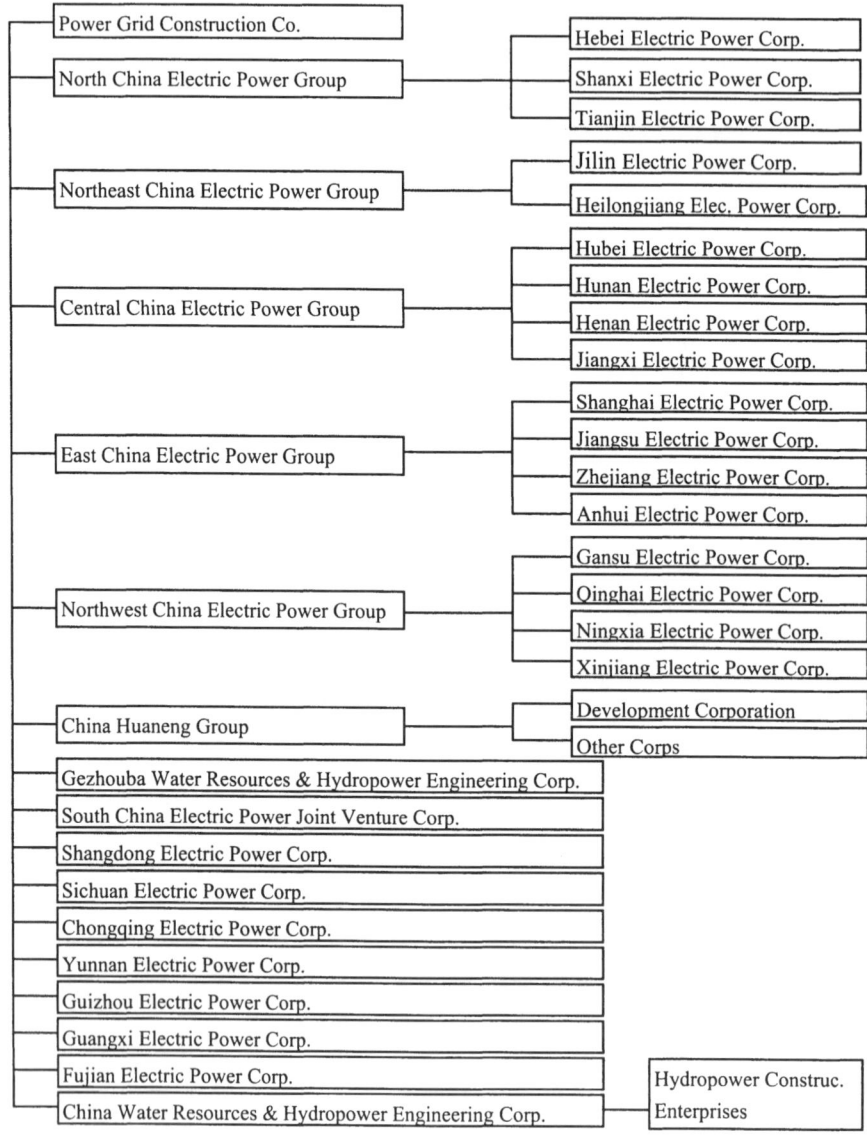

Figure 7.4 Organisation of the SP

7.3.2 Philosophy and Strategy of the SP

As for the philosophy and strategy of power industry development, the SP will observe the principle of sustainable development by relying on technical progress, further deepening reform and widening open policy.

In addition to focusing on fundamental research and staff training, the SP has actively launched five cross-century technology pilot projects, namely clean coal power generation, regional network interconnection, energy conservation and electricity saving, exploration of new energy resources, as well as a computerised information system. Since the beginning of the operation, the SP has focused on both international and domestic markets, and both international and domestic financing sources.

The development of China's power industry is still an arduous task. The mitigation in expanding the power sector is difficult due to the increase in electricity consumption. The per capita generating capacity in 1997 in China was only 0.21 kW, ranking about 80th in the world, while the per capita electricity consumption accounted for only one-third of the world average. It is planned that the nation's total installed capacity will reach 290 GW by 2000 and 500 GW to 550 GW in 2010, the nationwide power network interconnection with the Three Gorges Project being at the centre. In order to achieve the goals, SP will pursue a policy of

Optimising thermal power development, vigorously exploiting hydropower, appropriately developing nuclear power, tapping new energy power in line with the local conditions, synchronously constructing power networks, enhancing environmental protection and equally stressing exploitation and saving, but placing emphasis on savings.

An important change that must take place is that the electric power sector should be ready to be re-oriented to the market mechanism. The temporary balance in supply provides a good opportunity for the power sector to make adjustments optimise the sector itself. These include:

- *Regional adjustments*: Power development should keep pace with regional economic development. Power development plans should be worked out in accordance with the economic area that has been formed in the course of economic development instead of the administratively divided area. Encourage investors to develop power projects for the inter-regional economy. Support the economic development in mid-western China and give priority to the development of energy resources in the area. Priority should also be given to this area in using foreign funds for power projects.
- *Optimisation of power sources structure*: Optimise the development of thermal power in light of improving efficiency and protecting the environment. Give priority to the development of hydro power and develop nuclear power properly, while making great efforts developing new and renewable energy power sources.
- *Promoting large-size generators*: Encouraging the development of large-size generating units with high efficiency, in particular those above 600 MW. Strictly restrict the expansion of small-size condensing-type thermal power plants.
- *Strengthening the construction of the network*: With the rapid development of power plants, the construction of the network, including the facilities from trunk transmission lines to medium- and low-voltage distribution networks, lags behind the construction of

power sources and produces a bottleneck in the power industry. Great efforts should be made to expand and reinforce the power network.
- *Improving energy efficiency*: More and more funds including foreign funds will be spent on retrofitting existing older power plants (including both coal-fired power plants and hydro power stations)
- *Clean coal technology*: The use of clean coal technology and gas-fired combined-cycle plant is encouraged.

7.4   Power Market in China

The electric power sector is simultaneously a commodity and a public goods. As such, it is a sensitive area where conflicts are encountered between regulation and deregulation, between market value and controlled profit, and between government management and local and private initiatives. The officials and managers at the various levels understand the necessity and importance of transition to market economy, but they are not well informed about the mechanisms and approaches to realise the transition. This phenomenon is reflected by hesitation in both the manner and extent to which the electricity market is opened up to the outside, as well as in the relationship between the highly centralised dispatching system and the economic interests of the localities and electricity producers.

Electricity pricing reform lies at the heart of China's response to meeting future energy demand. Starting in 1979 measures were put in place that were designed to deregulate the electric power sector partially. The primary driving elements in this tariff reform included: allowing electricity prices to vary according to fuel and transportation prices; effecting a multiple-tier pricing policy for electricity generated outside the 'plan'; fee collection for guaranteed supply to firm power sources; and providing for both new and large-use customers.

In 1985, the State Council approved the Provisional Rules on the Encouragement of Investment for Electricity Generation and Implementation of Multiple Tariffs. These established the basic principles for power generation with foreign funds, including the establishment of multiple-tier tariffs that provide for a reasonable profit. Generally, however, parties interested in constructing power plants have had to develop legal and regulatory frameworks to support the investment on a case-by-case basis.

The Eighth Five Year Plan also calls for development in the energy sector to address both energy efficiency and the environment. The government is committed to stepping up a market-oriented reform of the country's pricing system in a bid to improve the competitiveness of Chinese industrial products and minerals. The Ministry of Materials and Equipment has worked in close collaboration with other government departments to sort out the outdated 'two-tier' pricing system (official and market prices) for key raw materials such as coal. This will include raising the state coal price to market levels. Prior to entry into the market economy, the pricing policy prevailing in the centrally planned economy of China was very simple: the government built and managed energy producers (oil fields, coal mines, power stations, etc.), took the profit away (if any was generated) and provided subsidies if entities operated poorly. However, as economic reforms are implemented, energy pricing becomes more and more complex and, in many instances, sensitive. Many energy experts and economists, both national and foreign, have offered proposals such as dual-track tariff pricing, comparative pricing and long-term marginal cost pricing:

however, none of these has been standardised as national policy. Such policies are inevitable; however, there is a growing realisation that the establishment of a rational pricing structure has become vital to the development of a sustainable energy industry. Since the mid 1980s an increasing number of enterprises, particularly joint ventures, have adopted cost-plus pricing structures that base the price of energy product on production costs (including the recovery of construction capital and interest, operation costs and labour costs), tax paid to the government, and profit. This is a considerable improvement over the administratively fixed price, but it still results in several ambiguities, including how to calculate costs in an environment where inflation is often double digit, and how to regulate the profits of enterprises.

Time-of-day pricing was introduced in 1987, along with seasonally adjusted rates where hydro power is a major component of base load. However, the rate of return on capital continued to decline, owing to the inability of rates to cover rising costs and of regional pricing structures to capture pricing differences, investment diversion to small, inefficient out-of-plan capacity, and the inability to collect user fees.

The regional and grid electricity tariffs, which are jointly fixed by the state, are administrated by the Ministry of Power. The limits of united pricing consist of the power grid prices and out-grid prices which are managed by the Ministry of Power and the provinces respectively. The prices of mid- to small-size power plants managed by regions and counties are fixed by local government, checked and ratified by the department responsible for price, as well as the power bureaux. The wholesale prices of grid and out-grid power plants which are priced by the state are checked and ratified by power grids bureaux and provincial power bureaux respectively. The base price reflects the fixed cost in the selling price, and has no relationship to consumption. The circulating price mirrors the variable production cost. The electricity tariff structure is divided into six categories, based on uses and voltages. The categories include:

- electricity rates for lighting;
- electricity rates for non-industry and ordinary industry;
- electricity rates for the larger industry;
- electricity rates for agriculture production;
- wholesale prices;
- electricity rates for grid co-supply.

Primarily because nationalised and standardised electricity pricing policies have not been fully implemented, the classification of electricity tariffs does not reflect the characteristics of different electricity consumption; for example, electricity rates in some areas (e.g. commerce, services and hotels) continue to be subsidised, as are the preferential electricity prices in industry and agriculture. At present, tariffs fixed by the state are significantly lower than the marginal production cost, i.e. electricity prices are being subsidised. Even pricing conventions were to be rationalised, and increased power generation costs and decreased tax revenues have offset revenue streams from the (much smaller) readjustments in state pricing, with system-wide effects of the resulting financial losses throughout the power industry.

7.4.1   Motivations for Reformation

Establishing a power market in China will introduce a market-oriented economy, promoting substantial development in the power industry. It is expected to solve the following problems:

1. Power resource location problem
   In the situation of an electric power shortage, the main issue that the electric power industry must face is to speed up construction of new power plants. Diversifying investment channels and ownership of power plants will help to achieve the expected goals. At the same time, however, it will also bring problems such as the inappropriate power mix structure, air pollution and non-synchronous construction of the power networks.
   In the past, the above problems were concealed by the power shortage situation. When power supply exceeds demand, these problems become the main consideration. After establishing the power market, power projects are to be decided according to market demand, not administrative order. Under regulation of the market mechanism power resources allocation will be more efficient and stable.

2. Low administration efficiency
   During the past 20 years, reform of the Chinese economy has undergone a rapid development. Supplies of commodities have become greatly abundant, while prices have decreased. To compete in the market, manufacturers spare no effort to improve their administration and promote their service qualities. However, China's power industry does not face such pressure. It still operates according to planning economy modules and has a monopoly electric power in selling.
   The generation cost has been increasing year by year. The generation and transmission indices are very low: for instance, the national net consumption rate is about 400 g/kWh (standard coal); the line loss rate is about 7%. These indices are far behind the world's average level.

3. Pricing problem
   Under the traditional monopolies of the power industry, to gain more benefits, the operators strived to maintain a higher rate of electricity. At the same time, the central government encouraged investment in the power plant by shortsighted polices such as 'new plant new rate', 'one plant one rate'. Thus the electricity rate continuously increased every year with new power plants put into operation. In January 2000, the average rate in urban areas was about 0.47 yuan/kWh and that in rural areas was about 0.67 yuan/kWh. Compared with the average income of Chinese people and prices of other commodities, the price of electricity in China is quite high.
   If the power shortage was an obstacle to developing China's economy in the past, the electricity rate gradually became a new barrier to the growth of China's economy. To maintain a sustainable development of the national economy thorough reform of the electric power industry is urgently needed.

7.4.2 Reform Plan of the SP

To implementing reform, the SP has set forth a 'four-step' restructuring framework. The period from the establishment of the SP in 1997 to the termination of the Ministry of Electric Power was the first step in realising corporate restructuring. From 1998 to 2000, the second step, the SP will continue to intensify restructuring, during which period the main tasks are to separate government functions from those of enterprises as follows:

1. Reorganisation of capital assets according to the strategy of *'establishing the corporation, commercialising the operation and legalising the management.'* The SP together with power group corporations and provincial power corporations will position themselves accurately among each other, and according to the requirement of the State Council will be gradually reorganised into stockholding companies, entity companies and grouped companies.
2. Pursuant to the 'Corporation Law' and the requirements of the modernised enterprise system, all enterprises under the SP will build normalised capital linkages among each other according to the respective hierarchical level, and establish perfect legal person management structures.
3. Taking the construction of the Three Gorges transmission project as a turning point, inter-regional power network construction will be speeded up to realise nationwide interconnection, optimise allocation of power resources, and prepare for the national integrated power market.
4. Based on the principle of *'the State controls power networks, while diversifies ownership of power plants'*, electric power networks will be separated from power plants. The independent power producers which are actively growing will become the main body of the competitive power market where access to power networks can be guaranteed on the principles of *'same network with same quality on same price'* and *'open, fair and equal'* basis.
5. To improve the power resource structure and speed up urban and rural network construction, and expand power consumption, the government will abolish obsolete, small thermal power units 6810 MW of total capacity. The power market will play an important role in introducing technology for technical progress and power industry development.

In the decade from 2001 to 2010, the third step of power industry reform, the SP will strive to realise nationwide network interconnection. In the meantime, networks shall be separated from power plants, and a well-regulated, technically advanced and orderly competed power market will be open to all power plants. The SP and all regional and provincial power corporations will run the power network as their main business, performed entirely on an enterprise/legal person and economic entity basis. In the period after 2020, the fourth step of the reform, the Chinese power industry management system will then approximate the international advanced level, moving towards a modernised transnational corporation at the international top level.

7.4.3 Obstacles in Establishing the Power Market in China

How to accelerate the pace of reform and smoothly make the transition from existing conditions to a market-oriented electric power industry are the questions of today. The obstacles that must be removed to achieve the reform goals are as follows.

Firstly, many power plants are not real companies. In fact, they are just shadows of true enterprises, with many of their necessary powers held by other higher administrative echelons belonging to the SP. Thus many of the key functions were out of their own control, being run directly or indirectly by the SP and its subsidiaries. In such conditions, the real power market cannot be established, because it requires open competition, fair dealing, justice and reasonableness.

Here we face two major problems. The first one is the property right problem. The 'property right' issue is based on the observable fact that under power market conditions, the enterprises in the power industry must be involved with the jurisdiction, definition, allocation and operation of their property. In fact, it is a problem of how the enterprise will be able to manage and operate itself. The second one is the problem of separating government and enterprise functions. Up to now, the traditional administrative function of the government remains basically unchanged, although the SP was established and the Ministry of Electric Power was terminated. Clearly, with such basic issues remaining unsolved, then the structure of property rights, optimum organisation form and the market mechanism will also remain unresolved [1].

Secondly, the electricity price is quite confused and firmly controlled by the government. Since economic reform began in China, relaxation of price controls over most commodities has been implemented. But the electricity price is still firmly controlled by the government because the electric industry is considered to be one of those spheres that affect the overall economy and standard of living. Therefore, in the initial stage, reformation of the power market is rigorously limited to the generation side. Under the policy of diversifying investment, different types of power plants have been built. At present, power plants can be economically classified as follows:

1. Plants that are run directly by the regional or provincial power company. These plants were basically constructed by government financial allocation in the past. They are internal accounting units of power companies.

2. Independent power plants constructed by government financial allocation. They sell electricity to the grid according to the prices audited by the government. Thus each plant has its own selling price.

3. Independent power plants constructed by investment. Most of new plants belong to this category. The investors include the regional and provincial power companies, local (provincial or manipulation) energy investment corporation, etc. They sell electricity to the grid also according to the prices audited by the government. Because the factor of repaying their investment must be taken into account, each plant also has its own selling price.

4. Small hydro power or thermal power plants constructed by local governments. They are usually run by local power companies and sell electricity according to the price audited by local governments.

5. Thermal power plants sold to foreign enterprises. To get funds to construct new power plants in some regions, several thermal power plants have been sold to foreign enterprises. According to the agreement of sale regional power companies guarantee that this kind of thermal power plant will sell a certain amount of electricity to the grid each year at a certain price.

Prices for electric power varied considerably across provinces and even within small regions. Power purchase agreements, which define how much electricity that the grid guarantees to buy from the power plant, are not being honoured. Old plants usually sell the most power since they need to cover only fuel, operation and maintenance costs and need not pay their capital cost.

From the above classification of power plants one can see that the price system on the generation side is very complicated and it is very difficult to form a normalised competitive market. The Electric Power Law approved by the People's Congress of China on 28 December 1995 is out of date. It is almost useless in restructuring China's electric power industry. The establishment and improvement of the electric power market must be standardised and backed by a complete legal framework. Therefore, new legislation must be in place as a precondition for the necessary reformation of the electric power.

The matters that must be dealt with are:

1. Property ownership: This subject occupies a very important and critical position in China's electric power industry reform. Without a thorough clarification and firm legal definition concerning the property right of regional, provincial power corporations and independent power plants, the transmission to a true power market will be greatly hampered.

2. Market relations: In today's SP management system, temporary regulations or administrative orders are still the main measures used to manage and control economic affairs. This situation shows the lingering characteristics of a planned economy. Therefore, equal completion must be defined as the basic principle of the power market.

3. Social relations: The provincial power corporations and independent power plants support the reform. But they often show more concern about their benefits from the reform. Therefore, ensuring distribution of social benefits is an important issue in promoting the development of the power market.

In summary, China's electric power industry is at its initial stage of reformation. There are many challenges to be overcome to establish a fair and efficient power market.

7.5 Electricity Pricing

The characteristics of electricity pricing in China originated from the requirement of a large amount of generation sinking fund and investment. The PPPs used in the UK are not suitable for China [8]. Many methodologies have been suggested to design electricity price systems, including a two-part pricing system. This section will discuss a one-part pricing system that can cope with the above problem.

Costs of electricity include operation costs and investment costs or the fuel costs and capacity costs. Determination of electricity costs involves optimisation of system operation and reliability, because operation optimisation is the basis of calculating fuel costs, while reliability is the basis of determining capacity investment. Therefore, a probabilistic production simulation of the power system becomes one of the most powerful tools to predict electricity cost [7].

7.5.1 Basic Theory of Predicting Electricity Costs

In order to analyse electricity costs, we should run a probabilistic production simulation hour by hour. Then we can obtain fuel costs $F(t)$ and loss of load probability $LOLP(t)$, $(t = 1,2,\cdots,8760)$. This data is the basis for calculating electricity costs.

Variable costs of electricity consist of fuel costs and capacity costs. Capacity costs can be represented by

$$I_G = W_b K_b + W_P K_P \tag{7.1}$$

where W_b : generation capacity for base load of the system
W_P : generation capacity for peak load of the system
K_b : annual rate per-unit capacity for base load
K_P : annual rate per-unit capacity for peak load

When predicting generation costs for each hour, the annual rate of base-load capacity should be evenly distributed among 8760 h, while, the annual rate of peak-load capacity should be distributed according to $LOLP(t)$ for each hour. Therefore, the cost of electricity for hour t is

$$C(t) = F(t) + W_P K_P \frac{R(t)}{R_A} + \frac{W_b K_b}{8760} \tag{7.2}$$

where $F(t)$: fuel cost of the power system in hour t
$R(t) : LOLP(t)$, the risk level in hour t
R_A : the risk level in the investigated year

$$R_A = LOLP_A = \sum_{t=1}^{8760} LOLP(t) \tag{7.3}$$

Hence, the average cost of electricity for hour t, $\overline{\rho}(t)$, is

$$\overline{\rho}(t) = \frac{C(t)}{P(t)} \tag{7.4}$$

where $P(t)$ is the system load at hour t.

The marginal cost $\rho(t)$ of electricity for hour t is

$$\rho(t) = \frac{\partial C(t)}{\partial P(t)}$$

Substituting equation (7.2) into the above equation, we have

$$\rho(t) = \frac{\partial F(t)}{\partial P(t)} + \frac{\partial (W_P K_P R(t)/R_A)}{\partial P(t)} \qquad (7.5)$$

$\frac{\partial F(t)}{\partial P(t)}$ in equation (7.5) can be found by running a probabilistic production simulation. To find the second term of equation (7.5), we can use the following two methods.

1. Maintain generation capacity W_P unchanged, and increase 1 unit load for each hour, then run a probabilistic production simulation in this situation, $LOLP(t)$.

2. Maintain $LOLP(t)$ unchanged, and increase 1 unit load in this situation we must increase peak load capacity W_P.

Because it is difficult to get the cost of loss load, the second way is preferred. Under the above condition equation (7.5) can be rearranged as

$$\rho(t) = \frac{\partial F(t)}{\partial P(t)} + K_P \frac{R(t)}{R_A} \cdot \frac{\partial W_P}{\partial P(t)} \qquad (7.6)$$

$\frac{\partial W_P}{\partial P(t)}$ can be approximately found by the following procedure as shown in Figure 7.5. LDC in the figure is the load duration curve formed from $P(t)$. After running a probabilistic production simulation, the equivalent load duration curve ELDC is formed. The risk level of the whole year, $LOLP_A$, is determined by the abscissa $W_b + W_P$. $\overline{LDC}$ in Figure 7.5 is the load duration curve formed by $P(t) + \Delta P$; here ΔP is an increment of the load. After running a probabilistic production simulation, the respective equivalent load duration curve become $\overline{ELDC}$. With the same $LOLP_A$ we can find a point in $\overline{ELDC}$, the abscissa of which is OB. Therefore, the section of line AB represents the capacity increment ΔW. Thus we can substitute the following equation

$$\frac{\partial W_P}{\partial P(t)} \approx \frac{\Delta W}{\Delta P} \qquad (7.7)$$

into equation (7.6) and further calculate the marginal cost of electricity for each hour.

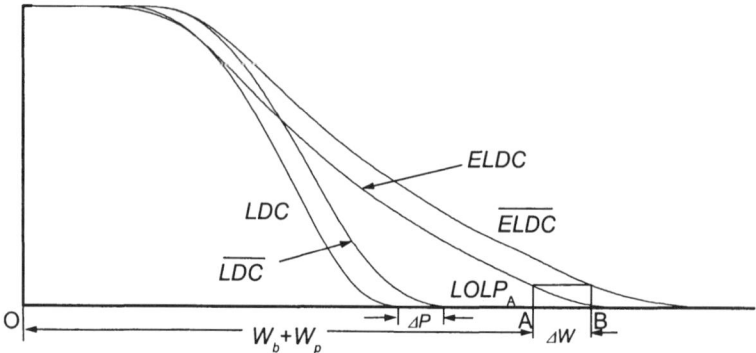

Figure 7.5 Equivalent duration load curve

In order to maintain the necessary capacity reserve and reliability level, besides increasing generation capacity, we can draw up contracts with some consumers. When the generation capacity of the system is not enough, the power supplied to consumers can be interrupted; in return these consumers will settle their bill by lower price. For these kinds of consumers the, generation cost should not include the capacity cost and the marginal cost become,

$$\rho'(t) = \frac{\partial F(t)}{\partial P(t)} \tag{7.8}$$

Some industrial enterprises have their own power units and sell electricity to the grid. If they do not contribute respective reserve, their electricity price will be calculated by equation (7.8) instead of equation (7.6).

7.5.2 Electricity Cost Derivation

When we design a price system of electricity, the renew period is one of the most important parameters that should be determined first. The shorter the renew period, the more regulation effect there is on the electricity price. The renew period of electricity price for a spot market can be as short as half an hour, even a quarter of an hour. To predict the change pattern of the electricity cost, we can use the theory and formulae presented in the above section. As an example, we calculate the electricity cost for a real power system in the northwest power grid. Figures 7.6 a and b demonstrate the change patterns for typical days in September and November respectively. In the figure, the thin lines represent average costs and the thick lines represent marginal costs.

From Figure 7.6 we can see that both the average and marginal costs of electricity dramatically increase at peak times. This is due to the allocation of peak capacity investment.

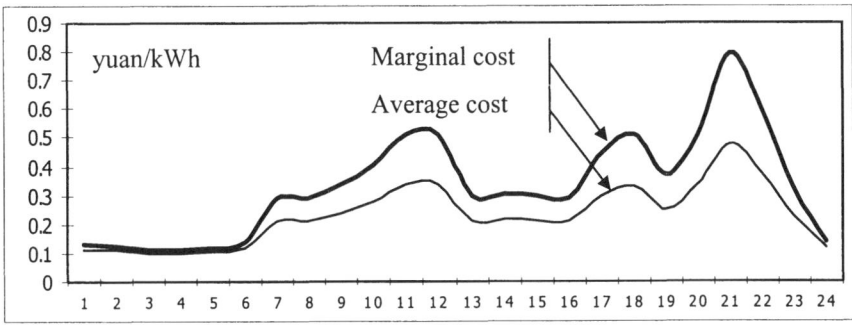

(a) September

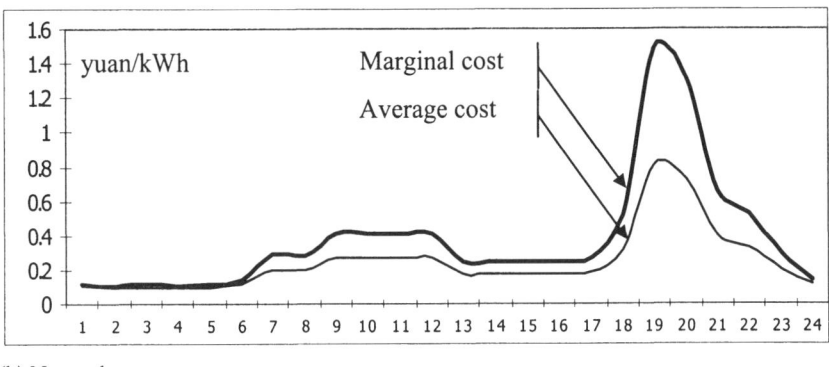

(b) November

Figure 7.6 Typical day's electricity cost curve (yuan/kWh)

In order to supervise the market price of electricity for the government easily, we can calculate several characteristic costs (or prices) for an interval of a certain time, say one week, one month or one year.

We now derive the formulae to determine the characteristic costs for an interval of one year. Assume that the set of peak-load time is defined by T_p, the set of valley load time is T_v and the set of shoulder load time is T_s. The number of hours are t_p, t_v and t_s respectively, and

$$t_p + t_v + t_s = 8760$$

Electricity consumed in the peak, shoulder and valley load periods are A_p, A_s and A_v, respectively, and are calculated as follows:

$$A_p = \sum_{t \in T_p} P(t) \quad A_s = \sum_{t \in T_s} P(t) \quad A_v = \sum_{t \in T_v} P(t) \tag{7.9}$$

If the cost of electricity generation in the peak, shoulder and valley load periods is C_p, C_s and C_v, respectively, then they can be found from equation (7.2) as

$$C_p = \sum_{t \in T_p} C(t) \quad C_s = \sum_{t \in T_s} C(t) \quad C_v = \sum_{t \in T_v} C(t) \qquad (7.10)$$

Hence we can determine the average cost of electricity for the peak, shoulder and valley load periods as

$$r_p = C_p/A_p \quad r_s = C_s/A_s \quad r_v = C_v/A_v \qquad (7.11)$$

The average cost of electricity for the year is thus

$$\bar{r} = \frac{C_p + C_s + C_v}{A_p + A_s + A_v} = \frac{C_\Sigma}{A_\Sigma} \qquad (7.12)$$

where C_Σ and A_Σ are the cost and amount of electricity for the whole year.

The average marginal costs for the peak, shoulder and valley load period are

$$\rho_{pa} = \sum_{t \in T_p} \rho(t)/t_p \quad \rho_{sa} = \sum_{t \in T_s} \rho(t)/t_s \quad \rho_{va} = \sum_{t \in T_v} \rho(t)/t_v \qquad (7.13)$$

where ρ_{pa}, ρ_{sa} and ρ_{va} are average marginal costs for the peak, shoulder and valley load periods respectively, and are calculated from equation (7.6). The average time used costs of electricity for a real power system are shown in Table 7.3.

Table 7.3 Average time used costs (yuan/kWh)

	Peak load	Shoulder load	Valley load
Average energy cost	0.4021	0.2014	0.1133
Marginal energy cost	0.5554	0.2299	0.0795

7.5.3 Electricity Pricing of Inter-provincial Power Market

Since 1997 the nationwide electric power shortage which lasted for 25 years began to be alleviated. As a consequence, many power plants suffered a limitation of generation for the first time. Therefore the income and benefits of the plants were decreased accordingly. Although there is the electric power policy of making the provincial power corporations into real entities, developing united networks and unified dispatch, the power flows along inter-province tie lines gradually died off. Owing to the emphasis on making the provincial companies into real entities, most provincial dispatching centres refused to use electricity from neighbouring provincial grids even in the case of power shortages.

This operation obviously contradicts with the principle of optimal allocation of resources. The cause of such a performance is due to the pricing of electricity exchanged among provinces.

Let us illustrate this problem by a real example of the northwest power system. In Northwest China, four provincial power systems are interconnected. These provinces include Shaanxi, Gansu, Ningxia and Qinghai. The Xinjiang Electric Power System is an isolated system. The power source mix of the northwest power system at the end of 1997 is shown in Table 7.4 and Figure 7.7. We can see that in Shaanxi and Ningxia provinces, electricity is mainly supplied by coal-fired power plants; in Gansu and Qinghai provinces, more than half the electricity is supplied by hydro power plants. Therefore, unified dispatching in the northwest power system can make a significant profit.

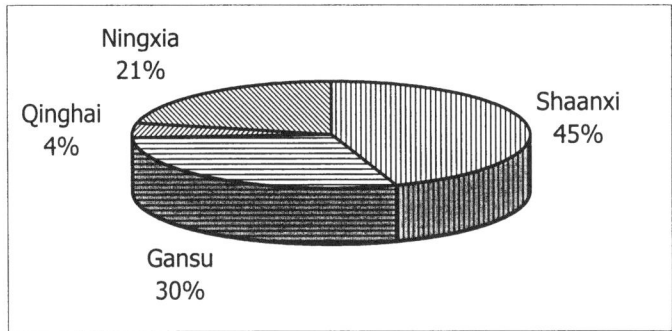

(a) Thermal power

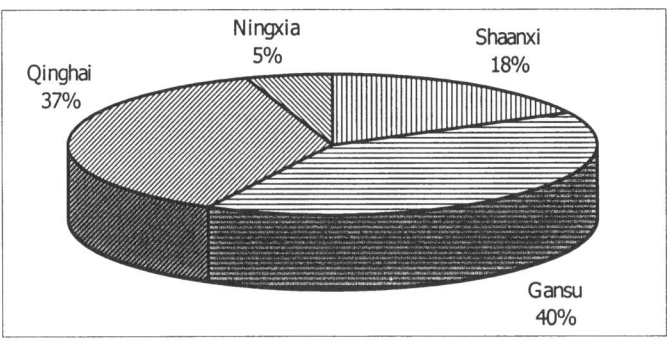

(b) Hydro power

Figure 7.7 Weight of power capacity installed in four provinces

To investigate the difference between a unified dispatching operation and as isolated operation, we run probabilistic production simulations for a typical day in February. The simulation results are shown in Figure 7.8, where one can see histogram curves of the interchanged power among the provinces for unified dispatching.

Table 7.4 Power source mix in the northwest power system (MW)

Province	Thermal power		Hydro power		Total	
	Capacity	%	Capacity	%	Capacity	%
Shaanxi	4025	44.6%	988	17.5%	5013	34.2%
Gansu	2668	29.7%	2285	40.4%	4953	33.7%
Qinghai	400	4.4%	2080	36.8%	2480	16.9%
Ningxia	1924	21.3%	302	5.3%	2226	15.2%
Total	9017	100%	5655	100%	14671	100%

Table 7.5 Economic benefits of interconnection, MWh, million yuan

		Separate Operation	Interconnected Operation
Shaanxi	Load Energy	63568	63568
	Generating Energy	63600	81458
	Exchange Energy	32	17890
	Operation Cost	3.26887	4.34856
	Fixed Cost	8.7768	11.24120
Gansu	Load Energy	64139	64139
	Generating Energy	64139	39318
	Exchange Energy	0	-24821
	Operation Cost	3.46376	1.78287
	Fixed Cost	9.71239	5.42588
Qinghai	Load Energy	18747	18747
	Generating Energy	18747	8669
	Exchange Energy	0	-10078
	Operation Cost	1.04226	0.19944
	Fixed Cost	3.37173	1.19632
Ningxia	Load Energy	30079	30079
	Generating Energy	30951	47088
	Exchange Energy	872	17009
	Operation Cost	1.68593	2.60171
	Fixed Cost	4.27124	6.49814
Total	Total Operation Cost	35.59297	33.29414
	Interconnect Benefits	0	2.29884

The economic comparisons between unified dispatching and isolated operation are shown in Table 7.5. For a typical day in February 1997, a profit of 1.742 million yuan can be made by unified dispatching, when Shaanxi exports 14238 MWh, Gansu imports 18043 MWh, Qinghai imports 14273 MWh and Ningxia exports 18078 MWh.

Distribution of the above profit is directly affected by electricity pricing in the inter-provincial power market. Table 7.6 shows the relationships between electricity price and profits allocated to different provinces. Obviously, the provincial power market can only be established when the electricity price is in an interval of 0.25-0.29 yuan/kWh.

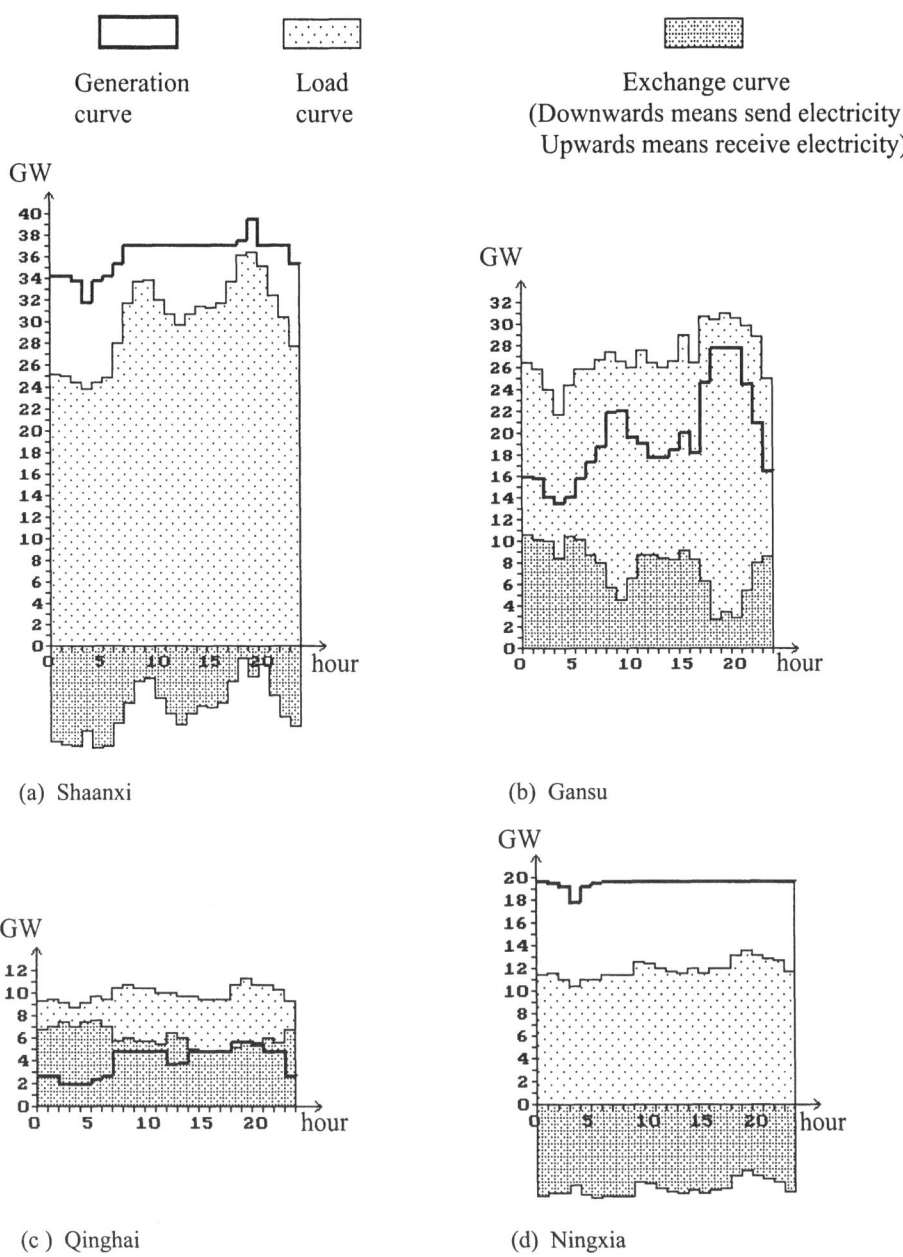

Figure 7.8 Electricity exchange for interconnection

Table 7.6 Effects of prices on interconnecting benefit distribution

Electricity Price of Exchange (yuan/kWh)	Interconnection Benefit (million yuan/day)			
	Shaanxi	Gansu	Qinghai	Ningxia
0.35	1.82291	-1.47891	-0.00517	1.96001
0.34	1.64401	-1.23070	0.09561	1.78992
0.33	1.46511	-0.98249	0.19639	1.61983
0.32	1.28621	-0.73428	0.29717	1.44974
0.31	1.10731	-0.48607	0.39795	1.27965
0.30	0.92841	-0.23786	0.49873	1.10956
0.29	0.74951	0.01035	0.59951	0.93947
0.28	0.57061	0.25856	0.70029	0.76938
0.27	0.39171	0.50677	0.80107	0.59929
0.26	0.21281	0.75498	0.90185	0.42920
0.25	0.03391	1.00319	1.00263	0.25911
0.24	-0.14499	1.25140	1.10341	0.08902
0.23	-0.32389	1.49961	1.20419	-0.08107
0.22	-0.50279	1.74782	1.30497	-0.25116
0.21	-0.68169	1.99603	1.40575	-0.42125
0.20	-0.86059	2.24424	1.50653	-0.59134

7.6 Transmission pricing

Transmission of electricity is becoming a separate industry player. A viable transmission business is critical to a successful competitive electricity market. In the past, the inappropriate trends of 'emphasising generation, ignoring transmission' in the electric power industry of China made adequacy transmission very poor. Since the basic business of the SP and its subsidiaries is transmission they have a duty to provide enough transmission capacity to satisfy the requirement of the power market. This is a massive undertaking that will cost billions of yuans. Where will the money come from?

The reform of the power industry brings transmission pricing into a new focus and there is a growing need to identify the costs of transmission services. In such a situation, one should answer questions such as how much is this 'generator or load' making use of this transmission line? Or 'what proportion of the network losses is allocated to this generator (or load)'? Solutions of these problems are very important to measure the services provided by transmission systems, and has a direct influence on transmission cost.

This section presents a comprehensive investigation of load flow analysis in wheeling costing. Two current decomposition axioms are first introduced as the fundamentals of load flow analysis in wheeling costing. Then rigorous mathematical models of the distribution factor problem and loss allocation problem are established. To solve these problems, we introduce a series of theorems based on graph theory and a very simple and efficient algorithm is developed. Finally, case studies are introduced to illustrate the usefulness of the proposed theory and algorithm [18].

7.6.1 Current Decomposition Axioms

Market-driven transactions have become the new independent decision variables defining the behaviour of the power system. Understanding the impact of bilateral transactions on system losses is important in order to allocate a corresponding loss component to each individual transaction and improve economic efficiency. One essential piece of information that the bilateral market needs in order to improve economic efficiency is knowledge of the transmission losses associated with each proposed bilateral transaction. This knowledge permits buyers and sellers to incorporate the level and cost of losses into their negotiations. The essence of the proposed loss allocation method is that given a path, along which the transactions vary with time, it is possible to find for each infinitesimal incremental transaction an associated unique and separable loss component. This leads then to a loss allocation component for each transaction. A number of current proposals for calculating associated contractual losses have been proposed [19-24]. When considering the problems of wheeling cost, we need to identify the power (or current) components of each branch and allocate the effects such as losses to its components. To solve this kind of problem it is not enough to use only Kirchhoff's laws of electric circuits. Therefore, in this section, we introduce two axioms.

Assume the current of branch k, $I_{(k)}$, consists of L current components $I_{(k)l} (l = 1,2,\cdots,L)$ supplied by L generators,

$$I_{(k)} = \sum_{l=1}^{L} I_{(k)l} \quad (7.14)$$

where $I_{(k)}$ and $I_{(k)l}$ are the effective or r.m.s. values of the currents, which can be either 'active' or 'reactive' components. Similarly, in the following description, the term 'power' can also be replaced by either 'active power' or 'reactive power' [25].

Axiom 1 The components of current in a branch are conservative.
The axiom states that each component $I_{(k)l}$ is the same at the initial and terminal node of a branch.

Definition The use of branch k made by generator l denoted by $f_{(k)l}$ is called the *distribution factor*, and defined by

$$f_{(k)l} = I_{(k)l}/I_{(k)} \quad (7.15)$$

Corollary 1 Distribution factors are the same at the two nodes of a branch.
Proof This is obvious when we define $f_{(k)l}$ by currents as shown in equation (7.2) because both $I_{(k)l}$ and $I_{(k)}$ maintain the same values at the two nodes of branch k.

However, in power system analysis we usually use power instead of current. Thus we should prove this statement is also true when we use power to define distribution factors.

Assume the voltages at the initial and terminal nodes of branch k are U_k' and U_k''. Thus the respective powers are

$$P_{(k)}' = U_{(k)}' I_{(k)} \, , \quad P_{(k)}'' = U_{(k)}'' I_{(k)} \quad (7.16)$$

The powers at the two nodes supplied by source l are

$$P'_{(k)} = U'_{(k)} I_{(k)}, \quad P''_{(k)l} = U''_{(k)} I_{(k)l} \tag{7.17}$$

Hence we have

$$f_{(k)l} = I_{(k)l} / I_{(k)} = P_{(k)l} / P_{(k)} \tag{7.18}$$

where $P_{(k)}$ and $P_{(k)l}$ can be either $P'_{(K)}$ and $P'_{(k)l}$, or $P''_{(k)}$ and $P''_{(k)l}$.

Corollary 2 The loss of a branch should be allocated proportionally to the current (or power) component.

Proof From equation (7.16), the loss of branch k, $\Delta P_{(k)}$, can be represented by

$$\Delta P_{(k)} = P'_{(k)} - P''_{(k)} = (U'_{(K)} - U''_{(k)}) I_{(k)} \tag{7.19}$$

According to equation (7.4), the loss caused by current component $I_{(k)l}$ is

$$\Delta P_{(k)l} = P'_{(k)l} - P''_{(k)l} = (U'_{(k)} - U''_{(k)}) I_{(k)l} \tag{7.20}$$

Combining equation (7.19) and equation (7.20) we obtain

$$\Delta P_{(k)l} = \Delta P_{(k)} \frac{I_{(k)l}}{I_{(k)}} = \Delta P_{(k)} f_{(k)l} \tag{7.21}$$

This concludes our proof.

The principle of loss allocation on the basis of demand squared was also suggested in [7], i.e. the loss allocated to component current $I_{(k)l}$ should be calculated according to

$$\Delta P'_{(k)l} = \Delta P_{(k)} \frac{I^2_{(k)l}}{\sum_{l=1}^{L} I^2_{(k)l}}$$

This principle is not *economically* reasonable, because it can cause inefficient resource allocation.

The next axiom is the proportional sharing assumption commonly used in [25].

Axiom 2 The current components in the outgoing lines of an injected current at a node are proportional to the currents of the outgoing lines.

Assuming that total current injected at node $P_{(i)}$ is I_i, this axiom states that when the current injected by generator $P_{(i)}$ at node i is $I_{i,l}$, its component current $I_{(k)l}$ in outgoing line k is

$$I_{(k)l} = I_{i,l} \frac{I_{(k)}}{I_i} = I_{i,l} a_{i(k)} \tag{7.22}$$

where $a_{i(k)}$ is called *allocation factor* of line k,

$$a_{i(k)} = I_{(k)} / I_i \tag{7.23}$$

Definition The whole loss caused by transmitting energy from generators to a node is called *loss of the node*. We will denote the loss of node $P_{(i)}$ by δP_i.

Obviously, when electricity of node $P_{(i)}$ is directly transmitted through its incoming lines from generators, the loss of node i, δP_i is equal to the total loss of these incoming lines. To allocate δP_i, to the outgoing lines of node i we have the following corollary.

Corollary The factor of node loss allocated to an outgoing line is equal to its allocation factor.

Proof Assume that node i has L_i incoming lines all directly connected with the generators; then the loss of node i is

$$\delta P_i = \sum_{m=1}^{L_i} \Delta P_{(m)}$$

According to Corollary 2 of Axiom 1 and Axiom 2, the loss allocated to outgoing line k of $\Delta P_{(m)}$ (the loss of incoming line (m)) is

$$\Delta P_{(m)(k)} = a_{i(k)} \Delta P_{(m)}$$

Therefore, the total loss allocated to outgoing line k, $DP_{(k)}$, can be calculated by

$$DP_{(k)} = \sum_{m=1}^{L_i} \Delta P_{(m)(k)} = \sum_{m=1}^{L_i} a_{i(k)} \Delta P_{(m)} = a_{i(k)} \delta P_i \qquad (7.24)$$

When the incoming lines of node i are not all connected to the generators, equation (7.24) can be proved by the recursive reasoning method.

7.6.2   Mathematical Models

As mentioned above, there are two problems related to load flow analysis in wheeling costing, namely the distribution factor problem and the loss allocation problem. We first present the distribution factor problem.

For a specified operating condition of a power system, one can obtain the power flow along each branch (transmission line or transformer) by a load flow study. The problem to be solved now is to calculate the distribution factors of each generator for each branch. Assuming that a power system has N nodes, N_G generators and N_B branches, we need to determine $N_B \times N_G$ distribution factors defined by equation (7.18), which represents the usage of branch k by generator l at the specified operating condition.

From Axiom 1, we know that $f_{(k)l}$ of generator l for branch k is the same at both nodes of branch k. Axiom 2 states that the current components in outgoing lines at a node of an injected current are proportional to their currents, as shown in equation (7.22). Multiplying the two sides of equation (7.22) by the voltage at node i yields

$$P_{(k)l} = P_{i,l} \frac{P_{(k)}}{P_i}$$

where $P_{i,l}$: injection power of generator l at node i

P_i: total injection power of node i

Substituting the above equation into equation (7.18), we have

$$f_{(k)l} = \frac{P_{(k)l}}{P_{(k)}} = \frac{P_{i,l}}{P_i} \qquad (7.25)$$

Therefore, if we find the injection power $P_{i,l}$ of each generator l in each node i, then usage of each outgoing line of node i by generator l can be calculated by equation (7.21). To do so, we first establish the following load flow distribution relationship:

$$P_g = BP_n \quad (7.26)$$

where

$$P_g = [P_1^G, P_2^G, \cdots, P_n^G]^t$$

is the vector of generator powers and

$$P_n = [P_1, P_2, \cdots, P_n]^t$$

is the vector of total node injection powers. B in Equation (7.26) is an $N \times N$ matrix, elements of which are defined by

$$b_{ji} = \begin{cases} 1 & i = j \\ P_{ji}/P_i & ij \in \Gamma_+(i) \\ 0 & \text{otherwise} \end{cases} \quad (7.27)$$

where $\Gamma_+(i)$ is the set of the outgoing lines of node i, and P_{ji} is the power flow in branch ji from j to i.

After a load flow run, the total injection power of each node and load flow in each branch is known; hence the elements of B can easily be calculated by equation (7.27). Let us illustrate this with a simple power system as shown in Figure 7.9a. For this system we have the following relationship:

$$\begin{bmatrix} P_1^G \\ P_2^G \\ 0 \\ 0 \end{bmatrix} = \begin{bmatrix} 1 & 0 & 0 & 0 \\ -59/400 & 1 & 0 & 0 \\ -218/400 & 0 & 1 & -82/283 \\ -112/400 & -17/173 & 0 & 1 \end{bmatrix} \begin{bmatrix} P_1 \\ P_2 \\ P_3 \\ P_4 \end{bmatrix} \quad (7.28)$$

The above mathematical model is rigorous, and does need not to have the assumption of a lossless branch as adopted in [25].

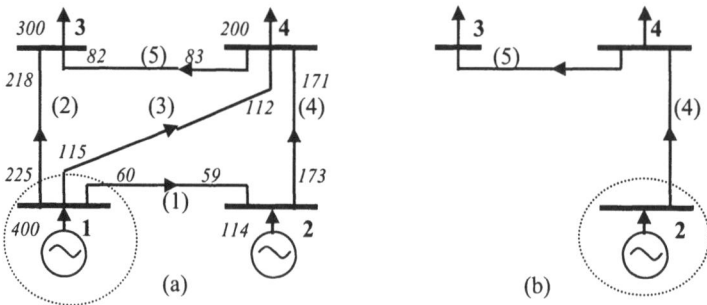

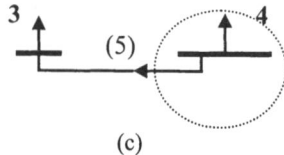

(c)

Figure 7.9 A circuit diagram for simple power systems

By solving equation (7.26) we can obtain the contribution of each generator to the total injection power at each node from the following equation:

$$P_n = B^{-1} P_g \qquad (7.29)$$

Thus the contribution factors can be readily calculated by equation (7.25).

After getting $f_{(k)i}$, we can further allocate the loss of the transmission network to each generator by the following equation according to equation (7.21):

$$DP_l = \sum_{k=1}^{N_B} \Delta P_{(k)} f_{(k)l} \qquad (7.30)$$

where DP_l is the loss allocated to generator l.

However, the loss allocation problem can be an independent problem. Therefore, we need a mathematical model for the problem to allocate loss to each load or each generator. To allocate the loss to each load, the key step is to calculate the losses of nodes $\delta P_j (j = 1,2,\cdots,N)$. δP_j consists of two parts, as follows:

Firstly, the sum of loss ΔP_{ij} in the incoming line $ij \in \Gamma_-(j)$, where $\Gamma_-(j)$ denotes the set of the incoming lines of node j. Secondly, the loss of δP_i allocated to line ij, $DP_{(k)}$, which can be calculated according to equation (7.24). Note that $k \Leftrightarrow ij \in \Gamma_-(j)$. The loss balance equations are as follows:

$$\delta P_j = \sum_{ij \in \Gamma_-(j)} (\Delta P_{ij} + a_{i(k)} \delta P_i) \qquad j = 1,2,\cdots,N \qquad (7.31)$$

where $a_{i(k)}$ is the allocation factor defined in equation (7.23). For convenience we can use the following form to determine $a_{i(k)}$:

$$a_{i(k)} = \frac{P_{ij}}{\sum_{k=ij \in \Gamma_+(i)} P_{ij} + P_i^L} \qquad (7.32)$$

Here P_i^L is the load power at node i.

Equation (7.31) is a linear equation system including N unknown variables δP_j, which can be solved by a conventional algorithm. After solving equation (7.31) for $\delta P_i (i=1,2,\cdots,N)$, loss allocation to the load at node j is then calculated by

$$DP_j = \delta P_j \frac{P_j^L}{\sum_{ji \in \Gamma_+(j)} P_{ji} + P_j^L} \qquad (7.33)$$

We can use a similar approach to formulate the problem of allocating the loss to generators. Based on the discussion above, we may conclude that to solve the distribution factor or loss allocation problem, we should first build and solve the linear equations, equation (7.26) or equation (7.31). However, this approach is not efficient and not flexible. We will develop a very simple and efficient algorithm by means of graph theory in the next section.

7.6.3 Methodology of Graph Theory

A load flow distribution graph is a directed graph. At this stage, the direction of each branch is determined by the direction of its active power flow. Each branch has its initial node and terminal node, while each node has its outgoing lines and incoming lines. The number of outgoing lines at node i is denoted by $d_+(i)$, the number of incoming lines by $d_-(i)$. As mentioned above, the set of outgoing lines is denoted by $\Gamma_+(i)$ and the set of incoming lines by $\Gamma_-(i)$. A directed path is formed along the direction of branches. When the initial node and terminal node of a directed path are identical, we have a directed circuit.

In a load flow graph, we use $R_{(k)}, X_{(k)}, P_{(k)}, Q_{(k)}$ to denote the resistance, reactance, active and reactive power flow of branch k, and we have the following theorem.

Theorem 1 If the following relationship holds for each branch along the direction of its active power in a load flow graph,

$$P_{(k)}X_{(k)} > Q_{(k)}R_{(k)}$$

then there exists no directed circuit in the graph.

Proof We use the methodology of reduction to absurdity. If there exists a directed circuit C, then the following relationship holds:

$$\sum_{k \in C} \Delta\theta_{(k)} = 0 \tag{7.34}$$

where $\Delta\theta_{(k)}$ is the phase angle difference between the two nodes of branch k, and can be expressed by

$$\Delta\theta_{(k)} = \arctan\left\{\frac{(P_{(k)}X_{(k)} - Q_{(k)}R_{(k)})/U_{(k)}}{U_{(k)} + (P_{(k)}R_{(k)} + Q_{(k)}X_{(k)})/U_{(k)}}\right\} \tag{7.35}$$

$P_{(k)}, Q_{(k)}, U_{(k)}$ in equation (7.35) take the values at the terminal node of branch k. When $P_{(k)}X_{(k)} > Q_{(k)}R_{(k)}$, then $\Delta\theta_{(k)} > 0$, and thus equation (7.34) cannot hold. Therefore, the directed circuit cannot exist in this situation.

It should be noted that $P_{(k)}X_{(k)} > Q_{(k)}R_{(k)}$ is a sufficient condition, which is satisfied for most load flow distributions. In case there is a branch not satisfied by the condition, the power flow along the branch is certainly negligible. So we can assume the branch is open.

Theorem 2 When a directed graph has no directed circuit, there are at least two nodes, i and j, that satisfy $d_+(i) = 0$, and $d_-(j) = 0$ respectively.

Proof Assume $d_+(i) > 0$ holds for all nodes, i.e. each node has at least one outgoing line. Thus setting out from any node n_1, we can travel to the next node n_2 along its outgoing line. And from n_2 we can travel further to n_3 by similar reasoning, and so on. Thus there are only two possible outcomes: one is that we have infinite travel but this is impossible for a finite graph; the other is that there exist directed circuits, but this contradicts with the condition of the theorem. Hence we can conclude that there is at least a node with $d_+(i) = 0$. Similarly, we can prove the other half of the theorem. Combining Theorems 1 and 2, we obtain the next corollary.

Corollary On a load flow graph, there exists at least one node without an outgoing line and one node without an incoming line.

Definition Assume i is a node with $d_-(i) = 0$ on a load flow graphs. The process of eliminating node i and its outgoing lines $\Gamma_+(i)$ is called the *eliminating process* for node i.

Theorems 3 In a load flow graph, all branches can be eliminated through a recursive node-eliminating process.
Proof Denote the node set of a load flow graph by V, and the branch set by U. Because $D(V,U)$ has no directed circuit, there exist at least exists one node i_1 with $d_-(i_1) = 0$. Carrying out an eliminating process for node i_1, we get subgraph $D'(V\backslash i_1, U\backslash \Gamma_+(i_1))$. Because $D \supset D'$, D' also has no directed circuit. Hence, there exists at least one node, say i_2, with $d_-(i_2) = 0$ in D'. Then we can carry out an elimination process for node i_2, and so on. Thus we can eliminate all branches by a finite (less than N steps) recursive elimination process.

Now we explain the elimination process by a simple example, as shown in Figure 7.9a. This directed load flow graph has no directed circuit, and $d_-(1) = 0$. Thus we can first eliminate node 1 and its outgoing lines 1, 2 and 3. After eliminating node 1, the subgraph D' is formed as shown in Figure 7.9b, in which $d'\_(2) = 0$. Therefore we can further eliminate node 2 and its outgoing line 4. After node 2 is eliminated, respective subgraph D'' is formed, as shown in Figure 7.9c, where $d''\_(4) = 0$. Hence we can eliminate node 4 and its outgoing line 5, and thus we complete the elimination process.

In the above elimination process, the node with $d\_(i) = 0$ is successively eliminated. This process can also be carried out by successively eliminating the node with $d_+(i) = 0$ and its incoming lines $\Gamma_-(i)$. The corresponding definitions and theorems are similar to the discussion above.

7.6.4 Algorithms and Case Studies

The following algorithm can be used in both problems of load flow analysis within a slight difference. For simplicity, in the algorithm below, we will use PDF and PLA to denote the distribution factor problem [26] and loss allocation problem respectively.

Algorithm Based on Eliminating Outgoing Line Process
Step 1 Preparative calculation, including a load flow run and forming
$\Gamma_+(i)$, $d_+(i)$, $d_-(i)$, $(i = 1,2,\cdots,N)$.
Step 2 Search node i with $d_-(i) = 0$ as the node to be eliminated.
PDF: the power injected by each generator at node i, $P_{i,l}$, is known.
PLA: the node loss δP_i is known.
Step 3 PDF: Calculate distribution factors $f_{(k)l}(k \in \Gamma_+(i))$ for the outgoing lines of node

i by equation (7.25).

PLA: Calculate loss allocation to the load of node i by equation (7.33).

Step 4 Do the following for all j, $ij \in \Gamma_+(i)$.

PDF: Transfer the power of each generator at node i to node j

$$P_{j,l} = b_{ji} \times P_{i,l}$$

Here b_{ji} is defined by equation (7.27).

PLA: Cumulate the loss of node j, δP_j, according to Equation (7.31).

Decrease $d_-(j)$ of node j by 1.

Step 5 Set $d_-(i)$ as -1, indicating that the node has been eliminated.

Step 6 Return to Step 2. Search for the next node without incoming lines, until all nodes are eliminated.

We can similarly introduce an algorithm based on eliminating incoming lines. In this case, the calculated results for PDF are the distribution factors of branches used by loads; for PLA the losses are allocated to generators.

The northwest power network is calculated by the proposed algorithm to evaluate the wheeling cost. The data used in the case study is real exchange power data between Shaanxi and Qinghai provinces dated 16 January 1998 as shown in Figure 7.10 a. The calculation results are the wheeling cost paid by Gansu province as shown in Table 7.7 and Figure 7.10 b. The calculation results are the wheeling cost paid by Gansu province as shown in Table 7.7 and Figure 7.10 b. In Table 7.7, sum of wheeling costs includes three parts: 'Energy' in the table refers to the energy loss cost, 'Capacity' refers to the cost of additional generator capacity to compensate power loss, while 'Line' refers to transmission line usage cost. The average wheeling cost of the day is 0.0247 yuan/kWh.

7.7 Conclusions

This chapter has described the Chinese power market that is an embryonic free market in which the state retains ownership of the generators and some of the transmission infrastructure, but is opening up the market to limited competition. Electricity pricing and transmission loss methods have been proposed and examples of a simplified Chinese power system have been used to demonstrate the advantages derived from such methods.

7.8 Acknowledgements

This work is partially supported by the Key Project of National Science Foundation of China. The authors would also like to thank IEEE for granting permission to reproduce the materials contained in reference [18].

Table 7.7 Wheeling cost for Shaanxi and Qinghai power exchange

Hour	Exchange Power (MW)	Loss of Wheeling (MW)	Line Using Cost (yuan)	Wheeling Cost (yuan / kWh)			
				Energy	Capacity	Line	Sum
1	273	10.29	4410	0.011	0.006	0.016	0.033
2	339	13.29	4510	0.012	0.005	0.013	0.030
3	279	7.68	2410	0.008	0.004	0.009	0.021
4	331	11.63	3630	0.011	0.004	0.011	0.026
5	325	11.83	4200	0.011	0.005	0.013	0.029
6	335	14.80	5640	0.013	0.006	0.017	0.036
7	268	7.52	2920	0.008	0.004	0.011	0.023
8	101	0.11	1080	0.001	0.001	0.017	0.019
9	-122	3.37	2160	0.008	0.004	0.018	0.030
10	-226	5.58	2170	0.007	0.003	0.010	0.020
11	-172	5.25	1870	0.009	0.004	0.011	0.024
12	-260	5.84	3760	0.007	0.003	0.014	0.024
13	-165	1.71	2370	0.003	0.002	0.014	0.019
14	-189	3.34	2660	0.005	0.003	0.014	0.022
15	-143	2.15	2400	0.005	0.001	0.017	0.023
16	-197	2.66	3280	0.004	0.002	0.017	0.023
17	-293	7.66	4090	0.008	0.003	0.014	0.025
18	-305	8.29	3130	0.008	0.004	0.010	0.022
19	- 30	0.83	420	0.008	0.004	0.014	0.026
20	0	0.00	0	0.000	0.000	0.000	0.000
21	-116	1.88	1220	0.005	0.002	0.011	0.018
22	-210	3.00	2700	0.004	0.002	0.013	0.019
23	0	0.00	0	0.000	0.000	0.000	0.000
24	154	0.75	1100	0.001	0.001	0.007	0.009

7.9 References

[1] S.Q. Gao and F.L. Chi, *Several Issues Arising During the Retracking of the Chinese Economy*, Foreign Language Press, 1997, Beijing.
[2] J.P. Sun, Electric Power Industry in China 1999, China Electric Power Information Center.
[3] W. Sweet and M. Hood, 'Can China consume less coal?', *IEEE Spectrum*, Vol.36, No.11, November 1999, pp.39-47.
[4] M Hood and W Sweet, 'Energy policy and politics in China', *IEEE Spectrum*, Vol.36, No.11, November 1999, pp.34-38.

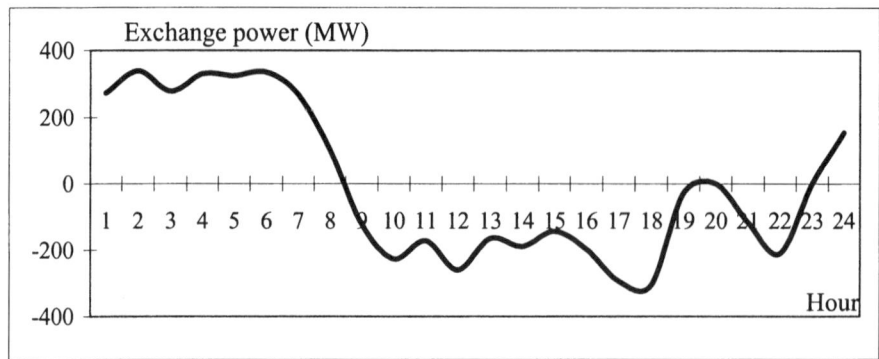

(a)

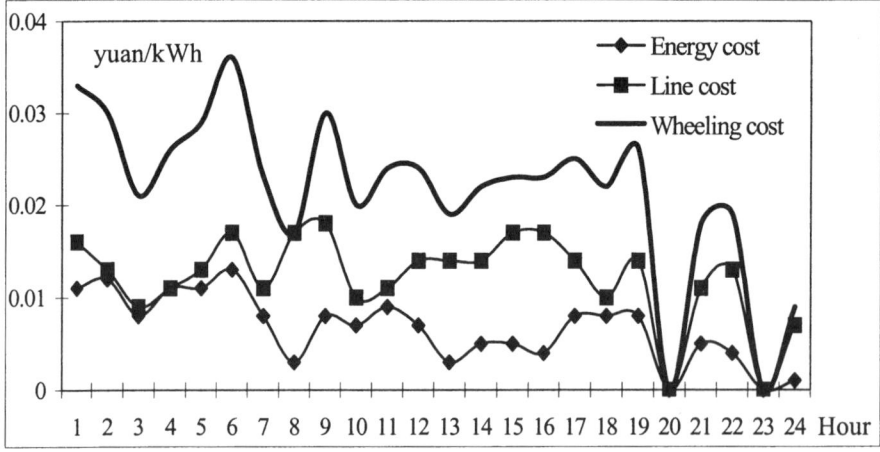

(b)

Figure 7.10 Wheeling cost for Shaanxi and Qinghai power exchange

[5] Y. Wang and K. He, 'The air pollution picture in China', *IEEE Spectrum*, Vol.36, No.12, December 1999, pp.55-58.

[6] Robert H. Williams, 'The Need for Research on Modernised Biomass in the Global Energy Economy', Renewable Energy for Development, Newsletter of the Energy, Environment & Development Programme of the Stockholm Environment Institute, Vol.9, No.3/4, December 1996.

[7] X.F. Wang, 'Preliminary Study of electricity pricing', *China Electric Power*, Vol.32, No.6, 1999, pp.1-3.

[8] M.S. Yan and J.Q. Xin, 'Power market pricing in developing countries, part I, pool purchase pricing and operation with deterministic electricity value analysis', *Proceedings of the Australia Research Council Workshop on Emerging Issues and Methods in the Restructuring of the Electric Power Industry*, 1998, pp.136-141.

[9] Shi Yubo, 'Take vigorous action to promote power industry's reform and development', *China Power Enterprise Management*, No.1, 1999, pp.7-8.

[10] Zhang Shaoxian, 'Clear up reform ideas and initiate a new chapter of professional management', *China Power Enterprise Management*, No.11, 1998, pp.4-5.
[11] Gao Yan, 'On the second step reform of the State Power Corporation of China', *China Power Enterprise Management*, No.2, 1998, pp.4-5.
[12] Lu Yanchang, 'A re-understanding about the simulated power market practice', *China Power Enterprise Management*, No.4, 1998, pp.8-9.
[13] Wang Yongjian, 'Strive to accomplish two reform in three years and basically realize equal power rate in five years', *China Power Enterprise Management*, No.12, 1998, pp.16-18.
[14] J. Logan, 'Balancing the books on energy pricing', *IEEE Spectrum*, Vol.36, No.12, December 1999, pp.59-63.
[15] Xuehao Hu, 'Electricity sector decision making in China', *IEEE Winter Power Meeting*, January 2000.
[16] Lu Yanchang, 'China power industry development and market-oriented reform', *Proceedings of the Fifth International Conference on Advances in Power System Control, Operation and Management*, Vol. 2, IEE, Publication Number CP478, October/November 2000, pp.1-9.
[17] Cha Keming, 'Speed up the nationwide interconnection of power grid', *China Power Enterprise Management*, No.1, 1999, pp.16-18.
[18] X.F. Wang, X.L. Wang and B. Jia, 'Power tracing analysis in wheeling costing', *Proceedings of the International Conference on Power Utility Deregulation, Restructuring and Power Technologies 2000 (DRPT2000)*, City University, London, IEEE, April 2000, pp.173-178.
[19] M.E.E. Baran and V. Banuarayanan, 'A transaction assessment method for allocation of transmission services', IEEE Power Meeeting Paper No. PE-410-PWRS-0-1-1998.
[20] D. Shirmohammadi and P.R. Gribik, 'Evaluation of transmission network capacity use for wheeling transactions', *IEEE Transactions on Power Systems*, Vol.4, No.4, October 1989, pp.1405-1413.
[21] CIGRE Task Force 38.04.03, 'Methods and tools for transmission costs', Electra, No. 174, October 1997.
[22] M. Ilic and R. Cordero, 'On providing interconnected operations services by the end user Case of transmission losses', *National Science Foundation Workshop*, November 1996.
[23] Francisco D. Galiana and Mark Phelan, 'Allocation of transmission losses to bilateral contracts in a competitive environment', *IEEE Transactions on Power Systems*, Vol.15, No.1, February 2000, pp.143-150.
[24] Autouio Gomei Exposito, Jesus Manuel Riquelme Santos, Tomas Gonzalez Garcia, and Enrique A. Ruiz Velasco, 'Fair allocation of transmission power losses', *IEEE Transaction on Power Systems*, Vol.15, No.1, February 2000, pp.184-188.
[25] J. Bialek, 'Topological generation and load distribution factors for supplement charge allocation in transmission open access', *IEEE Transactions on Power Systems*, Vol.12, No.3, 1997, pp.1185–1193.
[26] J. Nanda, L.L. Lai, J.T. Ma, N. Rajkumar, A. Nanda, and M. Prasad, 'A novel approach to computational efficient algorithms for transmission loss and line flow formulations', *International Journal of Electric Power and Energy Systems*, Elsevier Science Ltd, November 1999, pp.555-560.

8

Flexible AC Transmission Systems (FACTS)

Prof. Vijay K. Sood
Hydro-Quebec (IREQ)
Canada

8.1 Introduction

In recent years, major changes have been introduced into the structure of electric power utilities all over the world. The reason for this was to improve efficiency in the operation of the power system by means of deregulating the industry and opening it up to private competition. This is a global trend and similar structural changes have occurred elsewhere in other industries, i.e. in the telecommunications and airline transportation industries. The net effect of such changes will mean that the transmission, generation and distribution systems must now adapt to a new set of rules dictated by open markets. In particular for the transmission sector of the power utility, this adaptation may require the construction or modification of interconnections between regions and countries. Furthermore, the adaptation to new generation patterns will also necessitate adaptation and require increased flexibility and availability of the transmission system. Adding to these problems has been the growing environmental concern and constraint upon the rights-of-way for new installations and facilities. Yet further demands are continually being made upon utilities to supply increased loads, improve reliability, delivery energy at the lowest possible cost and with improved power quality. The power industry has responded to these challenges with the technology of flexible AC transmission systems or FACTS [1,2,3]. This term encompasses a whole family of power electronic controllers, some of which may have achieved maturity within the industry whilst some others are as yet in the design stage. FACTS have been defined by the IEEE [4] as:

A power electronic based system and other static equipment that provide control of one or more ac transmission system parameters to enhance controllability and increase power transfer capability.

For manufacturers of electrical equipment, this challenge provides an opportunity to build equipment that is reliable, flexible and relocatable since planners now demand rapid adaptation to changing system requirements.

FACTS rely, to a large degree, upon advances made in power electronics (PE) and microprocessors. The PE technology, well known in low-power industrial applications, has now migrated to high-power utility applications because of the economical availability of reliable high-power switching devices (i.e. thyristors, GTOs and IGBTs). Note that developments in other related areas such as communication systems (using fiber-optics etc.), super conducting materials for energy storage and metal oxides for surge arrestors will also play important roles in the continuing growth of FACTS applications. This technology will impact on all aspects of power system operations, for example, in:

- generation systems (i.e. from hydro, thermal, wind or photovoltaic means),
- storage systems (i.e. by conversion of energy from AC to DC, DC to AC, DC to DC),
- transmission systems (i.e. by the rapid control of system parameters such as voltage, current, impedance and phase angle),
- distribution systems (i.e. by the rapid circuit or current interruption for protection purposes), and
- consumer systems (i.e. by the power conditioning of consumable energy).

Particularly for transmission systems, FACTS technology offers the following possibilities:

- Greater control of power, so that it flows on the prescribed transmission routes.
- Secure loading (but not overloading) of transmission lines to levels nearer their thermal limits.
- Greater ability to transfer power between controlled areas, so that the generation reserve margin - typically 18% - may be reduced to 15% or less.
- Prevention of cascading outages by limiting the effects of faults and equipment failure.
- Damping of power system oscillations.

Static var compensators (SVC) is an example of a mature FACTS application. Other more novel applications (i.e. STATCOM, UPFC) are being developed and tested to provide increased flexibility, enhance stability and transmission capacity in the operation of power systems. The present environment of deregulation and constraints on building of more transmission facilities provide compelling reasons to develop FACTS controllers. The improvement of a deteriorating power quality will be an additional focus for FACTS controllers of the future.

8.1.1 Benefits of FACTS Technology

The two main objectives of FACTS controllers are:

- to increase the power transfer capability of transmission networks, and

- to provide direct control of power flow over designated transmission routes.

The flexible AC system owes its tighter transmission control to its ability to manage the inter related parameters that constrain today's AC systems, including series/shunt impedance, phase angle and the occurrence of oscillations at various frequencies below the rated frequency.

8.2 Transmission System Limitations

Power flow over a transmission system is limited by one or more of the following [4]:

- system stability,
- loop flows,
- voltage limits,
- thermal limits of either lines or terminal equipment, and
- high short circuit level limits.

Such limitations on power transfer are primarily due to the lack of high-speed control of inter related electrical parameters including voltage, current, impedance, phase angle, reactive and active power. High-speed control of any one or more of these parameters with PE controllers will enhance the value of AC transmission assets. Preliminary studies of several systems have shown that FACTS controllers can provide economic solutions to some of these problems. A discussion of each of the above-mentioned limitations is provided next.

8.2.1 System Stability

This requires the power system to retain a margin of power to ride through a perturbation on the system and still maintain synchronism. Since FACTS controllers operate at high speed and are able to control the power transfer to tighter margins, there is considerable opportunity for avoiding the addition of new generation or transmission facilities. Stability-related constraints [4] could be further split into the following subdivisions:

- Transient stability concerns the ability of a power system to maintain synchronism for the initial few seconds after a major perturbation. A number of examples exist to improve the performance by the use of, say, controlled series capacitors, high response excitation systems and the implementation of braking resistors.
- Ambient damping concerns the ability of a power system to damp out power swing oscillations once initiated by a small disturbance. Approaches that are used to improve ambient damping include power system stabilisers (PSS) and modulation controls on HVDC or SVC systems.
- The voltage stability limit describes the situation when the next increment of load causes a voltage collapse in the power system. This voltage reduction is generally a slow decay occurring over time periods ranging from many seconds to minutes. Approaches that are used to improve VSL include operator action, adding series compensation, generators or synchronous condensers and SVCs.

- Sub-synchronous resonance (SSR) is due to interactions between the series-compensated AC power transmission system and torsion vibration modes of turbine generator units. This issue is dealt with by constraining the level of series compensation permitted to safe limits; usually this level is below that desired for power system security. Approaches that are used to improve SSR conditions include bypass of series capacitors during unsafe operation, passive series blocking filters, active control of generator excitation or SVC on the generator bus. Generator-based protection is usually applied to cover any unexpected contingencies.

8.2.2   Loop Flows

Loop flows occur as an unwanted result of the operation of the interconnected transmission grid and are dictated by electrical circuit laws (i.e. Ohm's and Kirchhoff's laws). These flows are of concern at steady state where the undesired loading affects the voltage levels, losses or the reduction of thermal or stability limits. These effects are addressed by phase-shifting transformers or by series capacitors. The new FACTS controllers also achieve these functions. However, since speed of operation is not a major concern in this load flow problem, controllers will be justified only if frequent adjustments are required.

8.2.3   Voltage Limits

Voltage control is accomplished by a combination of generator reactive power adjustment, fixed or mechanically switched reactors/capacitors and mechanical tap-changers on transformers. Shunt reactive equipment is used for coarse control while the generators provide vernier control.

8.2.4   Thermal Limits

Thermal limits are inherent in transmission systems owing to both line conductors and series equipment (i.e. transformers, reactors and series capacitors). Transmission lines are operated below these limits to provide security in the event of a major disturbance. The role of FACTS controllers will be to use this inherent thermal capacity in a more efficient and secure manner.

8.2.5   High Short-circuit Level Limits

The problem of excessive short-circuit level can be quite difficult and expensive to correct and arises when a new addition is made to the transmission system. This can result in short-circuit levels creeping up in sub-transmission equipment.

8.3 FACTS Technology

The IEEE definition of a FACTS controller is:

A power electronic based system and other static equipment that provides control of one or more ac transmission system parameters.

The technology concerning FACTS is well known in the low-power industrial applications field, but is relatively less well known in the utility power field. This technology is intimately concerned with developments in the following two areas [5]:

- Power electronic switching devices and pulse width modulated (PWM) converters.
- Control methods using digital signal processor (DSP) and microprocessor technology.

Developments in both areas are advancing rapidly, and need to occur further before applications in the power utility field appear economically attractive. Applications of PE in the power utility field still need further research in the following areas:

- active harmonic filtering and reactive/active power support,
- single-node or area-wide application,
- compensation of non-linear loads, and
- transient performance of the controller.

8.3.1 Power Switching Devices and PWM Inverter

Of the switching devices presently and potentially available within the near future (next five years), the gate turn-off (GTO) thyristor and IGBT are the most promising. However, in the longer future (10 years), competition for these switching devices will occur from MOS-controlled thyristor (MCT) devices. A comparison of the various power-switching devices is presented in Table 8.1.

However, owing to the higher switching losses in GTO devices, the maximum switching frequency operation is limited to less than about 1 kHz. Furthermore, owing to the switching and drive characteristics of the device, it has not been feasible to operate devices in parallel for high power applications. Some limited success in the series operation of devices has been reported, but again this remains a limitation.

Hence, for increasing the rating capability of a FACTS converter deploying GTOs, the practical option appears to be the use of several converters operating in parallel. The apparent switching frequency presented to the total filter can also be increased by phase shifting the switching functions of individual inverters, and by the use of phase-shifting converter transformers. A new possibility exists with the use of multi-level converters.

Table 8.1 Comparison of power semiconductor devices

	Thyristor	GTO* Thyristor	IGBT*	SI* Thyristor	MCT*	MOSFET*
Max. voltage rating (V)	8000	6000	1700	2500	3000	1000
Max. current rating (A)	4000	6000	800	800	400	100
Voltage blocking	Sym./ Asym.	Sym./ Asym.	Asym.	Asym.	Sym./ Asym.	Asym.
Gating	Pulse	Current	Voltage	Current	Voltage	Voltage
Conduction drop (V)	1.2	2.5	3	4	1.2	Resistive
Switching frequency (kHz)	1	5	20	20	20	100
Development target max. voltage rating (V)	10000	10000	3500	5000	5000	2000
Development target max. current rating (A)	8000	8000	2000	2000	2000	200

*   GTO   : Gate Turn-Off thyristor
     IGBT   : Insulated Gate Bipolar Transistor
     SI   : Static Induction thyristor
     MCT   : MOS-controlled Transistor
     MOSFET : MOS Field-effect Transistor.

Two versions of switching converters are feasible depending on whether the DC storage device utilised is an inductor or a capacitor. When the storage device is an inductor, the converter is called a current source converter (CSC); when the storage device is a capacitor then the converter is called a voltage source converter (VSC). A noticeable change in converter topology usage will be the increasing use of VSCs instead of CSCs used in traditional HVDC transmission. The VSC will find applications in advanced static var compensators (ASVCs), active filters, STATCOMs, etc. The main reasons for this change are that VSCs are smaller and less expensive than CSCs; furthermore, VSCs are expandable in parallel for increased rating. A brief comparison between VSCs and CSCs is given in Table 8.2.

Table 8.2 Comparison of current source versus voltage source converters

Current source converters	Voltage source converters
Use inductor L for DC-side energy storage	Use capacitor C for DC-side energy storage
Constant current	Constant voltage
Fast accurate control	Slower control
Higher losses	More efficient
Larger and more expensive	Smaller and less expensive
More fault tolerant and more reliable	Less fault tolerant and less reliable
Simpler controls	Complexity of control system is increased
Not easily expandable in series	Easily expanded in parallel for increased rating

Traditional power converters used line-commutated thyristors as their active switching elements, but next-generation converters will exploit self-commutated GTO thyristors in the near-term future, and will probably exploit IGBT and/or MCT devices in the long-term future. The basic PE building blocks will comprise either the:

- anti-parallel thyristors which will be used to control inductive/capacitive impedances, or
- six-pulse CSC or VSC unit, employing multi-level operation (with or without multi-phase transformers) to increase the pulse number (up to 48 pulses), to reduce harmonic generation. The basic switching elements will be the anti-parallel GTO-diode or IGBT-diode unit.

8.3.2   Control Methods and DSP/Microprocessor Technology

Control methods based on either the time or frequency domain are feasible. These require instantaneous monitoring techniques and complex computation of switching functions for the firing of the converter switches. A comparison of the control methods in the two domains is made in Table 8.3.

Table 8.3 Comparison of time domain versus frequency domain compensation

Time domain	**Frequency domain**
Fast response	Slower response
Easy to implement	Complex measurements and analysis
Computational burden is low	Computational burden is high
Ignores past periodic characteristics	Depends on periodic characteristics of distortion

Owing to the complex switching functions required and the computational burden necessary, extensive use of DSPs and microprocessor technology will be required in a power system environment. Utilities have some experience with HVDC technology, SVCs and digital protection relays which use microprocessor-based controls. However, the application of FACTS devices is likely to be at a greater level of complexity than anything known previously within the utility environment. This will require careful considerations of reliability and ease of use within the utility environment.

8.3.3 Present Status on FACTS Activities

EPRI of the USA has been promoting a program (EPRI Project 3022) on FACTS for some years [1]. A number of special conferences on this topic have been organised by EPRI and these conferences comprise, by far, the largest effort on FACTS-related literature. Since the last five years or so, IEEE and CIGRE working groups have also become involved and publications are being reported in their literature also.

FACTS have been with the power industry for many decades in the form of SVC and other applications. However, it is only recently that these applications have become classified under the broad-based heading of FACTS controllers of the power system.

FACTS technology is not a single, high-power electronic controller, but rather a collection of controllers, which can be applied individually or collectively in a specific power system to control the interrelated parameters that constrain today's systems. The thyristor (either line or self commutated) is their basic switching element; however, in one particular application called the interphase power controller, no active switching device is used.

8.4 Solution Options with FACTS

8.4.1 Fundamental Concepts of Transmission

A simplified example of power flow in a loss-less transmission line, with inductive impedance X_L, connecting two ac systems with voltages V_s and V_r is shown in Figure 8.1. The transmitted power P is given by equation (8.1) and also shown in the figure. From equation (8.1) it is evident that power flow can be controlled by varying V_s, V_r, X_L or the angles δ_1 and δ_2:

$$P = (V_s \cdot V_r / X_L) \sin(\delta_1 - \delta_2) \tag{8.1}$$

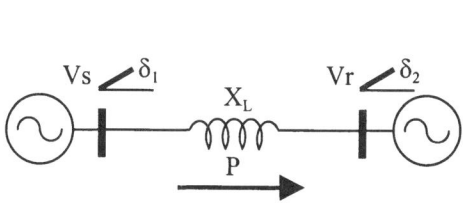
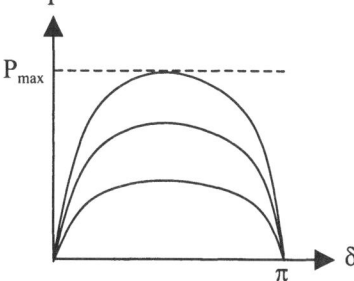

Figure 8.1 Fundamentals of AC power transmission

Transmitted power P can be regulated by control of any system parameter by a FACTS controller, or any combination of controllers, as indicated in Table 8.4.

Table 8.4 Control of system parameters by FACTS controllers

System Parameter	Controller Type	Examples of FACTS Controllers
Voltages V_s and V_r	Shunt	SVC, STATCOM
Impedance X_L	Series	TCSC, IPC
Angles δ_1 and δ_2	Phase angle regulator	TCPAR

The FACTS applications have been split into the following categories depending upon their mode of operation:

- Controllers which act in shunt to the transmission system.
- Controllers which act in series to the transmission system.
- Controllers which act in a series/shunt combination.
- Controllers which alter the phase angle between voltage and current.
- A special category which encompasses HVDC controllers and any remaining controllers.

Details of these various categories are provided in the following sections.

8.4.2 Shunt Controllers

Static Var Compensator (SVC)

The SVC [6,7] has been used for reactive power compensation since the mid 1970s, firstly for arc furnace flicker compensation and then in power transmission systems. One of the first 40 MVAr SVCs was installed at the Shannon Substation of the Minnesota Power and Light system in 1978. At present some 300 SVCs with an installed capacity of 40,000 MVArs are in service all over the world. The SVC results in the following benefits [8]:

- voltage support,
- transient stability improvement, and
- power system oscillation damping.

Although many versions of SVCs exist [9] (i.e. variants are TSR, TCR, TSC) the most common one (Figure 8.2a) usually employs (either thyristor or mechanically) switched capacitors and thyristor-controlled reactors (TCRs). With an appropriate coordination of the capacitor switching and reactor control (Figure 8.2b), the var output can be varied continuously and rapidly between capacitive/inductive values. It maintains the steady state and dynamic voltage at a bus within bounds, and has some ability to control stability, but not much to control active power flow.

Flexible AC Transmission Systems (FACTS)

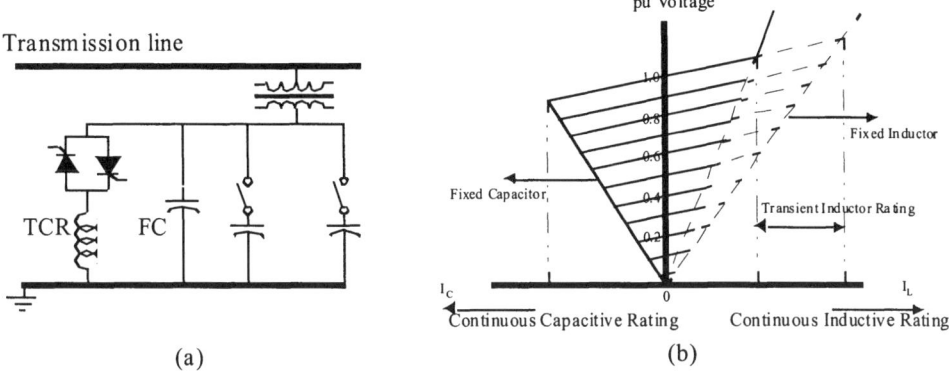

Figure 8.2 (a) The SVC and (b) its *V-I* characteristic

Static Compensator (STATCOM)

The STATCOM, a solid-state voltage source inverter coupled with a transformer, is tied to a transmission line. A STATCOM injects an almost sinusoidal current, of variable magnitude, at the point of connection. This injected current is almost in quadrature with the line voltage, thereby emulating an inductive or a capacitive reactance at the point of connection with the transmission line. The functionality of the STATCOM model is verified by regulating the reactive current flow through it. This is useful for regulating the line voltage.

An advanced static var compensator (ASVC) [10] using a voltage source inverter (VSI) is shown in Figure 8.3a and its *V-I* characteristic is shown in Figure 8.3b. The VSI is operated from a DC storage capacitor to generate an output AC voltage V_o. When V_o equals the voltage V of the AC bus, the VSI draws no current; when $V_o > V$, the current drawn by the leakage impedance of the transformer is purely capacitive. On the other hand, when $V_o < V$ then the current drawn is purely inductive. The functional performance of the ASVC is superior to the traditional SVC.

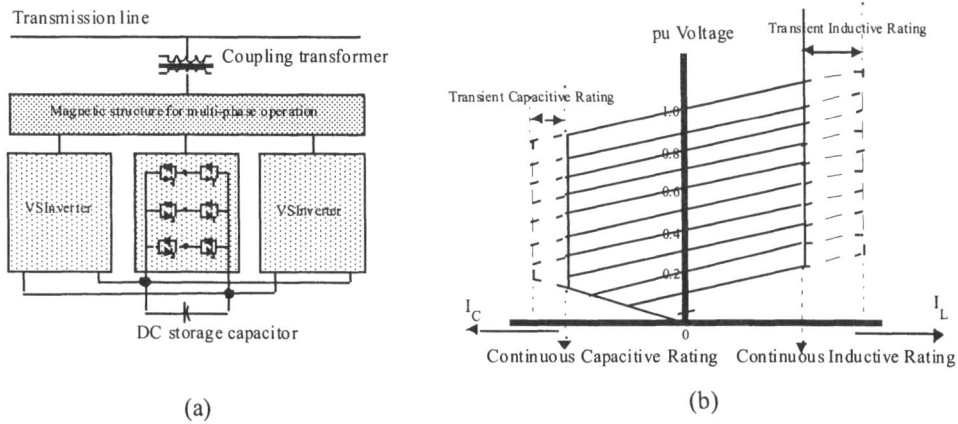

Figure 8.3 (a) STATCOM application and (b) its *V-I* characteristic

The ASVC is also superior to the conventional SVC for the following reasons:

- Reduction in outdoor area requirement, since it replaces the voluminous capacitor/reactor banks associated with a conventional SVC.
- Improved dynamic performance and enhanced stability due to its ability to increase transiently the var generation.
- Improved performance at low operating voltages down to about 0.15 p.u. (limited only by transformer leakage).
- Reduced need for AC filters.

Previous papers have referred to the GTO-based system as an 'Advanced Static Var Compensator'. The functional operation of this device is, however, more similar to that of a rotating synchronous condenser, but without the slow response time and mechanical inertia, and so it was briefly known as the 'Static Synchronous Condenser' (STATCON) [11]. However, current practice is to refer to these as STATCOMs - static compensators.

The STATCOM generates a three-phase voltage source with controllable amplitude and phase angle behind a reactance. When the AC output voltage from the inverter is higher (lower) than the bus voltage, the current flow is caused to lead (lag), and the difference in the voltage amplitudes determines how much current flows. This allows the control of reactive power.

The STATCOM (Figure 8.3a) is implemented by a six-pulse VSI comprising GTO thyristors fed from a DC storage capacitor. Multi-pulse circuit configurations are employed to reduce the harmonic generation and to produce practically sinusoidal current. The V-I characteristic of the STATCOM is shown in Figure 8.3b. The STATCOM is able to control its output current over the rated maximum capacitive or inductive range independently of AC system voltage, in contrast to the SVC that varies with the ac system voltage. Thus the STATCOM is more effective than the SVC in providing voltage support and stability improvements.

One difference between the STATCOM and the SVC is the performance at the limits of equipment capability. The SVC characteristic is a function of the voltage while the STATCOM can continue to produce capacitive current independent of voltage. In addition, the output current can temporarily exceed the steady-state rating. The amount and duration of the overload capability is dependent upon the thermal capacity of the GTO heat sinks and the minimum turn-off current of the GTO. With converter designs, the transient rating of the STATCOM is likely to vary from 120% to 180% of the steady-state rating. Studies on the comparison of performances of the SVC and STATCOM (Table 8.5) are the subject of EPRI Project RP 3023-4 [12].

Flexible AC Transmission Systems (FACTS)   269

Table 8.5 Comparison between a STATCOM and SVC

#	STATCOM	SVC
1	acts as a voltage source behind a reactance	acts as a variable susceptance
2	insensitive to transmission system harmonic resonance	sensitive to transmission system harmonic resonance
3	has a larger dynamic range	has a smaller dynamic range
4	lower generation of harmonics	higher generation of harmonics
5	faster response (within ms) and better performance during transients	somewhat slower response
6	both inductive and capacitive regions of operation possible	mostly capacitive region of operation
7	can maintain a stable voltage even with a very weak AC system	has difficulty operating with a very weak AC system
8	can be used for small amounts of energy storage	
9	temporary overload capability translates into improved voltage stability	

A STATCOM version, based on IGBT switches, and capable of operating at switching frequencies up to 2 kHz has been developed. The core parts of the plant, comprising the IGBT valves, DC capacitors, control system and the valve cooling system, are fitted into a container with a footprint of 10×20 m. The outdoor equipment is limited to heat exchangers, air-cored commutation reactors and the power transformer. A rating of ± 100 Mvar per converter is available; in case of increased rating, multiple units can be operated in parallel. The modular design makes it easily relocatable to another site when desired to meet changing system needs. The response time of this unit is very fast (about one-quarter cycle). As a result of its high switching frequency, the plant can operate without harmonic filters, or may only require a small high-pass filter. The risk for resonant conditions is therefore negligible. Furthermore, the possibility of active filtering of harmonics already present on the network makes this an attractive choice.

Thyristor-controlled Braking Resistor (TCBR) or Dynamic Brake
The stability limits of synchronous generators can be improved by matching the turbine mechanical power and the generator electrical power during system faults. This can be done by introducing either a series or shunt braking resistor. Shunt resistors are preferable because they are less expensive and easier to coordinate in a system with many generators and lines. Moreover, a shunt-connected thyristor-controlled resistor with a radial transmission line can be used effectively to damp power swing oscillations [13] in a transmission system.

These systems are designed to provide post-fault AC system speed control by compensating for fault accelerating power by dissipation in a shunt resistor. A pair of back-to-back thyristors (Figure 8.4) does the application of the shunt resistor. The application of braking resistors should take place as soon as possible after fault detection and they should not be switched out until the derivative of the swing curve becomes negative. The rating of the resistor should be such that the kinetic energy injected by the fault should be dissipated before the generators slip the first pole.

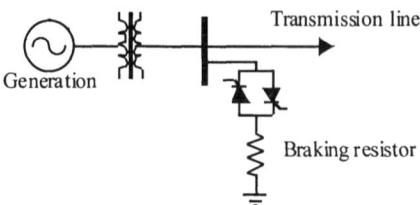

Figure 8.4 Dynamic brake application

The reliability and effectiveness of braking resistors have been demonstrated in three different projects:

1. BPA's Chief Joseph substation, 1400 MW, 3 seconds, 230 kV system;
2. BC Hydro's G.M. Shrum substation, 600 MW, 20 seconds, 138 kV system;
3. Argentina El Chocon Project.

Load-tap Changers
Basically LT changers regulate the output voltage when subjected to variations in the input voltage due to changing system conditions. Mechanical versions were used widely in the industry for many years. These mechanically operated load-tap changing transformers can now have thyristor-operated switches (Figure 8.5) to do the same function faster [14]. This permits the improvement of system stability and damping of the power system oscillations.

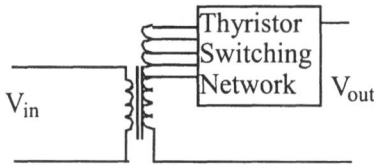

Figure 8.5 High speed static tap changer

Super conducting Magnetic Energy Storage Systems
Super conducting Magnetic Energy Storage (SMES) systems use a super conducting coil (Figure 8.6) to store energy in a magnetic field. The SMES acts as a buffer between the power generation and load consumption and aids in the load levelling and transient matching (within a few cycles) of the two, enabling a greater control and flexibility of the power system [15]. The benefits of energy storage systems are offset by the round-trip losses of storing energy. The round-trip efficiency of a SMES system is claimed to be greater than 90%. The SMES coil is fed by a current source GTO inverter from the AC system. When required, the SMES can supply transient active or reactive power to the AC supply to support it. The technology holds promise for improved energy storage, power quality and flexibility to meet peak utility system loads. A multi-terminal SMES has been proposed to act as a power flow control device also [6,7]. A fairly recent publication [16]

suggested the use of a SMES system for SSR damping of turbine generator units. A SMES unit has been in commercial use on the BPA system.

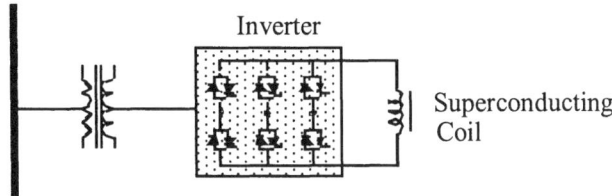

Figure 8.6 SMES operating principles

Battery Energy Storage System (BESS)
A BESS system for utility application is similar to a large uninterruptible power supply. A VSC connects the DC battery to the AC system. Such applications provide load-levelling benefits and act as a spinning reserve on islanded networks. Modulation of the 10 MW BESS at Chino has increased the transfer capability from Arizona to California by several hundred megawatts.

Battery storage has been applied at a number of locations including:

- an 85 MW/30 minute system in Berlin in 1986,
- a 10 MW/4 hour station commissioned in Chino, S. California, in 1988 [17], and
- a 20 MW/4 hour station commissioned in Puerto Rico.

8.4.3 Series Controllers

Thyristor-controlled Series Capacitor (TCSC)
The importance of adding series capacitors to long AC lines for increasing line loadability has been known for a long time. Adding a thyristor-controlled series capacitor (TCSC), however, is a more recent phenomenon and provides greater flexibility in power transmission. A TCSC can vary the transmission line impedance continuously to levels below and up to the line's natural impedance to force power flow along a 'contract path'. The advantages of the TCSC are:

- ability to mitigate sub-synchronous resonance (SSR),
- ability to balance three-phase power flow,
- ability to control power flow flexibly,
- ability to reduce short-circuit currents by rapidly controlling the capacitive to inductive impedance, and
- ability to damp power system oscillations.

The controlled series compensation installation will likely have two key components (Figure 8.7). One element will be the mechanically switched portion, and the second will be the thyristor-controlled portion. The relative sizes of the fixed and controlled mode portions will vary with application. The TCSC portion is made up of a number of small series-connected modules. Each module is either inserted (with the thyristors blocked) or bypassed (with the thyristors fully conducting). In this manner, a stepwise control is achieved with minimal losses and harmonics. There also exists the possibility of operating in a vernier mode where partial conduction of the thyristor path during each half-cycle is used to circulate inductive current through the capacitor and boost its effective ohmic value. One advantage of such small-signal modulation is the control of SSR oscillations.

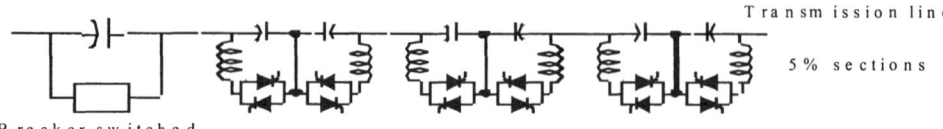

Breaker switched

Figure 8.7 Thyristor-controlled series capacitor

A new control scheme with a TCSC [18] indicates that a method of modulating the firing angle can be used to boost the series capacitor voltage and virtually eliminate the possibility of SSR oscillations. A phase-locked loop (PLL) is used for synchronising the thyristor firing with the line current rather than capacitor voltage for a more stable operation.

Interphase Power Controller (IPC)
Although the IPC [19,20] does not contain any PE equipment, it is included here as a FACTS device that can aid in the management of power flow between two synchronous systems. The basic IPC consists of a series-connected device comprising two susceptances, one inductive and the other capacitive, subjected to properly phase-shifted voltages. Thus, whatever the angle δ at the IPC terminals, some of the components are always subjected to a certain voltage. By adjusting the value of these components, it is always possible to force a current in each of the networks even if the angle at the terminals is zero. When all components are energised, the amplitude and phase angle of the current are set in one of the two buses to which the IPC is connected. This current control thus enables the control of active and reactive power through the device.

Many types of IPC are possible and each type can have different configurations. In one type called the IPC 120 (Figure 8.8), the voltage phase shifts are achieved with a cross-connection between phases using an inverting transformer to reduce the voltage magnitude applied to the reactive components. One practical application of such an installation has appeared in Vermont, USA.

Operating Principle
The IPC 120 uses a group of three-phase reactors and capacitors, each installed in series between two AC systems. The IPC is different from other series compensation devices in the way the series elements are connected. For example, the phase A reactor and capacitor of the sending end system is connected to phases B and C of the receiving end system. Thus, whatever the angle δ at the IPC terminals, some of the components are always subjected to a certain voltage. By adjusting the value of these components, it is always possible to force a current in each system even if the angle is zero. When all components are energised the amplitude and phase angle of the current are set in one of two buses to which the IPC is connected. This current control thus enables the power carried by the IPC to be set, as well as the reactive power absorbed or generated at one of the buses.

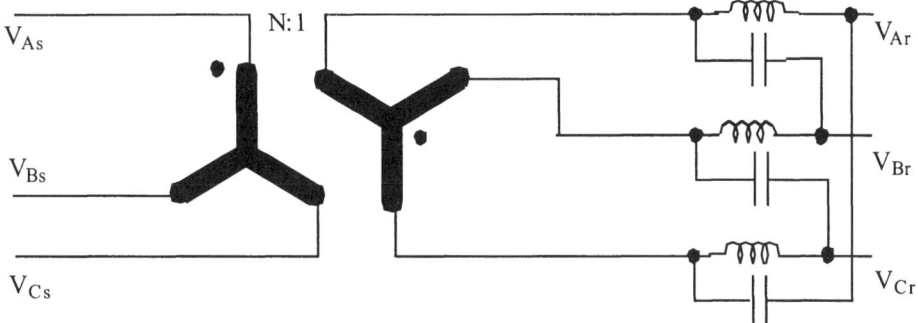

Figure 8.8 Three-phase diagram of the IPC 120

Static Synchronous Series Compensator (SSSC)

The SSSC, a solid-state voltage source inverter [21,43,44], coupled with a transformer, is connected in series with a transmission line. An SSSC (Figure 8.9) injects an almost sinusoidal voltage, of variable magnitude, in series with a transmission line. This injected voltage is almost in quadrature with the line current, thereby emulating an inductive or capacitive reactance in series with the transmission line. This emulated variable reactance, inserted by the injected voltage source, influences the electric power flow in the transmission line.

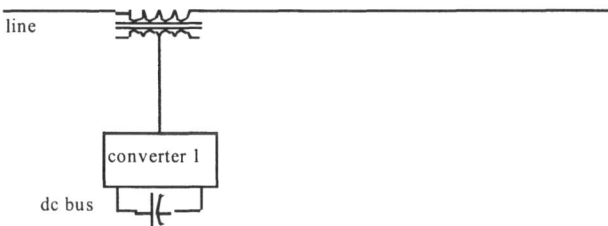

Figure 8.9 Static synchronous series compensator

SSR Dampers

NGH Damper

Another controller currently in use is the NGH-SSR damper [22] to counter SSR which was first observed at the Square Butte Project. SSR instabilities are at times an undesirable side effect of using mechanically controlled series capacitors to a transmission line [23]. The benefits of adding series capacitors are to lower the line's impedance, increase power flow and expand stability limits.

The NGH-SSR damper consists of back-to-back thyristors connected in series with a small inductor and resistor across the series capacitor (Figure 8.10). The operation of the damper is based on two principles. One is to fire the switch 8.33 ms after each zero crossover of the capacitor's voltage, or half a cycle (or 180 degrees) at 60 Hz. But if the voltage wave contains other frequencies, some half-cycles will be longer than 8.33 ms. In this case, the valve firing at 8.33 ms causes some current to flow during the extended part of the half-cycle and damps the oscillations. The second principle is to fire the switch somewhat earlier than 8.33 ms or less than 180 degrees following the voltage zero crossover. Earlier firing causes the impedance of the combined circuit to be more negative than that with the capacitor alone, thus de-tuning the circuit. Furthermore, by modulation of the firing angle, the impedance can have a powerful damping effect at any unwanted frequency below the main frequency. Similar effects can be achieved with HVDC controls. Alternatively, active filters can also be used.

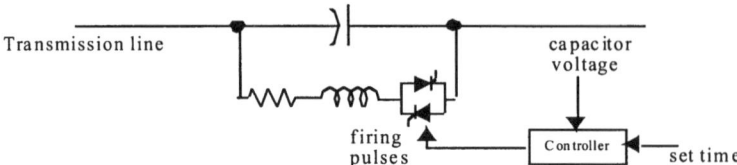

Figure 8.10 NGH resonance damper

SVR Damper
A new FACTS controller for using a TCSC to damp out SSR-related problems was presented in [18]. The new method controls the amount of voltage boost of the TCSC that makes it exhibit a virtually inductive impedance in the frequencies from 15 to 45 Hz where SSR problems may exist. Basically, the TCSC firing angle is modulated to provide damping at SSR frequencies.

ASC Damper
A study by the team at the Kayenta ASC [24] showed similar results that the TCSC exhibits an inductive impedance at sub-synchronous frequencies, and the danger from SSR problems was alleviated. However, the main SSR danger resulted from the uncontrolled portion of the series capacitance in the transmission line.

Current Limiters and Circuit Breakers
PE switches (either thyristors or GTO thyristors) can be used to interrupt AC currents. The thyristor depends on current interruption at the natural zero crossover point of the fault current, whereas the GTO thyristor may interrupt at a specified current setting (which is below its interruption capability). Such static switches have been applied mainly to distribution systems where the switch ratings are lower [25]. The static breaker can have two parts in it, one a static switch and another a current limiter (Figure 8.11). When a fault is experienced, the current-limiting switch is firstly triggered to take over the fault current, and the main static switch is opened. This forces the fault current into the current-limiting path owing to the series inductive element. The non-linear arrestor across the static switch is used to contain the overvoltages [26].

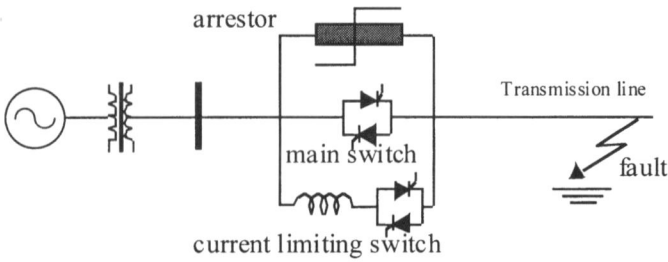

Figure 8.11 Solid-state breaker and current limiter

It is possible to consider the switching capability of thyristors to use as current limiters in the application of TSCS in the future [27]. The increasing interest in FACTS controllers in particular seems to indicate that fault-current-limiting functions can be economically added onto TSCS units. Furthermore, these additional features lend themselves to be retrofitted to existing facilities.

8.4.4 Combined Series/Shunt Controllers

Unified Power Flow Controller (UPFC)
The UPFC [28,29] is able to control both the transmitted real power and, independently, the reactive power flows at the sending and receiving ends of the transmission line. The UPFC consists of two GTO-based converters connected together by a DC link having a storage capacitor. This arrangement functions as an ideal AC to AC power converter in which real power can flow in either direction. Each converter can either generate or absorb reactive power at its own AC terminal.

Converter 2 of the UPFC (Figure 8.12a) injects an AC voltage V_{pq} of variable magnitude and angle in series with the line voltage thereby allowing the control of the phase angle between the resultant voltage and the line current. This injected voltage can be considered as a synchronous AC voltage source. The line current flows through this voltage source exchanging real and reactive power between it and the AC system. The real power exchanged is inverted into DC power and is stored in the DC link. The reactive power exchanged is generated internally by the converter.

Converter 1 supplies or absorbs the real power required by converter 2 through the DC link. Inverter 1 can also generate or absorb reactive power as a shunt device from the line. Converter 1 can be operated independently of converter 2.

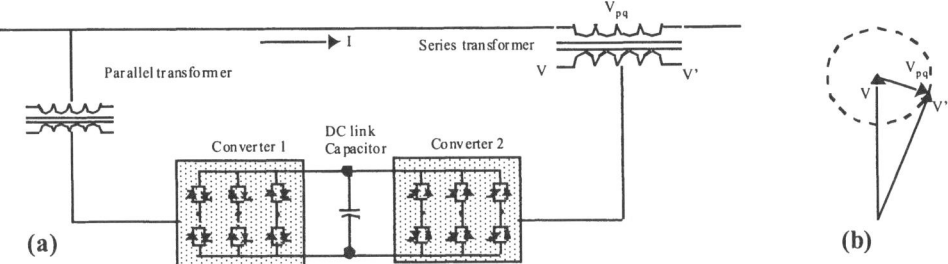

Figure 8.12 Unified power flow controller (UPFC)

The operation of the UPFC can fulfil the multiple functions of reactive shunt compensation, series compensation and phase shifting by injecting a voltage V_{pq} with appropriate amplitude and phase angle (Figure 8.12b). Comparisons between the UPFC and TCSC, and between the UPFC and TCPAR, are made in [28]. Results from transient network analyser (TNA) simulations and computer studies are also shown. An application of this technique is presently underway at WAPA, located at Mead, and is rated for 1060 MVA (series injection) and 475 MVA (shunt compensation) capability.

Interline Power Flow Controller (IPFC)

The IPFC (Figure 8.13) proposed is a new concept [30] for the compensation and effective power flow management of multi-line transmission systems. In its general form, the IPFC employs a number of converters with a common DC link, each to provide series compensation for a selected line of the transmission system. Because of the common DC link, any inverter within the IPFC is able to transfer real power to any other and thereby facilitate real power transfer among the lines of the transmission system. Since each inverter is also able to provide reactive compensation, the IPFC is able to carry out an overall real and reactive power compensation of the total transmission system. This capability makes it possible to equalise both real and reactive power flow between the lines, transfer power from overloaded to under loaded lines, compensate against reactive voltage drops and the corresponding reactive line power, and to increase the effectiveness of the compensating system against dynamic disturbances. In its simplest form, the IPFC becomes an SSSC.

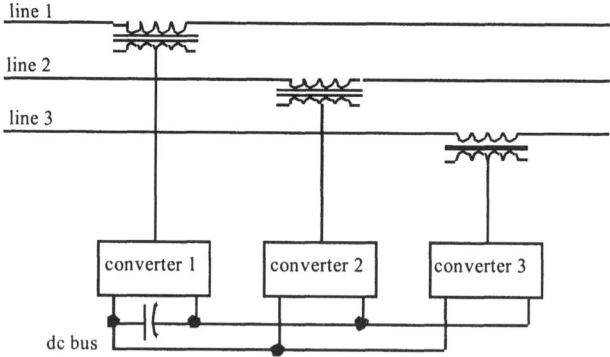

Figure 8.13 Interline power flow controller (IPFC)

8.4.5 Phase Angle Controllers

Phase Shifters
A schematic diagram of a phase shifter [31] is shown in Figure 8.14. A phase shift is accomplished by adding or subtracting a variable voltage component that is perpendicular to the phase voltage of the line. This perpendicular voltage component is obtained from a transformer winding connected between the other two phases.

In the scheme shown [32], the three secondary windings have voltages proportional to 1:3:9. Thyristor switches, one per winding, allow each winding to be included or excluded in the positive or negative direction. The choice of 1, 3 and 9 - along with the plus or minus polarity for each winding - yields a switchable voltage range of -13 to +13, thus giving a variable high-speed control of the perpendicular voltage component.

Flexible AC Transmission Systems (FACTS)

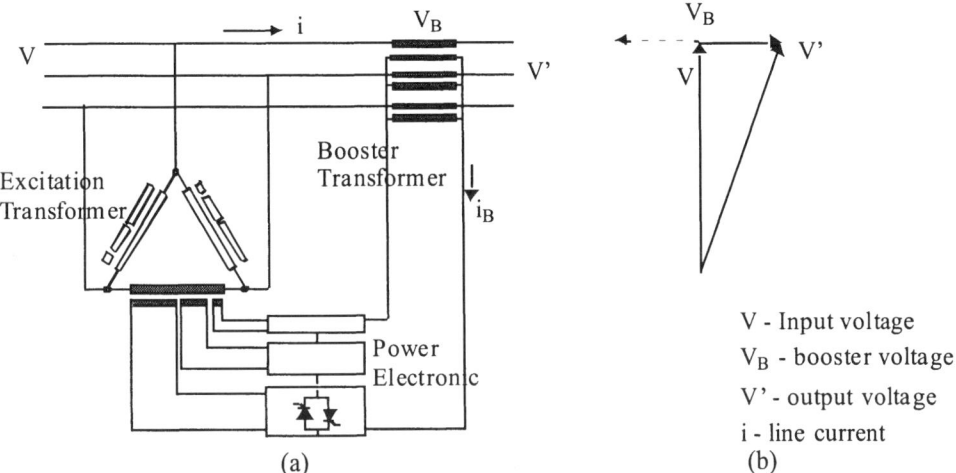

Figure 8.14 Schematic diagram of a phase shifter

The principles of a phase-shifting transformer (Figure 8.14a) with a thyristor tap-changer are discussed in [31]. Similar to a conventional phase-shifter with a mechanical switch, a continuously variable, quadrature voltage is injected in series with the transmission line voltage (Figure 8.14b). It uses three different transformer windings (in proportions of 1:3:9), with switch arrangements that can by-pass a winding or reverse its polarity. It can produce a total of 27 steps using only 12 thyristors (of 3 different voltage ratings) per phase. There is no thyristor-controlled phase shifter in service presently. The conventional phase shifter does not have the ability either to generate or absorb reactive power. The reactive power it absorbs or supplies must be supplied or absorbed by the AC system. Consequently, the phase-shifting transformer must be located close to a generation or load site to avoid large voltage drops due to reactive power transfer.

As part of EPRI Project RP 3022-13, Minnesota Power has developed a novel and economic version of the phase shifter [33] called the MP single core/single tank 'bang-bang' type TCPAR which uses mechanical and thyristor switches.

An advanced phase shifter employing voltage source inverters (VSIs) using GTOs was shown in Figure 8.12 in an earlier section. The converter 2 is used to inject voltage V_{pq} in series with the line. The phase relationship of this voltage V_{pq} to the line voltage is arbitrary, as shown in the phasor diagram. Thus the injected voltage can be used for phase shifting, voltage regulation or both. Furthermore, the VSI can itself generate or absorb all of the reactive power resulting from the compensating voltage injection. On the other hand, any active power supplied or absorbed must be provided by the AC source (unless an additional DC source is available). Switching converter 1 supplies to or absorbs from the dc link capacitor the real power involved in the overall compensation. Since converter 1 handles only real power, and as its AC side is in shunt with the transmission line, it is largely immune from the effects of surge currents during any line faults. Converter 2, however, has to handle its total injection VA as well as any surge currents during faults. Consequently, the rating of converter 1 is smaller than converter 2. The phase shifter of this type is economical to a total angle variation of 120 degrees. Above this value, the rating of the injection converter becomes larger than the power transmitted through the line. In such a case, it might be economical to consider the approach of the HVDC back-to-back configuration considered in Section 8.4.6. The advanced phase shifter has the ability to control all three parameters affecting power transmission: phase angle, voltage and impedance. For this reason, it has also been called unified power flow controller (UPFC) [28].

8.4.6 HVDC Transmission Controllers

Strictly speaking, HVDC transmission does not fit in with the definition provided for FACTS controllers. However, HVDC systems have been a dominant player for such a long time in the usage of PE controllers for transmission that their role in promoting high PE controllers cannot be overlooked. With the latest developments in PE technology, HVDC systems will play an even greater role embedded in AC systems. Traditionally, HVDC transmission is used only for special situations and applications:

- long-distance bulk power transmission where it was cheaper than the AC alternative,
- back-to-back asynchronous interconnections, and
- interconnections using a submarine (or underground) cable.

Long-distance HVDC Transmission
With traditional HVDC transmission, power is electronically controlled, and hence an HVDC line can be used to its full thermal capacity if the converters are adequately rated. Furthermore, owing to its high-speed control, an HVDC line can help a parallel AC line to maintain stability (as long as the HVDC converters do not sustain commutation failures). However, owing to its expensive implementation, HVDC transmission is used only for special situations and applications. An alternative arrangement with a controlled series capacitor in an owing transmission line may provide similar advantages at a lower cost. However, in integrated AC-DC systems, it is now possible to have a DC link in parallel with an ac link. In such systems, and there are a number of such instances (i.e. Pacific Intertie, Chandrapur-Padghe tie, etc.), the DC link can be used to increase the power transmitted over the AC system and provide additional damping when required for stability purposes.

With the availability of GTO/IGBT converters, it is feasible to consider HVDC inverters feeding into very weak and even dead AC systems [34], which have no synchronous machines at all. Some of the problems previously associated with multi-terminal HVDC systems using conventional thyristors may now also be addressed with parallel taps using force-commutated converters. This means that planners may now consider multi-terminal HVDC systems more sympathetically. Before multi-terminal HVDC systems can materialise, however, one additional device that may need further development will be the HVDC breaker; the prospects for this are excellent.

The practical difficulty of implementing GTO-based converters for high-voltage applications has been the problem of operating GTO devices in series. Some techniques have been suggested to build up the high voltage required for DC transmission by using multi-converters in series [35], or the use of multi-level converters; in either case, capacitors are used to equalise the voltages across the multi-converters. The economic viability of such techniques for high-voltage applications is far from clear at present.

Back-to-back (BB) HVDC Converters

Up till now both converters have been line-commutated and therefore have had control only over the direction of active power flow. With the use of self-commutated GTO converters (Figure 8.15), reactive and active power flow can now be controlled in any one of four quadrants, since there is no restriction from the commutation voltage of the valves. Additionally, use of PWM techniques will assist in the minimisation of harmonics generated by the converters and lowering the overall cost of the terminal equipment. We can expect further applications of BB ties at lower cost and improved performance.

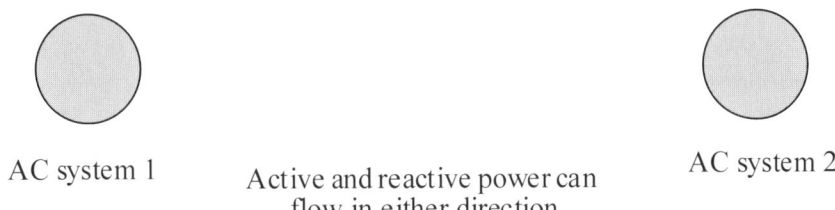

AC system 1   Active and reactive power can flow in either direction   AC system 2

Figure 8.15 Force-commutated BB link

In this respect, two recent developments that will have significant repercussions for future ties are:

- capacitor commutated converters (CCC) [36], and
- controlled series capacitor converters (CSCC) [37].

Both these techniques rely on utilising capacitors in series with the converter with the net effect that the reactive power demanded by the converter is effectively compensated for by the series capacitors. This is a fundamental departure from the previous HVDC converter practice of employing shunt capacitors for reactive power compensation. The beneficial impacts of the series capacitor are as follows:

- The capacitor voltage assists in the commutation voltage for the converter which allows operation with a very weak AC system.
- Since the reactive power flow through the converter transformer is reduced, the dimensions of the converter transformer can be reduced.
- The valve short circuit current is reduced to about 50% when compared with a conventional converter.
- Since the AC filter is reduced in size, the load rejection overvoltages are much smaller.

Coupled with these trends, manufacturers are now offering more efficient, continuously tuned AC filters, active AC and DC filters, compact and modular outdoor valves and fully digital controls. These new concepts are going to reduce the cost of converters and improve reliability.

HVDC Interconnections Using Submarine or Underground DC Cables

A new generation of DC cables is available based on polymeric insulating material instead of the classic paper-oil insulation. The mechanical strength, flexibility and low weight of the cables make them suitable for severe installation conditions. The cables use copper conductors for submarine usage and aluminium conductors for land usage. Land cables may either be installed underground by ploughing techniques or go overhead with aerial cables.

The development of IGBT valves combined with the use of newly designed DC cables has now led to new applications. Using PWM techniques with switching frequencies up to 2 kHz, these new IGBT-based, VSCs are self-commutated and can control active and reactive power flow. This reduces the size of components required appreciably. Hence these converters are constructed in a modular concept and are enclosed in relocatable containers. Their rating range can vary from 7- 600 MW over distances of 0-100 km.

The following applications scenarios are envisaged with this concept:

- Bulk power transmission.
- Reactive power controller, coupled with an active filtering role.
- Small-scale generation from wind, photovoltaic, offshore oil platforms or hydro plants.
- Feeding remote local loads.
- City centre infeed where adding of new rights-of-way may not be available.
- Multi-terminal DC grid.
- Power quality control by isolating disturbing loads such as smelters.

A number of applications have already been reported with this concept (Table 8.6) and future prospects are excellent.

Table 8.6 Applications of HVDC Light technology

#	Project	Rating		Distance	Application	Commissioned
		MW	kV	(km)		
1	Hellsjon	3	± 10	10	AC-DC conversion	Mar. 1997
2	Gotland	50	± 80	70	Feed from wind power generation	June 1999
3	Tjaereborg	7	± 10	4	Feed from wind power generation	Aug. 1999
4	Directlink	180	± 140	65	Asynchronous interconnection	Dec. 1999

Other Configurations

A possible extension would be to install a back-to-back in the middle of a radial AC line. This has the effect of segmenting the long-distance line into two shorter lines and thereby improving the stability of this radial line. The back-to-back can operate in any of four quadrants controlling both active and reactive power flows. The converters would use GTO devices to be able to provide full four-quadrant operation. Additional flexibility can be obtained by having a shorting switch across the back-to-back station. However, such an application will only be considered for special situations because of the high cost of implementation.

Recent considerations of using DC systems with transformer-less converters [38] indicate possible cost savings could result. These techniques could extend the range of applications for DC systems. These new concepts, however, still foresee considerable developments occurring in the ratings of electronic switches before practical applicability.

8.4.7 Other Controllers

These can comprise

(a) Thyristor -controlled Voltage Limiter (TCVL)
In this application a thyristor switch can be connected in series with a part of a gapless arrestor to lower the voltage limiting level dynamically.

(b) Thyristor-controlled Voltage Regulator (TCVR) [1]
This could be a regular transformer with a thyristor-controlled tap-changer or with a thyristor-controlled AC-AC voltage converter of variable AC voltage, in series with the line. Such a relatively low cost controller can be used for controlling the flow of reactive power between two AC systems.

8.5 FACTS Applications

8.5.1 SVC

Interconnection Between Canada and the USA
Northern States Power Co. (NSP) of Minnesota, USA, has installed an SVC in its 500 kV power transmission network, a part of the Manitoba-Minnesota Transmission Upgrade Project, the purpose of which is to increase the power interchange capability between Winnipeg and the Twin Cities on existing transmission lines. This solution was chosen instead of building a new line as it was found to be superior with respect to increased asset utilisation as well as minimised environmental impact. The main purpose of the SVC is to improve the generation and transmission system's dynamic response to network disturbances. It also provides improvement during steady-state conditions by supplying adequate reactive power support. With the SVC in operation, the power transmission capability of the system has increased by some 200 MW. Without the SVC the power transmission capacity of the NSP network would be severely limited, owing either to excessive voltage fluctuations following certain fault situations in the underlying 345 kV system, or to severe overvoltages at loss of feeding power from HVDC lines coming from Manitoba.

The system has a dynamic range of 450 Mvar inductive to 1000 MVAr capacitive at 500 kV, making it one of the largest of its kind in the world. It consists of a SVC and two 500 kV, 300 MVAr mechanically switched capacitor banks (MSC). The large inductive capability of the SVC is required to control the overvoltage during loss of power from the incoming HVDC at the northern end of the 500 kV line. The SVC consists of two thyristor-switched reactors (TSRs) and three thyristor-switched capacitors (TSCs). The short-time ratings are utilised only during severe disturbances in the 500 kV network. Additionally, the SVC has been designed to withstand overvoltages up to 150 % of rated voltage for short periods (< 200 ms).

8.5.2 STATCOM

Prototype Installation
EPRI and the Tennessee Valley Authority teamed up with Westinghouse to install a 100 MVAr STATCON at TVA's Sullivan 500 kV substation [10,11] in Johnson City. This application was demonstrated in 1995. The selection of this site was made to:

- Test the full range of reactive power of the STATCON.
- Aid in damping the oscillations in the TVA system fed in from the neighbouring AEP system.
- Regulate the 161 kV bus voltage during the daily load build-up so that the 500/161 kV transformer bank can be used less often.
- Regulate the 500 kV bus at Sullivan during off-peak periods.

EPRI and GE are considering STATCON applications on the Commonwealth Edison and Pacific Gas and Electric Co. Some cost evaluations have been reported at the EPRI FACTS 3 Conference [12].

8.5.3 TCSC

Prototype Installations
Three installations with different manufacturers are presently being tested in North America with the support of EPRI.

In 1991, AEP of Columbus, Ohio, with the manufacturer ABB began testing of the world's first prototype switch of a single-phase series capacitor bank at its 345 kV Kanawha River substation in W. Virginia. Following successful tests, a 788 MVAr, 2500 A, 42 ohm series three-phase capacitor bank was installed. Each phase of this installation consists of two platforms, one with both a 10% (7 ohm) and 20% (14 ohm) segment and the other with the remaining 30% (21 ohm) segment. This allows compensation from 0% to 60%, in steps of 10% [39].

In 1992, Western Area Power Authority (WAPA) with the manufacturer Siemens dedicated the first three-phase TCSC 230 kV, 330 MVAr installation at Kayenta substation in Arizona [40]. For the requirement of increased power flow on the single 230 kV transmission line between Shiprock Substation (New Mexico) and Glen Canyon (Arizona), WAPA and Siemens/Nokia jointly agreed to install an advanced series compensation (ASC) at the Kayenta substation. In addition to the benefit of adjustable impedance, the thyristor-controlled reactor can provide high-speed protection of the 15 ohm capacitor section.

EPRI with manufacturer GE successfully managed the installation of a 208 MVAr TCSC in September 1993 on a 500 kV line at Slatt substation of the BPA system [41]. The TCSC consists of six series capacitor modules. Each module has a capacitance of 1.33 ohms at 60 Hz, in parallel with a thyristor-controlled inductor of 0.2 ohms. Smooth control of the module is achieved by firing angle control.

8.5.4 UPFC

Application at Inez Station of AEP
The first UPFC application was commissioned in mid 1998 at the Inez station of AEP in Kentucky for voltage support and power flow control.

Voltage Control
The UPFC regulates the substation 138 kV bus voltage by controlling six capacitor banks of 330 MVArs to reduce daily and seasonal voltage fluctuations to within acceptable limits. The controllable reactive power range of the shunt converter is from -160 to +160 MVArs to compensate for dynamic system disturbances.

Power Flow Control (PFC)
The PFC is maintained at a level of 300 MW on the line between Big Sandy and Inez to minimise system losses. Under severe contingency conditions, the UPFC controlled line is capable of transferring 950 MVA.

In order to increase the system reliability and provide flexibility for future system changes, the UPFC installation allows the operation of the shunt converter as an independent STATCOM and the series converter as an independent SSSC. It is also possible to couple both converters together either in shunt or series over a double control range.

Power Circuit
Each GTO converter is rated at ± 160MVA. The converter output is a three-phase voltage set of nearly sinusoidal (48-pulse) quality. Each converter feeds an intermediate transformer that is coupled to the transmission line via a conventional three-winding transformer. The rating of the intermediate transformer is 50 % of the main transformer.

The converters are constructed from three-level poles, each composed of four valves. This arrangement assists in waveform construction to facilitate harmonic elimination. Each converter employs 48 valves in 12 three-level poles with a nominal dc voltage of + 12kV and –12 kV with respect to the mid point. The mid point voltage is maintained by means of a split capacitor and diode arrangement.

8.6 Concluding Remarks

With the deregulation of the power industry [42], FACTS controllers will be required by power systems to manage power flow to utilise transmission lines nearer to their thermal limits. The ability to transmit at higher transfer limits will necessitate greater co-ordination to balance reliability and economy of operation of the power system. Before wide-scale adoption of FACTS controllers, the following concerns of the power industry need to be addressed:

- Transient overvoltages.
- System restoration.
- Generator torsion behaviour.
- Power quality.
- Economic considerations and cost benefits.

To evaluate these concerns, study tools are required to test the FACTS controllers themselves (either as concepts, prototypes or before they enter commercial service). The study tools based on electromagnetic transient program (EMTP) and the real-time power system simulator are presently available. However, models of these controllers are still being developed or enhanced.

Owing to the capital costs involved, FACTS designers will seek to add features to make FACTS controllers more viable, such as the feature of fault current limiting with the TSCS. For the applications of STATCONs, value may be added if features such as power conditioning (i.e. harmonic cancellation) can also be provided along with the reactive power.

8.7   Acknowledgements

The author pays tribute to the many pioneers whose vision of the FACTS controllers has led to the rapid evolution of the power industry. Although it is impossible here to name all of them individually, the contributions of Drs N. Hingorani and L. Gyugyi are noteworthy. The author also thanks his wife Vinay for her considerable assistance in the preparation of this manuscript.

8.8   References

[1]   N. Hingorani, 'FACTS - flexible ac transmission systems', *IEE International Conference on AC-DC Power Transmission*, 1991, pp.1-7.

[2]   N.G. Hingorani and L. Gyugyi, *Understanding FACTS - Concepts and Technology of Flexible AC Transmission Systems*, IEEE Press, 2000

[3]   L.Gyugyi, 'Solid-state control of electric power in AC transmission systems', *International Symposium on Electric Energy Converters in Power Systems,* Invited Paper No. T-IP. 4, Capri, Italy, 1989.

[4]   FACTS Overview, IEEE PES Working Group Report and *CIGRE International Conference on Large High Voltage Electric Systems*, Chairmen: E.Larsen and T. Weaver, April 1995 TP108.

[5]   V.K. Sood, Position Paper on FACTS Technology, Canadian Electrical Association, Contract CEA ST-460, March 1995.

[6]   A. Erinmez, Ed, 'Static Var Compensators', Working Group 38-01, Task Force No.2 on SVC, CIGRE, 1986.

[7]   L. Gyugyi, 'Fundamentals of thyristor controlled static var compensators in electric power system applications', IEEE Special publication 87TH0187-5-PWR, presented in 1987.

[8]   A. Hammad, 'Analysis of power system stability enhancement by static var compensators', *IEEE Transactions on Power Systems*, Vol.PWRS-1, No.4, November 1986.

[9]   R. Varma and R. Mathur, Supplement to a bibliography for static VAR compensators (SVC) and related flexible ac transmission system (FACTS) devices [1988-1994].

[10]   C.Schauder et al., 'Development of a 100 MVAr static condenser for voltage control of transmission systems', *IEEE Transactions on Power Delivery*, Vol.10, No.3, July 1995, pp.1486- 1496.

[11]   H. Mehta, et al., 'Static condenser for flexible ac transmission systems', *EPRI FACTS Conference* 2: TR 101784, Meeting in May 1992, Proceedings December 1992

[12] A. Ekstrom et al., 'Studies of the performance of an advanced static Var compensator, STATCON, as compared with the conventional SVC - EPRI Project RP-3023-4', *EPRI FACTS3 Conference*, Baltimore, Maryland. October 1994.

[13] W. Mittlestadt, 'Four methods of power system damping', *IEEE Transactions on Power Apparatus and Systems*, Vol.PAS-87, No.5, May 1968.

[14] P. Wood, 'Study of improved load-tap-changing for transformers and phase angle regulators', EPRI EL-6079, Project 2763-1, Final Report November 1988.

[15] M. Gavrilovic and G. Begin, 'SMES systems for transient stability and damping improvement of power systems', *American Power Conference*, Chicago, Ill. April 13-15, 1993.

[16] C. Wu and Y. Lee, 'Application of simultaneous active and reactive power modulation of super-conducting magnetic energy storage unit to damp turbine-generator sub-synchronous oscillations', *IEEE Transactions on Energy Conversion*, Vol.8, No.1, March 1993, pp.63-70.

[17] B. Bhargava and G. Dishaw, 'Application of an energy source power system stabilizer on the 10 MW battery energy storage system at Chino substation', *IEEE Transactions on Power Systems*, Vol.13, No.1, February 1998

[18] L. Angquist, G. Ingestrom, and H. Othman, 'Synchronous voltage reversal (SVR) scheme - A new control method for thyristor controlled series capacitors', *EPRI FACTS3*, Baltimore, Maryland. October 1994.

[19] M. Gavrilovic, G. Roberge, P. Pelletier, J-C. Soumagne, 'Reactive and active power control by means of variable reactances', *11th Pan-American Congress* (COPIMERA), Montreal, November 1987.

[20] J. Brochu, P. Pelletier, F. Beauregard and G. Morin, 'Interphase power controller - A new concept for managing power flow within ac networks', *IEEE Transactions on Power Delivery*, Vol.9, No.2, April 1994, pp.833-841.

[21] K.K. Sen, 'STATCOM - Static synchronous compensator: Theory, modeling and applications', *IEEE PES Winter Meeting*, 1999, pp.1177-1183.

[22] N.G. Hingorani, 'A new scheme of sub-synchronous resonance damping of torsional oscillations and transient torque - Parts I and II', *IEEE Transactions on Power Apparatus and Systems*, Vol.PAS-100, No.4, April 1981, and *IEEE PES Summer Meeting* 1980. Papers 80 SM 687-4 and 80 SM 688-2.

[23] J.W. Ballance and S. Goldberg, 'Subsynchronous resonance in series compensated transmission lines', *IEEE Transactions on Power Apparatus and Systems*, Vol.PAS-92, No.5, September/October 1973, pp.1649-1659.

[24] R. Hedin, S. Weiss, D. Mah and L. Cope, 'Thyristor controlled series compensation to avoid SSR', *EPRI FACTS3*, Baltimore, Maryland, October 1994.

[25] T. Ueda et al., 'Solid state current limiter for power distribution system', *IEEE Transactions on Power Delivery*, Vol.8, No.4, 1993, pp.1796-1801

[26] R.K. Smith, P.G. Slade, M. Saarkozi, E.J. Stacey, J.J. Bank and H. Mehta, 'Solid state distribution current limiter and circuit breaker: Application requirements and control strategy', *IEEE Transactions on Power Delivery*, Vol.8, No.3, July 1993, pp.1155-1164.

[27] G. Karady, 'Concept of a combined short circuit limiter and series compensator', *IEEE Transactions on Power Delivery*, Vol.6, No.3, July 1991, pp.1031-1037.

[28] L. Gyugyi et al., 'The unified power flow control controller for independent P and Q flow control in transmission systems', *EPRI FACTS3*, Baltimore, Maryland, October 1994.

[29] K.K. Sen and E. Stacey, 'UPFC - Unified power flow controller: Theory, modeling and applications', *IEEE Transactions on Power Delivery*, Vol.13, No.4, October 1998, pp.1453-1460.

[30] L. Gyugyi, K. Sen and C. Schauder, 'The interline power flow controller concept: A new approach to power flow management in transmission systems', *IEEE Transactions on Power Delivery*, Vol.14, No.3, July 1999, pp.1115-1123

[31] R. Mathur and R. Basati, 'A thyristor controlled static phase shifter for ac power transmission', *IEEE Transactions on Power Apparatus and Systems*, Vol.PAS-100, No.5, May 1981, pp.2650-2655.

[32] R. Baker, G. Guth, W. Egli and P. Elgin, 'Control algorithm for a static phase shifting transformer to enhance transient and dynamic stability of large power systems', *IEEE Transactions on Power Apparatus and Systems*, Vol.PAS-101, No.9, September 1982.

[33] J. Kappenman et al., 'Thyristor controlled phase angle regulator applications and concepts for the Minnesota-Ontario Interconnections', *EPRI FACTS3 Conference*, Baltimore, Maryland, Oct 1994.

[34] V.K. Sood, Position paper for Canadian Electrical Association on Artificially Commutated HVDC Inverters, March 1989, Contract No. ST-174B.

[35]. Z. Zhang, J. Kuang, X. Wang and B. Ooi, 'Force-commutated HVDC and SVC based on phase-shifted multi-converter modules', *IEEE Transactions on Power Delivery*, Vol.8, No.2, April 1993, pp.712-718.

[36] T. Jonsson and P. Bjorklund, 'Capacitor Commutated Converters for HVDC', Paper SPT PE 02-03-0366, *IEEE/KTH Stockholm Power Tech Conference*, Stockholm, Sweden, June 1995.

[37] K. Sadek, M. Pereira, D. Brandt. A. Gole and A. Daneshpooy, 'Capacitor commutated converter circuit configurations for DC transmission', *IEEE Transactions on Power Delivery*, Vol.13, No.4, October 1998, pp.1257-1264.

[38] J. Vithayathil, P. Bjorklund and W. Mittlestadt, 'DC systems with transformerless converters', IEEE Transactions on Power Delivery, Vol.10, No.3, July 1995, pp.1497-1504.

[39] A. Keri, A. Mehrbahn and P. Halvarsson, 'AEP experience with the 788 Mvar series capacitors and the controlled thyristor switch', *EPRI FACTS3*, Baltimore, Maryland. October 1994.

[40] N. Christl, et al., 'Advanced series compensation with variable impedance', *EPRI Conference 1 on FACTS*, Cincinnati, Ohio, November 1990. Proc. March 1992, EPRI TR-100504, Project 3022,

[41] J.Urbank et al., 'Thyristor controlled series compensation prototype installation at the Slatt 500 kV substation', *IEEE Transactions on Power Delivery*, Vol.8, No.3, July 1993, pp.1460-1469.

[42] M. Henderson, 'Operating issues for FACTS devices - An operations planning perspective', *EPRI FACTS3*, Baltimore, Maryland. October 1994.

[43] K.K. Sen, 'SSSC - Static Synchronous Series Compensator: Theory, modeling and applications', *IEEE Transactions on Power Delivery*, Vol.13, No.1, January 1998.

[44] L. Gyugyi, C. Schauder and K. Sen, 'Static synchronous series compensator: A solid state approach to the series compensation of transmission lines', *IEEE Transactions on Power Delivery*, Vol.12, No.1, July 1997, pp.406-417.

9

Asset Management

Kevin Morton
London Electricity plc
UK

Cliff Walton
London Electricity plc
UK

9.1 Introduction

Asset management has been one of the most debated topics over the past decade, yet often those words are used to label some very different processes. Asset management can range from the maintenance and renewal regime associated with a specific individual or group of assets to the management of a multi-billion-pound international portfolio of networks of assets spanning a range of industries. This introduction explores the drivers of the development of asset management from a UK electricity distribution perspective. The drivers for change have most often arisen from regulatory initiatives or from the financial position of new owners, with asset management evolving to meet each new challenge.

Understanding the drivers gives an insight as to why asset management means different things to different players depending on where they are in the restructuring of their business.

9.2 Pre-privatisation (1990): The Public Purse

In the years immediately before privatisation, the electricity industry's finances and investments were very much Treasury driven to meet the public sector borrowing requirements. Competing demands for government investment meant that most electricity companies were required to curtail capital investment and were given annual targets to return cash to the Treasury.

At this stage of development asset management was normally considered synonymous with time-based planned maintenance. However, the constraint on the capital expenditure (Capex) investment meant that as little in the way of reinforcement or renewal was possible and this brought about a focus of improving asset utilisation. Unsatisfactory assets were removed and wherever possible not replaced, whilst underutilised plant was recovered and relocated to meet load growth.

Small operational areas with local policies and/or interpretation of policy meant that there was frequently suboptimal allocation of resources resulting in widely differing staffing and work levels, unit costs and performance of networks.

Competition between business operating units was used to drive lower operating costs and improved resource utilisation but the availability of comprehensive and verifiable data about the costs and performance proved to be a limitation and a constant source of disputes about the accuracy of the statistics between rivals.

9.3 Post-privatisation (1990): Freedom

Privatisation enables governments to realise the considerable capital tied up in utilities and at the same time free industries from the constraints of public sector borrowing requirements. Prices in the UK were initially set via RPI + x with most companies receiving a positive x, thereby enabling the perceived need for significant infrastructure investments to be funded by the new investors.

But the new shareholders brought about a strong profit driver for staff numbers and costs to fall, which in turn brought about wholesale organisational changes and the flattening of structures to reduce operating costs and improve profits.

A key part of these changes was to begin to change the culture of the previously state-owned monopolies into something more appropriate for a profitable private sector. Large numbers of existing technical managers were retrained at university business schools and in doing so mixed freely with colleagues in a variety of other industries and cultures. It quickly became apparent that there were other ways of managing large capital asset bases, and a whole array of statistical and quality control techniques that could be harnessed by a new breed of asset manager if freed from the day-to-day operational obligations and provided with the right information and tools.

9.4 Early-mid 1990s: Getting the Same for Less

Asset manager/service provider business models began to be adopted by many companies but in a variety of forms. Often initially with service level agreements (SLAs) between a relatively small asset management group and internal service provider contractors, some companies moved to adopt formal contracts between the parties and outsource non-core activities to external companies. Key benefits were seen to be:

- Focus on reducing operating expenditure (Opex).
- Development of coherent investment decisions (Opex and Capex).
- Separating the decisions from the doing.
- Analysis and decisions required a different skill set from the execution.
- Enabling companies to force changes in practices and local customs.
- Allowing the doing to be contracted out to the most efficient competitive operator.
- Only what is specified gets done (or paid for).

These changes meant that companies had to acquire additional skills in business and contractor management. Centralising the decision making enabled a new focus on business objectives with less reliance on field technical expertise enabling large-scale reductions of technical staff.

The asset manager/service provider model has met with mixed success. It could easily become confrontational with the drivers of the service provider not necessarily aligned with those of the asset manager. Both sides need experts, one to specify and one to do.

9.5   1994/5+: Getting More for Less

The distribution price control brought major power cuts and successive year-on-year RPI-x reductions in regulatory income, with the promise of unspecified penalties and rewards at end of the five year review period for the achievement of (company-set) performance targets. The response of the companies typically involved:

- World class studies, benchmarking and business process re-engineering.
- Consideration of a whole life approach towards investment.
- Organisations moving towards a three-layer model as they begin to separate asset ownership from operational management:

a   Strategy
b   Asset management
c   Service provider

- A more accountable set of relationships specifying what needs to be done but leaving the doing to the accountable unit.
- Strategic asset management approach to understanding where value is created and destroyed, summarised in the question – 'where best to invest the next pound?'
- Scenario analysis to identify investment strategies that are most immune from regulatory and external risks.

9.6   Late 1990s: Capital Efficiency

The second half of the 1990s saw the lapse of the UK government's Golden Share in the ownership of regional electricity companies (RECs) and the rapid advent of multi-national ownership, alliances and mergers. This brought in an enlarged set of asset management options as owners had to consider in which country, utility and network to invest the next available dollar for maximum return. Regulatory uncertainty as to how investment choices and changes in performance would be rewarded or penalised remained unclear, particularly to overseas investors who appeared unused to dealing with high levels of regulatory ambiguity. New owners withdrawing cash to fund their acquisitions, or to get better returns on their investment outside the UK, forced a renewed drive for capital efficiency.

Merging the management of the two power or more distribution networks potentially allows the sharing of expensive resources such as IT, control, strategic analysis and research whilst applying combined best practice.

The focus on meeting performance targets with severely reduced capital investment brought about a new wave of interest in asset management techniques including:

- Data mining, fault causation analysis and targeting of worst-served customers and most expensive to operate networks.
- Condition monitoring to inform selective refurbishment or renewal.
- Reliability-centred design, engineering and maintenance.
- Removing dormant and problem assets and improving asset utilisation.
- Turnkey capital project management for smaller and smaller projects.
- Value-based procurement.
- Innovation in technology and processes.

9.7 August 1999 Interim Report: All Change?

The regulator's first thoughts on penalties and rewards for past investment and performance were set out based on simplistic assessments made some 9 months ahead of end of the five year review period. At the same time indications of future income caps and performance targets were published with 50% of the savings from mergers clawed back.

The immediate reaction by PESs to the regulator's initial thoughts depended upon the robustness of their asset management scenario planning and their long-term strategic intent. Some continued much as before but overall the publication of the initial review results created a dramatic fall in capital investment orders and in the asset replacement contracts being placed.

With limited rewards for excellent performance and prudent capital investment, the PESs typically switched fixed resources onto those targets they saw as having a good prospect of achieving without additional investment, whilst soft-pedalling on those targets that required investment and additional resources.

9.8 The 1990/2000 Regulatory Settlement and a Major Challenge

The final distribution price control review (DPCR) resulted typically in:

- DUoS prices cut by 20-35% per-unit from April 2000.
- Further 3% per-unit price reductions for 4 years for all PESs from April 2001.
- Proposed efficiency savings of 19%-29%.
- Data collection and aggregation costs transferred to supply.
- The role of distribution redefined and almost all customer service costs transferred to supply.
- A 6.5% allowed rate of return (good asset management can deliver more).
- A metering competition from April 2000.
- Agreement on a business separation compliance plan.
- Performance targets set by the regulator.
- Information and incentives project to come.

It is a major challenge to deliver the DPCR price savings whilst improving customer service with an uncertain incentive mechanism. The suggestion from the regulator is that companies will be placed within an incentive framework intended to mimic a competitive market with companies that do least well in meeting their agreed targets financially rewarding those companies that do best by an exchange of penalty payments.

The uncertainty posed by Ofgem's Information and Incentives Project in terms of what will be incentivised, how performance will be defined and measured has for many companies effectively extended the moratorium in investment.

Companies need to consider how the required scale economies can be effected whilst at the same time delivering improving performance. Some companies may choose to defer major new investment commitments and perhaps organisational changes until there is greater clarity about the rules of the next round of the regulatory game, but this bring its own risks of failure to deliver required improvements sufficiently quickly. The uncertainty highlights the need for a robust frameworks for modelling and valuing the impacts of the various organisational and investment opportunities against a range of scenarios.

The scope of asset management has developed with each previous stage of the restructuring of the distribution business and is therefore set to do so again.

For companies already recognised by the regulator as being frontier efficient or as leaders in effective asset management, but still being presented with a very significant cut in regulated income, a further radical change is essential to achieve the required step change in results and still remain at the frontier.

Combining the management of the two power distribution networks by creating a Joint venture company (24seven) is LE's and TXU's innovative response to the pricing and performance challenge. Creating an outsourcing arrangement with the transfer of staff, vehicles and tools, etc., allows the sharing of expensive resources, such as expert staff, offices, IT, control, strategic analysis and research, applying best practice, determining optimum solutions and delivering a range of services at best value for money - whilst allowing each company to retain is ownership, distribution licence and to develop its own unique competitive market position should this be appropriate.

Such an approach creates the driver for the next evolutionary phase of asset management and requires the separating out and future competitive assignment of the responsibilities between asset owner, asset governor, asset manager and operators.

9.9 Asset Ownership

The owner of major sets of utility assets, whether it be a government, multinational corporation, publicly quoted company or municipal cooperative, will normally have a relatively small set of strategic objectives it is seeking to achieve by its ownership, e.g. profit, service expansion etc. It will not normally wish to concern itself with the detailed financial, regulatory or technical management of the assets but merely to satisfy itself that they are in the hands of an effective governor who can reliably deliver its strategic objectives with frontier efficiency and effectiveness.

9.10 Asset Governance

Separating out the responsibilities of governance from those of ownership, asset management and operations to an organisation dedicated to the creation and release of value through the effective management and exploitation of the assets.

The asset governance concept provides for even a non-technical organisation to benefit from the ownership of a world class set of distribution assets and services with limited regulatory risk and with a minimal staff.

The concept is new to the electricity industry but similar opportunities exist in other capital-intensive industries such as rail and airports where the owners see the need to improve operational efficiency and returns significantly but have other major and more profitable commercial opportunities to commit their management time to.

Regulatory compliance, supply business satisfaction, income maximisation and value generation require a different skill set from the management of individual assets or sets of assets, typically involving:

- Promoting the use, availability, capacity and income generation from the existing system.
- Understanding and actively managing the portfolio of risks.
- Strategically developing network assets to match new markets for differentiated products and services.
- Determining the optimum market position with respect to frontier efficiency, regulatory risks and connections pricing policy.
- Load and economic forecasting.
- Seeking out new opportunities arising from the removal of geographical and competitive constraints.

Asset governance allows economies of scale in asset management and operations whilst retaining and maximising the returns from substantial capital assets and developing a separate and possibly unique market position and set of commercial products. It enables the asset management and operations to be contracted out to the most efficient competitive operator. The principal role of the asset governor is therefore as an informed purchaser. To procure the most effective and efficient asset management and network operations services the governor will require frontier understanding of:

- Innovation in the strategic development and deployment of technology.
- Contract and risk management.
- Appropriate content and volumes of services to be provided within a network management plan.
- The quantity and quality of work done and the value added.
- The strategies, policies and standards necessary to deliver the vision.
- Expertise to develop and renew contracts in an intensely competitive and rapidly developing market.
- When to change, withdraw services or terminate contracts.

The asset governor may also discharge a range of statutory, regulatory and financial planning issues on behalf of the owner that cannot effectively be contracted out, including:

- Liaison with customers and regulatory groups.
- Health, safety and quality audits.

- Publication of quality of supply, environmental and network asset and performance statements.
- Statutory and regulatory accounts.
- Tax and treasury.
- Business and financial planning.
- Legal and some properties issues.

Asset governance provides a solution for the uncertainty of incentive regulation and increased competition as follows:

- Treating asset ownership as a business in its own right and managing inputs and outputs without day-to-day operational distractions enable focus on exploiting the assets and managing the risks associated with incentive regulation and increased competition.
- The economies of scale in asset management and operations are clear and quickly recoverable. Service providers only need one set of IT facilities, control and support processes regardless of how many clients they serve and these can be located advantageously in terms of recruitment and salaries.
- The contract with the asset manager can be structured to require him/her to deliver frontier efficiency performance and costs on each of a generic set of services and projects.
- The contractor can be incentivised to deliver regulatory performance requirements so that the risks associated with regulatory incentives and penalties are effectively transferred to the contractor.
- The governor can withdraw any or all unsatisfactory or uncompetitive services and seek new tenders. The contractor will wish both to retain and to grow the business, to do so he/she will need to demonstrate convincingly to existing and potential clients frontier efficiency in all areas of the business.

It is anticipated that a range of asset management and service Operator contractors will emerge over the next review period to establish a competitive market for these services. This matches well with the regulator's stated objective of advancing and promoting competition in the electricity markets and conducting effective regulation where competition is not fully effective, and of being in favour of experimentation in corporate restructuring with no prescription as to which format may be best.

The asset governor could act for several owners (X and Y), be an owner him/herself, or provide independent advice to an owner (Z), governing one or more discrete and perhaps very different networks and regulatory regimes. There may be several competing network asset managers to choose from.

Many competing network operators, providing individual or groups of services, or perhaps competing in different geographical regions of the same franchised area may supply the services. Figure 9.1 shows such a relationship.

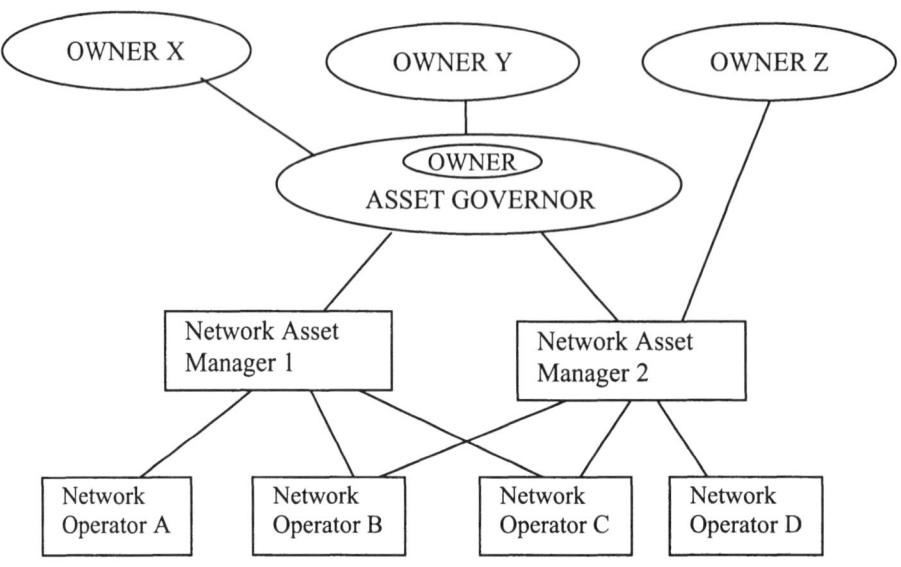

Figure 9.1 Asset governor

9.11 Asset Management

As has been seen, asset management is given a wide variety of interpretations throughout industry and even within the electricity supply industry. Even within a single company the interpretation may change with time - particularly as the company learns from its mistakes!

Typically, asset management has been seen as the core of the distribution business, being primarily responsible for the strategy of the network and the business, albeit that both are derived through teamwork and cooperation throughout the business. The other main areas of focus are asset and network performance, policy and standards, investment and operating costs. The focus on the latter is through work reduction and avoidance, with the operational groups focusing on productivity issues.

9.12 Asset Information and the Ageing Process

Data is the essential ingredient to effective asset management. The asset management process adds value by converting this data into decisions, which reduce the overall lifecycle cost of the network.

In service lifecycle costs can be broken down into three distinct areas: installation, operations & maintenance and decommissioning. However, one of the major factors in determining the overall lifecycle cost is the actual life of the asset. Installation and decommissioning costs have traditionally been evaluated at the conceptual stage of a project, with operating & maintenance costs being considered over a fixed lifecycle for the particular asset. Asset managers today are faced with decisions on where and when to invest, but also have a responsibility for the much wider issue of the existing, ageing assets. For all these existing assets, decisions must be taken which reduce the cost of keeping the asset in service and extending the period for which the asset provides satisfactory service - this is the essence of asset management.

The important question Asset Managers must ask themselves is:

> **'Is the age of an asset the prime indicator of its**
> **ability to remain satisfactorily in service?'**

If the answer is yes then the task is straightforward - we simply record the age of our assets and replace them at the correct time to prevent them becoming a safety hazard to staff, disrupting service to our customers, or becoming expensive to maintain.

Life just isn't that simple. Assets 'age' at different rates depending on the nature of the duty imposed on them, the environment which they inhabit, the way they were installed, as well as a whole host of other contributory factors. Even if we are able to 'normalise' this ageing process it is still necessary to be able to predict the life span for each asset type if we are to avoid replacing plant too early, or allowing our service level to deteriorate.

Tools exist to enable us to determine the condition of some of our plant. The actual condition of the asset is far more important than its age. The present condition is a measure of how well the asset has aged over time.

9.13   Condition Monitoring

Condition monitoring has become commonplace in a number of asset-intensive industries. It has the potential for complicated information systems to capture precise information about particular aspects of an asset's performance and present them in a user-friendly way to facilitate decision making on maintenance regimes and replacement periods. In its simplest form it provides the opportunity for an operator to inspect visually, or put hands on, a piece of equipment and report whether anything has changed since the last visit. Taken to its other extreme, it could mean a fully automated monitoring and reporting system complete with early-warning alarms for an indication of wear or the need for maintenance.

The degree of complexity is a major factor in whether the cost of condition monitoring can be recovered by reduced maintenance costs, higher utilisation, extended life or better management of risk. There is little point in just monitoring the power factor of a transformer on a regular basis, via expensive analytical equipment, if a simple and inexpensive analysis of an oil sample will provide a more reliable indication of ageing, wear or potential failure. If neither of these techniques enable accurate prediction of failure, or reduction in maintenance, then we must question their usefulness in the asset management process.

The major advantage of condition-based monitoring is that it allows the asset manager to have a greater degree of confidence in how the assets are performing and reduces the need to rely on simple time-based preventative maintenance. In short, it provides the opportunity to reduce the cost of maintenance and extend the life of the asset. If regular monitoring is carried out and records collated, a 'footprint' for each item of plant can be established and trends monitored.

This can be useful for predicting potential failures and enabling corrective maintenance to take place, which is normally less expensive than repair following a catastrophic failure. Even when guidelines for unacceptable performance are unavailable, collation of data for a relatively large population enables 'outliers' to be identified and examined in more detail.

Whatever the degree of complexity of the monitoring, the underlying criteria must remain the same - decide on the criteria for performance, or ageing, identify a reliable indicator of potential variation from this standard and then monitor it in the most cost-effective manner. There is little point in continuously monitoring the temperature of a circuit breaker if the real indicator of wear is the time it takes to operate, or the level of electrical discharge activity.

A wide range of monitoring equipment and techniques are currently available to the asset manager. The following sections detail a selection of some of those employed within London Electricity.

9.13.1 Transformers

There are many aspects of a transformer which can be monitored. One simple but effective monitor is the level of moisture in the insulating oil. In its simplest form this is achieved by visually inspecting the colour of the silica gel in the breather. An alternative is the use of refrigerated breathers, which actively reduce the level of moisture in the oil and can be set to trigger alarms if the moisture level should rapidly increase.

This 'ageing' process can effectively be slowed down by proactive treatment of the transformer. London Electricity has effectively employed molecular sieves to reduce the level of moisture within the transformer insulating oil.

A more detailed picture of the condition of the transformer can be gained by carrying out dissolved gas analysis (DGA) on a sample of the oil. This provides an indication of the gases present and what activity is likely to have caused their presence. This is now possible with on-site equipment to provide a coarse indication, by use of a permanent hydrogen and carbon monoxide monitor. This provides an indication of the presence of the major gases associated with overheating, electrical discharge and faults and can be set to alarm at pre-set levels. This provides the opportunity for preventative action in advance of further damage to the plant arising and actually causing the plant to fail in service. Electrical strength, acidity and moisture content tests can also be carried out to help build up a complete picture of the condition of the plant. We can even monitor the condition of the winding insulation and estimate the remaining useful life by carrying out furfuraldehyde measurements on the oil. The level of furfuraldehyde can be related to the degree of polymerisation of the winding insulation papers. The use of fuzzy dissolved gas analysis method for the diagnosis of multiple incipient faults in a transformer is given in the Appendix.

9.13.2 On-load Tap Changers

London Electricity has installed a number of tap changer monitors which record, remotely, not only the number of operations of each device, but the voltage and current at the time of each operation. A widespread data collection operation is in progress to validate the hypothesis that it is not simply the number of operations which determines the need for maintenance, but the duty of the contacts at the time of operation. These monitors transmit data, via modem, back to a central PC, which collates the information and presents it in a user-friendly format. Early experience with this type of monitoring has identified bandwidth problems and unequal numbers of operations between multiple tap changers on the same site. The monitoring of voltage and current also enables a more accurate classification of duty type for different sites, as opposed to the traditional split between continuously loaded transformers in Central London and those with a more varied load profile in Outer London.

Accelerated ageing trials are now in progress to simulate many years' worth of duty over the space of a few months. Inspection of the oil and contacts at various intervals within these trials should enable us to confirm our initial assertion and develop an algorithm for determining maintenance intervals for differently loaded transformers.

9.13.3 Switchgear

Apart from the routine oil condition tests mentioned previously, circuit breaker timers allow the amount of wear on the operating mechanism to be monitored. London Electricity makes use of a simple and inexpensive electronic timer when carrying out operational trip checks on circuit breakers. The timing of each breaker is also monitored by the SCADA system each time it is operated, thus providing a continuous monitoring regime capable of indicating when breaker mechanisms are moving out of tolerance.

Relative humidity monitors are installed on the compressed-air systems associated with air blast circuit breakers. This allows the statutory inspection (Pressure Vessel Regulations) intervals to be extended and hence avoid unnecessary outages.

9.13.4 Other Plant

London Electricity makes extensive use of infrared detectors and thermovision cameras to identify hot spots caused by loose connections or worn couplings on exposed busbars or cable boxes.

Machine health check monitors are used to check for excessive vibration and hence monitor mechanical wear on the bearings for transformer cooling system pumps and fans. The monitor is not a vibration monitor in its truest sense since it is basically an acoustic device, which assimilates noise with wear.

Radio frequency interference detectors are simply carried by the operative around a substation compound to monitor discharge activity within substation plant.

Portable transient earth voltage (TEV) discharge locators are used to pinpoint the source of the discharge activity, or compare similar arrangements within the same substation. Continuous monitoring is also available by installing TEV equipment to monitor 'leaky' plant. This can be moved along a switchboard over a period of time to compile a footprint for a particular substation. Where a high level of leakage has been identified, permanent monitoring equipment with pre-set alarms can be installed to warn of increased discharge levels and potential failure.

9.13.5   Understanding Long-term Asset Costs

If we are to understand the long-term costs of employing assets, then we must have a good understanding of how they perform in service and what techniques can be employed to extend asset life or reduce the level of maintenance required, to keep them in service. As indicated in the previous section, purely time-based maintenance inevitably leads to equipment being maintained too early or too late. In both cases the result is likely to be unnecessary expenditure.

We therefore need to develop a data model of the asset, which can accurately reflect its condition, maintenance requirements and life span. In many cases this can be provided by a wealth of historical data, coupled with on-line indication of performance. Unfortunately this is not always the case and we are left with the problem of developing a model based on assumptions and very little feedback from the asset itself.

Continuous monitoring of assets is not always feasible owing to a number of factors, such as the shear size of the population, accessibility, cost of monitoring, etc. It is therefore necessary to develop sampling techniques to enable prediction of performance across the whole population based on tests performed on a representative group of assets. One such area for London Electricity is the underground cable network.

9.13.6   Underground Cables

The secondary network consists of 8500 km of HV and 17500 km of LV underground cable. The environment in which it exists makes it difficult to monitor, vulnerable to third-party damage, particularly with the high level of excavation activity within London, and expensive to replace.

Faults on the HV network account for nearly two-thirds of the interruptions, which make these assets strategically important to the company. Owing to the high cost of replacing underground cables, investment must be targeted at those assets which are reaching the end of their useful life.

It is essential, therefore, that we are able to derive a measure of the performance of individual circuits and even localised sections within those circuits. The number of faults on a circuit has traditionally been accepted as a measure of its ability to perform effectively. The assumption has been that the performance follows the 'bath tub' curve of failures against asset life, or at least the middle and end portions. The main problem with this approach is knowing when the particular asset has reached the point of ever-increasing failures, without the volume of these failures seriously affecting network reliability. We need to consider the generic model of the bath tub curve in more detail, with particular focus on the bottom portion of the curve.

Figure 9.2 demonstrates a series of curves with varying rates of failure at the end of their life span. The ideal situation would enable us to identify small increases in failure rate as the beginning of a steep increase in the failure rate predicted by the latter portion of the model. The diagram indicates that the slope on many of the small variations is similar to the initial slope of this unacceptable failure rate.

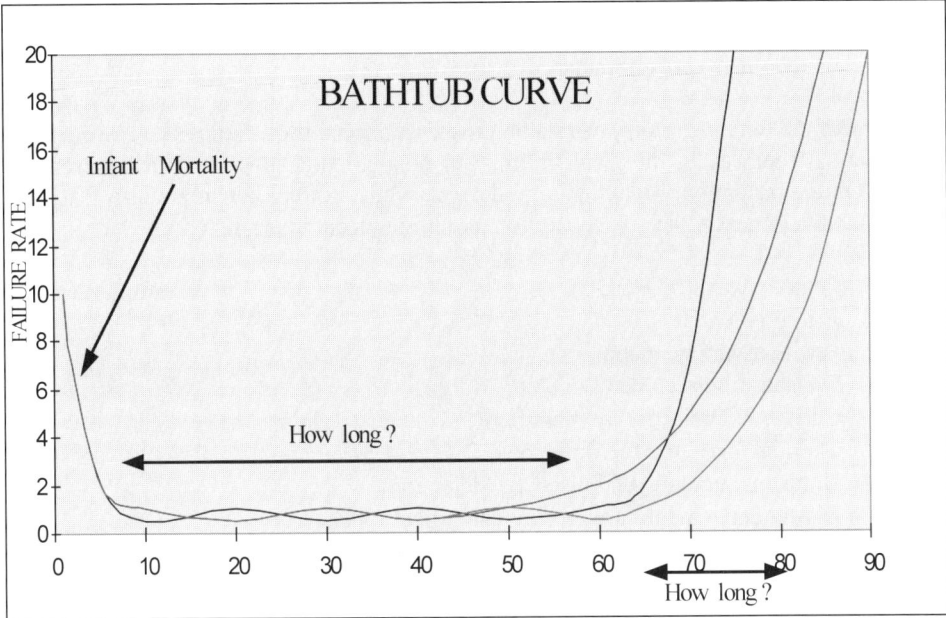

Figure 9.2 Bath tub curve predictive failure mode

We clearly need to identify some other means of predicting failure if we are to avoid extensive and rapidly declining standards of network reliability.

9.13.7   HV Cables

London Electricity's approach to this problem is to analyse the faults, which do occur in greater detail. Each HV fault results in the cable or joint being analysed to identify if there are any generic problems which are likely to lead to similar failures elsewhere on the same cable. The analysis of cable failure statistics provides the crucial key to understanding the types of faults that occur as well as when and where they are most likely to occur.

Analysis of MV cable failure indicates two primary modes of failure:

- *External*: Failure of the cable sheath or cable joint water barrier leading to the ingress of moisture and electrical failure. Failures typically arise from damage during installation or during streetworks, cracking of lead plumbs on cable joints due to thermal cycling, corrosion of armour wires or lead sheaths which leads to arcing under fault conditions and secondary failure. Analysis indicates that some combinations of cable types and environments have significantly higher distributed risk than normal and in these cases a selective replacement programme may be appropriate. Elsewhere the application of targeted condition monitoring techniques can identify individual circuits with high risk of failure. Some of the condition monitoring techniques include:

- Tan δ and delta tan δ
- Zero sequence impedance
- Partial discharge mapping
- time domain reflectrometry.

- *Internal*: Cable failure from overloading itself is rare but most failures can be attributed to thermal runaway in the insulation due to poor jointing practice or the presence of voids in the insulation. Manufacturing defects, whilst not unknown, are also thankfully comparatively rare in the UK. Condition monitoring techniques include:

  - Pressure tests (AC & DC)
  - Fall of potential
  - Partial discharge mapping
  - Dielectric Loss angle (Tan δ)
  - Change in Dielectric Loss angle with voltage (Delta tan δ)
  - Thermal imaging of terminations
  - Ultrasonic imaging and discharge detection
  - Distributed temperature sensing using fiber optics

A series of tests have been devised, for recommissioning faulty circuits, to construct a more detailed footprint of the circuit.

These tests involve the simple, traditional tests for dampness in the papers and DC pressure test as an indication of the circuit's ability to withstand service voltage, although this test has been extended to include a fall in potential measurement as an indication of the quality of the insulation.

More extensive tests have also been included to carry out partial discharge and zero sequence impedance measurement. The objective of these other tests is to gain an appreciation of the performance of the remainder of the circuit, seeking to identify the presence and location of electrical discharge and the quality of the bonding and earth continuity.

9.13.8 Partial Discharge

Partial discharge mapping provides a fairly detailed indication of where the next faults are likely to occur. Figure 9.3 indicates a typical discharge map against cable length and joint positions, the high points on the map indicate areas of high discharge. This information can be used to tackle potential problems on a proactive basis, thus avoiding the interruption to customers or the need to respond to an emergency situation. Whilst it cannot be used to predict precisely when a fault will occur, it has proved invaluable in identifying whether it is necessary to replace an entire circuit or simply a localised section.

Asset Management

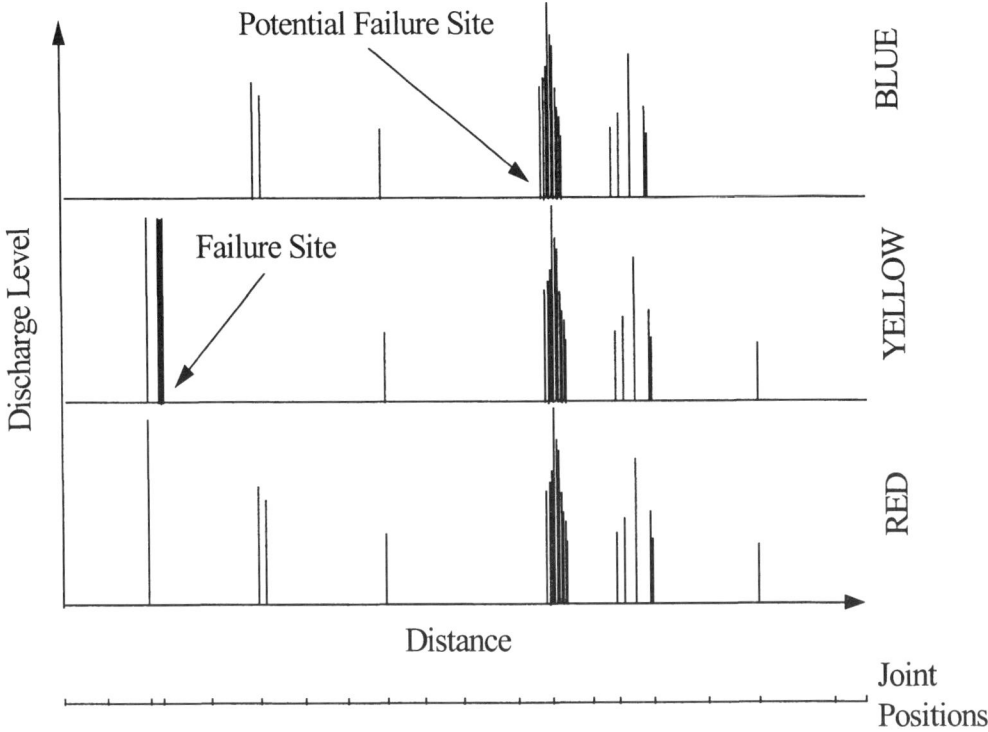

Figure 9.3 Partial discharge map of 11 kV circuit

Techniques have been developed at London Electricity to enable live line monitoring of paper-insulated cables over the entire range of feeders from a substation. This technique has proved invaluable in prioritising the actual partial discharge mapping of individual circuits some of which can now be mapped with the circuits in commission.

9.13.9 Zero Sequence Impedance

Many faults can be attributed to problems with the earth continuity of the circuit, such as corroded armour bonds, cracked or damaged lead at the ends of the joint, etc. Zero sequence impedance measurements provide the opportunity to identify weaknesses in the earth continuity circuit and point the way towards potential failures. Figure 9.4 indicates a scatter plot of zero sequence impedance measurements for a variety of circuits against circuit length. Precise values against circuit length are difficult to predict, owing to the variation in age, environment, etc. However, the information on the chart is an obvious place to target proactive investigations. Partial discharge can then be used to pinpoint potential failures.

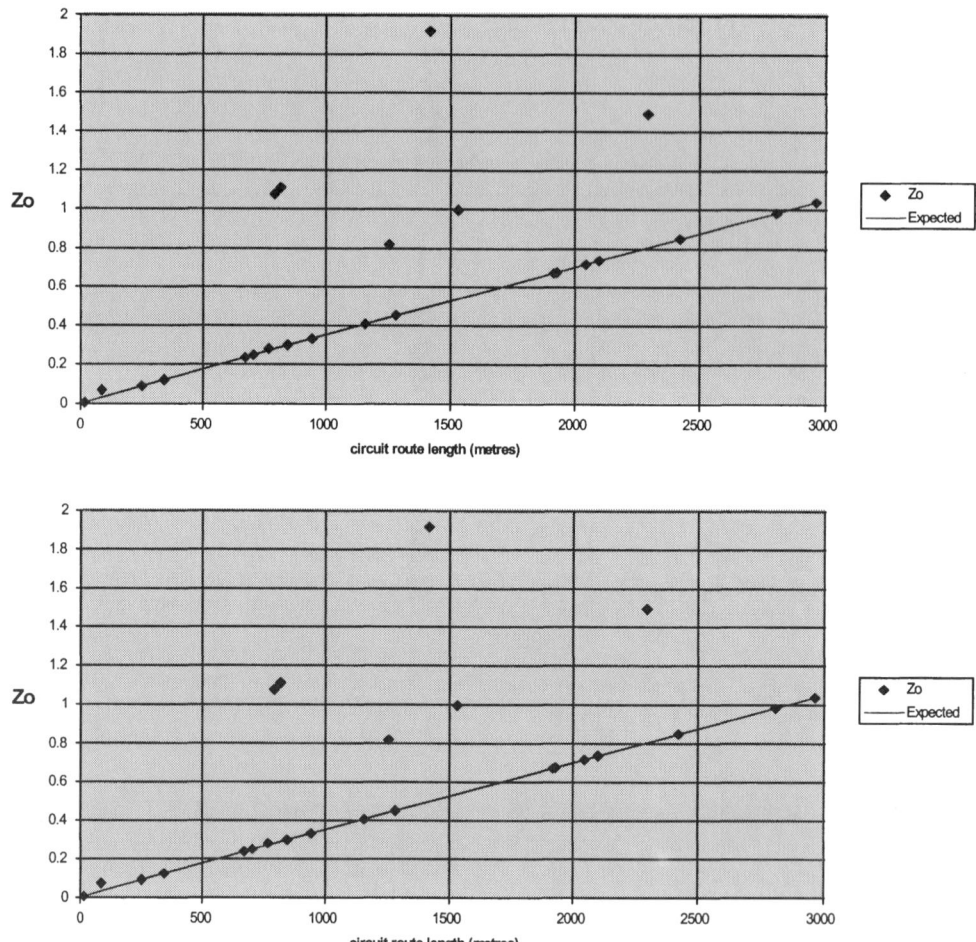

Figure 9.4 Zero sequence impedance values for 11 kV circuits

9.14 Asset Replacement Analysis

Customers' expectations for the reliability of electricity supply have significantly increased in the last 30 years and this trend is likely to continue. Reflecting these expectations, the regulator monitors closely the performance of the electricity distribution companies and strongly encourages them to reduce the number and the duration of service interruptions.

Some of these interruptions are the unavoidable consequence of essential maintenance or repair work. A few results from operating errors while a significant number are caused by accidental or intentional damage to the equipment. However, a large majority of these outages is caused by equipment failures.

The rate of occurrence of interruptions caused by premature ageing or deterioration could be reduced if all the installed equipment were replaced by new equipment.

Considering the enormous investment that such a replacement would represent and the relatively slow growth in the demand for electricity, this refurbishment must be spread over a relatively long period of time. To optimise this replacement programme, it is essential to know where investments in new equipment are likely to have the largest effect on the reliability of service, i.e. to know which equipment is most likely to fail soon and ought to be replaced first.

If failures occurred on a purely random basis, replacing any piece of equipment would have the same effect on system reliability.

On the other hand, if it was possible to show that a single factor (e.g. the type of insulation used for cables) has a much stronger negative influence on the failure rate than any other factor, the replacement policy would be simple: all cables with that type of insulation should be replaced first. A review of the existing literature on this subject suggests that the actual situation is considerably more complex than either of these extremes.

For example, while it is clear that cable failures do not occur on a purely random basis, a number of factors seem to contribute to their probability of failure. These factors include (among others) the type and the age of the cable, the method of installation, the type and condition of the soil in which the cable is buried, the instantaneous and historical loading of the circuit and the previous occurrence of faults in a particular cable section.

Faults are comparatively rare given the asset base and have multiple causes. As a result, chance is the scourge of fault research. The same unsafe behaviour may in one instance go unpunished yet in another result in a catastrophic fault. All sorts of external factors may influence the outcome: weather, co-workers, loading, mechanical failure, making the prediction of large amounts of variance in fault likelihood extremely difficult.

Future research, having demonstrated a relationship between an unsafe behaviour and faults, should then focus on the investigation of factors that predict that unsafe behaviour. This change of focus has already happened to some extent in relation to driving accidents. It is well established that driving above the posted speed limit is predictive of road traffic accidents in the long run. However, any attempt to demonstrate a direct link between speeding as measured in a single study and the occurrence of accidents within that study is unlikely to meet with success. Most speeding goes unpunished by negative consequences. However, that does not mean that speeding is not important in accident causation. Therefore, much research is now dedicated to determining the characteristics that are associated with this dangerous driving behaviour. This approach could also be adopted in fault causation analysis.

Cracking down on relatively small numbers of repetitive faults may have only limited effectiveness in changing overall performance (though it is vital in terms of meeting specific repetitive failure targets). What is required are countermeasures directed at the whole population. Weather-related faults would appear to be such a group where the fault problem may not be located at the extremes of the normal distribution. The problem of weather-related faults may require an approach which focuses on fault causation more broadly conceived, rather than maintaining a rather narrow interest in individual differences in fault liability. It is recommended that future research also consider this perspective

So far, researchers into fault liability have focused almost exclusively on those factors that predict inclusion in the fault group, which in most populations is much smaller than the no-fault group and subject to a high chance factor. Perhaps future research will also devote attention to those networks which manage over a long period of time to avoid faults and determine the factors that promote fault avoidance. It would be possible to target interventions aimed at encouraging such factors. Another sensible strategy would be to shift the focus of interventions towards the positive benefits of avoidance, and away from the negative effects of faults.

9.14.1 Benchmarking

The electricity supply industry has grown up with a fairly risk-averse culture. Network reliability in the UK is above, or at least on par with, levels achieved elsewhere in the world. The industry adopted the traditional approach to maintenance - standard time-based intervals based on the most troublesome plant with little or no drastic changes to the intervals being introduced. A purely time-based maintenance programme tends to lead to a risk-averse approach. It is many years before the impact of a change in frequency is fully understood and, if everyone else is doing much the same, it is unlikely that any one company will be courageous enough to implement radical change.

The advent of privatisation put in place a number of restrictions to sharing information. The competition between the RECs, real or imaginary, meant that companies were seeking to gain competitive advantage by reducing costs in whatever way possible. This inevitably led to some companies adopting a more radical approach to plant maintenance - but how were they to assess the risks?

Utility benchmarking programmes had been popular for a number of years in other parts of the world, notably America and Australasia, and these offered the opportunity to learn from other company's techniques, practices and even their mistakes.

Benchmarking programmes provide the opportunity for companies to see how other similar companies are performing and open up a wide network of contacts to discuss the suitability of new test equipment, or the latest condition monitoring trends. London Electricity has learned a valuable lesson from its involvement in maintenance benchmarking - to question the reasoning behind the entire maintenance programme. This is obviously something which could be carried out without being part of a formal programme, but the information on what other companies are doing, or not doing, has proved invaluable on making the questions more informed and more challenging.

Successful companies do not join a benchmarking programme and simply copy what the best performers are doing. They listen, and learn of strategies which, have worked in other environments and with different plant, and they seek to identify those best practices which are appropriate for their own plant and environment.

Figure 9.5 indicates a typical analysis chart which plots performance against service level. Performance is measured as the unit cost of a particular activity, whilst service level is a combination of measures such as defect rate, customer interruptions, safety performance, etc., for the particular activity.

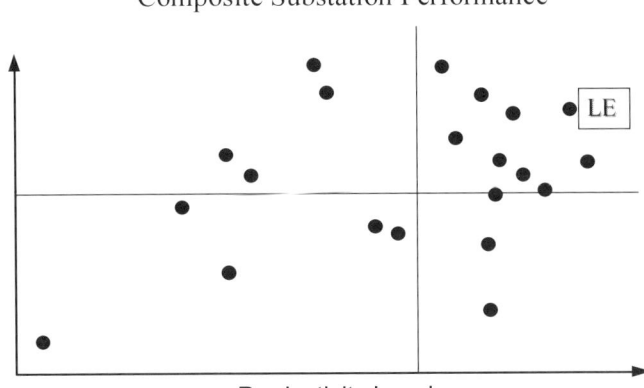

Figure 9.5 Benchmarking performance matrix for substation maintenance

9.14.2 Asset Lifecycle

Management of a large portfolio of assets also necessitates the management of risk. Historically, the growth in usage of electricity has not been linear and we should not be surprised to find that our asset base has not been constructed at a continuous rate. Figure 9.6 details the approximate age profile of London Electricity's major assets, indicating peaks of investment during the 1960s.

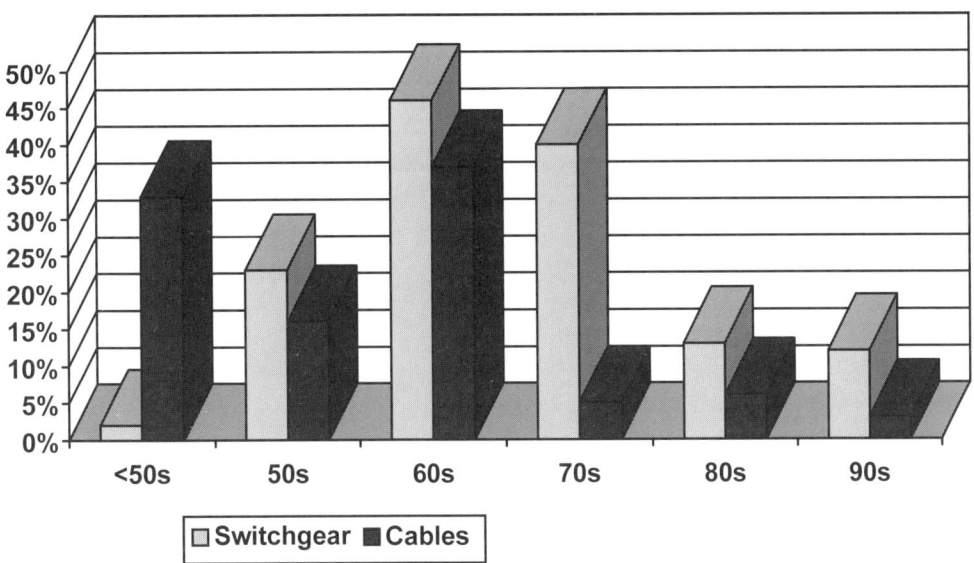

Figure 9.6 Asset age profile

Age-related replacement of assets will clearly lead to similar peaks in investment in the future. Asset management techniques, such as condition based monitoring (CBM), can be used to extend the life of individual assets - assuming that they are in good condition. CBM can similarly warn of the need for early replacement without the need for failure to occur. Another useful technique which is available to companies with dynamic networks is to use other work as a driver for replacement.

This is best illustrated by the following example. A 'typical' substation constructed in the peak investment period of the 1960s would be a 4×15 MVA transformer site with 16 11 kV feeders. Its modern-day equivalent would be a 3×60 MVA double secondary transformer site with 36 feeders. Reinforcement of one substation in an area can normally enable a further two similar substations to be removed, thus avoiding the need for replacement. Extensive use of this technique normally requires an element of load growth.

Even if we do opt for an age-related replacement programme, we need to plan for a more gradual replacement programme. The easy option is to replace assets before they reach the end of their useful life. Our task as asset managers is to manage the risks associated with pushing assets closer towards the end of their useful life by introducing alternative options, or devising ways of closely monitoring their performance.

The actual life in service of assets may frequently be observed to be lower than the accounting life of plant, as used for depreciation by companies, or the much higher assigned service life. This difference has normally been driven by reasons other than replacement needs such as: upgrade for load growth, faults, change of building occupancy, diversions, etc

Figures 9.7.and 9.8 show examples of actual life in service where this has been less than the assigned service life. The data represents all secondary transformers and secondary switchgear removed from the London network since 1991.

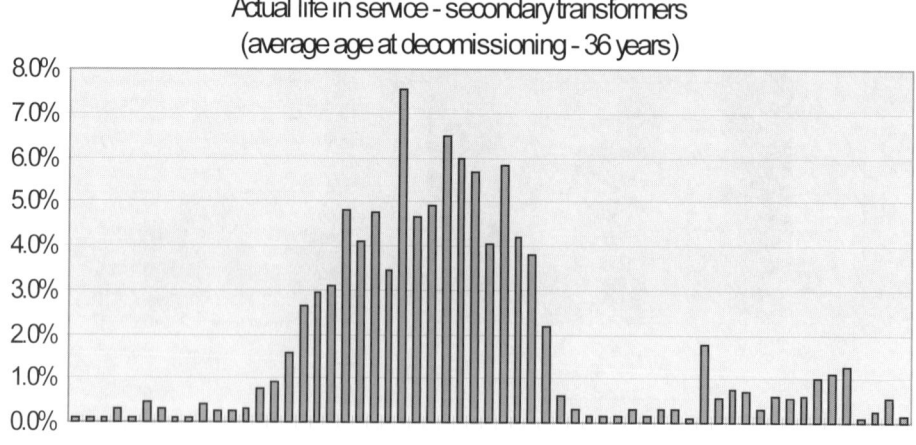

Figure 9.7 Actual life in service – secondary transformer

Asset Management 307

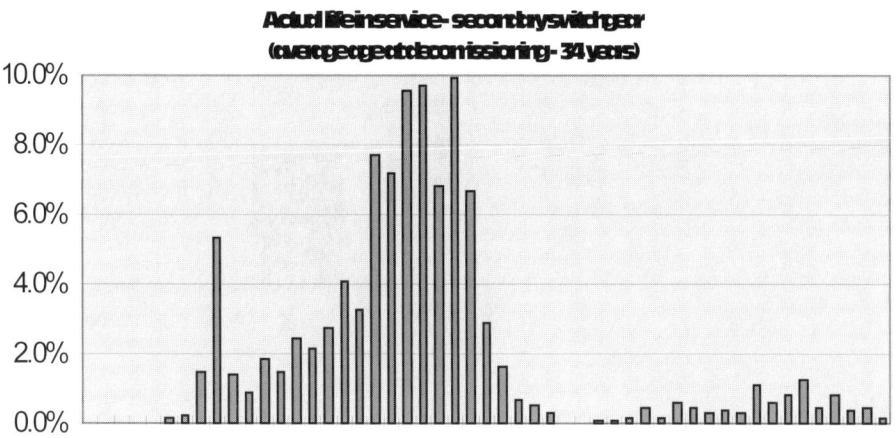

Figure 9.8 Actual life in service – secondary switchgear

9.14.3 Asset Replacement Models

Condition monitoring and assessment provides a very useful guide to the specific assets that need attention within the next review period but are less useful at present for long-term investment planning.

Several models have been used by London Electricity to assess possible asset replacement requirements in the long term (e.g. beyond the next price review period). These use a number of techniques for projecting the current profile of assets using different replacement regimes.

The most elementary replacement profile model is one that targets replacement of all assets in the year they reach the end of the assigned service life. This will have the effect of recreating the same age profile curve as the present population of assets.

The models used by London Electricity apply a spread of replacement ages centred around the assigned service life. This is intended to represent a more realistic view of the range of ages at which assets will be replaced, caused by the impact of the widely varying drivers for replacement such as safety, obsolescence, environment and reliability.

The shape of the replacement profile can be selected to represent how wide the variation from the average service life is likely to be. The most simplistic approach is to take a flat profile, which replaces an equal proportion of the asset population over a given period of time. Figure 9.9 shows 7.5% of the population replaced each year over a 15-year period. The effect of this is to create a new profile of assets which is smoothed and has a wider spread by 15 years.

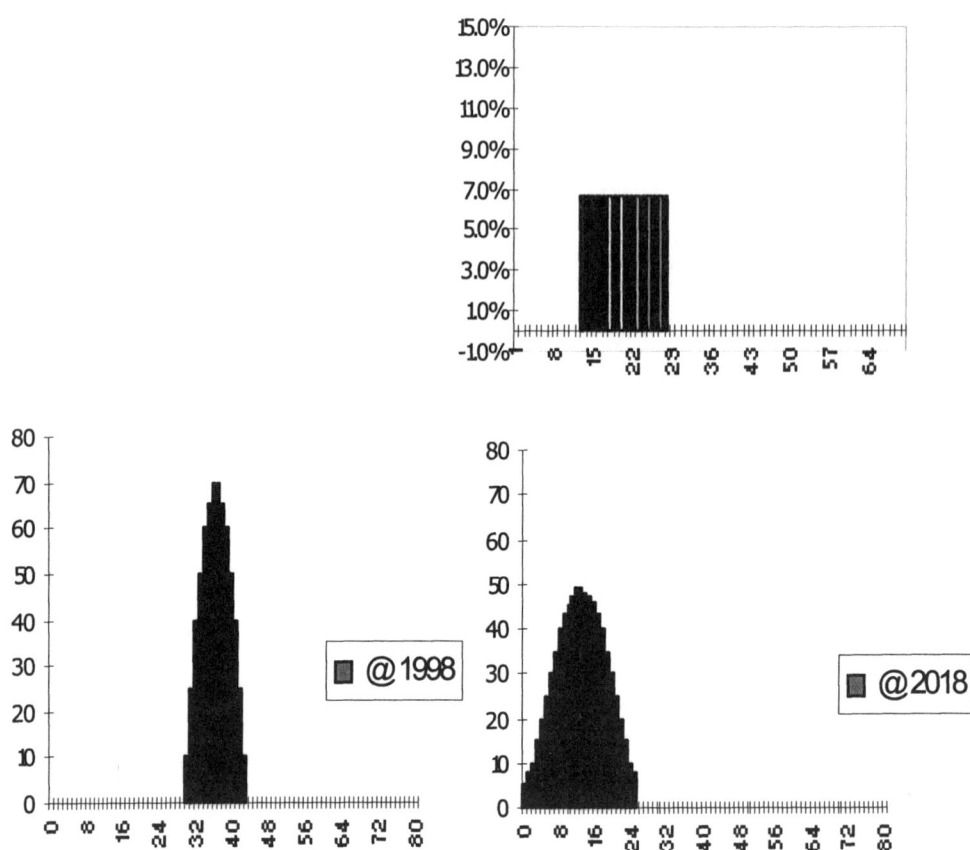

Figure 9.9 Example: effect of flat replacement profile on asset age profile

Alternative replacement curves used for some assets are shown in Figures 9.10 and 9.11. A flat-top curve represents a targeted programme of work to replace a given population of assets over a given period (e.g. ring main unit replacement programme). In cases where replacement is driven by a large number of different drivers a probabilistic curve has been used, with an adequate spread to enable the majority of assets to be replaced within acceptable age limits.

Asset Management

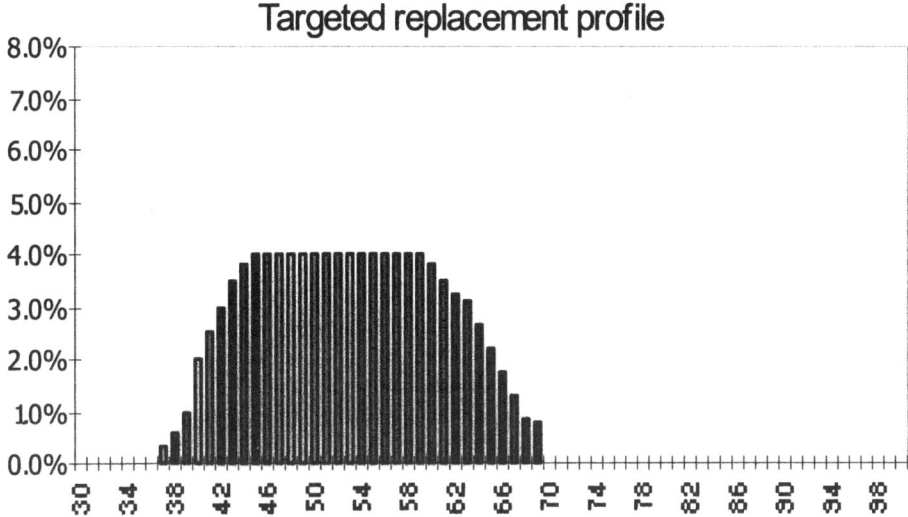

Figure 9.10 Targeted replacement profile

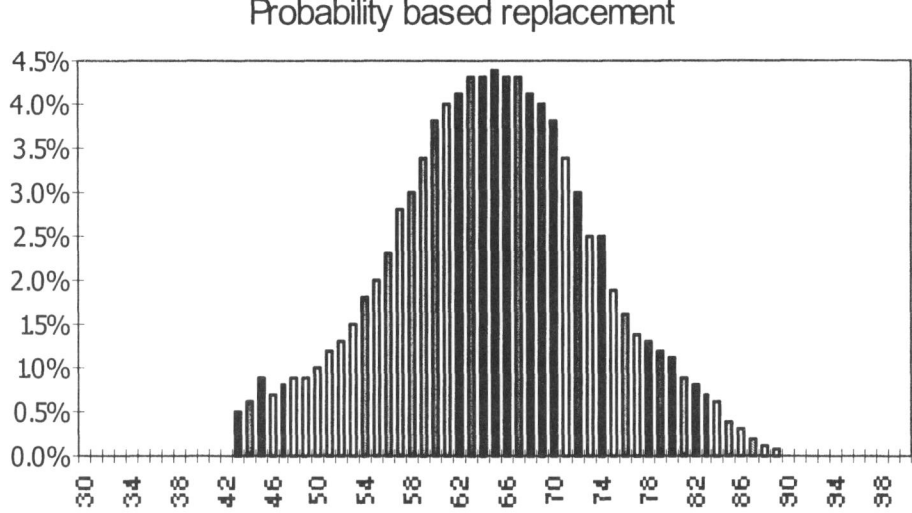

Figure 9.11 Probability-based replacement profile

9.14.4 Investment Prioritisation

Maintenance is only one aspect of effective asset management. The ability to extend the useful life of an asset can be based on the amount and quality of the maintenance carried out, but it can equally be affected by the ability to balance benefits against costs in the overall replacement and investment strategy. Wholesale replacement of assets is an expensive business and we need to ensure that our investment is always targeted at those areas which provide the most benefit.

London Electricity has been deriving a methodology for developing an understanding of what adds value to the stakeholders in the company and putting in place suitable techniques to determine where best to invest in order to maximise stakeholder value. This involves assigning values to non-monetary benefits as well as estimating potential cost savings attributable to certain actions.

The technique involves constructing a model of the project or process and creating an influence diagram to ensure that all internal and external influences are taken into account in evaluating the benefits of a particular project and the way in which it is implemented. Figure 9.12 indicates the cumulative cost/benefit analysis of a portfolio of investment projects. From this we can judge which projects provide the greatest return for the investment and allow us to prioritise within budget or cash flow constraints.

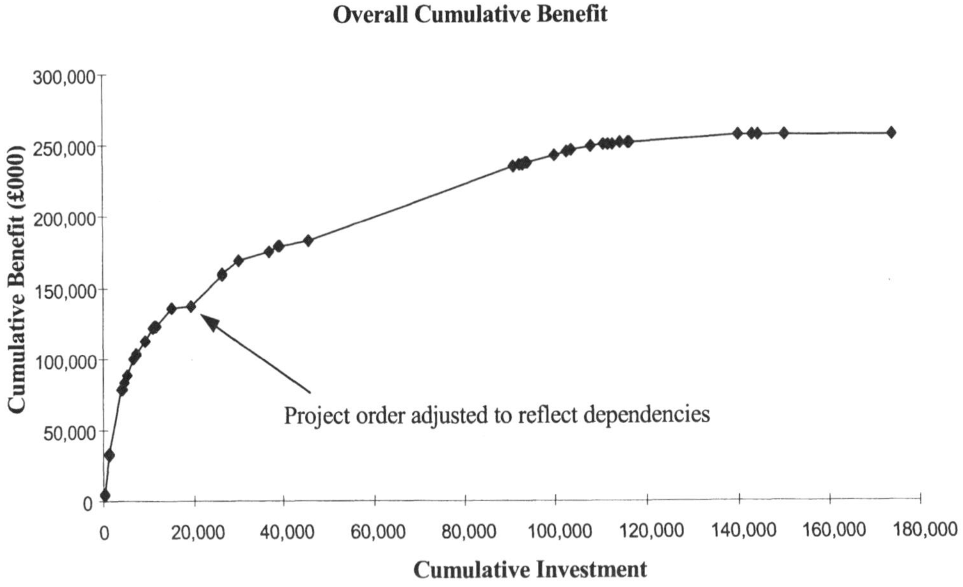

Figure 9.12 Cumulative cost/benefit analysis of project portfolio

The steep slope at the beginning of the curve indicates that the projects at this end exhibit the greatest benefit to cost ratio, whilst those at the other end appear less attractive. Occasionally there is a need to recognise the importance of a less beneficial project, due to the dependency of another, more beneficial, project upon it. This is illustrated by the step in the curve.

Each of these individual projects can similarly be evaluated against a variety of options, such as: do more or quicker, do less or slower, do nothing, etc. This enables evaluation replacement/refurbishment/maintenance decisions to be calculated as well as the prediction of optimal replacement times.

Deferring replacement, refurbishment or maintenance always has a risk associated with it. The use of a formal methodology, which evaluates costs against benefits, provides a useful risk management tool for all the staff associated with and affected by the decisions taken.

The variability of elements within each project are also assessed for criticality to eliminate statistical uncertainty associated with those elements which do not significantly impact on the overall project. This allows us to concentrate on those elements where we need to be more accurate in assessing probabilities or variabilities.

9.14.5 Technology Strategy

London Electricity has been developing a technology strategy to ensure that all potential areas of development for the network are complementary to each other. The main objectives of the technology strategy has been to ensure that state-of-the-art technology and information systems are evaluated and potential operating cost savings are identified.

The various strands of the technology strategy all need to build towards a common objective of providing the degree of network performance required. Each element or task can be evaluated on its own merits but, in general, those projects which also 'add value' to other future projects are the most likely to be included in the investment portfolio.

A good example of this is the functionality of the remote terminal unit (RTU) currently being installed to facilitate remote control of the 11 kV network. Additional features have been built into these units to facilitate the transfer of data from the LV system when suitable devices have been manufactured to obtain the required information. This kind of specification would not be possible without such a cohesive strategy.

Much of the monitoring work is experimental, but it is already possible to install power outage disturbance sensors (PODS) in the premises of a customer who has suffered repetitive supply interruptions which will contact the control centre in the event of a supply failure via a telephone line. Fault passage indicators installed at various points on the LV network provide more localised information about the position of the fault, which will eventually be relayed back to the office via the RTU. These RTUs also have the ability to provide on-line loading and status information for the substation, which can provide invaluable information to the network planners and analysts.

Other work has concentrated on ensuring that many of the independently developed information systems, for control, network design and analysis, etc., are able to share information via a 'data hub'.

Another major task has been the development of a more proactive version of the partial discharge mapping technique mentioned previously. Continuous discharge monitoring of HV and EHV feeders is economic and this, coupled with the ability to switch the HV network remotely, could facilitate the isolation of potentially faulty sections without customer interruption.

9.15 Refurbishment and Replacements

The cost of replacing plant can be very high especially where space to install new equipment before decommissioning the old is not available and load transfers have to be made before work can begin. The costs of delivering enhanced performance and functionality can often be minimised by considering the four Rs:

1. Removal
2. Retrofit
3. Refurbishment
4. Replacement.

Frequently networks have changed considerably since original installation and a critical design review can often result in a considerable quantity of less satisfactory plant being able to be removed entirely without significantly affecting network performance.

Where enhanced functionality such are remote control, telemetry, condition monitoring facilities, faster fault clearance, higher fault levels or lower maintenance levels are required then the solution may be to utilise kits that have been developed by specialist suppliers to retrofit existing, basically sound equipment. This technique has been extensively used in the UK to replace withdrawable oil circuit breakers with modern vacuum equivalents whilst retaining the existing fixed portion of the switchgear. Similarly many thousands of oil and SF_6 ring main units have been retrofitted with actuators and condition monitoring equipment at a fraction of the price of complete replacement.

Some switchgear, transformers and tap changers may have known type defects or design limitations that make them unsuitable for long-term retention or immediate retrofit although their basic performance is sound. In these cases more complete refurbishment may be appropriate in order to increase their serviceable life significantly. Typically refurbishment may include the replacement of inadequate switchgear mechanisms, bushings and seals and the treatment of oil to remove moisture and contaminants. Normally such programmes of work can be carried out by specialist teams *in situ*, but for some items of switchgear a rolling factory exchange programme may be more suitable. *In situ* refurbishment of transformers is normally limited to the reconditioning of oil and the removal of moisture and contaminants.

Programmes to refurbish and retank smaller transformers have been utilised with considerable success in the past but the price of replacement units currently makes refurbishment uncompetitive. There may still be a refurbishment case for large transformers with more serious defects being returned to the factory for the removal of cores, desludging, core tightening and corrosion repairs.

Where the condition of the plant is very poor, the population too small to warrant the development of standard refurbishment or retrofit kits, or simply that a modern equivalent is cheaper to install and maintain then a replacement programme may be most appropriate. For replacement within existing substations the costs of installation, cabling and jointing may exceed the cost of the new plant several times over. A critical look at the suitability of the replacement equipment for the physical environment and the installation techniques used may indicate that a collaboration between the switchgear manufacturer and the installation contractor on a major replacement programme can reduce the overall outage times and costs considerably.

'Cablecure' injection of gas and silicone fluid into the circular stranded core of single-core XLPE cables to remove moisture and fill voids is established practice in the USA where direct-laid HV cables can be refurbished for typically less than a third of the replacement cost. Until recently most HV cables in the UK have generally been three-core paper insulated with lead or corrugated aluminium sheaths which are unsuitable for this technique. At present there are no viable refurbishment techniques for these types of HV cables but preliminary research work is under way to establish the viability of similar refurbishment techniques.

At EHV some targeted refurbishment of oil and gas pressure cable joints is undertaken to extend the useful life of these circuits where thermal expansion and ground movement have resulted in the cracking of joint plumbs and the movement of ferrules within joints. Oil sections can also be redesigned to reduce pressures and real-time monitoring can enable lower operating pressures to be employed. The presence of oil or gas transmission paths through pressurised EHV opens up the possibility of limited reconditioning although in practice electrical failure of these assets is extremely rare under normal operating conditions.

9.16   Risk Management and Insurance Consequences

Active asset management will include an assessment of the relative network risks and of their financial and public relations consequences to the whole business to allow the implementation of appropriate risk control measures.

For power systems risk may be considered as the product of three independent variables:

risk = population × probability × consequence

Each system will have its own mix of the components of risk and so it is necessary for the asset manager to draw up a prioritised assessment of the risks faced by each of the networks under his/her management.

9.16.1   Risk Control

HV faults are rare but the consequences can be high. Effective asset management bounds the risks of major events by reducing:

The probability by
- Identifying circuits and parts of circuits with higher that normal risks of damage, thermal stress, fire, ground movement, vibration, etc., and where appropriate taking remedial action, e.g. additional mechanical protection for circuits on bridges, in tunnels, alongside railway lines, motorways, etc.
- Routine automatic assessment of network design for normal and outage conditions against long-term changes in loading patterns such as the following:

    - Regularly exercising switchgear reduces the risks of failure.
    - Targeted condition monitoring with timely remedial action.

- Automated securement and/or remote restoration of supplies with real-time telemetry, which can significantly reduce the possibility of overloading and secondary failure which if sustained may cause far more extensive damage than the initial, perhaps simple, failure.

The consequences by

- Minimising the number of customers affected through active risk management design of the systems and the use of appropriate protection zones.
- Knowing the status of the network and keeping customers advised.
- Providing restoration in seconds or minutes, not hours.
- Ensuring adequate supplies of spares and skilled resources.
- Establishing contingency plans and regularly exercising them.
- Laying off some of the financial risks, contract exclusions and insurance.

Some events that may initially be considered improbable, such as the coincidental loss of multiple independent circuits or of substations, may nevertheless be worth considering with respect to the physical, political and economic environment. Some rare events such as aircraft crashing from their flight paths, failure of flood defences, earthquakes, terrorism, major fires, industrial disputes, computer viruses, etc., may be of particular of concern when combined with supplies to central business districts, continuous process industries, the media, security and transport services.

Asset management strategies for these situations might include multiple redundancy from outside the environmental zones, fall-back or manned reserve control centres, local generation or just a targeted set of contingency plans with regular exercises to validate them.

9.16.2 Major Incidents

Major incidents are fortunately comparative rare but typically just a handful of events each year will account for around 10% of customer interruptions. At first sight these major incidents often seem to arise from a unique set of circumstances. However, analysis of these disparate types of events using large populations of components and over many years may reveal underlying patterns and trends. Such analysis can produce significant surprises: for example, that perhaps a high proportion of such failures are associated with stuck circuit breakers and that even these are most often associated with equipment that has been recently been installed or maintained, or that a certain type of equipment has a low but catastrophic failure fate.

Failure at major substations can result in large numbers of customers being affected for prolonged times whilst repairs are effected, and also follow-up outages for inspections and remedial works on other similar equipment can result in the overall security of the system being compromised for prolonged periods with a significant loss of resource opportunity.

Regular trip testing is a key performance indicator that the breaker will operate when called upon to do so. With the increase in remote control facilities this activity can be most economically carried out on both primary and secondary systems from control centres and will increasingly be performed and reported automatically, releasing maintenance staff to tackle other activities.

Correct installation and commissioning of plant and equipment is critical to both life cycle costs and system reliability:

- Check protection operation/operative after modifications and circuit outages.
- Primary system and busbar modifications post-commissioning inspection after 3 months with thermal and discharge surveys.
- Exercise MSS circuit breakers remotely regularly, e.g. twice a year.
- Visual inspection of outdoor installations and precautions against flying debris.
- Minimise repair time on first circuit outage.

9.16.3   Type Failures

The economics of purchasing often means that large quantities of the same type of switchgear, transformers or ancillary equipment are purchased and installed in close proximity on networks that are being built, extended or refurbished at the same time. Experience has shown that whilst the widespread catastrophic failure of new or refurbished equipment is rare, problems that could lead to longer term failure or maloperation are identified considerably more often. Health and safety considerations mean that such defects may require live operation of the plant to be restricted until after it can be taken out of service for inspection and modification.

Such type failures can present the operator with very significant risks, as in the extreme, large sections of networks could be rendered inoperable under routine and fault conditions. This is particularly the case with wholly underground HV cable networks where the opportunity to undertake live line overhead work does not exist. Large-scale and repeated outages could also be necessary to effect the necessary remedial inspections and modifications.

The risks posed by type failure can be managed by:

- Selecting equipment with a proven excellent performance record.
- Actively maintaining a diversity of makes, types and versions of equipment throughout networks so that a type failure of one type does not result in widespread paralysis of networks.
- For new types of equipment and changes to existing designs, encouraging the manufacturer to participate in formal and independent failure mode and criticality analysis that can identify and address the probability and consequences of a type failure in the design.

9.16.4   Common Mode Failure

Common mode failure can occur where a single incident places a number of system components at risk at the same time. Typical causes of common mode failures are:

- Widespread storm damage to overhead lines.
- Mechanical excavator damage to several cables in the same trench.

- Damage to protection wiring or pilot cables affecting multiple circuits.
- Failure of tripping batteries and their chargers.
- Change of environment affecting the ambient temperature or the cooling capacity of backfill or ventilation equipment.
- Smoke from fire in a switchboard contaminating other switchboards or accessories.
- Flooding of low-lying or underground substations.
- Damage to poles, towers or foundations on duplicate circuit overhead lines.

Again the principles of risk assessment can be used to rank and prioritise the probability and consequences of failure. Mitigation measures can include physical separation or segregation of routes and switchgear, provision of ventilation, flood defences, alternative supplies, condition alarms and contingency plans. Often it is found that it is the simple and often relatively low-cost items that contribute some of the largest unmanaged risks.

9.16.5 Financial Risk Management

Today's new environment brings new uncertainties and risks [1-14]. Different stakeholders see different risks. Much talk about risk today is focused on financial risks, but it is remarkable that many important risks are not measured in dollars. We see more uncertainties and risks under deregulation. System planning, engineering and operating procedures reduce the risks of widespread or local service interruptions. Transmission, generation, distribution and system protection engineers develop techniques and advanced software for reducing risk.

Uncertainty introduces risk, which is associated with decisions. There is no risk when there are no uncertainties. Decisions made can affect these uncertainties and can reduce or eliminate hazards. Most risk analysis in power and elsewhere is based on either probabilistic or unknown-but-bounded models [1,2]. Total risk can be decreased by selecting investments with negative correlations, i.e. one can hedge against an undesirable outcome by betting on the opposite direction.

As electric industry restructuring brings more and more players into energy markets around the world, the number of opportunities for power producers and marketers is increasing as well. We know the electric power markets to be extremely volatile. Already, many companies have learned the hard way that risk assessment and financial management are crucial to the success of their businesses. Participants in deregulated markets engage in trading to hedge against the risks arising from uncertain price movements and other uncertainties. Perhaps the most attractive attribute of trading in deregulated markets is flexibility along with limitless opportunities.

The financial risks of failure faced by distribution companies are considerable. At the lowest level these will include the routine costs of repairing or replacing equipment that is lost owing to electrical, mechanical or weather-related damage.

Because of the enormous potential for financial loss to customers arising from the failure of electricity supply, consequential loss is normally explicitly excluded by clauses contained with the contract for supply with the supplier or end use customer. The option exists for customers to insure these consequential risks separately, although there may be some advantages for network managers in marketing commercial power insurance services. However, the network owner faces an increasing array of both very real and potential penalties from regulatory and competitive regimes that are increasingly placing companies in multiple jeopardy from failure to deliver required levels of service.

Despite exclusions on consequential loss arising from loss of supply, a fire in a substation located within a major commercial tower block could present a significant financial risk if the network operator could be found liable for the damage to the building and its effects on the occupier's business and staff.

Similar risks exist with respect to the heath and safety of employees, customers and members of the public injured or killed by incidents on the power system. The risks arise not only from the incidents themselves but also from any requirements for widespread remedial works to prevent repetition that are recommended by panels of inquiry or required by statutory or judicial reviews.

A major environmental disaster such as the tornadoes seen in the south of England in 1987 or by EdF in France at Christmas 1999 could present the owner with financial risks of a similar order of magnitude to the current value of the owner's network assets or annual income.

Because of the scale of the potential financial risks it is normal for owners to act as their own insurers for all routine events and only to lay off risk with other insurers for the most rare and unusual events that individually cost more than a significant proportion of annual income. Effective asset management enables these risks to be economically managed and bounded and thus has a significant payback not only in terms of direct and consequential costs avoided but also in terms of limiting the costs of insurance and penalty payments incurred.

The introduction of increasingly competitive and incentivised operating regimes will inevitably mean that there must be a greater emphasis on active risk management to counterbalance the removal of risk underwriting provided by state or municipal ownership or regulated price regimes. Large multinational groups can be expected increasingly to use their financial size and new deregulated ownership opportunities, to balance their portfolio of risks across a variety of regulatory and physical environments in order to minimise the risks of any localised disaster seriously jeopardising their overall position.

9.17 Asset Information Acquisition

Asset information is the key to effective asset management; however, determining what data to capture, how often, how to store it and then how to use it effectively is far from simple or without cost. Sparse, undated or corrupted data can be seriously misleading whilst too much data inappropriately presented can cause information overload, slow IT system response to a standstill and create inappropriately large costs in data collection and maintenance.

There is a real cost to collecting and maintaining data and asset managers need to have a clear understanding about what information is required and the purpose to which it is to be put. Where data is required only on change of state, such as when work is undertaken on site, then it may be appropriate to routinely collect data manually where data is required; more frequently it may be more economical to automate the collection. Collecting trial data on a small sample of the assets, or for just a limited time, in order to explore the performance or to look for correlation with other factors can be a simple but effective technique to determine what is actually of benefit to the business.

Typically distribution companies have a numerous asset management IT systems which if not managed effectively can exhibit problems in the following areas:

- data duplication,
- repeated manual data entry,
- lack of electronic interfaces,
- unavailability of data for strategic analysis and business reporting, and
- islands of automation that cannot exchange information.

These problems are due to the lack of a strategic 'integration architecture' enabling the easy development and execution of electronic interfaces and the processes required to consolidate information for strategic analysis and reporting.

The IEC 61968 series 'System Interfaces for Distribution Management' is intended to facilitate inter-application integration of the various distributed software application systems supporting the management of utility electrical distribution networks.

Figure 9.13 clarifies the scope of IEC 61968-1 graphically in terms of business functions and shows a distribution management system with IEC-61968-compliant interface architecture.

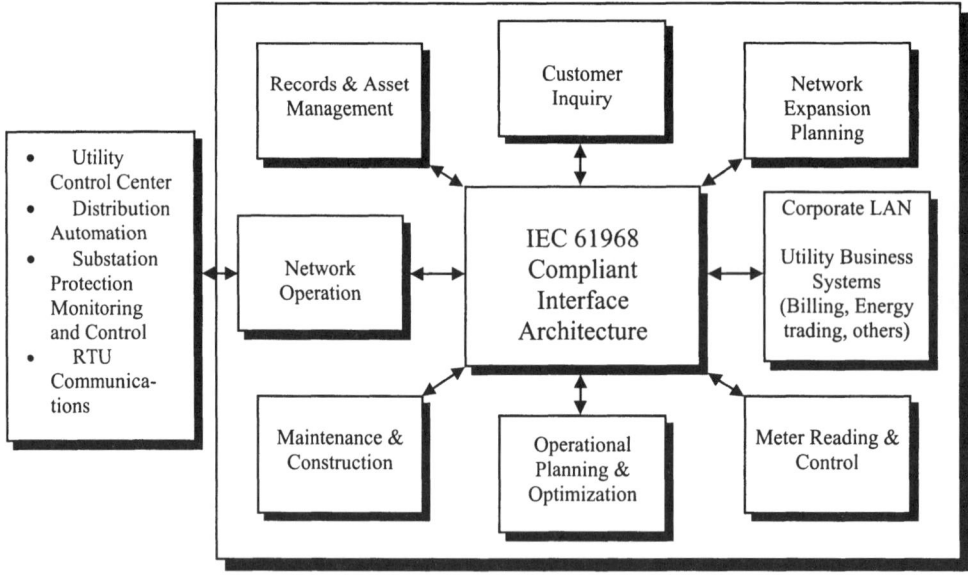

Figure 9.13 Distribution management system with IEC-61968-compliant interface architecture

9.17.1 Asset Management Systems

Asset management systems typically hold data, including ownership costs, on all the electrical assets, linking them together via parent/child relationships. These systems normally share a comprehensive power system model with other applications so that operational and planning tools and data can be employed as part of asset management. Integrated systems are comparatively new in many companies and the scale of the data collection validation can be immense, so many systems have not yet had a chance to prove their full worth in monitoring lifecycle costs etc. Proven uses at the moment are as plant database and maintenance schedulers utilising network analysis and contingency planning tools. Variable maintenance triggers can be set within the database and condition feedback and defects recorded from maintenance or inspection visits.

The performance and effective life of otherwise identical assets is largely driven by the duty they are required to perform and the environment in which they operate. An additional benefit of integrated information systems is the ability to download large sections of data for 'off-line' analysis and data mining to understand and exploit the relationships between performance, duty and environment.

9.17.2 Data Gathering

The data required by asset managers typically resides in several separate databases each of which has to be designed, populated and maintained by an effective set of business processes.

The equipment database contains information about the items of plant and circuits which make up the distribution network. The volume of the assets and the variety of information that is available about each and every type are very large indeed. Cable records are particularly complex, often requiring multiple spatial representations to enable all potential users of the data to access the information they need whether it be ownership, manufacturer, specification, age, installation method, condition, fault history, expenditure, loading, electrical parameters, etc. The recording of costs against individual circuits is far from straightforward as cables are continuously being cut into new circuits or being jointed together to form new ones.

A faults and interruptions database will typically describe the nature and exact geographical and circuit location of the faults which have taken place in the distribution network over at least the last three years and the time required to restore service to the customers.

The method of estimating the number of customers affected by incidents may vary widely from company to company. Tighter regulatory reporting and incentive arrangements will require consistency of reporting between competing companies. The explicit linking of end use customers to a full network connectivity database is often proposed as a solution to this issue but the establishment and maintenance of such a data base could be a major undertaking, particularly if it is required to establish real-time connectivity down to individual circuits and phases at LV within the distribution systems of individual blocks of flats and offices. The costs of such systems and their maintenance could not easily be justified when the consistent application of simple estimation rules could well suffice.

Experience shows that the most reliable data comes when it is collected in the field by trained staff who are familiar with the data collection systems, causes of failure and who are aware of the significance of accurate data collection both to them and to the company. Increasingly hand-held data collection terminals and radio telemetry enable near-real-time collection of data with instant querying of values or sequences of data that do not fall within expected limits or patterns.

Manual recording of time series data that is routinely required to assess the duty performed by the asset, such as transformer and feeder loading, temperatures and for some condition monitoring, is no longer cost effective in a tightly regulated market. RTUs at both primary and increasingly secondary substations, in addition to providing remote control and automation, are able to provide comprehensive control and telemetry facilities typically including:

- Voltage
- Load
- Temperatures
- Harmonics
- Out of balance
- Disturbance recording
- Security access control
- Condition monitoring
- Switching and status
- Alarm monitoring and reset.

9.17.3   Property Rights

Data will also be required concerning the property in, through or over which the assets operate, the charges payable and any constraints. This will usually include details of leases, wayleaves and freeholds together with agreed plans of access routes for plant and staff. For more complex substations or those integral with other buildings details of ventilation, fire, smoke and flood control systems may be required.

9.17.4   Environment

To be able to specify the most appropriate standards, installation and physical protection of assets it is necessary to know and understand the effects of the local environment on the performance of the assets. Generalised information, often at postcode or community level, about the nature of the environment will also be necessary to understand and then predict the effects on individual circuits and groups of circuits within localised areas. Examples of information that may be required are attributes and, where appropriate, histories of:

Terrain	Mountains, forestation, proximity to sea, altitude
Active environment	Temperature extremes, windspeed, storms, lightning, flooding, earthquakes, geomagnetic and sunspot activity, wildlife

Subterrain	Soil type, acidity, water table, shrink/swell potential, resistivity, stability
Built environment	Air pollution, corrosion, vibration, ground contamination, thermal sources, damage

The above list is by no means exhaustive.

Asset managers will also need to manage the effects of the assets upon the environment in order to control the risks of damage, prosecution and unfavourable media attention. To do so will require data concerning the leakage and disposal of oil, gases and waste material.

Spatial information concerning constraints such as:

- areas of outstanding natural beauty
- sites of special scientific interest
- archaeological and historic sites
- conservation and preservation areas
- areas with protective covenants
- military and security restricted sites

is also essential.

9.17.5 Data Cleaning

Databases created from input provided by people tend to include a significant proportion of erroneous or incomplete records. The data format is often not uniform, particularly if provided by many people who may have different interpretations of what data should be provided or how it should be entered. A significant amount of effort therefore needs to be devoted to cleaning up this data to a standard format and keeping it clean. This cleanup involves removing obviously incorrect or irretrievably incomplete records and harmonising formats.

Data provided by telemetry can also be subject to corrupted or missed data packets, blown potential fuses, faulty sensors or failed communications. Since it is unrealistic to expect a totally clean and complete database, a strategy for dealing with missing or noisy data fields must also be decided.

9.17.6 Confidential Information

Within an increasingly competitive industry there will be regulatory and energy business concerns that the distribution business has information that could be of significant benefit to other supply or generation businesses. Typically regulators will respond by imposing licence conditions that define as confidential any information relating to or deriving from the management or operation of the distribution business.

For example this may include:

- Supplier registration information, and particularly information that would facilitate customer segmentation analysis.
- Customer consumption data, and particularly half-hourly load shapes.
- Advance information regarding the setting of future DUoS prices.
- Information about the credit-worthiness of suppliers.

Asset managers will need to ensure that only those having appropriate rights and a need to know can have access to, copy or export information that is to be regarded as confidential.

9.17.7 *Quality of Information*

As operating margins become smaller and further efficiencies becomes more difficult to realise, increased quality of information becomes ever more essential for the effective management of assets.

For regulatory as well as asset management purposes it is essential that data is:

- Accurate
- Relevant
- Comparable across companies
- Consistent over time
- Collectable at reasonable cost.

Because the data collected will be used for asset regulatory price controls and used within incentive regimes, there will be a need to be able to demonstrate, to external auditors, the quality and timeliness of the data. One way of managing this may be though formal quality management processes such as those of the ISO 9000 series. Regular automated sequences of data quality audits that run structured queries and power system analysis studies, comparing results with those previously obtained and flagging abnormal changes for investigation, can also be used to significant advantage.

9.18 Conclusions

None of the above examples individually can provide the solution to the problem of how we manage our assets. A combination of all, or at least some, of the above provides a solution, which matches the point on the evolutionary curve which London Electricity has reached at this moment in time. The only thing that is certain is that 'nothing is certain' and this overall model will continue to change as more information becomes available, trends become more established or more varied, or if pressure from customers, the regulator or other stakeholders pushes investment decisions in a new direction. The challenge for the asset manager is to evaluate continually the benefits of individual techniques and equipment against the cost of installation and operation. We must not become over-enthusiastic with the principle of condition monitoring and data collection, lest we forget to recognise the potentially high cost associated with both the collection and analysis of data. It is worth bearing in mind that the most effective way of identifying when the mechanism of a manually dependent item of equipment, such as an isolator, requires maintenance is to ask the last person who operated it.

Asset Management

9.19 Appendix: Fuzzy DGA for Diagnosis of Multiple Incipient Faults

Large power transformers are probably the most important equipment in an electrical system. Correct diagnosis of their incipient faults is vital for the safety and reliability of an electrical network. An in-service transformer is subject to electrical and thermal stresses, which can break down the insulating materials and release gaseous decomposition products. Overheating, partial discharge and arcing are three primary causes of fault-related gases. There are many interpretative methods based on DGA to diagnose the nature of transformer deterioration, such as the IEC ratio codes which were developed from intensive investigations on gases generated from individual faults.

Although DGA has widely been used in the industry, in some cases, the conventional methods fail to diagnosis incipient faults. This normally happens for those transformers which have more than one type of fault. Actually, the conventional diagnostic methods are based on the ratio of gases generated from a single fault or from multiple faults but with one of dominant nature in a transformer. When gases from more than one fault in a transformer are collected, the relation between different gases becomes too complicated and may not match the pre-defined codes. For instance, the IEC codes are defined from certain gas ratios. When the gas ratio increases across the defined limits (boundaries), the code changes suddenly between 0, 1 and 2. In fact, the gas ratio boundary may not be clear (i.e. fuzzy), especially when more than one type of fault exists. Therefore, between different types of faults, the code should not change sharply across the boundaries. A new method has been developed to employ fuzzy boundaries between different IEC codes.

9.19.1 The IEC DGA Codes

In DGA, the IEC codes have been used for several decades and considerable experience accumulated throughout the world to diagnose incipient faults in transformers. The individual gases used to determine each ratio and its assigned limits are shown in Tables 9.1 and 9.2. Codes are then allocated according to the value obtained for each ratio and the corresponding fault characterised.

9.19.2 The Fuzzy IEC Code – Key Gas Method

The fuzzy IEC code-key gas method (FIK) developed is a combination of fuzzy diagnosis using IEC codes and key gases. This method produces nine fuzzy components from $F(0)$ to $F(8)$. These components are related to the fault types as detailed in Table 9.2.

Table 9.1 IEC ratio codes

	IEC codes		
Sharply defined ranges of the gas ratio	Codes of different gas ratios		
	$\dfrac{C_2H_2}{C_2H_4}$	$\dfrac{CH_4}{H_2}$	$\dfrac{C_2H_4}{C_2H_6}$
<0.1	0	1	0
0.1-1	1	0	0
1-3	1	2	1
>3	2	2	2

Table 9.2 Fault classification according to the IEC Gas Ratio Codes

No.	Fault type	$\dfrac{C_2H_2}{C_2H_4}$	$\dfrac{CH_4}{H_2}$	$\dfrac{C_2H_4}{C_2H_6}$
0	No fault	0	0	0
1	Partial discharges of low energy density	0	1 (not significant)	0
2	Partial discharges of high energy density	1	1	0
3	Discharges of low energy	1 or 2	0	1 or 2
4	Discharges of high energy	1	0	2
5	Thermal fault of low temperature < 150°C	0	0	1
6	Thermal fault of low temperature 150–300°C	0	2	0
7	Thermal fault of medium temperature 300–700°C	0	2	1
8	Thermal fault of high temperature > 700°C	0	2	2

9.19.3 Fuzzy Diagnosis Results

Using the FIK method, a number of 110-330 kV power transformers were diagnosed and some typical results are given in Table 9.3. It can be seen from sample No.1 that the new method is generally in agreement with the IEC method for transformers of a single or a dominant fault. Compared with IEC method, the FIK method also has some advantages. For example, owing to no matching codes, 13 transformers could not be diagnosed by the IEC method but are diagnosed by the FIK method, as shown in Table 9.3 by Nos 2-3 for some typical results. In some cases, the faults may be only at the early stage or intermittent which did not produce sufficient gases to give a stronger indication, such as F(2) in No.2 and F(6-8) in No.3. However, the information obtained should be useful for future trend analysis. Transformer No.4 was diagnosed by the IEC method to have a thermal fault of medium temperature (300-700°C), in comparison, the FIK method indicates that both high (>700°C) and medium temperature (300-700°C) faults existed. The likelihoods of each fault indicated by the fuzzy component are 0.477 and 0.431 respectively. The analysis results of another transformer (No.5) show that although the gas level is below the guide value, the FIK method can still be used and a low-energy discharge is diagnosed. The fuzzy vector ranges from 0 to 0.441, which could be useful for future trend analysis when the gas level increases.

The gas concentrations for the samples in Table 9.3 are given in Table 9.4. Most transformers in Victoria, Australia have shown medium or low levels of gases and lower increasing rates, such as transformers No.1, No.2 and No.5. Therefore they are closely monitored but not untanked. Transformer No.3 was dismantled and arc damage in the insulation was found in the core. In the inner inspection of another transformer, No.4, two locations were identified with high-temperature damage due to eddy currents and a bad contact. More transformers will be further investigated when certain criteria are met. Laboratory tests will also be carried out to fine-tune the fuzzy diagnosis technique.

Table 9.3 Comparison of diagnosis results from the FIK method and the conventional IEC method

No	Conventional IEC method		Fuzzy IEC code - key gas method (FIK)		COMMENTS
	IEC code	Remarks	Fuzzy fault components	Explanatory note	
1	020	Thermal fault (150-300°C)	F(0)=0.479 F(1)=0.006 F(2)=0.006 F(3)=0.000 F(4)=0.005 F(5)=0.000 **F(6)=0.496** F(7)=0.000 F(8)=0.009	Normal ageing PD of low energy PD of high energy Discharge of low energy Discharge of high energy Thermal fault (<150°C) Thermal fault (150-300°C) Thermal fault (300-700°C) Thermal fault (>700°C)	Both FIK and IEC diagnose a thermal fault (150-300°C), indicating a general agreement between these two methods. *Actual fault will be checked during the next overhaul.*

2	100	No match	F(0)=0.525 F(1)=0.053 **F(2)=0.231** F(3)=0.045 F(4)=0.050 F(5)=0.000 F(6)=0.047 F(7)=0.000 F(8)=0.050	Normal ageing PD of low energy **PD of high energy** Discharge of low energy Discharge of high energy Thermal fault (<150°C) Thermal fault (150-300°C) Thermal fault (300-700°C) Thermal fault (>700°C)	IEC cannot diagnose but FIK indicates a high-energy PD fault, which could be at an early stage. *Actual fault will be checked during the next overhaul.*
3	121	No match	F(0)=0.005 F(1)=0.052 F(2)=0.052 F(3)=0.000 **F(4)=0.408** F(5)=0.000 **F(6)=0.161** **F(7)=0.161** **F(8)=0.161**	Normal ageing PD of low energy PD of high energy Discharge of low energy Discharge of high energy Thermal fault (<150°C) Thermal fault (150-300°C) Thermal fault (300-700°C) Thermal fault (>700°C)	IEC cannot diagnose probably due to the existence of more than one fault. The fuzzy component of the early thermal fault indicated by FIK is useful for future trend analysis. *Actual fault was an arc damage to the core.*
4	021	Thermal fault (300-700°C)	F(0)=0.007 F(1)=0.026 F(2)=0.026 F(3)=0.000 F(4)=0.030 F(5)=0.000 F(6)=0.003 **F(7)=0.477** **F(8)=0.431**	Normal PD of low energy PD of high energy Discharge of low energy Discharge of high energy Thermal fault (<150°C) Thermal fault (150-300°C) Thermal fault (300-700°C) Thermal fault (>700°C)	IEC diagnoses medium-temperature fault but actually both medium- and high-temperature faults existed as indicated by FIK. *Two locations of overheating damages were found due to eddy currents and a bad contact.*
5	Low values	No diagnosis	F(0)=0.479 F(1)=0.005 F(2)=0.057 **F(3)=0.441** F(4)=0.013 F(5)=0.000 F(6)=0.000 F(7)=0.000 F(8)=0.005	Normal PD of low energy PD of high energy Discharge of low energy Discharge of high energy Thermal fault (<150°C) Thermal fault (150-300°C) Thermal fault (300-700°C) Thermal fault (>700°C)	Although the gas level is below the guide value, an early indication of low-energy discharge by FIK should be useful for trend analysis in the future. *Actual fault will be checked during the next overhaul.*

Table 9.4 Gas concentration for the transformers listed in Table 9.3 (in ppm)

No.	H_2	CH_4	C_2H_2	C_2H_4	C_2H_6	IEC ratio codes
1	95	110	<0.1	50	160	0 2 0
2	120	17	4	23	32	1 0 0
3	300	490	95	360	180	1 2 1
4	200	700	1	740	250	0 2 1
5	25	3	0.1	0.1	0.1	below guide level

9.19.4 Trend Analysis of Individual Faults

In FIK diagnosis, a fault can be more accurately determined by its fuzzy component that indicates the likelihood or dominance of the fault. Deterioration of the fault may therefore be closely monitored from trend analysis. This technique has been used for a transformer that was tested over a 15-month period. Thermal faults of medium- and high-temperature (300-700°C and >700°C) were diagnosed by the FIK method and the fuzzy components against the test time are plotted in Figure 9.14. The graph clearly shows the development of each thermal fault in this transformer. It can be seen that at the beginning of this monitoring period, the medium temperature thermal fault F(7) was the main problem of this transformer and the fuzzy component of the high-temperature thermal fault (>700°C) was very small, i.e. below 0.05. The high-temperature thermal fault F(8) was diagnosed from Day 114 onwards and then become stable until Day 406 when the oil was de-gassed. After de-gassing, because the thermal faults remained, the fuzzy components F(7) and F(8) went up again from Day 453. It took a few weeks for the gases to be released and dissolved in the oil to a sufficient level for accurate diagnosis. A small fluctuation of F(8) was recorded on Day 178, which might be due to the lighter load during the specific time period.

It must be noted that if a transformer has no fault, the fuzzy component F(0) always gives a large value in the range of 0.6-1. For example, the DGA results for a healthy transformer are (in ppm) H_2 - 95, N_2 - 73000, O_2 - 11000, CO - 1000, CH_4 - 20, CO_2 - 8400, C_2H_4 - 25, C_2H_6 - 45 and C_2H_2 - 2. The fuzzy component of no-fault F(0)=0.863 indicates that no fault exists in the transformer. The IEC codes are 0, 0, 0, also indicating no fault. From our experience, when the value of F(0) is between 0.3 and 0.6, an incipient fault may have occurred at earlier stage. When the fault is getting worse, F(0) will decrease to <0.1.

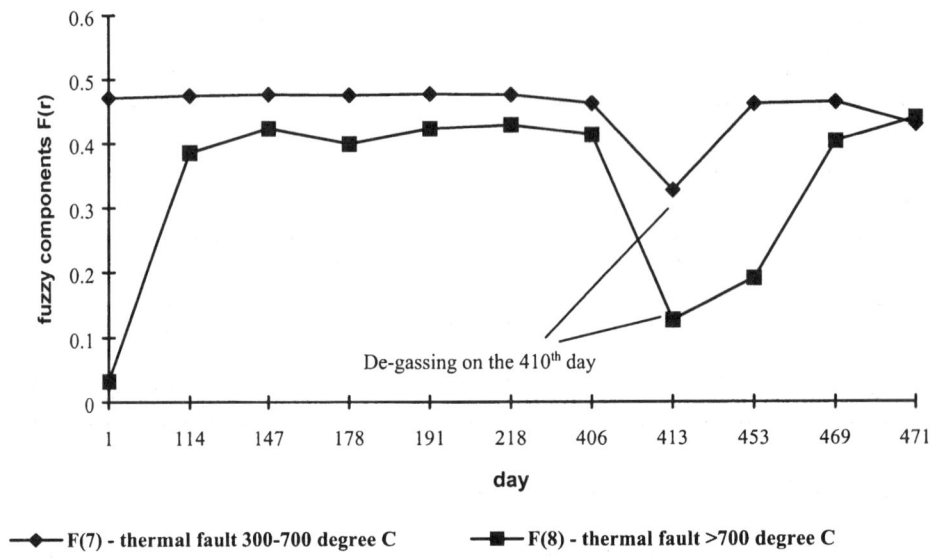

Figure 9.14 The trend of two types of thermal fault in a 330 kV transformer determined by the FIK method

9.19.5 Comments

The FIK method developed has been successfully used for the diagnosis of several transformers in Australia. It has been proved that, using the fuzzy diagnosis method, more detailed information about the faults inside a transformer can be obtained. This is an improvement over the conventional IEC code method, which may be due to the more realistic representation of the relationship between the fault type and the dissolved gas levels with fuzzy membership functions. Also, multiple faults can be diagnosed using this method, which may not be possible for any other methods available in the world.

Another advantage of the fuzzy diagnosis method is its quantitative indication of the fault likelihood/dominance by means of fuzzy diagnostic vectors. This has been used to identify the main faults and determine their severity in comparison with each other. The trend of development of each fault in a transformer can also be determined from its fuzzy diagnostic vector after a certain period of monitoring. This information is important for any decision regarding transformer replacement or/and refurbishment. In the current economic climate, there is an increasing demand to extend the service life and reduce maintenance costs of transformers. With the aid of more accurate insulation condition monitoring techniques, such as the FIK method, the maintenance schedule can be optimised and a longer service life could be achieved.

9.20 References

[1] Mario V.F. Pereira, Michael F. McCoy and Hyde M. Merrilli, 'Managing risk in the new power business', *IEEE Computer Applications in Power*, Vol.13, No.2, April 2000, pp.18-24.
[2] George Anders, Robert Entriken and Puica Nitu, Risk Assessment and Financial Management, Tutorial, IEEE Catalog Number 99TP137-0, 1999.
[3] B.G. Gorenstin, N.M. Campodonico, J.P. Costa and M.V.F. Pereira, 'Power system planning under uncertainty', *IEEE Transactions on Power Systems*, February 1993, pp.129-136.
[4] Teofilo De la Torre, James W. Feltes, Tomas Gomez San Roman, Hyde M. Merrill, 'Deregulation, privatization, and competition: transmission planning under uncertainty', *IEEE Transactions on Power Systems*, May 1999, pp.469-465.
[5] J.C. Hull, *Options, Futures and Other Derivatives*, Prentice Hall, New Jersey, 1998.
[6] J. Schwager, *A Complete Guide to the Futures: Fundamental Analysis, Technical Analysis, Trading, Spreads, and Options*, John Wiley & Sons, New York, 1984.
[7] Price Waterhouse LLP, *The Corporate Risk Management Handbook*, Risk Publications, London, 1996.
[8] P. Jorion, *Value at Risk: The New Benchmark for Controlling Market Risk*, Irwin Professional Pub., Chicago, 1997.
[9] J. Douglas, A. Altman, V. Niemeyer, R. Goldberg, and C. Clark, 'Navigating the currents of risk', *IEEE Power Engineering Review*, March 1998, pp.6- 10.
[10] D. Duffie and J. Pan, 'An overview of value at risk', *Journal of Derivatives*, Institutional Investor, Inc., New York, Spring, 1997.
[11] J.P. Morgan & Co. and Arthur Andersen & Co., *The JP Morgan/Arthur Andersen Guide to Corporate Risk Management*, Risk Publications, London, 1997.
[12] G.L. Gastineau, *Dictionary of Financial Risk Management*, Swiss Bank Corporation, New York, 1992.
[13] Enron Capital & Trade Resources, *Managing Energy Price Risk*, Risk Publications, London, 1995.
[14] R.L. Nersesian, *Computer Simulation in Financial Risk Management: A Guide for Business Planners and Strategists*, Quorum Books, New York, 1991.
[15] Q. Su, C. Mi, L.L. Lai and P. Austin, 'A fuzzy dissolved gas analysis method for the diagnosis of multiple incipient faults in a transformer', *IEEE Transactions on Power Systems*, Vol.15, No.2, May 2000, pp.593-598.

10

Power Quality

Prof. Jos Arrillaga
University of Canterbury
New Zealand

Dr N.R. Watson
University of Canterbury
New Zealand

10.1   Introduction

10.1.1   A General Overview

Prior to deregulation, electricity has been generally sold from one supplier to the consumer, with ownership changing hands at only one physical point. In contrast, after deregulation it is expected that the product will be exchanged at several points along the generation, transmission and distribution systems and there will be power quality (PQ) issues at each physical location where ownership is transferred.

PQ is, of course, an ambiguous term which in its broadest sense is interpreted as service quality including reliability of supply, waveform quality and provision of information [1].

In a deregulated environment, only national grids have the ability to acquire, analyse and act on the information necessary to provide system security and quality. In that privileged position, the grids can be unreasonably demanding in their connection codes for new generation plant. In the long term, however, the expectation is that conventional plant will find stronger competition from distributed generation, both of renewable energy (like micro-hydro, wind and solar) and non-renewable energy sources (such as gas-fired microturbines and fuel cells), the latter in the kilowatt rather than megawatt area. Much of the technology used in these energy sources involves power electronic devices. A variety of active equipment is now commercially available covering the full range of power ratings, from HVDC links and FACTS (flexible AC transmission systems) to custom power devices.

At the generation level, an increase in the connection of IPPs (independent power producers such as wind and gas-fuelled microturbines) with poorly controlled synchronisation will make PQ more difficult to control. The increase in embedded generation will cause further voltage magnitude variations as well as introduce additional voltage magnitude steps [2]. Wind power is known to lead to an increase in flicker severity. Solar power and the more advanced ways of connecting wind power will lead to an increase in harmonic distortion. At the transmission level, the need for system operators to transmit power according to contracts between the requested locations is likely to accelerate the demand for series-connected FACTS controllers. In the future, series compensation and unified power flow controllers are expected to be used extensively once they are shown to offer better technical features at reasonable costs.

Because system planning under deregulation will be more difficult owing to uncertainty in the generation and load locations, fast solutions will be needed to improve the operating conditions and FACTS controllers can offer such solutions with short delivery and installation times. The use of asynchronous grid interconnections, both national and international, is also likely to increase with deregulation. The controllability of asynchronous interconnectors is currently limited by the switching restrictions of the silicon-controlled rectifier, which only permits two-quadrant converter operation, i.e. bi-directional active power transfers. The availability of gate turn-off switching devices permits four-quadrant converter operation and considerable development is already going on to improve the efficiency and power handling capability of these devices. The reality is that whether in the form of two-quadrant or four-quadrant configurations, the asynchronous link is likely to be an important player in modern transmission systems planning and its impact on PQ needs to be carefully examined.

Power electronic devices, whether in the form of asynchronous interconnectors, FACTS or custom power, have the potential to improve various aspects of PQ [3]. Power electronic control at distribution level may mitigate voltage variations, harmonics and voltage sags. But the increased use of power electronic controllers may introduce new problems like additional harmonic voltage distortion, especially in the form of higher order harmonics.

In a competitive environment there will be reluctance to expand the distribution system, leading to increased customer interaction. And, at the loads themselves, increased awareness of power costs will create an emphasis on local compensation with corresponding passive or active components. Some of these changes tend to degrade PQ. Fast local compensation will create loads of a constant-power type. Thus, the controller will draw more current when the voltage drops causing additional voltage drops for non-compensated customers. Widespread use of compensation equipment may even become a voltage stability issue.

Most of these problems are not exclusive to deregulation. In fact, there is a continuing increase in the use of non-linear loads, such as adjustable speed drives, office equipment, and high-efficiency fluorescent lighting. At the same time, sensitive information technology equipment, such as PCs, continues to be dispersed into power locations that previously were restricted to lights, motors and heaters. There is no reason to believe that this trend will reverse.

Following deregulation, the power exchanges should be subjected to close quality scrutiny on a continuous basis. This requires dynamic evaluation of the three-phase voltage and current waveforms, either by local measurements exclusively or by a combination of measurements and system simulation using harmonic state estimation techniques. The latter should provide more *intelligent* and economical solutions for the control of the distortion problem on a system-wide basis. Deregulation disperses responsibility for PQ. In the past it was clear, for the most part, that the utility was responsible for delivering some level of PQ to the customer. After deregulation, however, who is responsible for the customer's PQ? The generator? The energy supplier? The distributor? The retailer? The customer? These blurred PQ boundaries will lead to confusion, and possibly to an increase in disputes.

Some studies have shown that customer expectations rise unreasonably immediately after deregulation. Although most of these studies have concentrated on price expectations, the same effect may occur for quality expectations as well [4,5]. To many customers, the quality of power has become every bit as important as price and availability. The products that help correct PQ problems are part of a growing, billion-dollar industry. Obviously, no one can predict the future, but as the product of electricity unbundles, quality and reliability will become every bit as important as price. Customer choice will dictate that energy providers must look at power from the customer's viewpoint, and when they do, they will see that customers want value, and those who can provide the best power value will be positioned the best. Power value will be determined by the price, availability, quality and usefulness. It will also be customer dependent. Some will be quite content to continue buying from the local distribution company. Others will need higher quality and reliability. Some may even be willing to pay a higher price.

Many local electric utilities have had some experience in providing PQ consulting services. Activities are generally restricted to monitoring the quality of the power (voltage anomalies) at the meter. Some utilities go beyond the meter to investigate 'power problems' for their customers as part of a complaint resolution process. Several utilities have found unexpected benefits from monitoring PQ at key customers, and sharing real-time PQ data with individuals at those customers via pagers. A sense of partnership develops, and that sense of partnership gives a competitive advantage to that utility. This sense of partnership is a somewhat surprising benefit of deregulation on power quality. On the downside, in regions where utilities have been deregulated, there has generally been a reduced exchange of information and cooperation between utilities. This is an inevitable result of competition, but it has an unfortunate effect on the development of PQ technology, tools and standards.

The rest of the chapter discusses the main aspects of PQ affecting or being affected by deregulation and restructuring as well as the impact that modern equipment, such as FACTS and large power converters, is having on the quality of the voltage and current waveforms. A general feeling among experts is that momentary voltage sags constitute the most pressing PQ problem. This view is influenced by the extensive disruption of industrial processes caused by low voltage conditions.

Voltage sags and interruptions at a transmission substation will affect all customers in the distribution system, but a sag originating at distribution level will not affect other distribution systems. However, faults cleared by overcurrent relays or by fuses may lead to sag durations of 1 second or more, which have been shown to be of serious concern even after propagation through the transmission network [6,7]. Deregulation will not change the technical issues behind this; but the responsibility for the problem may become rather complicated, especially as a local load may affect the severity of voltage sags and a short-circuit fault inside customer premises will cause voltage sags elsewhere.

10.1.2 PQ Issues During System Disturbances

In the context of PQ, a disturbance is a temporary deviation from the steady-state waveform caused by faults of brief duration or by sudden changes in the power system. The disturbances considered by the International Electrotechnical Commission include voltage dips (sags), brief interruptions, voltage increases (swells), and impulsive and oscillatory transients. These are illustrated in Figures 10.1 and 10.2.

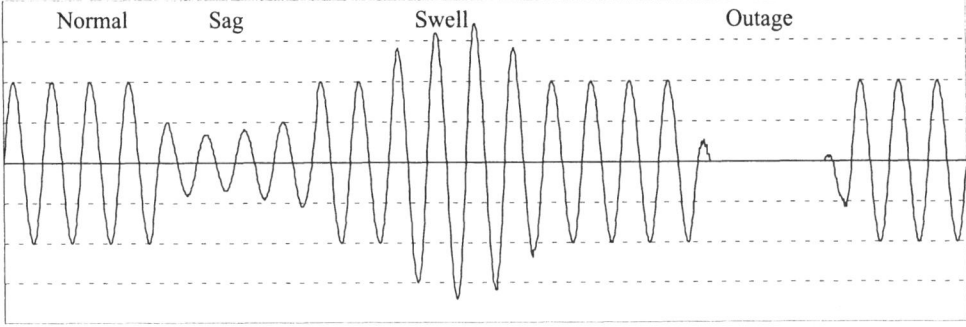

Figure 10.1 Voltage disturbances

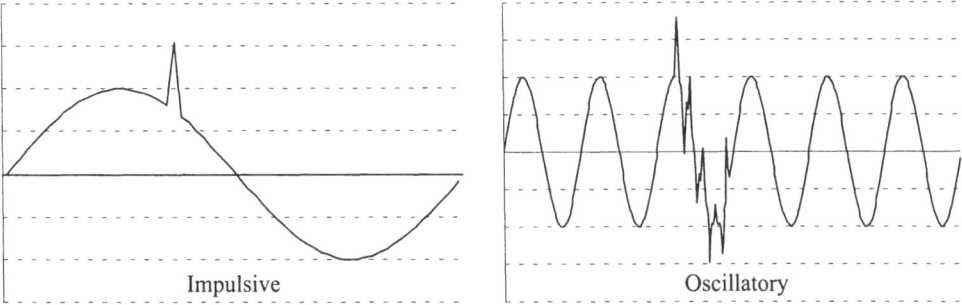

Figure 10.2 Voltage transients

A voltage dip (or sag) is a sudden reduction (between 10 and 90%) in the voltage magnitude and lasting for 0.5 cycles to several seconds. They may be caused by switching operations associated with a temporary disconnection of supply, the flow of heavy current associated with the start of large motor loads or the flow of fault currents. These events may emanate from customers' systems or from the public supply network. The main cause of momentary voltage dips is probably the lightning strike.

Possible effects of voltage dips are: extinction of discharge lamps; incorrect operation of control devices; speed variation or stopping of motors; tripping of contactors; computer system crash; or commutation failure in line commutated inverters. The effect of a voltage dip on equipment depends on both its magnitude and its duration; in about 40% of the cases observed to date, they are severe enough to exceed the tolerance standard adopted by computer manufacturers. Brief interruptions can be considered as voltage sags with 100% amplitude. The cause may be a blown fuse or breaker opening and the effect an expensive shutdown. For a given system design and fault location, a certain number of customers will be affected and there is no way to prevent this process without major system structural changes.

However, interruptions due to overload are somewhat more predictable. These include overload of the whole system (due to lack of generation) as well as individual lines and cables. Voltage collapse can also be viewed as an overload situation, but in this case load shedding can alleviate it. In the pre-deregulation era, load shedding took place according to utility rules. Deregulation allows utilities to offer interruptible and non-interruptible supply. During times of overload or overload risk, utilities may decide to increase the incentive for customers to be interrupted [8,9]. At present, this action only covers a very small fraction of the interruptions but this will obviously change if the congestion in the system increases.

Voltage swells are brief increases in r.m.s. voltage that sometimes accompany voltage sags. They appear on the unfaulted phases of a three-phase circuit that has developed a single-phase short circuit. They also occur following load rejection. Swells can upset electric controls and electric motor drives, particularly the adjustable-speed drives, which can trip because of their built-in protective circuitry. Swells may also stress delicate computer components and shorten their life. Voltage disturbances shorter than sags or swells are classified as transients and are caused by sudden changes in the power system [10].

According to their duration, transient overvoltages can be divided into switching surges (duration in the range of milliseconds), and impulse spikes (duration in the range of microseconds). Surges are high-energy pulses arising from power system switching disturbances, either directly or as a result of resonating circuits associated with switching devices. They also occur during step load changes. In particular, capacitor switching can cause resonant oscillations leading to an overvoltage some three to four times the nominal rating, causing tripping or even damaging protective devices and equipment. Electronically based controls for industrial motors are particularly susceptible to these transients. Impulses result from direct or indirect lightning strikes, arcing, insulation breakdown, etc.

10.1.3 Voltage Sags

In the present stage of deregulation, no serious consideration is given to full competition at transmission and distribution levels and, therefore, there is little incentive to design for an overall reduction in the frequency of sags. Although there are indications that sag events will increase in the future, some customers are likely to demand a reduction in their number.

One option is to introduce '*power quality guarantees*' whereby the customer receives compensation for each event exceeding a certain severity (in magnitude, duration or frequency). Such an additional service may be offered by the (monopolised) distribution company, by the supplier, or by any other player in the market (e.g. an insurance company). Alternatively, a regulatory body may decide to enforce a basic compensation scheme for all customers as part of the connection fee [11]. However, some customers may not be satisfied with any compensation scheme, safety being their main consideration. The option in this case is for the utility to offer high-quality power to a small group of customers. These customers will experience less voltage sags than similar customers elsewhere. This special service will require the installation of mitigation equipment, which may be offered by the distribution company, by the supplier, or by any other player in the market. Additional regulations are needed to guarantee a minimum level of compatibility between equipment and supply:

- Requirements for equipment immunity must be produced by standard-setting organisations. The IEC is obviously the best platform for the development of such a standard. In the USA, the IEEE may take the lead. Standards for equipment testing, like IEC 61000-4-11 [12], are also needed to obtain and verify equipment immunity.
- As a complement to equipment immunity requirements, voltage characteristics for the supply must be made available to the customers. The European standard EN 50160 should be extended with voltage characteristics for voltage sags and other events. Equivalent documents should be written for other parts of the world as well as local standards for individual countries [13].
- Regulatory bodies should publish statistics on the PQ performance of utilities. Such a scheme is already in place in the UK for long interruptions [14].
- Voltage sag characterisation is an important basis for the above standards and regulations. At the time of writing, standardisation on this issue is under development both in the IEC [4] and in the IEEE [15]. However, current activities concentrate on sags experienced by single-phase equipment.

A technique has been proposed for the characterisation of voltage sags [16] experienced by three-phase equipment. It enables the characterisation through one complex voltage, without significant loss of information. The method is based on the decomposition of the voltage phasors into symmetrical components. An additional characteristic is introduced to enable the exact reconstruction of the three complex voltages. The mathematics behind the method and additional examples is described in references [2,17-20].

The ITIC (Information Technology Industry Council) curve [21] shown in Figure 10.3 can be used to evaluate the voltage quality of a power system with respect to voltage interruptions, sags or undervoltages and swells or overvoltage. This curve was originally produced as a guideline in the design of the power supply for computer and electronic equipment for use in the 60 Hz, 120 V distribution voltage system. By noting the changes of power supply voltage on the curve, it is possible to assess if the supply is reliable for operating electronic equipment, which is generally the most susceptive equipment in the power system.

The curve shows the magnitude and duration of voltage variations on the power system. The region between the two sides of the curve is the tolerance envelope within which electronic equipment is expected to operate reliably. Rather than noting a point on the plot for every measured disturbance, the plot can be divided into small regions with a certain range of magnitude and duration. The number of occurrences within each small region can be recorded to provide a reasonable indication of the quality of the system.

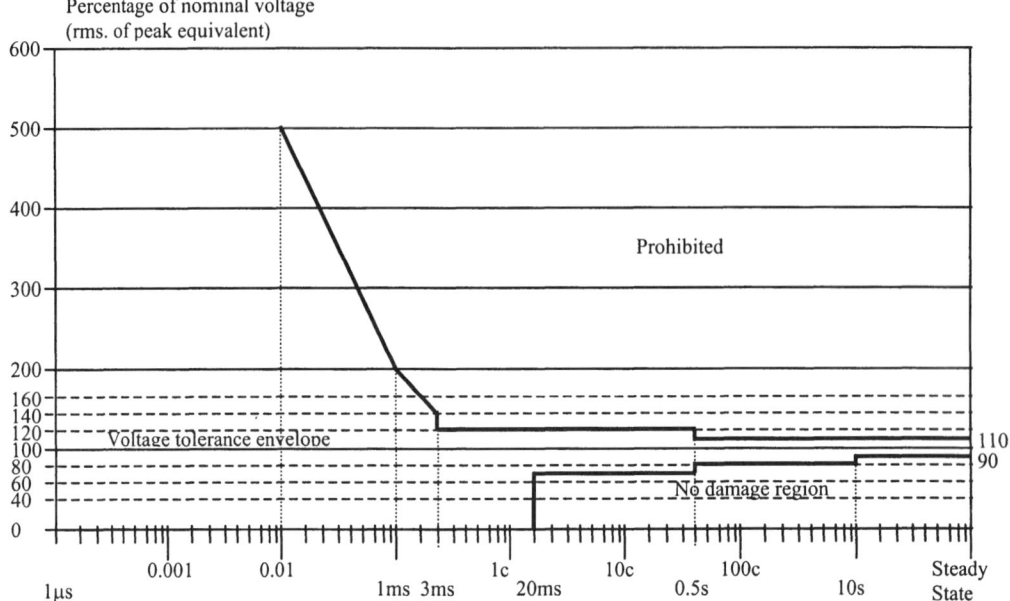

Figure 10.3 ITIC curve

10.2 Disturbance Assessment

10.2.1 The Wavelet Transform

The wavelet transform (WT) provides a fast way of analysing non-stationary voltage and current waveforms because, unlike the Fourier transform, the wavelet can tailor the frequency resolution.

Wavelets use short time intervals for high-frequency components and long intervals for low-frequency components thus improving the analysis of signals containing impulses and oscillations, particularly in the presence of a fundamental and low-order harmonics. A 'mother' wavelet, such as shown in Figure 10.4, is the product of an oscillatory function and a decay function [22], i.e.

$$g(t) = e^{-at^2} e^{j\omega t} \tag{10.1}$$

Power Quality

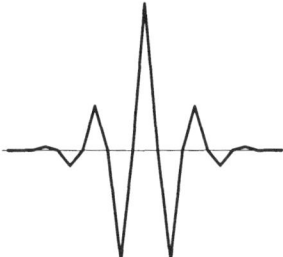

Figure 10.4 A sample mother wavelet

The one-dimensional signal can be transformed to the two-dimensional function of a (scale) and b (translation) to constitute derived wavelets from the mother wavelet, i.e.

$$g'(a,b,t) = \frac{1}{\sqrt{a}} g\left(\frac{t-b}{a}\right) \quad (10.2)$$

The WT of a continuous signal $x(t)$ is defined as

$$WT(a,b) = \frac{1}{\sqrt{a}} \int_{-\infty}^{\infty} f(t) g\left(\frac{t-b}{a}\right) dt \quad (10.3)$$

The time extent of the wavelet $g((t-b)/a)$ is expanded or contracted in time depending on whether $a > 1$ or $a < 1$. A value of $a > 1$ ($a < 1$) expands (contracts) $g(t)$ in time and decreases (increases) the frequency of the oscillations in $g((t-b)/a)$. Hence, as a is ranged over some interval, usually beginning with unity and increasing, the input is analysed by an increasingly dilated function that is becoming less and less focused in time.

For digital implementation, a discrete wavelet transform (DWT) is used where the scale and translation variables are discretised but not the independent variable of the original signal. A DWT gives a number of wavelet coefficients depending upon the integer number of the discretisation step in scale and translation, denoted by m and n, respectively. So, any wavelet coefficient can be described by two integers, m and n. If a_0 and b_0 are the segmentation step sizes for the scale and translation, respectively, the scale and translation in terms of these parameters will be $a = a_0^m$ and $b = nb_0 a_0^m$.

In terms of the new parameters a_0, b_0, m and n, equation (10.2) becomes,

$$g'(m,n,t) = \frac{1}{\sqrt{a_0^m}} g\left(\frac{t - nb_0 a_0^m}{a_0^m}\right) \quad (10.4)$$

or

$$g'(m,n,t) = \frac{1}{\sqrt{a_0^m}} g\left(ta_0 - nb_0\right) \quad (10.5)$$

and the discrete wavelet coefficients are given by

$$\mathrm{DWT}(m,n) = \int_{-\infty}^{\infty} \frac{1}{\sqrt{a_0^m}} f(t) g(a_0^{-m} t - nb_0) dt \qquad (10.6)$$

Although the transformation is over continuous time, the wavelets representation is discrete and the discrete wavelet coefficients represent the correlation between the original signal and wavelets for different combinations of m and n.

The inverse DWT is given by:

$$f(t) = K \sum_{m=0}^{\infty} \sum_{n=0}^{\infty} W_g f(m,n) \frac{1}{\sqrt{a_0^m}} g(a_0^{-m} t - nb_0) \qquad (10.7)$$

where $K = (A + B)/2$, and A and B are the frame bounds (maximum values of a and b).

10.2.2 Wavelet Analysis

Wavelet analysis is normally implemented using multi-resolution signal decomposition (MSD). High- and low-pass equivalent filters, h and g respectively, are formed from the analysing wavelet. The digital signal to be analysed is then decomposed (filtered) into smoothed and detailed versions at successive scales, as shown in Figure 10.5 where (2↓) represents a down sampling by half.

Scale 1 in Figure 10.5 contains information from the Nyquist frequency (half the sampling frequency) to one-quarter the sampling frequency, scale 2 contains information from one-quarter to one-eighth the sampling frequency and so on. The decomposition can be halted at any scale, with the final smoothed output containing the information of all of the remaining scales, i.e. scales 8,16,32, if it is halted at scale 4; this is one of the desirable properties of MSD. The choice of mother wavelet has a significant effect on the results obtained. The orthogonality of wavelets ensures that the signal can be reconstructed from its transform coefficients [23]. Wavelets with symmetric filter coefficients generate linear phase shift.

A large wavelet family derived by Daubechies [24] covers the field of orthonormal wavelets. It includes members ranging from highly localised to highly smooth. Daub4 and Daub6 wavelets are the best choice for short and fast transient disturbances, while for slow transient disturbances, Daub8 and Daub10 are the most appropriate. However, the selection of a mother wavelet without knowledge of the types of transient disturbances is a difficult task. A simpler solution is the use of one type of mother wavelet in the whole course of detection and localisation for all types of disturbances.

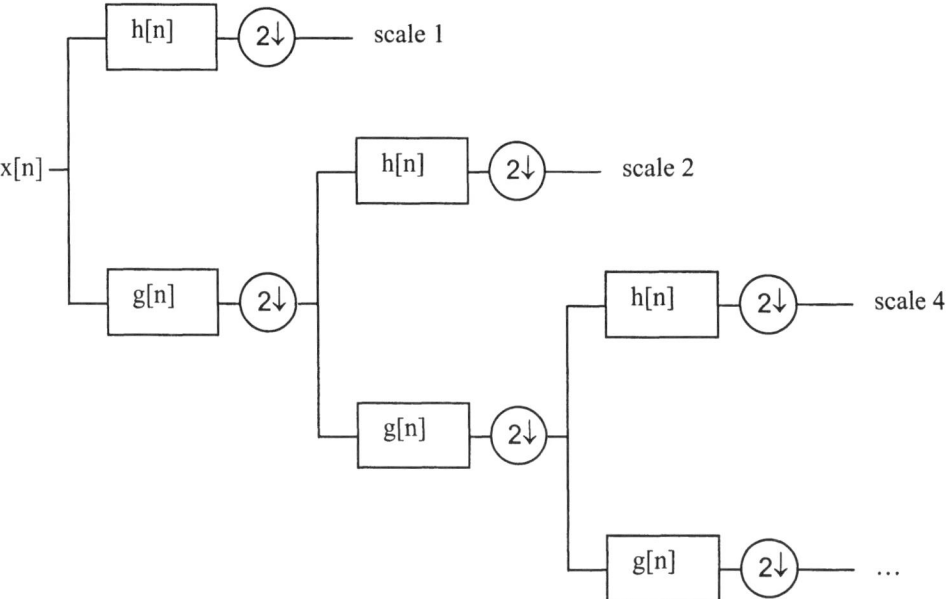

Figure 10.5 Multi-resolution signal decomposition implementation of wavelet analysis

In doing so, higher scale signal decomposition is needed. At the lowest scale the mother wavelet is most localised in time and oscillates rapidly within a very short period of time. As the wavelet goes to higher scales, the analysing wavelets become less localised in time and oscillate less owing to the dilation nature of the WT analysis. As a result of higher scale signal decomposition, fast and short transient disturbances are detected at lower scales, whereas slow and long transient disturbances will be detected at higher scales.

10.2.3 *Application to PQ [25]*

Figure 10.6a shows a sequence of voltage disturbances. To remove the *noise* present in the waveform, squared wavelet transform coefficients (SWTCs) are used at scales $m = 1, 2, 3$ and 4, respectively (shown in Figures 10.6b, c, d and e; these are analysed using the Daub4 wavelet. Figure 10.6a contains a very rapid oscillation disturbance (high frequency) before time 30 ms, and is followed by a slow oscillation disturbance (low frequency) after time 30 ms. The SWTCs at scales 1, 2 and 3 catch these rapid oscillations, while scale 4 catches the slow oscillating disturbance, which occurred after time 30 ms. Note that the high SWTCs persist at the same temporal location over scales 1, 2 and 4.

It must be pointed out that the same technique can be used to detect other forms of waveform distortion (like notches and harmonics) and other types of disturbance such as momentary interruptions, sags and surges. However, rigorous uniqueness search criteria must be developed for each disturbance for the WT to be accepted as a reliable tool for the automatic classification of PQ disturbances.

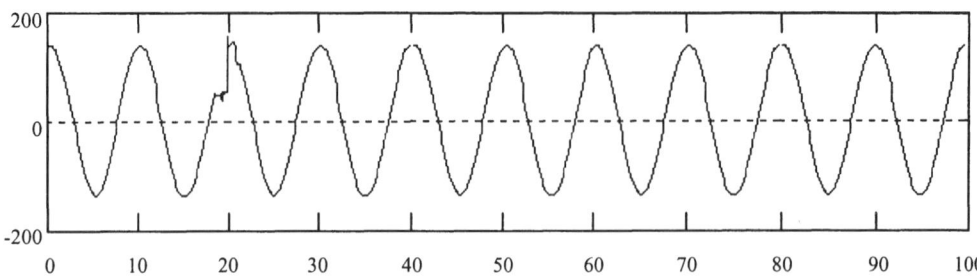

Figure 10.6a The voltage disturbance signal (© *1996, IEEE*)

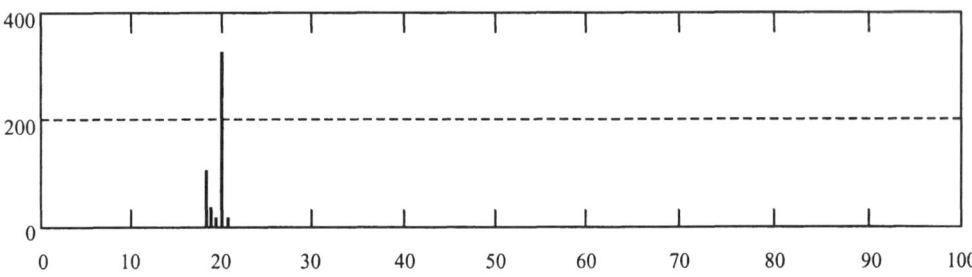

Figure 10.6b The SWTCs at scale 1 (© *1996, IEEE*)

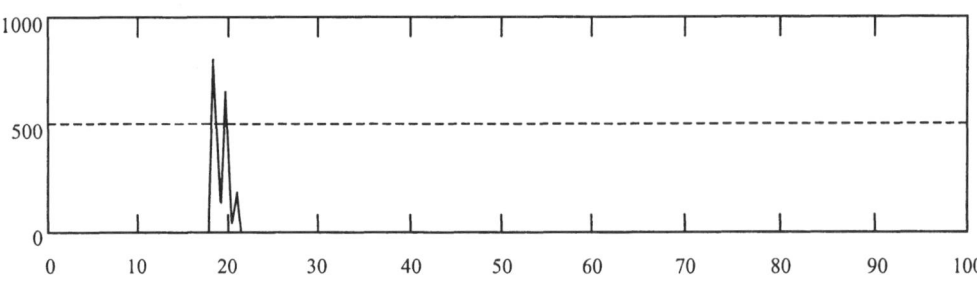

Figure 10.6c The SWTCs at scale 2 (© *1996, IEEE*)

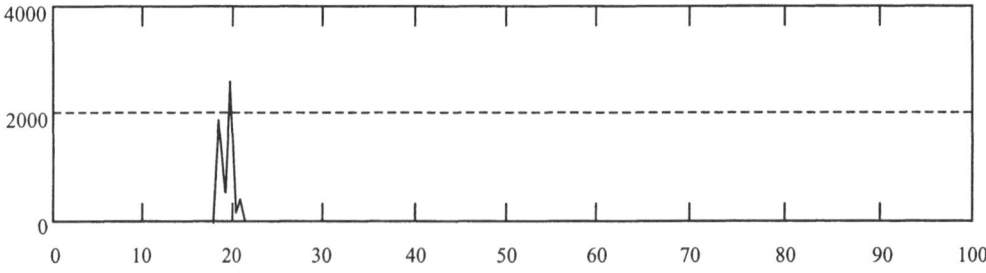

Figure 10.6d The SWTCs at scale 3 (© *1996, IEEE*)

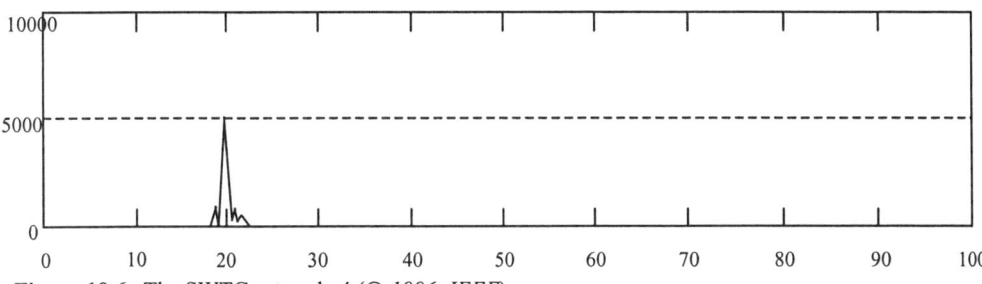

Figure 10.6e The SWTCs at scale 4 (© *1996, IEEE*)

10.2.4 Automated Disturbance Assessment

The process of disturbance recognition [26,27] can be automated to improve the speed, reliability and ease of data collection and storage. Such a scheme involves three separate stages as shown in Figure 10.7. These are a pre-processing stage to extract the disturbance information from the generated power signal; a main processing stage to carry out pattern recognition on the disturbance data; and a post-processing stage to group the output data and form decisions on the possible nature and cause of the disturbance.

The WT is an obvious candidate to extract the disturbance information owing to its greater precision and speed over Fourier methods. A collection of standard libraries of wavelets can be developed to fit specific types of disturbance or transient. Artificial neural networks can then be used in the main processing stage to perform pattern recognition. The neural network can be trained to classify the preliminary information extracted in the pre-processing stage.

Finally, fuzzy logic [28,29] is well suited in the post-processing stage to make decisions on the disturbance category. It is simple and fast to compute. The output of the disturbance recognition system is produced as one or a list of disturbance categories with an associated degree of belief. A list of disturbance categories with belief degree is necessary, as pattern recognition systems are inexact by nature. The system should, however, produce high belief degrees for only disturbance categories that are likely causes.

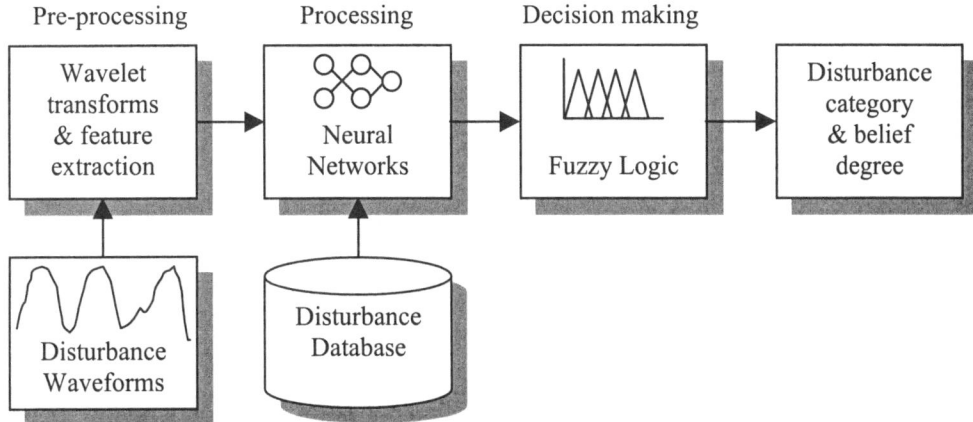

Figure 10.7 Block diagram of the automatic disturbance recognition system

10.3 Waveform Distortion

Waveform distortion is generally discussed in terms of harmonics, which are sinusoidal voltages or currents having frequencies that are whole multiples of the frequency at which the supply system is designed to operate (e.g. 50 Hz or 60 Hz). When the frequencies of these voltages and currents are not an integer of the fundamental they are termed interharmonics.

Both harmonic and interharmonic distortion is generally caused by equipment with non-linear voltage/current characteristics. In general, distorting equipment produces harmonic currents which in turn cause harmonic voltage drops across the impedances of the network. Harmonic currents of the same frequency from different sources add vectorially. It is believed that, in general, harmonic levels tend to be influenced primarily by local and immediately adjacent conditions rather than wider zonal effects.

The main detrimental effects of harmonics are [30]:

- maloperation of control devices, mains signalling systems and protective relays,
- extra losses in capacitors, transformers and rotating machines,
- additional noise from motors and other apparatus,
- telephone interference, and
- the presence of power factor correction capacitors and cable capacitance which can cause shunt and series resonances in the network producing voltage amplification even at a remote point from the distorting load.

As well as the above, interharmonics can perturb *ripple control signals* and at sub-harmonic levels can cause flicker. To keep the harmonic voltage content within the recommended levels, the main solutions in current use are:

- the use of high pulse rectification (*e.g.* smelters and HVDC converters),
- passive filters, either tuned to individual frequencies or of the band-pass type, and
- active filters and conditioners.

10.3.1 Harmonic Sources

Lower order odd harmonics are the most prolific among consumer electronic systems. However, the third harmonic (of zero sequence) is usually prevented from entering the high voltage system by the use of appropriate transformer connections. The fifth harmonic (in the UK) has been identified as the harmonic order exhibiting the highest peak levels of high voltage systems, with values between 2.5% and 3.0% at some locations. The fifth also most frequently presents the highest mean harmonic levels, a characteristic which has been found to be consistent both geographically and with time.

Power Quality

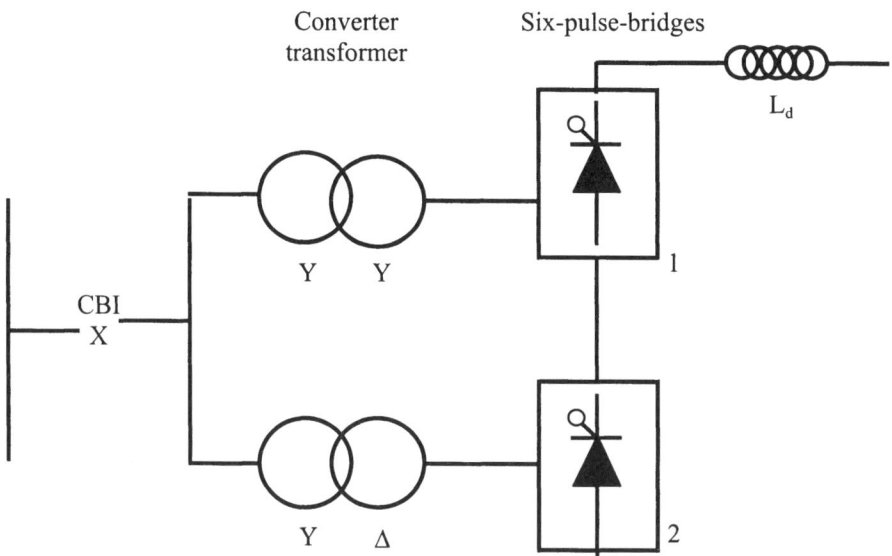

Figure 10.8 12-pulse converter

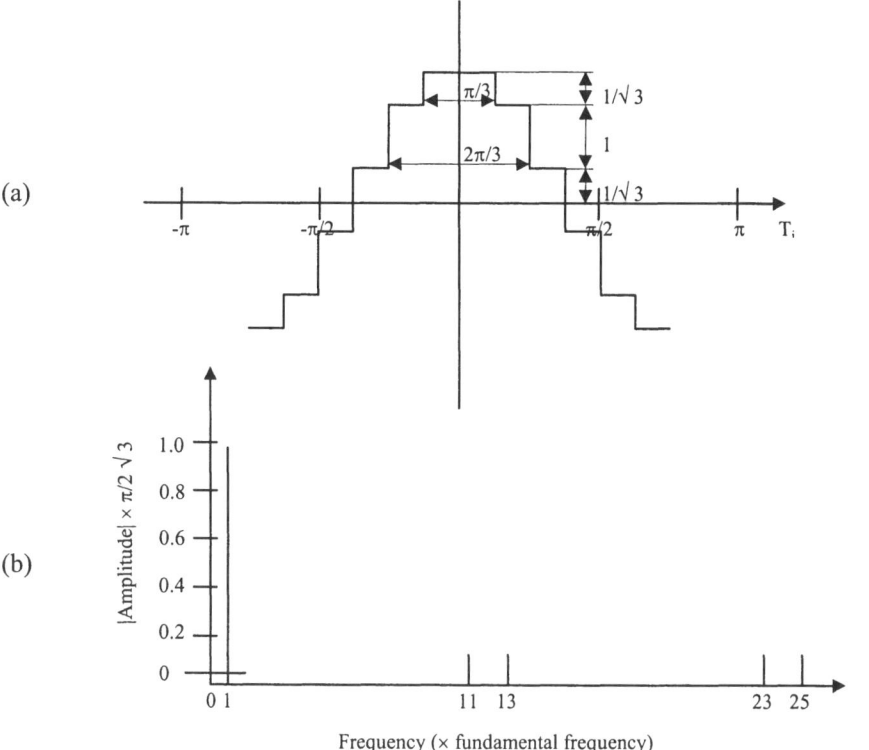

Figure 10.9 12-pulse converter current: (a) waveform, (b) harmonic spectrum

The standard configuration for industrial and HVDC applications is the 12-pulse converter, shown in Figure 10.8. The *characteristic* harmonic currents for the configuration are of orders $12k+1$ (of positive sequence) and $12k-1$ (of negative sequence) and their amplitudes are inversely proportional to the harmonic order, as shown by the spectrum of Figure 10.9b which corresponds to the time waveform of Figure 10.9a. These are, of course, maximum levels for ideal system conditions, i.e. with an infinite (zero impedance) AC system and a perfectly flat direct current (i.e. infinite smoothing reactance). When the AC system is weak and the operation not perfectly symmetrical *uncharacteristic harmonics* appear.

While the characteristic harmonics of the large power converter are reduced by filters, it is not economical to reduce in that way the uncharacteristic harmonics and, therefore, even small injection of these harmonic currents can, via parallel resonant conditions, produce very large voltage distortion levels. An example of uncharacteristic converter behaviour is the presence of fundamental frequency on the DC side of the converter, often induced from AC transmission lines in the proximity of the DC line, which produces second harmonic and direct current on the AC side.

Even harmonics, particularly the second, are very disruptive to power electronic devices and are, therefore, heavily penalised in the regulations. The flow of DC current in the AC system is even more distorting, the most immediate effect being asymmetrical saturation of the converters or other transformers with a considerable increase in even harmonics which, under certain conditions, can lead to *harmonic instabilities* [31].

Another common example of uncharacteristic behaviour is the appearance of triplen harmonics. Asymmetrical voltages, when using a common firing angle control for all the valves, result in current pulse width differences between the three phases which produces triplen harmonics. To prevent this effect, modern large power converters use the equidistant firing concept instead [32]. However, this controller cannot eliminate second-harmonic amplitude modulation of the DC current which, via the converter modulation process, returns third-harmonic current of positive sequence. This current can flow through the converter transformer regardless of its connection and penetrate far into the AC system.

Good harmonic prediction requires understanding of two closely related topics. One is the accurate location and characteristics of the harmonic sources and the other the interaction of these sources via the predominantly linear AC system that interconnects them. This task is made difficult by insufficient information on the composition of the system loads and their damping to harmonic frequencies. Further impediments to accurate prediction are the existence of many distributed non-linearities, phase diversity, the varying nature of the load, *etc.*

Simulation of the interaction between large static power converters and the AC systems is a complex issue considering the large size of the converter plant in many applications and its sophisticated switching control. The operation of the converter is highly dependent on the quality of the power supply, which is itself heavily influenced by the converter plant.

10.3.2 Characterisation of Harmonic Sources

Mathematical models with various levels of complexity are appearing in the literature to represent individual non-linear components, such as AC/DC converters, in the form of harmonic Norton equivalents. They involve iterative harmonic analysis to represent the interaction between the converter and the linear system. Further work is needed, however, to represent simultaneously the effect of multiple interconnected non-linear components. The system steady state is substantially, but not completely, described by the harmonic voltages throughout the network. In many cases, it is assumed that there are no other frequencies present apart from the fundamental frequency and its harmonics. This type of analysis can be viewed as a restriction of frequency domain modelling to integer harmonic frequencies but with all non-linear interactions modelled. Harmonic domain modelling may also encompass a solution for three-phase load flow constraints, control variables, power electronic switching instants, transformer core saturation, etc.

There are two important aspects to the harmonic domain modelling of the power system:

1. The derivation, form and accuracy of the non-linear equations used to describe the system steady state.
2. The iterative procedure used to solve the non-linear equation set.

Many methods have been employed to obtain a set of accurate non-linear equations, which describe the system steady state. After partitioning the system into linear regions and non-linear devices, the non-linear devices are described by isolated equations, given boundary conditions to the linear system. The system solution is then predominantly a solution for the boundary conditions for each non-linear device. Device modelling has been by means of time domain simulation to the steady state [33], analytic time domain expressions [34,35], waveshape sampling and FFT [36] and, more recently, by harmonic phasor analytic expressions [37].

In the past, harmonic domain modelling has been hampered by insufficient attention given to the solution method. Earlier methods used the Gauss-Seidel type fixed point iteration, which frequently diverged. Improvements made since then have been to include linearising *RLC* components in the circuit to be solved in such a way as to have no effect on the solution itself [35,38]. A more recent approach has been to replace the non-linear devices at each iteration by a linear Norton equivalent, chosen to mimic the non-linearity as closely as possible, sometimes by means of a frequency-coupled Norton admittance. The progression with these improvements has led towards Newton-type solutions, as employed successfully in the load flow for many years.

When the non-linear system to be solved is expressed in a form suitable for solution by Newton's method. The separate problems of device modelling and system solution are completely decoupled and the wide variety of improvements to the basic Newton method, developed by the numerical analysis community, can readily be applied.

10.3.3 Harmonic Flows [30]

In its simplest form the frequency domain provides a direct solution of the effect of specified individual harmonic or non-harmonic frequency injections throughout a linear system, without explicit consideration of the harmonic interaction between the network and the non-linear component(s).

The sources of harmonic injection, depending on the available information of the non-linear components, can be current sources or Norton or Thévenin harmonic equivalents. A common experience derived from harmonic field tests is the asymmetrical nature of the readings.

Asymmetry, being the rule rather than the exception, justifies the need for three-phase harmonic models. The basic component of a three-phase algorithm is the multi-conductor transmission line, which can be accurately represented at any frequency by means of an appropriate equivalent PI-model, including mutual effects as well as earth return, skin effect, etc. The transmission line models are then combined with the other network passive components to obtain three-phase equivalent harmonic impedances.

The system harmonic voltages are calculated by direct solution of the linear equation

$$[\mathbf{I}_h] = [\mathbf{Y}_h][\mathbf{V}_h] \tag{10.8}$$

where $[\mathbf{Y}_h]$ is a reduced system admittance matrix of order equal to (three times) the number of injection busbars.

10.3.4 Aperiodic Distortion

The voltage and current waveforms often have an aperiodic component. The most common and damaging load causing aperiodicity in the waveform is the arc furnace. The result is random variations of harmonic and interharmonic content, which are uneconomical to eliminate by conventional filters.

Aperiodicity also produces voltage fluctuations and light flicker. Connection to the highest possible voltage level and the use of series reactances are among the measures currently taken to reduce their impact on PQ. The conventional PQ indices do not take into account the aperiodic components.

For example, the total harmonic distortion (THD) is basically a ratio of the energy content in the harmonics to that in the fundamental component. It is possible to define a similar index for the aperiodic case by defining the power frequency (and thereby defining the fundamental component), and then using the remaining portion of the signal as the numerator of a THD-like index [39]:

$$\text{THD}' = \frac{E[i(t) - i_0 \cos(\omega_0 t)]}{i_0}$$

where the power frequency is denoted as ω_0 and $E[.]$ denotes the calculation of the energy of a time signal. The prime on the THD indicates that this is not quite the same as the conventional THD calculation. Of course, THD' degenerates to THD for the periodic case. With reference to the flicker disturbance, the measurement and frequency windows in which flicker exists is defined by international standards, mainly through the International Electrotechnical Commission (IEC). Generally, flicker is limited to 10 Hz and lower fluctuations in the supply voltage.

A problematic feature of this index is how the flicker is to be measured. As an example, should the flicker energy (i.e. sideband energy in the vicinity of the power frequency) be measured in root mean square amplitude, or zero to peak?

Also, is it more meaningful to integrate the sideband energy over a specified range of frequencies. The latter appears to have less physiological implications but better mathematical properties. Also, the integration of energy might be done using a physiological weighting factor as specified by the IEC standards. A windowed Fourier transform, short-time Fourier transform, and Fourier linear combiner have been suggested as possible solutions to the problem.

10.4   Need for Adequate PQ Indices and Standards

A variety of PQ indices have been produced with the intent of summarising the degree of distortion of the voltage and current waveforms of the active power loss caused by the distortion, and the interference on telephone and data communication circuits. Most of these indices have evolved from experience with power systems, from intuitive reasoning and from heuristics. However, with the advent of power electronics and other non-linear electronic devices, there are problematic cases in the general application of the traditional PQ indices.

For example, consider the simple use of the power factor index to minimise distribution system losses, with a time-varying load, such as a pulsating load on the shaft of a three-phase induction motor. The load can become a source over part of the cycle (e.g. a power stroke occurs followed by a regenerative period). Thus, the power factor of such an induction machine load may go leading and lagging. In this case, the issue of how to *correct the power factor* to minimise loss in the distribution supply creates a counter-intuitive result. The power factor index should be applied with caution in cases of time variation, unbalance and presence of non-power frequency signals.

The main motivation for using indices is the ease in calculation, the standardisation of definition, the simple application of the indices (and the simplified specification of the indices). However, in some cases, indices should not be used at all. Instead, it might be possible to utilise the time waveshape of voltages and currents directly. Some promising alternatives of PQ definitions include sojourn time, wavelet spectrum, Liapunov exponent, and Park's and Clarke's components.

The industry needs to establish uniform and complete PQ measurement standards so that data can be compared (over location, over time, etc.) and disputes resolved. Standards such as IEC 61000-4-7, which covers harmonics measurements, are a good model. IEC 77A Working Group 09 has made enormous progress in this area [40].

Utility-oriented PQ standards are under continuous development [41-68]. These types of standards can be used to set a common quality basis for competition between suppliers and they should create a minimum acceptable level of PQ. The European standard already contains some well-defined margins for harmonic distortion and other variations. Much work, however, still needs to be done to set acceptable levels for events like voltage sags and interruptions. The voltage characteristics by themselves are not sufficient for equipment immunity requirements and a maximum permissible number of equipment trips needs to be decided on.

More work is needed on PQ standards that can be used by equipment manufacturers. It is far less expensive to inform manufacturers about the real level of PQ that is available than it is to attempt to improve the level of power quality. Some industries, such as the semiconductor industry, have already developed their own standards [69]. This kind of power compatibility standard will ultimately minimise all PQ issues, including those introduced by deregulation.

Next to these industrial initiatives, a serious effort is needed from standard-setting organisations like the IEC and IEEE, to publish requirements for equipment immunity against voltage sags and short interruptions.

10.5 Need for Adequate PQ Monitoring [70,71]

The interdependence between PQ monitoring equipment and PQ standards is a chicken and egg situation. The latter are normally constrained by the characteristics of commercial instrumentation, whereas the manufacturers of the former are reluctant to invest in more advanced technology until it is required by the standards. The expectation, in this respect, is that deregulation will force a faster development of suitable instrumentation capable of assessing the agreed indices of PQ at points of energy interchange.

The limitations of present transducers (voltage and current transformers) to reflect accurately the variety of frequencies involved in PQ indices is the main concern, considering the impracticality of replacing them in the short term. Therefore, the recommendations of IEC 1000-4-7 are likely to provide the basis for transducers' suitability for a long time.

Regarding PQ instrumentation, there is no reason for accepting the present constraints placed on data acquisition, namely the use of snapshots, number of data channels, and processing capability.

Developments in distributed processing architecture (shown schematically in Figure 10.10), the use of DSPs, possible shifting of A/D conversion to the switchyard close to the transducers and use of digital communications with the central system (via fiber optics) can provide on-line practically unlimited processing capability as well as noise immunity.

Multi-channel three-phase real-time monitoring is now becoming available. One proposed system [72] includes remote data conversion modules (RDCMs), digital fiber optic transmission, GPS synchronisation, central parallel processing and ethernet-connected PCs for distance control and display.

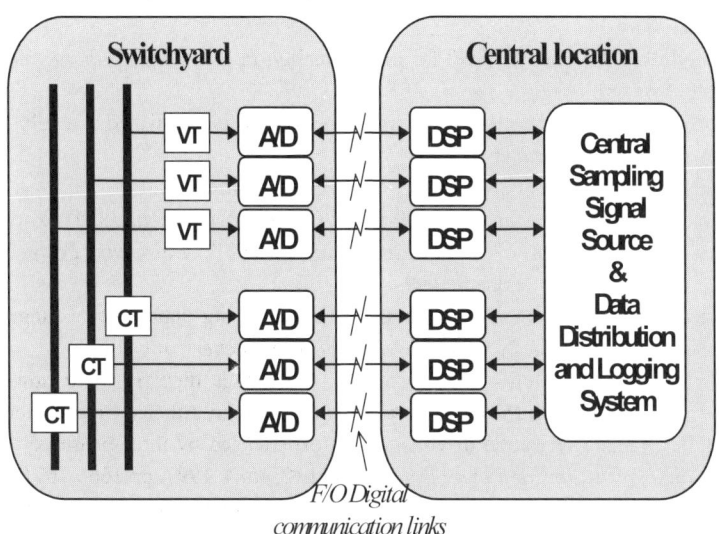

Figure 10.10 A possible distributed processing architecture

10.6 References

[1] J. Meeuwsen and W. Kling, 'The influence of different network structures on power supply reliability', *Quality of Power Supply, ETG Conference*, 1997.

[2] A. Larsson, 'Flicker and slow voltage variations from wind power', *International Conference on Harmonics & Quality of Power*, Las Vegas, 1996, pp.270-275.

[3] N.G. Hingorani, 'Introducing custom power', *IEEE Spectrum,* Vol.32, No.6, 1995, pp.41-47.

[4] C. Degand, P. Fauquembergue, M. Regnier, C. Levillain and E. Serres, 'EDF customized approach of power quality, and opportunity to meet customer's expectations', *ICHPQ '98*, Athens, 1998, pp.190-196.

[5] M. Shepard, 'Corporate energy managers', *Strategic Memo SM-96-6, E Source,* Boulder, USA, 1996.

[6] M.H.J. Bollen, *Understanding Power Quality Problems - Voltage Sags and Interruptions*, IEEE Press, New York, 1999.

[7] M.H.J. Bollen, 'Fast assessment methods for voltage sags in distribution systems', *IEEE Transactions on Industry Applications*, Vol.32, No.6, 1996, pp.1414-1423.

[8] C.S. Chen and J.T. Leu, 'Interruptible load control for Taiwan Power Company', *IEEE Transactions on Power Systems,* Vol.5, No.2, 1990, pp.460-465.

[9] K. Bhattacharaya and M.H.J. Bollen, 'Optimal interruptible tariffs and their impact on quality of power supply', *Power Systems Computation Conference,* Trondheim, Norway, 1999.

[10] J. Arrillaga, B.C. Smith, N.R. Watson and A.R. Wood, *Power System Harmonic Analysis,* John Wiley & Sons, Chichester, 1997.

[11] 'Report on customer services', published annually by Office of Electricity Regulation, Birmingham, UK.

[12] 'Voltage dips, short interruptions and voltage variations immunity tests', *IEC Standard Document* 61000-4-11.

[13] 'Basnivö för elkvalitet', (Basic level for power quality, in Swedish), *Götborg Energi Nät AB*, Gothenburg, Sweden, 1997.

[14] 'Report on distribution and transmission system performance', published annually by Office of Electricity Regulation, Birmingham, UK.

[15] IEEE Project Group 1159.2: Power quality event characterization.

[16] M.H.J. Bollen, J. Svensson and L.D. Zhang, 'Testing of grid-connected power-electronics converters for the effects of short circuits in the grid', *European Power Electronics Conference*, Lausanne, Switzerland, 1999.

[17] L.D. Zhang and M.H.J. Bollen, 'A method for characterizing unbalanced voltage dips (sags) with symmetrical components', *IEEE Power Engineering Letters*, July 1998.

[18] L.D. Zhang and M.H.J. Bollen, 'Characteristics of voltage dips (sags) in power systems', *International Conference on Harmonics and Quality of Power*, Athens, Greece, October 1998.

[19] M.H.J. Bollen, 'Characterization of voltage sags experienced by three-phase adjustable-speed drives', *IEEE Transactions on Power Delivery*, Vol.12, No.4, 1997, pp.1666-1671.

[20] L.D. Zhang and M.H.J. Bollen, 'A method for characterisation of three-phase unbalanced dips (sags) from recorded voltage waveshapes', *International Telecommunications Energy Conference (INTELEC)*, Copenhagen, Denmark, June 1999.

[21] ITIC (Information Technology Industry Council, formerly known as the Computer & Business Equipment Manufacturer's Association), *ITIC Curve Application Note*, available at http://www.itic.org/iss_pol/techdocs/curve.Pdf.

[22] P.F. Ribeiro, T. Haque, P. Pillay and A. Bhattacharjee, 'Application of wavelets to determine motor drive performance during power systems switching transients', *Power Quality Assessment*, Amsterdam, 1994.

[23] C.K. Chui, *An Introduction to Wavelets*, Academic Press, 1992, 6-18.

[24] I. Daubechies, 'Orthonormal bases of compactly supported wavelets', *Communications in Pure and Applied Mathematics*, Vol.41, 1988, pp.909-996.

[25] S. Santoso, E.J. Bowers, W.M. Grady and P. Hoffmann, 'Power quality assessment via wavelet transform analysis', *IEEE Transactions on Power Delivery*, Vol.11, No.2, 1996, pp.924-930.

[26] M. Ringrose and M. Negnevitsky, 'Automated disturbance recognition in power systems', *Australasian University Power Engineering Conference (AUPEC 98)*, Hobart, 1998, pp.593-597.

[27] P.F. Ribeiro and P. Celio, 'Advanced techniques for voltage quality analysis; unnecessary sophistication or indispensable tools', *Paper A-206, Power Quality Assessment*, Amsterdam, 1994.

[28] L. Zadeh, 'Fuzzy sets', *Information and Control*, Vol.8, No.3, 1965, pp.338-354.

[29] K. Tanaka, *An Introduction to Fuzzy Logic for Practical Applications*, Springer, 1997.

[30] J. Arrillaga, D. Bradley and P.S. Bodger, *Power System Harmonics*, John Wiley & Sons, Chichester, 1985.

[31] S. Chen, A.R. Woods and J. Arrillaga, 'HVDC converter transformer core saturation instability: A frequency domain analysis', *IEE Proceedings - Generation, Transmission, Distribution*, Vol.143, No.1, 1996, pp.75-81.

[32] J.D. Ainsworth, 'The phase-locked oscillator. A new control system for controlled static converters', *IEEE Transactions on Power Apparatus and Systems*, Vol.PAS-87, 1968, pp.859-865.

[33] J. Arrillaga, N.R. Watson, J.F. Eggleston and C.D. Callaghan, 'Comparison of steady state and dynamic models for the calculation of a.c./d.c. system harmonics', *IEE Proceedings*, Vol.134C, No.1, 1987, pp.31-37.
[34] R. Yacamini and J.C. Oliveira, 'Harmonics in multiple converter systems: a generalised approach', *IEE Proceedings*, B, Vol.127, 1980, pp.96-106.
[35] G. Carpinelli, et al., 'Generalised converter models for iterative harmonic analysis in power systems', *IEE Proceedings - Generation, Transmission and Distribution*, Vol.141, No.5, 1994, pp.445-451.
[36] C.D. Callaghan and J. Arrillaga, 'A double iterative algorithm for the analysis of power and harmonic flows at ac-dc converter terminals', *IEE Proceedings*, Vol.136, No.6, 1989, pp.319-324.
[37] B.C. Smith et al., 'A Newton solution for the harmonic phasor analysis of ac-dc converters', *IEEE PES Summer Meeting '95*, SM 379-8.
[38] C.D. Callaghan and J. Arrillaga, 'Convergence criteria for iterative harmonic analysis and its application to static converters', *ICHPS IV*, Budapest, 1990, pp.38-43.
[39] G.T. Heydt, 'Problematic power quality indices', *Panel Session on Harmonic Standards, IEEE Winter Power Meeting*, Singapore, 2000.
[40] R. Ott (Chairman), IEC 77A Low Frequency Phenomena, Working Group 9, 'Power quality measurements', *Draft in progress,* R.ott@edf.fr,1999.
[41] IEEE 141:1986, Recommended Practice for Electric Power Distribution for Industrial Plants.
[42] IEEE 1159: 1995, IEEE Recommended Practice on Monitoring Electric Power Quality.
[43] IEC 61000-2-5: 1995, Electromagnetic Compatibility (EMC), Part 2: Environment, Section 5: Classifications of Electromagnetic Environments.
[44] IEC 61000-2-1: 1990, Electromagnetic Compatibility (EMC), Part 2: Environment, Section 1: Description of the Environment – Electromagnetic Environment for Low-Frequency Conducted Disturbances and Signalling in Public Power Supply Systems.
[45] IEC 61000-2-2: 1990, Electromagnetic Compatibility (EMC), Part 2: Environment, Section 2: Compatibility Levels for Low-Frequency Conducted Disturbances and Signalling in Public Power Supply Systems.
[46] IEEE c62.41: 1991, IEEE Recommended Practice on Surge Voltages in Low-Voltage AC Power Circuits.
[47] IEC 816: 1984, Guide on Methods of Measurement of Short Duration Transients on Low Voltage Power and Signal Lines.
[48] UIE-DWG-2-92-D, UIE Guide to Measurements of Voltage Dips and Short Interruptions Occurring in Industrial Installations.
[49] Federal Information Processing Standards Publication 94: Guideline on Electrical Power for ADP Installations, National Technical Information Service, 1983.
[50] D.L. Brooks, R.C. Dugan, M. Waclawiak and S. Sundaram, 'Indices for assessing utility distribution system R.M.S. variation performance', *IEEE Transactions Power Delivery*, PE-920-PWRD-1-04, 1997.
[51] IEEE 519: 1992, IEEE Recommended Practices and Requirements for Harmonic Control in Electric Power Systems (ANSI).
[52] IEC 61000-4-7, 1991, Electromagnetic Compatibility (EMC), Part 4: Limits, Section 7: General guide on harmonics and inter-harmonics measurements and instrumentation, for power supply systems and equipment connected thereto.

[53] Directives concerning the Protection of Telecommunication Lines against Harmful Effects from Electricity Lines, International Telegraph and Telephone Consultative Committee (CCITT) published by the International Communications Union, Geneva, 1963.
[54] IEEE/ANSI C57.110-1986, Recommended Practice for Establishing Transformer Capability when Supplying Non-sinusoidal Load Currents.
[55] IEEE/ANSI Std 18-1980 (Reaff 1991), IEEE Standard for Shunt Power Capacitors.
[56] IEEE Inter-harmonic Task Force, CIGRE 36.05/CIRED 2 CC02 Voltage Quality Working Group, Inter-harmonics in Power Systems, January 1997.
[57] IEC 868: 1986, Flickermeter – Functional and design specifications.
[58] IEC 868-0: 1991, Flickermeter – Evaluation of flicker severity.
[59] IEC 61000-4-15 Ed. 1, 1997, Electromagnetic Compatibility (EMC), Part 4: Limits, Section 15: Flickermeter – Functional and design specifications.
[60] IEC 38: 1983, IEC Standard Voltages.
[61] ANSI C84.1: 1982, American National Standard For Electric Power Systems and Equipment – Voltage Ratings (60 Hz).
[62] UIE-DWG-3-92-G, Guide to Quality of Electrical Supply for Industrial Installations – Part 1: General Introduction to Electromagnetic Compatibility (EMC), Types of Disturbances and Relevant Standards.
[63] IEEE 100:1992, IEEE Standard Dictionary of Electrical and Electronics Terms.
[64] EN 50160: 1994, Voltage Characteristics of Electricity supplied by Public Distribution System, CENELEC.
[65] IEC 61000-3-2: 1994, Electromagnetic Compatibility (EMC), Part 3: Limits, Section 2: Limits for harmonic current emissions (equipment input current $\leq$ 16 A per phase).
[66] IEC 61000-3-4: 1994, Electromagnetic Compatibility (EMC), Part 3: Limits, Section 4: Limits for harmonic current emissions (equipment input current $\geq$ 16 A per phase).
[67] IEC 61000-3-3: 1994, Electromagnetic Compatibility (EMC), Part 3: Limits, Section 3: Limitation of Voltage Fluctuations and Flicker in Low-Voltage Supply System for Equipment with Rated Current $\leq$ 16 A.
[68] IEC 61000-3-5: 1994, Electromagnetic Compatibility (EMC), Part 3: Limits, Section 5: Limitation of Voltage Fluctuations and Flicker in Low-Voltage Supply System for Equipment with Rated Current greater than 16 A.
[69] SEMI Facilities Committee, Power Quality and Equipment Ride-through Standard SEMI-9803, http://www.semi.org, 1999.
[70] J. Arrillaga, N.R. Watson and S. Chen, *Power Systems Quality Assessment*, John Wiley & Sons, Chichester, 2000.
[71] 'An Assessment of Distribution System Power Quality', EPRI TR-106249, May 1996.
[72] A.J.V. Miller and M.B. Dewe, 'Multi-channel continuous harmonic analysis in real-time', *IEEE Transactions on Power Delivery*, Vol.7, No.4, 1992.

11

Information Technology Application

Prof. Gerald B. Sheblé
Iowa State University
USA

Prof. Kit Po Wong
University of Western Australia
Australia

Dr Loi Lei Lai
City University. London
UK

11.1 Introduction

Power utility deregulation and restructuring leads to the break-up of monolithic companies. It is now necessary for the power producers, transmission network controllers, distribution companies and service providers to cooperate within a competitive framework to supply power cheaply, securely and of a suitable quality. The separation of companies adds an extra dimension of uncertainty to the whole power system operation, control and planning.

Algorithms that have been developed to provide solutions to specific power system problems are mainly based on highly optimised numerical algorithms that rely on a number of assumptions. The reality of the situation is that the errors in making the assumptions can have a severe effect on the results of such numerically based algorithms.

Since the 1980s, many of the approaches adopted in power system analysis and design have turned away from the methodology of formal mathematical modelling which came from the fields of operational research, control theory and numerical analysis to the less rigorous computational intelligence techniques. Today the main computational intelligence techniques found in power system applications are artificial neural networks (ANNs) [1-4], evolutionary computing [5-12], fuzzy systems [13-14] and expert systems [15]. With the advance of information technology capabilities, there has been a significant increase in research and applications within computational intelligence techniques in power engineering. This chapter describes some of the latest work in research and applications in agents, evolutionary programming, complex neural networks and virtual reality to power engineering.

11.2 Software Agents

Software agents have evolved from multi-agent systems (MAS), which in turn form one of three broad areas which fall under distributed artificial intelligence (DAI), the other two being distributed problem solving (DPS) and parallel artificial intelligence (PAI) [16]. Hence, as with multi-agent systems, they inherit many of DAI's motivations, goals and potential benefits. For example, software agents inherit DAI's potential benefits including modularity, speed (due to parallelism) and reliability (due to redundancy). It also inherits those due to AI such as operation at the knowledge level, easier maintenance, reusability and platform independence. The concept of an agent can be traced back to the early days of research into DAI in the 1970s.

The study of multiple collaborative agents includes the interaction and communication between agents, decomposition and distribution of tasks, coordination and cooperation, conflict resolution via negotiation. These resulted in work such as DAI planning and game theories [17]. DAI's 'smartness' derives from the fact that the 'value' gained from individual standalone agents *coordinating* their actions by working in *cooperation* is greater than that gained from any individual agent. Application domains [18] in which agent solutions are being applied to or investigated include workflow management, network management, air-traffic control, business process re-engineering, data mining, information retrieval/management, electronic commerce, education, personal digital assistants (PDAs), e-mail, digital libraries, command and control, smart databases, scheduling/diary management, etc.

It is important to note that most agent-based systems are still *demonstrators* only: converting them into real usable applications would provide even greater challenges, some of which have been anticipated but, currently, many are unforeseen. The essential message of this section is that agents are here to stay, not least because of their diversity, their wide range of applicability and the broad spectrum of companies investing in them.

11.2.1 Types of Agents

An agent may be defined as a component of software and/or hardware which is capable of acting exactingly in order to accomplish tasks on behalf of its user. There are several factors to classify existing software agents.

Firstly, agents may be classified by their mobility, i.e. by their ability to move around some network. This yields the classes of *static* or *mobile* agents.

Secondly, they may be classed as either *deliberative* or *reactive*. Deliberative agents derive from the deliberative thinking paradigm: that is, the agents possess an internal symbolic, reasoning model and they engage in planning and negotiation in order to achieve coordination with other agents. Work on reactive agents originates from research carried out by Brooks [19]. These agents on the contrary do not have any internal, symbolic models of their environment, and they act using a stimulus/response type of behaviour by responding to the present state of the environment in which they are embedded [20]. Indeed, Brooks has argued that intelligent behaviour can be realised without the sort of explicit, symbolic representations of traditional AI [21].

Thirdly, agents may be classified along several ideals and primary attributes that agents *should* exhibit. At BT Labs, three main attributes, namely autonomy, learning and cooperation, have been identified. *Autonomy* refers to the principle that agents can operate on their own without the need for human guidance, even though this would sometimes be invaluable. Hence agents have individual internal states and goals, and they act in such a manner as to meet their goals on behalf of their user. A key element of their autonomy is their proactiveness, i.e. their ability to 'take the initiative' rather than acting simply in response to their environment [22]. *Cooperation* with other agents is paramount. In order to cooperate, agents need to possess a social ability, i.e. the ability to interact with other agents and possibly humans via some communication language [22]. Having said this, it is possible for agents to coordinate their actions without cooperation [23]. Lastly, for agent systems to be truly 'smart', they would have to *learn* as they react and/or interact with their external environment. Agents are (or should be) disembodied bits of 'intelligence'. With these three minimal attributes, Figure 11.1 was used to derive four types of agents, namely *collaborative agents, collaborative learning agents, interface agents* and *smart agents*.

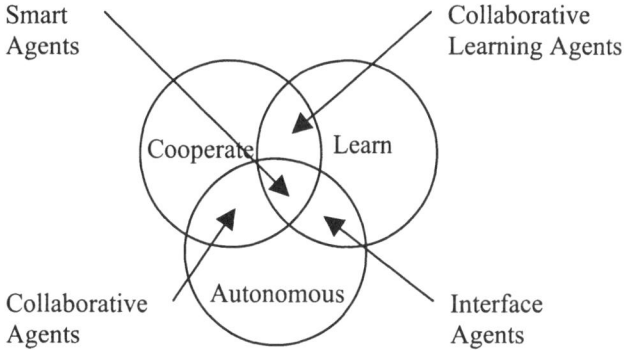

Figure 11.1 A part view of an agent typology

It must be emphasised that these distinctions are *not* definitive. For example, with collaborative agents, there is more emphasis on cooperation and autonomy than on learning; hence, it is not implied that collaborative agents never learn. Likewise, for interface agents, there is more emphasis on autonomy and learning than on cooperation. Anything else which lies outside the 'intersecting areas' is not considered to be agents. For example, most expert systems are largely 'autonomous' but, typically, they do not cooperate or learn.

Fourthly, agents may sometimes be classified by their roles (preferably, if the roles are major ones), e.g. World Wide Web (www) information agents. Again, information agents may be static, mobile or deliberative.

Fifthly, two or more agent philosophies are combined in a single agent that is termed a *hybrid* agent. There are other attributes of agents, which we consider *secondary* to those already mentioned. For example, is an agent versatile (i.e. does it have many goals or does it engage in a variety of tasks)? Is an agent benevolent or non-helpful, antagonistic or altruistic? Does an agent lie knowingly or is it always truthful (this attribute is termed veracity)? Can you trust the agent enough to (risk) delegate tasks to it? Does it degrade gracefully in contrast to failing drastically at the boundaries? Perhaps unbelievably, some researchers are also attributing emotional attitudes to agents - do they get 'fed up' being asked to do the same thing time and time again? What role does emotion have in constructing believable agents [24]? In essence, *agents exist in a truly multi-dimensional space*. It is quite possible that agents may be in competition with one another, or perhaps quite antagonistic towards each other. In agent-based systems generally, the communication involves high-level messages. The use of high-level messaging leads to lower communication costs, easy reimplementability and concurrency. Lastly, and perhaps most importantly, agent-based applications operate typically at the *knowledge level* [25].

Collaborative Agents
As shown in Figure 11.1, collaborative agents emphasise autonomy and cooperation (with other agents) in order to perform tasks for their owners. They may learn, but this aspect is not typically a major emphasis of their operation. In order to have a coordinated setup of collaborative agents, they may have to *negotiate* in order to reach mutually acceptable agreements on some matters

The *motivation* for having collaborative agent systems may include one or several of the following:

- to solve problems that are too large for a centralised single agent to do owing to resource limitations;
- to allow for the interconnecting and interoperation of multiple existing systems, e.g. decision support systems;
- to provide solutions to inherently distributed problems, e.g. distributed sensor networks;
- to provide solutions which draw from distributed information sources, e.g. for distributed on-line information sources;
- to provide solutions where the expertise is distributed;
- to enhance modularity (which reduces complexity), speed (due to parallelism), reliability (due to redundancy), flexibility (i.e. new tasks are composed more easily from the more modular organisation) and reusability at the knowledge level (hence shareability of resources);
- to research into other issues, e.g. understanding interactions among human societies.

Interface Agents
Interface agents emphasise autonomy and learning in order to perform tasks for their owners. Note the subtle emphasis and distinction between collaborating with *the user* and collaborating with *other agents*, as is the case with collaborative agents. Collaborating with a user may not require an explicit agent communication language as one required when collaborating with other agents. Essentially, interface agents support and provide assistance to use a particular application such as learning an operating system. The user's agent observes and monitors the actions taken by the user in the interface, learns new 'short-cuts', and suggests better ways of doing the task. Thus, the user's agent acts as an assistant, which cooperates with the user in accomplishing the task. As for learning, interface agents learn typically to assist their user better in the following four ways [26]:

1. By observing and imitating the user (i.e. learning from the user).
2. Through receiving positive and negative feedback from the user (learning from the user).
3. By receiving explicit instructions from the user (learning from the user).

4. By asking other agents for advice (i.e. learning from peers).

Their cooperation with other agents, if any, is limited typically to asking for advice, and not in getting into negotiation deals with them, as is the case with collaborative agents. The learning modes are typically by memory-based learning or other techniques such as evolutionary learning which are being introduced.

An interface agent is a quasi-smart piece of software that assists a user when interacting with one or more computer applications, where boring and laborious tasks could be delegated to interface agents, to eliminate the tedium of humans performing operations.

Mobile Agents

Mobile agents are computational software processes capable of roaming wide area networks (WANs) such as the WWW, interacting with foreign hosts, gathering information on behalf of their owner and coming 'back home' having performed the duties set by their user. These duties may range from a flight reservation to managing a telecommunications network. However, *mobility is neither a necessary nor sufficient condition for agenthood*. Mobile agents are agents because they are autonomous and they cooperate. For example, they may cooperate or communicate by one agent making the location of some of its internal objects and methods known to other agents. By doing this, an agent exchanges data or information with other agents without necessarily giving all its information away. Hence, mobile agents provide a number of *practical* advantages, which escape their static counterparts. They provide the following benefits:

- Reduced communication costs: there may be a lot of raw information that needs to be examined to determine its relevance. Transferring this raw information can be very time consuming. It is much more natural to get the agents to 'go' to that location, do a local search and only transfer the necessary information back across the network.
- Limited local resources: the processing power and storage on the local machine may be very limited (only perhaps for processing and storing the results of a search), thereby necessitating the use of mobile agents.
- Easier coordination: it may be simpler to coordinate a number of remote and independent requests and only collate all the results locally.
- Asynchronous computing: the mobile agents could do something and place the results back in the user's mailbox at some later time. They may operate when the user is even connected.
- A natural development environment for implementing 'free market' trading services. New services can come and go dynamically and much more flexible services may coexist with inferior ones, providing more choices for consumers [27].
- A flexible distributed computing architecture: mobile agents provide a unique distributed computing architecture which functions differently from the static setups. It provides an innovative way of doing distributed computation.

Mobile Agents: Some Challenges

Wayner [28] lists the major challenges. They include at least the following:

- Transportation: how does an agent move from place to place? How does it pack up and move?

- Authentication: how does the user ensure the agent is who it says it is, and that it is representing who it claims to be representing? How does the user know it has navigated various networks without being infected by a virus?
- Secrecy: how does the user ensure that the agents maintain privacy? How does the user ensure someone else has not read the personal agent and executed it for their own gains? How does the user ensure that the agent is not killed?
- Security: how does the user protect against viruses? How does the user prevent an incoming agent from entering an endless loop and consuming all the CPU cycles?
- Cash: how will the agent pay for services? How does the user ensure that it does not run up an outrageous bill on the user's behalf?
- Performance issues: what would be the effect of having hundreds, thousands or millions of such agents on a WAN?
- Interoperability/communication/brokering services: how does the user provide brokering/directory-type services for locating engines and/or specific services? How does the user publish or subscribe to services, or support broadcasting necessary for some other coordination approaches?

The *motivation* for developing information/internet agents is simply a need/demand for tools to manage such information explosion. Everyone on the WWW would benefit from them. In the future, agents are going to search the Internet, because no matter how much better the Internet may be organised, it cannot keep pace with the growth in information.

Hybrid Agents

Since each type has (or promises) its own strengths and deficiencies, the trick (as always) is to maximise the strengths and minimise the deficiencies of the most relevant technique for a particular purpose. Frequently, one way of doing this is to adopt a *hybrid* approach [26], which brings together some of the strengths of both the deliberative and reactive paradigms. Hence, hybrid agents refer to those whose constitution is a combination of two or more agent *philosophies* within a singular agent. These philosophies include a mobile philosophy, an interface agent philosophy, collaborative agent philosophy, etc.

The key *hypothesis* for having hybrid agents or architectures is the belief that, for some applications, the benefits accrued from having the combination of philosophies within a singular agent are greater than the gains obtained from the same agent based entirely on a singular philosophy. Otherwise having a hybrid agent is meaningless. Clearly, the *motivation* is the expectation that this hypothesis would be proved right; the ideal *benefits* would be the set union of the benefits of the individual philosophies in the hybrid. Consider the obvious case of constructing an agent based on both the collaborative (i.e. deliberative) and reactive philosophies. In such a case the reactive component, which would take precedence over the deliberative one, brings about the following benefits: robustness, faster response times and adaptability. The deliberative part of the agent would handle the longer term goal-oriented issues. For example, there is scope for more hybrids within a singular agent by combining the interface agent and mobile agent philosophies that would enable mobile agents to harness features of typical interface agents or some other combinations.

Heterogeneous Agent Systems

Heterogeneous agent systems refer to an integrated setup of at least two or more agents which belong to two or more different agent classes. A heterogeneous agent system may also contain one or more hybrid agents. Genesereth and Ketchpel [29] articulate clearly the *motivation* for heterogeneous agent systems. The essential argument is that the world abounds with a rich diversity of software products providing a wide range of services for a similarly wide range of domains. Though these programs work in isolation, there is an increasing demand to have them *interoperate* - hopefully, in such a manner that they provide 'added value' as an ensemble than they do individually. A new domain called *agent-based software engineering* has been invented in order to facilitate the interoperation of miscellaneous software agents. A key requirement for interoperation amongst heterogeneous agents is having an agent communication language (ACL) via which the different software 'agents' can communicate with each other. The potential *benefits* for having heterogeneous agent technology are several:

- Standalone applications can be made to provide 'value added' services by enhancing them in order to participate and interoperate in cooperative heterogeneous setups;
- The legacy software problem may be ameliorated because it could obviate the need for costly software rewrites as agents are given 'new leases of life' by getting them to interoperate with other systems. At the very least, heterogeneous agent technology may lessen the effect of routine software maintenance, upgrade or rewrites.
- Agent-based software engineering provides a radical new approach to software design, implementation and maintenance in general, and software interoperability in particular.

Genesereth and Ketchpel [29] note that agent-based software engineering is often compared to object-oriented programming in that an agent, like an object, provides a message-based interface to its internal data structures and algorithms. However, they note that there is a key distinction: in object-oriented programming, the meaning of a message may differ from object to object (this is the principle of polymorphism); in agent-based software engineering, agents use a common language with an agent-independent semantics. They highlight three important questions raised by the new agent-oriented software engineering paradigm. They include:

- What is an appropriate agent communication language?
- How are agents capable of communicating in this constructed language?
- What communication architectures are conducive to cooperation?

Once the agents are available, there are two possible architectures to choose from: one in which all the agents handle their own coordination or another in which groups of agents can rely on special system programs to achieve coordination. The disadvantage of the former is that the communication overhead does not ensure scalability, which is a necessary requirement for the future of agents. As a consequence, the latter federated approach (see Figure 11.2) is typically preferred.

In the above federated setup, there are five agents distributed in two machines, one with two agents and the other with three. The agents do not communicate directly with one another but do so through intermediaries called facilitators, which are similar to Wiederhold's *mediators* [30]. Essentially, the agents surrender some of their autonomy to the facilitators who are able to locate other agents on the network capable of providing various services. They also establish the connection across the environments and ensure correct 'conversation' amongst agents.

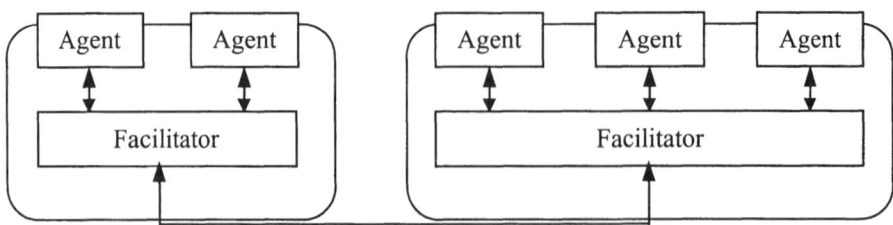

Figure 11.2 A federated system (adapted from [29])

11.2.2 General Issues and the Future of Agents

Apart from technical issues, as mentioned earlier, there is also a range of social and ethical problems, which are looming. They include the following:

- Privacy: how does the user ensure that agents maintain much needed privacy when acting on the user's behalf?
- Legal issues: imagine an agent offers some bad advice to other peer agents resulting in liabilities to other people; who is responsible?
- Ethical issues: agents must limit their searches to appropriate servers, share information with others and respect the authority placed on them by server operators.

11.3 Electricity Options Markets with Agents

Under deregulation, the formation of electricity markets is a topic of great interest in the power industry and in financial institutions worldwide. Using derivative financial instruments (including options) becomes important for hedging against uncertainty and managing risk - limiting exposure to adverse market conditions. Black and Scholes' equation is often used to value options, but its validity is questionable because of assumptions that may not hold for electricity, most notably the assumption of log-normally distributed prices for the underlying commodity. In this section, a put options market for electricity is modelled. Adaptive agents trade in this market to maximise profit. They are not forced to use an explicit economic or financial model (e.g. Black-Scholes) in their valuation. A genetic algorithm (GA) is used to find alternative valuations that are used to generate buy and sell signals. The results show that it is possible to evolve profitable valuations for use with buying and selling options in this simple model. Reasons for and implications of this finding (e.g. that Black-Scholes may not be a good method for pricing electricity derivatives) are discussed.

Throughout the world, the electricity industry is in the midst of major changes designed to promote competition. No longer vertically integrated with guaranteed customers and suppliers, electric generators and distributors will have to compete to sell and buy electricity. The traditional electricity utilities of the past will find themselves in a highly competitive environment. Some countries and regions of the USA (e.g. California) are already operating in a restructured environment. There does not yet appear to be a standardised final market structure that works for all areas, but each market that springs up adds to our experience and helps us make the next market implementation work a little better and more competitively. It is believed that to some degree, depending on the market implementation, regional commodity exchanges will play a key role in buying and selling electricity.

This section assumes a framework which has been described in detail by Sheblé [31,32]. Companies presently having both generation and distribution facilities would be divided into separate profit and loss centres. Power is generated by generation companies (GENCOs), transported via transmission companies (TRANSCOs) and distribution companies (DISTCOs), and is sold to energy service companies (ESCOs) representing the end-consumers. The North American Electricity Reliability Council (NERC) sets the reliability and security standards. Energy mercantile associations (EMAs) will emerge in this competitive electric industry. The EMAs will promote liquidity and as an intermediate partner to all multilateral trades, they will provide assurance to traders, that they need not worry about trading because a defaulting contract partner.

This framework allows for cash (consists of spot and forward markets), futures and planning markets. See Figure 11.3. The spot market allows for trading power each hour (or other duration, e.g. 30 minutes) in the next 30 days. Forward contracts allow energy traders to buy or sell firm electricity contracts as specified in the contract from 1 to 18 months. The futures market allows traders to purchase a non-firm electricity contract for a given month in the future (e.g. 1 to 18 months). Futures contracts provide a means for electricity traders to manage their risk. The planning market is a longer term market used to develop capital for building large items like new plants and transmission lines.

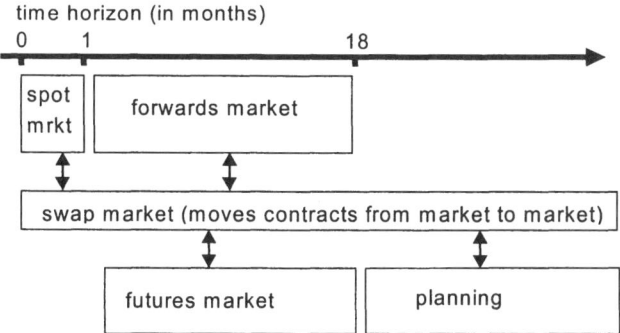

Figure 11.3 Interconnection between the markets

Options markets (for both futures and physicals) for electric energy are expected to be common and will be an important means of mitigating risk. An option contract gives its holder the right to buy or sell without the obligation to buy or sell. For this right, the holder must pay an up-front premium. The amount of the premium should reflect the value of the option to the potential holders. The worth of an option may vary from trader to trader owing to risk preferences, makeup of portfolios (collection of assets and contracts), etc. So the question is, how does one determine the value of an option? The Black-Scholes equation has been used in many markets to value options. Its usage assumes many things about the traded physical commodity that may not be true about electricity.

The approach taken in this research is to allow computerised agents to develop their own valuation formulae as they participate in a simulated option markets. The agents with proper options valuation should achieve higher profit than do other agents with poor valuations. The computerised agents evolve in a genetic algorithm. Those with poor valuations are replaced with new agents that are based on the successful ideas of the better agents.

11.3.1 Electricity Markets and Options

As mentioned previously, it is quite likely that regional commodity exchanges in which buyers and sellers participate in a double auction will soon exist. Such exchanges are utilised in other markets and are essentially an extension of the electricity market operating in California. A centralised exchange allows many and varied traders easily to trade a common commodity and derivatives based on that commodity for various periods.

In the cash (spot and forward) market, buyers and sellers interact through an independent contract administrator (ICA), who matches the bids subject to all operational constraints. GENCOs and ESCOs cooperate with the ICA, who is responsible for ensuring that the energy transactions resulting from the matched bids do not overload or render the electrical transmission system insecure. The ICA monitors and responds to the power system limits and transmission capacities.

The *spot market* is what we are most familiar with in the electrical industry. A seller and a buyer agree (either bilaterally or through an exchange) upon a price for a certain number of megawatts to be delivered sometime in the near future (e.g. 10 MW from 1.00 p.m. to 4.00 p.m. tomorrow).

An *options contract* is a form of insurance that gives the option purchaser the right, but not the obligation, to buy (sell) a contract at a given price. For each options contract, there is someone 'writing' the contract who, in return for a premium, is obligated to sell (buy) at the strike price. See Figure 11.4. Both the options and the futures contract are financial instruments designed to minimise risk. Although provisions for delivery exist, they are not convenient (e.g. the delivery point is not located where you want it to be located). The trader ultimately cancels his/her position in the futures market with either a gain or loss. The physicals are then purchased on the spot market to meet demand with the profit or loss having been locked-in via the futures contract.

'Long' denotes ownership; to go long means to purchase the item in question. In the figure, long indicates that the trader has purchased the option and now has the right to buy (call) or the right to sell (put) the future. A trader who writes the option is 'short'; to go short is to sell the item in question. Let's assume that the item in question is a MWh of electricity. In the long call diagram, the long trader has paid a premium (e.g. $1) to the option writer for the call option. This call option gives the trader the right to buy a MWh for the strike price (e.g. $7). At any price greater than the strike price, the option is exercised and the trader is said to be 'in-the-money' regardless of whether he/she gains or loses. If the price goes above the strike price plus the premium (e.g. $8), the trader has made a profit. The long trader has reduced risk by limiting his/her losses to the premium. On the other side of the figure is shown what happens from the option writer's point of view. The option writer receives the premium for assuming the risk and is obligated to sell the MWh at the strike price even though the market price is higher. The bottom half of the

Information Technology Application

figure shows how the 'put' works. The long trader pays a premium to lock-in a maximum price (exercise price) that he/she will have to pay for the MWh. The short trader takes that premium in return for promising to sell the MWh for that same exercise price.

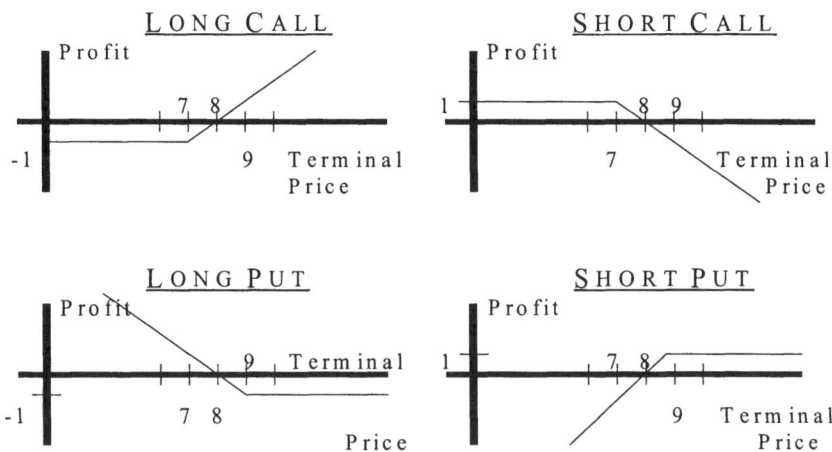

Figure 11.4 Using put and call options

Determining the value of the option has been the subject of some debate. A couple of decades ago, Black and Scholes put together their formula which has been widely used for valuing options in other commodity markets. Marshall [33] states that Black-Scholes requires that:

- The short-term interest rate is known and constant.
- The underlying asset pays no dividends.
- The underlying asset is efficiently priced.
- The option is of the European type.
- There are no transaction costs (for buying and selling).
- Any fraction of underlying asset value can be borrowed.
- No artificial restrictions on, or penalties for short selling.
- The Black-Scholes equation for valuing a put option is as follows [34]:

$$p = [X \cdot \exp(-r \cdot (T-t)) \cdot N(-d2) - S \cdot N(-d1)]$$

where:

X = strike price
S = spot price
r = risk free rate
$N(dn)$ = cumulative normal distribution
T = expiration date
t = current time

$$d1 = \frac{\ln\left(\frac{S}{X}\right) + \left(r + \frac{\sigma^2}{2}\right) \cdot (T-t)}{\sigma \cdot \sqrt{T-t}}$$

$$d2 = d1 - \sigma \cdot \sqrt{T-t}$$

11.3.2 Agent-Based Computational Economics

Market participants (suppliers and consumers) and their responses to prices can be quite complex, changing with time and with market conditions. Realistic models allow agents to modify their behaviour as time goes along, varying their strategies to optimise their position. Although some researchers model market agents with fixed rules, and model the market responses using control theory, it is quite likely that they make some unrealistic assumptions about the players or the market. In reality the solutions reached by a market are a function of the agents who are participating in the market at that point in time.

Agent-based computational economics (ACE) allows one to model each part of these complex systems without having to make possibly unrealistic assumptions in order to ensure that a solution to the model can be found. As suggested by its name, ACE uses interacting intelligent/adaptive agents to do economic computations. Agents are well suited to economic problems for many reasons. Firstly, an economy is essentially a collection of participants each acting in its own way; hence each player should be modelled independently. Secondly, the agents can be allowed to interact in myriad ways, without those interactions being constrained by the necessity of formulating them as a set of equations. Thirdly, the agents' behaviour need not be modelled as a differentiable function; it can be discontinuous and non-differentiable if that is what best models the system of interest. The power and flexibility afforded by agent-based computational models make an ACE approach very worthwhile.

With ACE, one can simulate thousands of possible systems and classify their behaviour. Because of advancements in computer technology and the computational power it gives researchers, it is possible to search for an optimal system of functions rather than trying to optimise a contrived system.

The number of different scenarios and possibilities is almost limitless. Starting with a simple model in which the producers and consumers interact via a simple pricing mechanism, one quickly begins to think of things to add and different things to test. The modular design that comes with using agents makes it possible to add to the model. One can model the spot market as well as the futures and/or options market. The agents then use these markets to maximise their utility. Trying to formulate an optimal futures market and complex interactions with the spot market and other derivative markets using linear algebra would be difficult if not impossible.

There are many different ways of programming adaptive agents (e.g. neural networks, genetic algorithms, etc.). Neural networks (NNs) are nice for problems that can be reduced to finding an optimal function for some task. They are well suited for problems in which the search space is relatively smooth and there are no localised suboptimal solutions. Genetic algorithms are another search mechanism in which the parts of the solution are coded as a 'gene' and one searches for the best collection of genes. They are less susceptible to being trapped in localised suboptima than are neural nets because one normally tests an entire population of genes all at the same time, which essentially examines discrete points in the search space and selects those grouping of genes that best solve the problem.

The basic genetic algorithm, as described by Goldberg [35], can be written as follows (see Figure 11.5 for a block diagram):

1. Randomly initialise a population and set the generation counter to zero.
2. Until done or out of time, do the following:

   - Calculate the fitness of each member of the population.
   - Select parents using some fitness bias.
   - Crossover the parents to create candidate offspring.
   - Mutate these new offspring.
   - Replace the less fit members with the offspring.
   - Increment the generation counter and go to step 2.

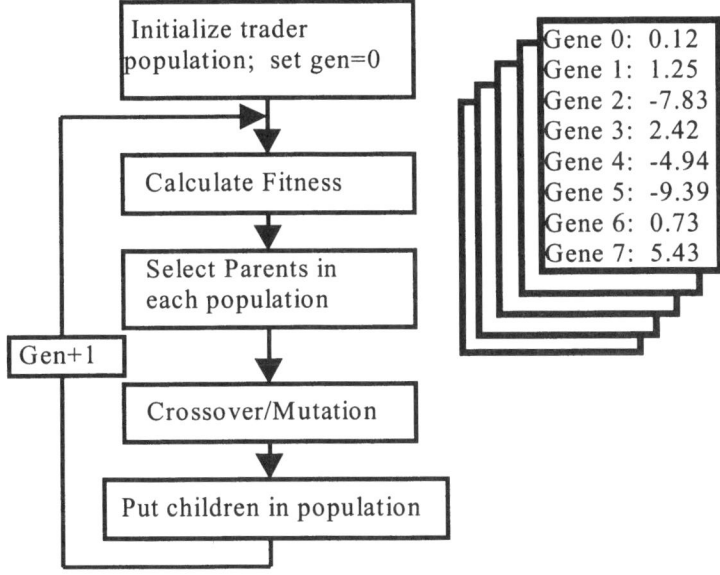

Figure 11.5. Genetic algorithm to evolve a population of trading agents

11.3.3 Valuing Options with Agents

General
A simple electricity market with four generators that provide power to an inelastic demand is modelled. Generators are dispatched to meet demand and a market price is determined from the aggregate marginal cost curves of the dispatched generators. Four put options with strike prices of $15, $20, $25 and $30 are offered with valuations determined using Black-Scholes and the market price data. GA-based agents then buy and sell the options at

these Black-Scholes prices. Implicit in the generation of buy and sell signals is a valuation of the put options by each of the agents.

Data Preparation
1. Demand data: Hourly demand data for an extended period was provided by a large Midwestern utility and was used as a source of realistic load data in this simulation. See Figure 11.6.
2. Electricity market price data: Before evolving strategies for buying and selling, price data was needed with which the put option prices were calculated using Black-Scholes.

The hourly demand data was used in conjunction with the generator parameters to determine the market price in an iterative procedure reminiscent of unit-commitment. Each of the suppliers has a unit that is modelled with a quadratic cost curve ($Cost = a + bP + cP^2$). See Table 11.1 for the values of the coefficients. The supplier produces power as long as the market price does not fall below the supplier's minimum marginal cost (which is determined by their minimum production level).

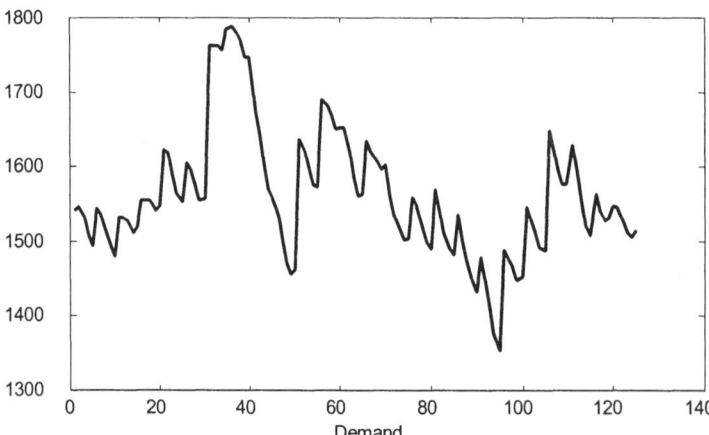

Figure 11.6 Demand on vertical axis (MW) versus time (hours)

Table 11.1 Generator parameters

Generator	a	b	c	P_{min}	P_{max}	λ_{min}	λ_{max}
1	100	6	0.005	100	600	7.0	12.0
2	150	7	0.004	120	700	8.0	12.6
3	200	9	0.006	150	750	10.8	18.0
4	250	8	0.007	200	800	10.8	19.2

The marginal cost is found by taking the derivative of the cost curve ($\lambda = b + 2cP$). The marginal cost curves for each generator are shown in Figure 11.7. Note that each generator has both a minimum and a maximum operating level. (Startup and shutdown costs, ramp rates, and minimum up and down time constraints were not considered in this simulation.) If the market price is below the minimum marginal price for a generator, that generator is removed from consideration and the market price recalculated. This process is repeated until demand is balanced by a set of generators for whom it is profitable to produce at the discovered price. If price discovery does not occur after 20 iterations, the market price from the 20th iteration is taken as the market price. (Under this simple scheduling scheme, it is possible that a unit could be forced to produce below its minimum marginal cost but a check showed that this never happened.)

A brief clarification at this point may be in order to prevent confusion in the use of the term 'spot'. The market price is referred to here as the spot price. This is in keeping with the terminology used in finance (i.e. options prices are determined by spot prices); this is not to imply that the hourly market here is the same as the spot electricity market (i.e. the 'spot' electricity market as the real-time electricity market). The spot price data for a typical week is shown in Figure 11.8.

3. Standard deviation of spot price: The standard deviation (sigma) of the spot price is used when calculating the Black-Scholes formula. For a given hour, sigma is calculated using a window of the last 25 peak-period hours prices. The standard deviation of the market price is shown in Figure 11.9.
4. Put options price data: There are four put options, which can be bought and sold, having strike prices of $15, $20, $25 and $30. The market valuation (price) of each of these is calculated using the Black-Scholes formula for put options, as presented earlier.

    Note that the risk-free rate is taken to be constant throughout the simulation and that T-t is a constant 90 days. This was done to prevent having to 'roll over' the options position because the expiration date was reached.

    Valuations for the put options are shown in Figure 11.10. One can see that they go up and down with swings in the underlying spot price of electricity and that the put options with higher strike prices have higher market valuations, as would be expected.

Evolving Trading Strategies with a GA

Each agent in the population buys and/or sells the four put options. These agents act according to internally generated buy and sell signals. These signals are generated using a GA to vary the coefficients in a modified Black-Scholes calculation. Options could be traded only for peak periods on weekdays, i.e. Monday-Friday, 11.00a.m.-4.00p.m.

1. GA valuation of options and buy/sell signals: The GA is coded as a string of real number genes. The number of genes is determined by the calculation being performed by the GA (described next). For these simulations each GA has eight genes, each of which is a real number.

    The equation currently used by the GA to generate a buy or sell signal is a modified Black-Scholes valuation. A signal to buy or sell an option will be generated if the GA valuation minus the market valuation is greater than some threshold. The terms $d1$ and $d2$ in the Black-Scholes formula are recalculated using a modified sigma called σ', where $\sigma' = (Gene2) \cdot \sigma$ and where σ is the 'standard' calculation of the standard deviation of the spot price. A buy signal is generated if $[Gene\,0 \cdot X \cdot \exp(-r \cdot (T-t)) \cdot N(-d1) - (Gene\,1) \cdot S \cdot N(-d2)] + (Gene\,2)$ is greater than the Market Price. Similarly, if a new $d1$ and $d2$ are calculated with $\sigma' = (Gene6) \cdot \sigma$ and the Market Price is greater than $[(Gene4) \cdot X \cdot \exp(-r \cdot (T-t)) \cdot N(-d1) - (Gene5) \cdot S \cdot N(-d2)] + (Gene7)$ then a sell signal is generated.

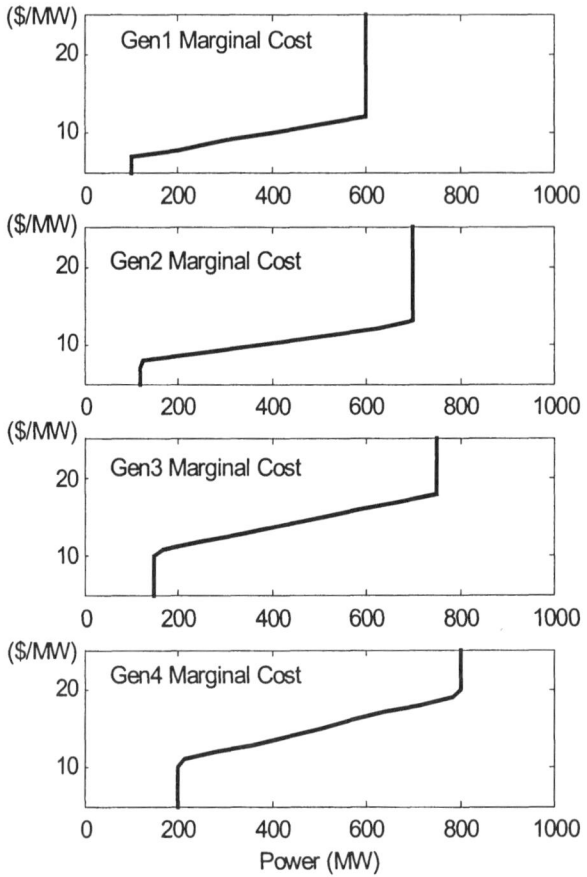

Figure 11.7 Marginal costs on vertical axes ($/MW) vs. MW for each generator

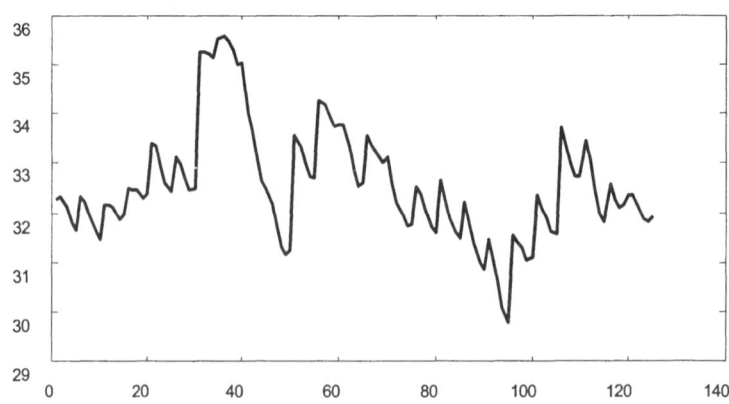

Figure 11.8 Spot price on vertical axis ($/MWh) versus time (hours)

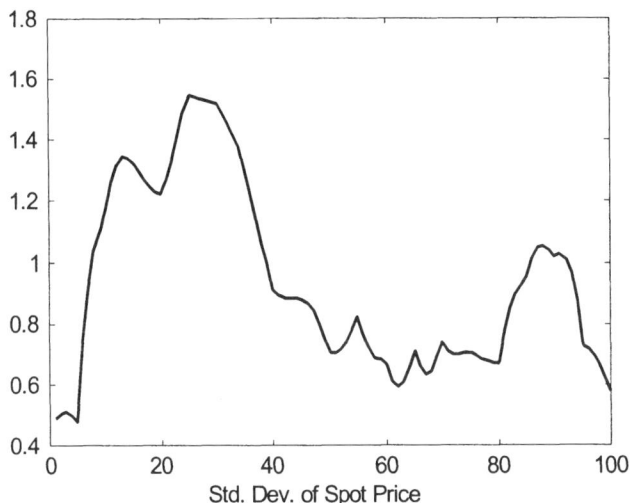

Figure 11.9 Standard deviation of spot price on vertical axis versus time

2. Fitness measure: Fitness is determined by the amount of profit made from playing options market many times. The agents are speculators who attempt to buy when options are undervalued by the market and sell when they are overvalued.

A generation consists of 100 hours of buying and selling. The agents have 100 chances or time steps to buy and sell options. A buy signal causes the agent to buy one option of the specified strike price, increasing its holdings of that option type by one and decrementing its bank account by the current market price of that option. A sell signal causes the agent to sell all its holdings of the type of option that generated the sell signal, causing its bank account to increase by (price * number of options held); its holdings for that type of option are then set to zero (since all options were sold). To repeat, only one option is bought in response to a buy signal, while all options are sold in response to a sell signal. This may seem like an oversimplification since real world agents decide how many options to buy or sell. However, determining the quantity of the transaction adds another dimension of complexity to the problem with which the GAs are presented. Note that agents earn (pay) interest at the risk-free rate when they have a positive (negative) bank account. Fitness is calculated as money earned (or lost) per hour. It is the agent's bank account at the end of the generation divided by the generation length.

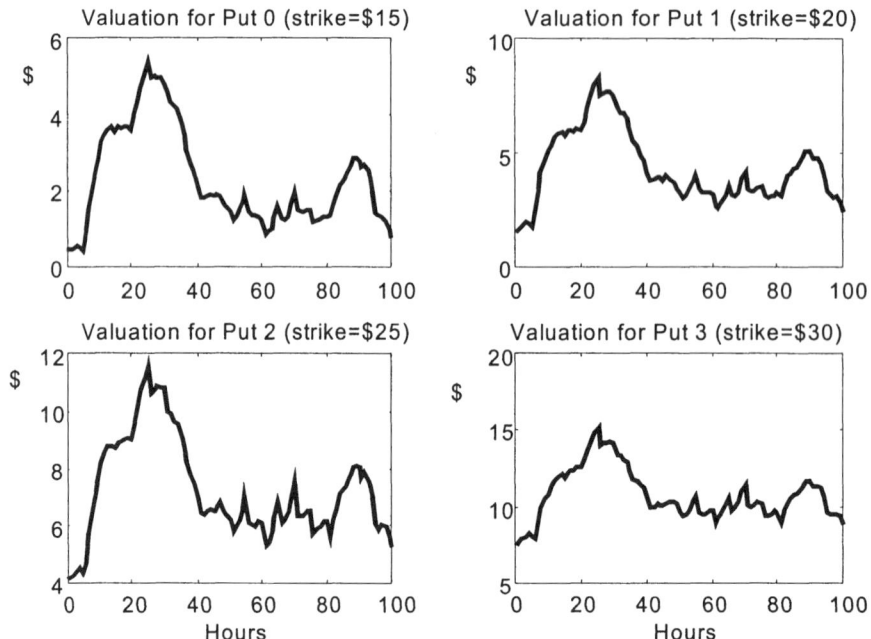

Figure 11.10 Market valuation for the put options ($ vs. hours)

3. Reproduction: After fitness is calculated, the agents are sorted according to their fitness. Reproduction is performed using single-point crossover of two parents selected from the best half of the population using rank selection. One child is created and replaces an agent in the worst half of the population.

Each child's genes can be mutated in four different ways (bearing in mind that the genes are real-valued):

1. 2% of the time the gene is replaced randomly.
2. 5% of the time the gene is multiplied or divided by 1.5.
3. 10% of the time the gene is multiplied or divided by 1.05.
4. 1% of the time the sign of the gene is changed.

Random genes are generated according to the relation:
$$NewGene = GeneMin + Random[0..1] \cdot (GeneMax - GeneMin)$$
where *GeneMin* and *GeneMax* are the max and min values of that gene over the entire population. (This was tried as a reasonable way of generating new genes without discarding what the population has collectively learned about the 'reasonable' range for a coefficient. This was devised because the space for real numbers is infinite.)

This process is repeated until every agent in the worst half of the population has been replaced. (A variation on this theme is to replace the worst quarter of the population with randomly generated agents, in an effort to introduce new genes into the gene pool and prevent stagnation.)

4. Results: The GA was able to evolve a strategy that consistently made a profit buying and selling put options in this market. As shown in Figure 11.11, the fitness of the best

agent is positive and improves over the generations, ultimately reaching a value of $1.5 per trade (with one trade allowed each hour).

Figure 11.11 also shows the fitness of the worst agent and the average fitness for the whole population. One can see that at the start of the run most agents actually lose money (make a negative profit) but by the end of the simulation the average fitness has risen to nearly zero. Figure 11.12 shows the best genes from 4 different runs.

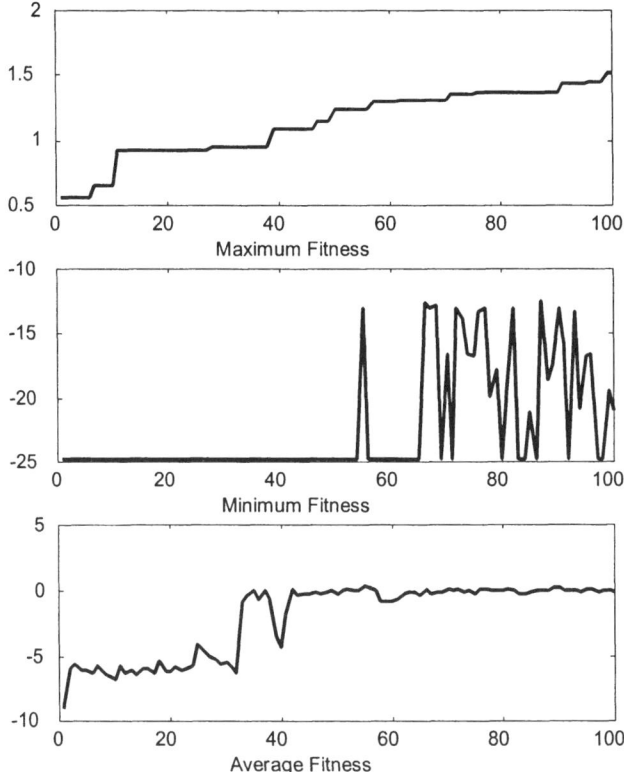

Figure 11.11 Maximum, minimum and average fitness over a typical run. The vertical axis measures profit per generation; the horizontal axis counts generations.

11.4 Evolutionary Programming-based Optimal Power Flow Algorithm

Solving the optimal power flow (OPF) problem is fundamental to the unbundling of transmission costs associated with transmission open access and is of increasing importance in power system operation under the deregulated environment of the electricity industry. It is a highly constrained and large-dimension non-linear optimisation problem, which is difficult to solve. The computational difficulties in solving the OPF problem have limited its use in power system operations.

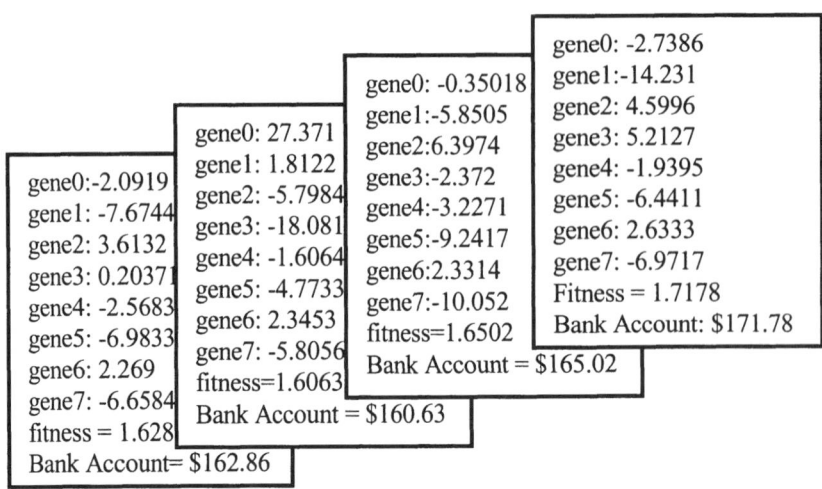

Figure 11.12 The best genes after 100 generations from 4 different runs

Many solution techniques have been applied to the OPF problem such as linear programming [36], non-linear programming [37] and interior point methods [38]. These methods rely on convexity to obtain the global optimum solution and as such are forced to simplify relationships in order to ensure convexity. However, the OPF problem is in general non-convex and, as a result, many local minima may exist. This non-convexity is further increased when valve-point loading effects of thermal generators have to be included [39,40] or FACTS devices are included on the network.

With a non-monotonic solution surface, classical optimisation methods are highly sensitive to starting points and frequently converge to local optimum solutions or diverge altogether. These methods are usually confined to specific cases of the OPF and do not offer great freedom in objective functions or the types of constraints that may be used. It is therefore important to develop new, more general and reliable algorithms, which are capable of incorporating new constraints arising from open access, non-convex solution surfaces and FACTS devices.

One such technique is that of evolutionary programming (EP) [41]. The EP technique is a stochastic optimisation method in the area of evolutionary computation, which uses the mechanics of evolution to produce optimal solutions to a given problem. It works by evolving a population of candidate solutions towards the global minimum through the use of a mutation operator and selection scheme. The EP technique is particularly well suited to non-monotonic solution surfaces where many local minima may exist.

This section reports an EP-based OPF (EP-OPF) solution algorithm that makes use of an EP load flow. Solution acceleration concepts in [42] are implemented which improve the basic EP algorithm. This acceleration is implemented using the gradient information obtained using the steepest descent method [43] to perform a local search. The method is capable of determining the global optimum solution to the OPF for a range of constraints and objective functions. The algorithm is not sensitive to starting points and is capable of handling non-convex generator cost curves. The performances of the algorithm when applied to the IEEE 30-bus test system under different generator input-output curves are presented.

11.4.1 OPF

The OPF problem seeks to optimise steady-state power system performance with respect to an objective f while subject to numerous constraints. For optimal active and reactive power dispatch, the objective function, f, is that of total generation cost. Other objectives may include minimisation of transmission losses and voltage level optimisation. Mathematically this may be stated as

$$\min f(\mathbf{x}, \mathbf{u}) \qquad (11.1)$$

subject to

$$g(\mathbf{x}, \mathbf{u}) = 0$$
$$h(\mathbf{x}, \mathbf{u}) \leq 0$$

where $\mathbf{u}$ is the vector of control variables (these include generator active power/voltage levels and transformer tap settings); $\mathbf{x}$ is the vector of dependent variables (load (PQ) node voltages, generator reactive powers); $f(\mathbf{x}, \mathbf{u})$ is the objective to be optimised; $g(\mathbf{x}, \mathbf{u})$ are the nodal power constraints; and $h(\mathbf{x}, \mathbf{u})$ are the inequality constraints on dependent and independent variables.

11.4.2 EP

EP seeks the optimal solution by *evolving* a *population* of candidate solutions over a number of *generations* or iterations. During each iteration, a second new population is formed from an existing population through the use of a *mutation* operator. This operator produces a new solution by perturbing each component of an existing solution by a random amount. The degree of optimality of each of the candidate solutions or *individuals* is measured by their *fitness*, which can be defined as a function of the objective function of the problem.

Through the use of a competition scheme, the individuals in each population compete with each other. The winning individuals form a resultant population, which is regarded as the *next generation*. For optimisation to occur, the competition scheme must be such that the more optimal solutions have a greater chance of survival than the poorer solutions. Through this the population evolves towards the global optimal point.

The EP technique is iterative and the process is terminated by a stopping rule. The rule widely used is either (a) stop after a specified number of iterations or (b) stop when there is no appreciable change in the best solution for a certain number of generations. Rule (a) is adopted in the present work. The main stages of the EP technique including initialisation, mutation and competition are shown in the flowchart of Figure. 11.13.

11.4.3 EP-OPF

Based on the EP methodology, an algorithm for solving the OPF problem can be established. The basic flowchart of the algorithm is shown in Figure 11.13 with its components described below and in Sections 11.4.2 and 11.4.3.

Representation of solution: An individual in a population represents a candidate OPF solution. The elements of that solution consist of the controllable and uncontrollable variables. Specifically the controllable variables are specified power generation at all generator (*PV*) nodes other than the slack node, the specified voltage magnitude at all *PV* nodes and tap positions for variable tap transformers. Each candidate solution also stores dependent variables such as the most recent load flow solution for subsequent use in initialising the load flow on the next iteration to reduce computation time within the loadflow algorithm.

Initialisation: Each of the controllable variables of an individual is initialised randomly using a uniform random number distribution within its feasible range. For example, for the specified active power generation for a *PV* node *i*, with active power limits of P_{min} and P_{max}, we have

$$P_i = U[P_{min}, P_{max}] \quad (11.2)$$

where $U[P_{min}, P_{max}]$ is a uniform random number between P_{min} and P_{max}. In addition to this, one candidate solution will have its specified active power generation for all *PV* nodes excluding the slack node set to the economic dispatch solution for the system with an active power load as the aggregate active power load of all nodes plus 2% to approximate transmission losses. This economic dispatch solution is obtained using the EP-based method of [44].

Fitness of candidate solutions: Each candidate solution is assigned a fitness to measure its optimality with respect to the objective being optimised. In the case of active and reactive power dispatch the fitness of individual *i* will be,

$$f_i = \frac{M}{C_i + \sum_j VP_j + SQ} \quad (11.3)$$

$$VP_j = \begin{cases} K_v(V_j - 1.0)^2 & \text{if } V_j > V_j^{max} \text{ or } V_j < V_j^{min} \\ 0 & \text{otherwise} \end{cases}$$

$$SQ = \begin{cases} K_q(Q_{slack} - Q_{slack}^{max})^2 & \text{if } Q_{slack} > Q_{slack}^{max} \\ K_q(Q_{slack} - Q_{slack}^{min})^2 & \text{if } Q_{slack} < Q_{slack}^{min} \\ 0 & \text{otherwise} \end{cases}$$

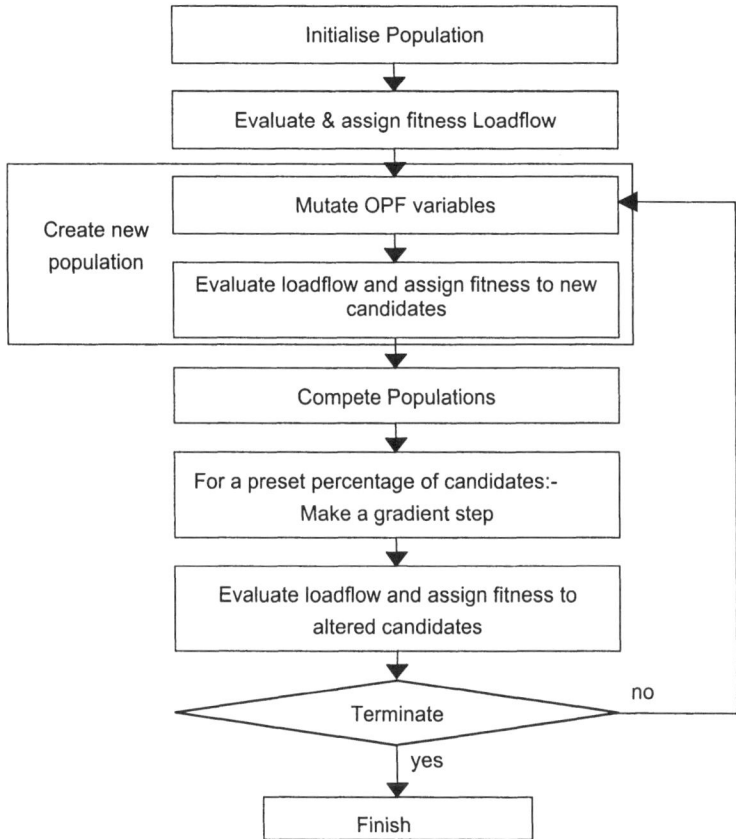

Figure 11.13 Flowchart of EP-OPF

In the above equation, M is the maximum possible cost of generation and C_i is the generation cost of individual i. The term VP_j denotes a penalty term on PQ or switched PV node j for violating preset voltage limits V_j^{min}, V_j^{max}. SQ represents a penalty on the slack node for violating a reactive power limit. K_v and K_q are penalty weighting constants. It is not necessary to impose a penalty on slack node active power violations as the mutation stage helps to satisfy this constraint. The EP-OPF algorithm seeks the solution with the maximum fitness.

Producing new solutions by mutation: A new population of OPF solutions is produced from the existing population through the mutation operator. A new individual p_i' is produced from each individual p_i, where the jth OPF variable in the new individual p_i' is calculated as

$$x_{ji}' = x_{ji} + N(0, \sigma_{ji}^2) \tag{11.4}$$

where x'_{ji} denotes the value of variable j in p'_i. x_{ji} is the value of variable j in the parent p'_i and $N(0, \sigma_{ji}^2)$ is a Gaussian random number with a mean of zero and a standard deviation of σ_{ji}. The expression designed for σ_{ji} is

$$\sigma_{ji} = (x_j^{max} - x_j^{min})((f_{max} - f_i)/f_{max} + a^r) \tag{11.5}$$

where f_i is the fitness of individual i; f_{max} is the maximum fitness within the population; x_j^{max}, x_j^{min} denote the upper and lower limits of variable j; a is a positive number constant slightly less than unity; and r is the iteration counter. The term a^r provides a decaying mutation offset the rate of which depends on the value of a [44]. It can be observed from (11.5) that a solution that has a much lower fitness than that of f_{max} will have a higher value for σ_{ji}; hence it will be moved further by mutation to a hopefully more optimal location.

Constrained mutation: To help in the satisfaction of the slack node active power constraints, all units other than the slack are assigned a loading according to (11.4). The total of their dispatches is then compared with the total generation found from the previous load flow of that individual. If the difference between them is within the operating limits of the slack unit, then the candidate is accepted. If not, the process is repeated for a total of five attempts. If within these attempts a feasible assignment is not found, the mutation is constrained to force satisfaction by sharing the excessive generation of the slack node among the remaining generators as follows.

Assuming the slack node active power in an individual has exceeded its upper bound and that the slack unit is unit 1, the total available capacity of units 2 to N in that individual is given by

$$C_2 = \sum_{i=2}^{N}(P_i^{max} - P_i) \tag{11.6}$$

and the excessive generation of the slack node is

$$E_2 = D_L - \left(\sum_{i=2}^{N} P_i + P_1^{max}\right) \tag{11.7}$$

where D_L is the sum of the active power demand and the transmission loss the value of which is set to that found in the immediately previous load flow solution of the individual. The loading of unit 2 is then modified according to

$$P'_i = P_2 + E_2 (P_2^{max} - P_2)/C_2 \tag{11.8}$$

If the modified loading exceeds the maximum loading of unit 2, it is set to the limiting value. The amount of excessive generation of the slack node left to be shared is

$$E_3 = E_2 - (P_2' - P_2)$$
(11.9)

The above procedure is repeated to modify the loadings of the units 3 to N. After all the loadings of the units are modified, the slack node active power will be on its upper limit. The same process is used when the lower active power limit is violated.

Selection of individuals by competition: In the competition stage, a selection mechanism is used to produce a new population from the two existing populations. For optimisation to occur, the fitter or more optimal solutions should have a greater chance of selection. The selection technique used is a *tournament* scheme described in the following.

The k parent solutions p_i, along with their corresponding offspring p_i' formed by mutation, $i = 1,\ldots,k$, each undergo a series of N_t tournaments with randomly selected opponents. Each individual i is assigned a score s_i according to

$$s_i = \sum_{j=1}^{N_t} n_j$$

$$n_j = \begin{cases} 1 & \text{if } f_i > f_r \\ 0 & \text{otherwise} \end{cases}$$
(11.10)

where f_i is the fitness of individual i. The opponent r, is chosen at random from the $2k$ individuals based on $r = \lfloor 2ku+1 \rfloor$. $\lfloor x \rfloor$ denotes the greatest integer less than or equal to x. u is a uniform random number in the interval $[0,1]$. The k highest scoring candidate solutions are taken as the individuals in the next generation.

11.4.4 Load flow Solution

During the solution of the OPF, it is necessary to solve the load flow problem numerous times. From Figure 11.13, the EP-OPF requires at most two load flows to be performed per individual, per iteration. The load flow is solved using an EP-based load flow developed using a similar methodology to that in [45] but incorporating a Jacobian acceleration stage. In order to keep computation time to a minimum during the optimisation the population size for the loadflow is set dynamically in the following way.

For the first attempt at solving a load flow the population size is set at 2. These two individuals are initialised with one being a flat start and if available the second will be the solution to the previous load flow solution for the OPF candidate in question. If a previous solution is unavailable, a random starting point is generated. If the load flow fails to converge within 10 iterations, the population size is increased to five and these candidates are initialised randomly. Following a failure of five consecutive load flows the OPF candidate is removed from the population and replaced with a randomly generated individual.

Many OPF solution techniques handle generation node reactive power limits through the use of penalty functions exclusively. This can lead to convergence problems due to distortion of the solution surface, or in some instances, it may be necessary to switch a node in order to obtain a load flow solution. The approach used in the EP-OPF is to handle the reactive power limits on all *PV* nodes other than the slack node by the conventional method of switching which is applied within the load flow stage. When a *PV* node has been switched to a *PQ* node, it is no longer possible to control the voltage at that bus and as a result the algorithm does not adjust the voltage of a switched *PV* node.

11.4.5 Gradient Acceleration

Owing to the large dimensionality of the OPF problem, evolutionary computation techniques such as EP can take an unacceptable number of iterations to converge. In order to improve the speed of convergence of the EP-OPF algorithm, acceleration techniques [42], which provide an intermediate re-mapping of candidates to a more optimal position, are implemented. To achieve this acceleration, a proportion of the population is moved in the direction of the negative gradient. This is achieved by using the steepest descent (SD) algorithm of [43]. As the gradient step forms only a part of the overall EP-OPF algorithm the choice of step size is not as critical as it is in a pure SD algorithm. It may simply be set to a constant small step size to ensure convergence.

The sensitivity of the solution to changes in different variables varies in the OPF. The solution is less sensitive to changes in active power than to changes in *PV* bus voltage magnitudes and transformer tap settings. As a result of this, different step sizes are used for each variable. Active power has a large step size while transformer tap and voltage magnitudes have much smaller step sizes. These variable-dependent step sizes provide a faster convergence in the SD.

With this methodology SD is providing a focused local optimisation, while EP is controlling the global optimisation. Reactive power penalty terms are not included in the SD formulation except for the slack node, which cannot be switched in the load flow process. The effect of generator node switching in the load flow routine creates disturbances in the solution process. These disturbances can cause purely gradient-based methods such as SD to diverge or converge to local optima. However, with EP as a global optimisation scheme these problems are avoided.

It is also necessary to reflect any penalties in the fitness function (11.3) in SD when it is used with EP. If these are not included within SD, EP will often discard solutions produced by the gradient step, as they will usually incur greater penalties in (11.3).

Non-convex Curves
The SD algorithm performs well on convex problems where there is only the global optimum; however, if the solution surface is multi-modal the SD algorithm will usually become trapped in a local optimum. This is the case when the input/output curves of generators are modelled by non-convex curves such piecewise quadratics or quadratics with sine components [39,46]. These curves present a problem to the SD method due to their non-convexity and discontinuities in the gradient.

As the step is based on this gradient, it is possible for the solution to cross the discontinuity where the gradient information is no longer valid. To prevent the SD from moving the candidate solutions beyond the local region, the SD formulation was modified such that if an active power loading of a unit crossed a discontinuity boundary, it would be set to the value at that boundary. These boundaries, while impassable to the SD routine, may be crossed by the global EP framework through mutation, so the solution space will still be traversed fully.

11.4.6 Application Studies

The EP-OPF algorithm was applied to the IEEE 30-bus test system. Three sets of generator cost curves were used to illustrate the robustness of the technique. The first case considered is where all curves are quadratic [47]; in cases (b) and (c) some of the cost curves are replaced with either piecewise quadratics or quadratics with sine components. Therefore in cases (b) and (c), there are many local optimal solutions for the dispatch problem and as a result the SD algorithm cannot determine the global optimum solution. The problem is therefore well suitable for validating the developed algorithm.

The EP-OPF algorithm was implemented using the C language and the software program was executed on a 200 MHz Pentium Pro computer. The specific settings for the algorithm and system data are summarised in the Appendix . In all cases, the standard IEEE 30-bus loading is used.

Case (a): Quadratic Cost Curves
In this case the unit cost curves are represented by quadratic functions from [47] and are summarised in Table 11.2. The program was run 100 times with the settings in the Appendix. The average cost of solution obtained was $803.51 with the minimum being $802.62 and maximum $805.61. The average execution time was 51.4 seconds. The solution details for the minimum cost are provided in Table 11.2.

For this case, a solution of $802.40 was reported in [47]. This was obtained using penalty functions for generator reactive power limits. The EP-OPF returned a solution with no PV nodes being switched. However, the solution from [47] violates the slack node lower Q-limit slightly by approximately 1.7 MVAr.

Table 11.2. Generator data and cost coefficients for base case (a)

Bus No.	P_G^{min} MW	P_G^{max} MW	Q_G^{min} MVAr	S_G^{max} MVA	Cost Coefficients		
					a	b	c
1	50	200	-20	250	0.00	2.00	0.00375
2	20	80	-20	100	0.00	1.75	0.01750
5	15	50	-15	80	0.00	1.00	0.06250
8	10	35	-15	60	0.00	3.25	0.00834
11	10	30	-10	50	0.00	3.00	0.02500
13	12	40	-15	60	0.00	3.00	0.02500

Generation input/output function $C_i = a_i + b_i P_i + c_i P_i^2$

Case (b): Piecewise Quadratic
In this study, units 1 and 2 cost curves were replaced by piecewise quadratic curves summarised in Table 11.3 to model different fuels or valve-point loading effects. In order to allow more precise control over units with discontinuities in cost curves, the unit with the simplest type of cost curve and largest capacity was selected to be the slack node. In this case, node 5 is treated as the slack bus. The average cost of solution obtained was $649.67 with the minimum being $647.79 and maximum $652.67. The average execution time was 51.6 seconds. The solution details for the minimum cost are provided in Table 11.5. When the pure SD method was applied to this problem, it failed to converge to the minimum above and the solution, which it provided, was approximately $850. This

demonstrates that the pure SD has difficulties with non-convex solution surfaces. It is possible to guide the SD method to the global optimum if the modifications described above in Section 11.4.5 are applied. However, the global optimum will only be found if the starting point is within the correct operating intervals for units 1 and 2. The developed EP approach is able effectively to search the entire solution space unlike the SD method and hence provides the minimum solution. The voltage profile at the solution is shown in Figure 11.14 for comparison in case (d).

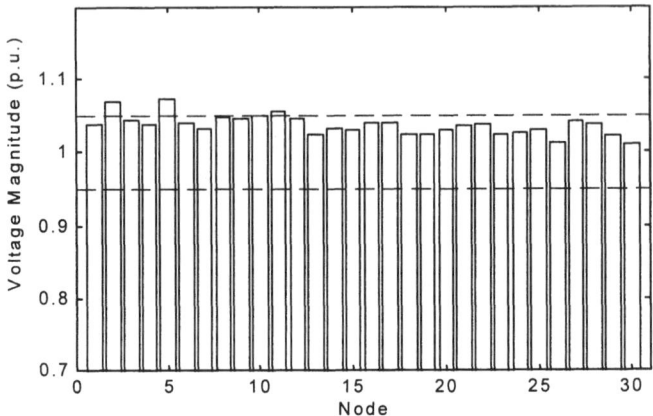

Figure 11.14 Voltage Profile Solution in Case (b)

Table 11.3 Generator cost coefficients in case (b)

Bus No.	From MW	To MW	Cost Coefficients		
			a	b	c
1	50	140	55.0	0.70	0.0050
	140	200	82.5	1.05	0.0075
2	20	55	40.0	0.30	0.0100
	55	80	80.0	0.60	0.0200

Generation input/output function $C_i = a_i + b_i P_i + c_i P_i^2$

Case (c): Sine Components

In this case the unit cost curves of the generators connected to buses 1 and 2 were quadratics with a sine component superimposed upon them. The sine component was used to represent the valve-point loading effects [39, 46]. The data for these curves is provided in Table 11.4. As in case (b) above, node 5 was taken to be the slack bus for the studies. The program was run 100 times with the average solution cost being $921.45. The minimum and maximum costs found were $919.89 and $926.68 respectively. The solution details for the minimum cost are provided in Table 11.5. To illustrate the convergence of the algorithm, the average statistics of the population over the 100 trials are plotted in Figure. 11.15. From this figure, it can be seen that the EP-OPF converges rapidly towards the solution.

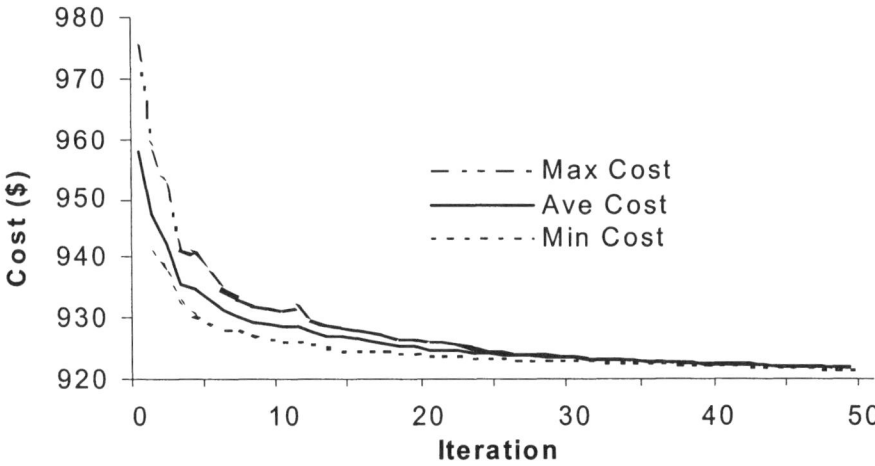

Figure 11.15 Convergence of the EP-OPF algorithm in case (c)

Again when the SD algorithm is applied to this case its ability to converge to the global optimum solution is strongly dependent on the starting point provided. With the inclusion of non-convex cost curves, the ability of the SD method to find the global optimum solution is greatly reduced. The parallel search mechanisms of the EP-OPF using the SD method for a directed local search, however, perform well in these cases.

Table 11.4 Generator cost coefficients in case (c)

Bus No.	P_G^{min} MW	P_G^{max} MW	Cost coefficients				
			a	b	c	d	e
1	50	200	150.00	2.00	0.0016	50.00	0.0630
2	20	80	25.00	2.50	0.0100	40.00	0.0980

Generation input/output cost function $C_i = a_i + b_i P_i + c_i P_i^2 + \left| d \sin(e(P_i^{min} - P_i)) \right|$

Case (d): Voltage Profile Optimisation
With the objective of the OPF problem set to an almost purely cost-based objective, the solution which is found, while being feasible, may have a less desirable voltage profile as shown in Figure 11.14. With a change to the objective function of the EP-OPF and a corresponding change in the SD penalty functions, it is possible to produce solutions which, while slightly more expensive, provide a flatter voltage profile.

Ideally all load nodes will have a voltage magnitude of 1 per unit. To achieve this the fitness function (11.3) was modified to

$$f_i = \frac{M}{C_i + \sum_j VP_j + \sum_k VF_k + SQ} \qquad (11.11)$$

$$VF_k = \begin{cases} K_f(V_k - 1.0)^2 & \text{if } V_k \neq 1.0, \; k \text{ a PQ node} \\ 0 & \text{otherwise} \end{cases}$$

The term VF_k denotes a penalty term on a load node k and K_f is a constant penalty weighting. The penalties within the SD formulation for voltage violation were also changed to the form of VF_k above. With this penalty the EP-OPF will attempt to minimise the cost of generation while trying to maintain the load flow voltages close to 1.0 p.u.

To demonstrate the effect of this change, case (b) was optimised with these settings. The voltage profile achieved is shown in Figure 11.16. This profile provides a much more satisfactory voltage level to load nodes than in case (b). The cost of this solution (flattest profile) is $651.54, which is close to that in (b). The solution details for the flattest profile are provided in Table 11.5. Of the 100 trials run, the average cost was $655.38 while all solutions returned a better profile than that found in (b). Once again the SD algorithm had difficulty in providing adequate solutions.

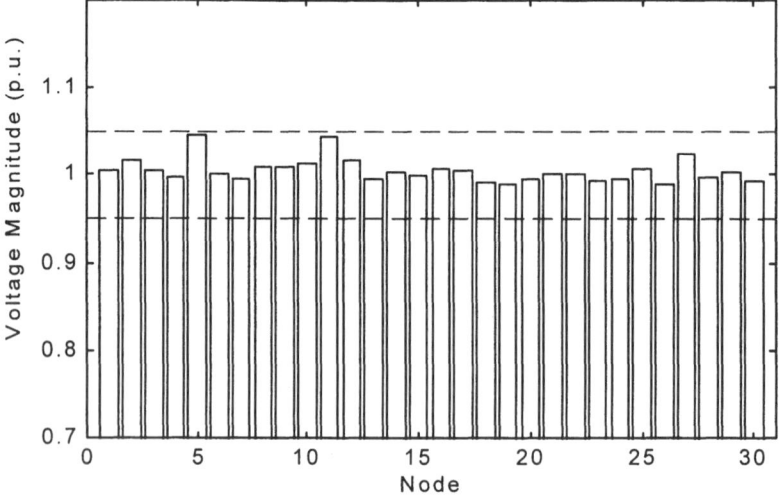

Figure 11.16 Voltage profile solution in case (d)

Table 11.5 Minimum solution found by EP-OPF in case (c)

	Case (a)	Case (b)	Case (c)	Case (d)
P_1	173.848	140.000	199.600	140.000
P_2	49.998	55.000	20.000	55.000
P_5	21.386	24.165	22.204	24.458
P_8	22.630	35.000	24.122	33.849
P_{11}	12.928	18.773	14.420	14.518
P_{13}	12.000	17.531	13.001	23.322
V_1	1.050	1.019	1.050	1.045
V_2	1.036	1.068	1.061	0.952
V_5	1.005	1.038	1.043	1.004
V_8	1.016	1.055	1.036	1.027
V_{11}	1.069	1.055	1.100	1.044
V_{13}	1.055	0.980	1.038	0.990
t_{11}	1.020	1.010	1.030	1.030
t_{12}	0.900	0.930	1.050	0.940
t_{15}	0.950	0.930	1.010	0.910
t_{36}	0.940	0.970	0.980	0.940

11.5 Complex Artificial Neural Networks for Load Flow Analysis

Artificial neural networks (ANNs) have been widely used in the power industry in fault classification, protection, fault diagnosis, relaying schemes, load forecasting and power generation. At present, most ANNs are built upon the environment of real numbers. However, it is well known that in computations related to electric power systems, such as load flow analysis and fault level estimation, complex numbers are extensively involved. The reactive power drawn from a substation, the impedance, busbar voltages and currents are all expressed in complex numbers. Therefore, ANNs in the complex domain must be adopted for these applications although it is possible to use ANNs in the conventional way by breaking up a complex number into two real numbers representing both the real and imaginary parts. In this section, it will be shown, by illustrating with a simple complex equation, that the behaviour of a 'real' ANN simulating complex numbers is inferior to that of an ANN which is intrinsically 'complex' by design. The structure of the 'complex' ANN and the numerical approach in handling back propagation for on-line training under the complex environment are described. The application of this newly developed ANN to load flow analysis in a simple six-bus electric power system is used as an illustrative example to show the merits of incorporating 'complex' ANNs in power system analysis.

ANNs have been proved to be capable of learning from raw data. They can be used to identify internal relations within raw data not explicitly given or even known by human experts and there is no need to assume any linear relationship between the data. ANNs represent the promising new generation of information processing networks [48]. Advances have been made in applying such systems to problems that have been found to be intractable or difficult for traditional computation. ANNs can supplement the enormous processing power of the von Neumann digital computer with the ability to make sensible decisions and to learn by ordinary experience. ANNs have widely been used in electric power engineering [49]. For energy management, load flow and OPF problems were solved by ANNs [50,51]. However, most existing ANNs for electric power applications have been designed using real numbers. In power engineering, applications such as load flow analysis, phasor evaluation, signal processing and image processing mainly involve complex numbers. Although conventional ANNs are able to deal with complex numbers by treating the real parts and the imaginary parts independently, it will be shown in this section that their behaviour is not so satisfactory. A new approach is introduced in this chapter where a computational ANN, particularly designed for manipulation of complex numbers in electric power systems, is described. It will be shown that this new 'complex' ANN has a superior performance on operations and computations of complex numbers as compared with the conventional 'real' counterparts. The 'complex' ANN is implemented to estimate busbar voltages in a load flow problem.

11.5.1 Conventional ANN for Real Numbers

Figure 11.17 shows a typical ANN for real numbers where there are n number of input nodes, m number of hidden nodes and l number of output nodes, totalling three layers. Of course, this network is freely extensible to any number of layers. All the x and the w in the network are real numbers and all the outputs o are real numbers within an interval [0, 1]. The pre-superscript of each w identifies the layer to which that w belongs. A set of desirable outputs, d_k, for k=1,...,l, corresponding to a set of inputs, x_j, j=1,...,n, is used as a training set. The standard sigmoid function is employed and the following equations hold:

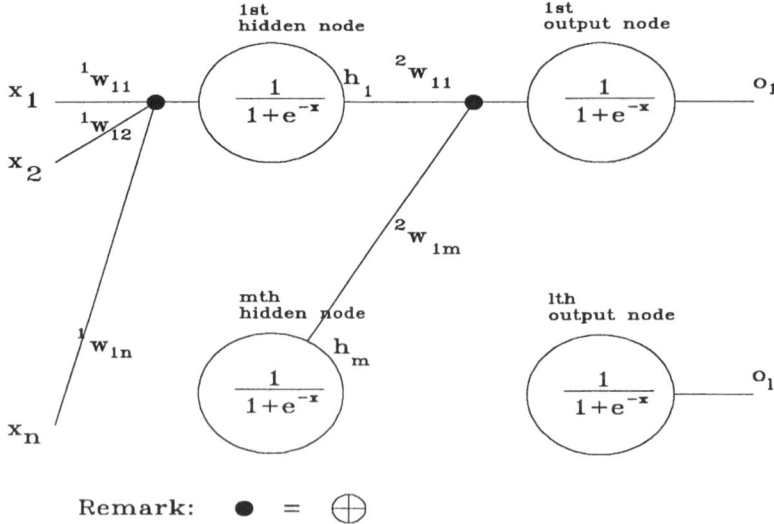

Figure 11.17 A typical ANN for real numbers

$$O_k = \frac{1}{1+e^{-\sum_{i=1}^{m} {}^2w_{ki} h_i}} \quad k=1,\ldots,l \tag{11.12}$$

$$h_i = \frac{1}{1+e^{-\sum_{j=1}^{n} {}^1w_{ij} x_j}} \quad i=1,\ldots,m$$

The following energy function, E, is being minimised

$$E = \frac{1}{2}\sum_{k=1}^{l} [O_k - d_k]^2 \tag{11.13}$$

to obtain an optimal set of values of w using the 'hill-climbing' algorithm so that at the pth iteration step, the following holds:

$$[w]^{p+1} = [w]^p - \lambda \nabla E$$

where λ = step size = 1.5

$$\Delta E = \left[\frac{\partial E}{\partial\, {}^2w_{ki}} \Big|_{i=1,..,m}^{k=1,..,l} \quad \frac{\partial E}{\partial\, {}^1w_{ij}} \Big|_{j=1,..,n}^{i=1,..,m} \right]^T \tag{11.14}$$

$$\frac{\partial E}{\partial\, {}^2w_{ki}} = (O_k - d_k) O_k (1 - O_k) h_i$$

$$\frac{\partial E}{\partial\, {}^1w_{ij}} = \sum_k (O_k - d_k) O_k (1 - O_k)^2 \, {}^2w_{ki}\, h_i (1 - h_i)\, x_j$$

11.5.2 New ANN for Complex Numbers

Figure 11.18 shows the basic elements of the newly designed ANN for complex numbers. For the operation of a basic function, say $z = w\, x$, where x is the input complex number, w is the weighting and z is the output complex number, the upper part of Figure 11.18 is referred to and the operation is shown by equation (11.15). For the addition of two complex numbers, x_1 and x_2, the operation is clearly shown in the lower part of Figure 11.18.

$$z_r + j z_i = (w_r + j w_i)(x_r + j x_i)$$
$$= (w_r x_r - w_i x_i) + j(w_i x_r + w_r x_i) \quad (11.15)$$

$$\text{where} \quad j = \sqrt{-1}$$

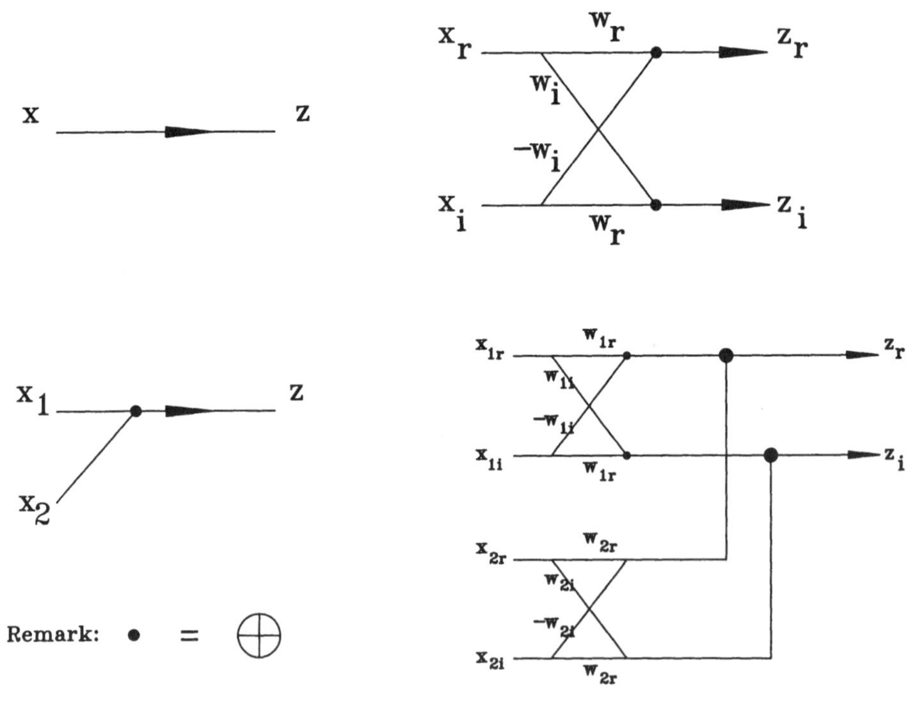

Figure 11.18 Basic elements of the newly designed 'complex' ANN

These basic elements form the foundation of our newly designed ANN for complex numbers. This new ANN was first proposed in [52]. The full configuration is shown in Figure 11.19 where it basically follows the format of Figure 11.17 but utilising new basic elements. Similarly, there are n number of input nodes, m number of hidden nodes and l number of output nodes. It should be noted that all nodes and weights are complex, a subscript r indicating the real part and a subscript i indicating the imaginary part. The sigmoid function is similar to that used in the real ANN but it is a complex operation, as shown by equation (11.16).

$$\frac{1}{1+e^{-(z_r+jz_i)}} = \frac{1}{1+e^{-z_r}\cos z_i - j e^{-z_r}\sin z_i}$$

$$= \frac{\left[1+e^{-z_r}\cos z_i\right]+j e^{-z_r}\sin z_i}{\left(1+e^{-z_r}\cos z_i\right)^2+\left(e^{-z_r}\sin z_i\right)^2} \qquad (11.16)$$

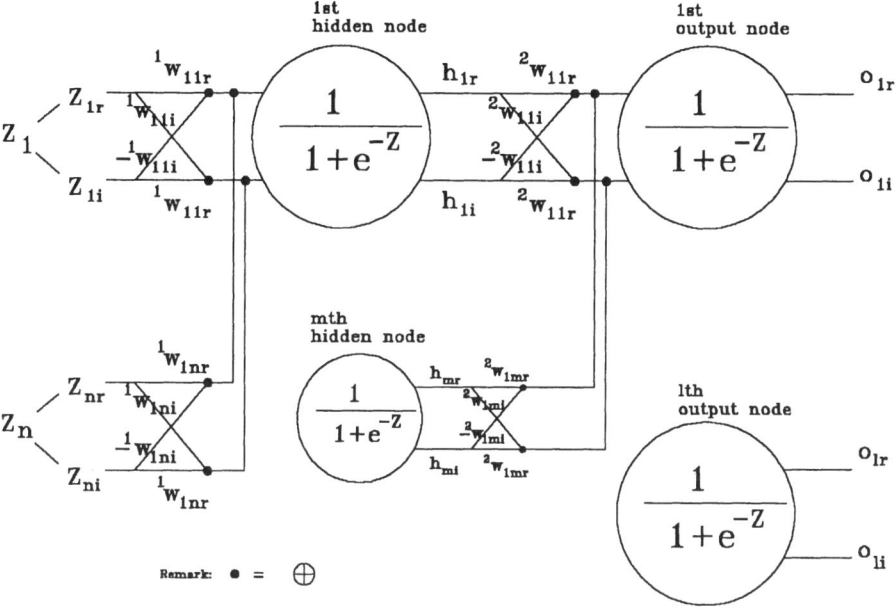

Figure 11.19 The new ANN for the complex number format

As this sigmoid function is highly non-linear and complicated, ∇E needs to be calculated numerically. The method is to perturb each w by a very small amount while all other values of w are kept constant. A new value E is then evaluated. The ratio of the difference of the new E from the old E, due to the perturbation, gives the corresponding element in ∇E. E itself now becomes a complex number and the gradient function refers to its magnitude, i.e. $\|E\|^2$, as defined by the following equation:

$$E = \frac{1}{2}\sum_{k=1}^{l}(O_{kr}-d_{kr})+j(O_{ki}-d_{ki})$$

$$\|E\|^2 = \frac{1}{2}\sum_{k=1}^{l}\left[(O_{kr}-d_{kr})^2+(O_{ki}-d_{ki})^2\right] \qquad (11.17)$$

11.5.3 Comparison of the two ANNs by Computer Simulation

In order to test the performance of the newly developed 'complex' ANN versus the conventional 'real' ANN in handling complex numbers, a simple function shown in equation (11.18) is used. A data set with nine training examples are available and shown in Table 11.6. During the training process, we continuously keep track of the total squared error of the output from the nine training sets.

$$O = x + \frac{1}{x}$$

(11.18)

$$\text{or} \quad O_r + j O_i = x_r + j x_i + \frac{1}{x_r + j x_i}$$

For the implementation on the conventional 'real' ANN, there are two input nodes, six hidden nodes and two output nodes. For the implementation on the new 'complex' ANN, there are one input complex node, three hidden complex nodes and one output complex node. All values of w for both the conventional and new ANNs are set to one initially before training, i.e. a fair initial guess, and the step size, λ, is arbitrarily set to 1.5. The history of squared error of the nine training examples of the two ANNs during training is shown in Figure 11.20. It can be seen that it takes 23,000 iterations for the 'complex' ANN to arrive at a total squared error of 10^{-3} or below while the 'real' ANN can only achieve a total squared error of 3.8×10^{-2} after 23,000 iterations. After the two ANNs have been trained up, a value of $x = 0.25 + j0.25$ is fed into the two networks while the correct answer expected should be $2.25 - j1.75$. The 'real' ANN gives an output of $1.85 - j1.4$, i.e. 19 % error, while the 'complex' ANN gives an output of $2.3 - j1.75$, i.e. 1.8 % error. From this illustrative example, it can be concluded that it is better to use a 'complex' ANN to handle systems involving complex numbers instead of using a 'real' ANN to handle the real parts and imaginary parts separately.

Table 11.6 Sample values of the complex test function

O	x
5.1 - j4.9	0.1 + j0.1
2.1 - j3.8	0.1 + j0.2
1.1 - j2.7	0.1 + j0.3
4.2 - j1.9	0.2 + j0.1
2.7 - j2.3	0.2 + j0.2
1.74 - j2	0.2 + j0.3
3.3 - j0.9	0.3 + j0.1
2.61 - j1.34	0.3 + j0.2
1.97 - j1.37	0.3 + j0.3

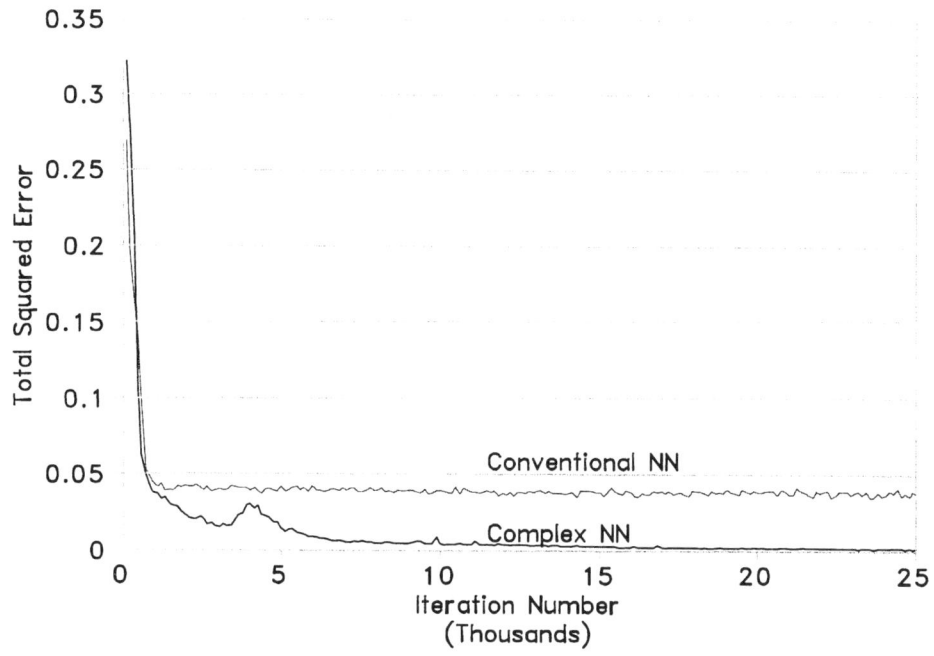

Figure 11.20 Error history of two ANNs under training

In order to make a fair comparison, the computer simulation has been carried out again by using three different network configurations. The same function and training examples as shown in equation (11.18) and Table 11.6 have been used. The first configuration consists of two separate real NNs, each consisting of one real input node, one real hidden node and one real output node, thus termed 'Two Separate NNs'. The second configuration consists of one real NN with two real input nodes, two real hidden nodes and two real output nodes, thus termed 'Conventional NN'. The third configuration consists of one complex NN with one complex input node, one complex hidden node and one complex output node, thus termed 'Complex N'. The objective of this simulation is for detailed comparison and, hence, the step size has been reduced by 10 times compared with the previous simulation. The results are shown in Figure 11.21. It can be seen that the behaviour of two separate NNs is very poor, as expected, because there is no cross-information between the two real NNs. For the conventional NN, it can still achieve a low error although it takes more iteration steps to arrive at the squared error of 0.05. Once again, it has been verified that the behaviour of the newly developed complex NN is superior. Regarding the speed of convergence, it can be seen from both Figures 11.20 and 11.21 that for the same level of accuracy, the complex NN can always reach the target earlier than the conventional real NN. However, the time duration of each iteration step of the complex NN is slightly longer than that of the conventional real NN. Therefore, from the speed of convergence point of view, both NNs have more or less the same performance. As mentioned in the introduction of this section, complex numbers are widely used in electric power systems and, thus, the 'complex' design should be adopted whenever ANNs are applied to electric power systems. One typical example of applying the 'complex' ANN to load flow analysis is shown in the following section.

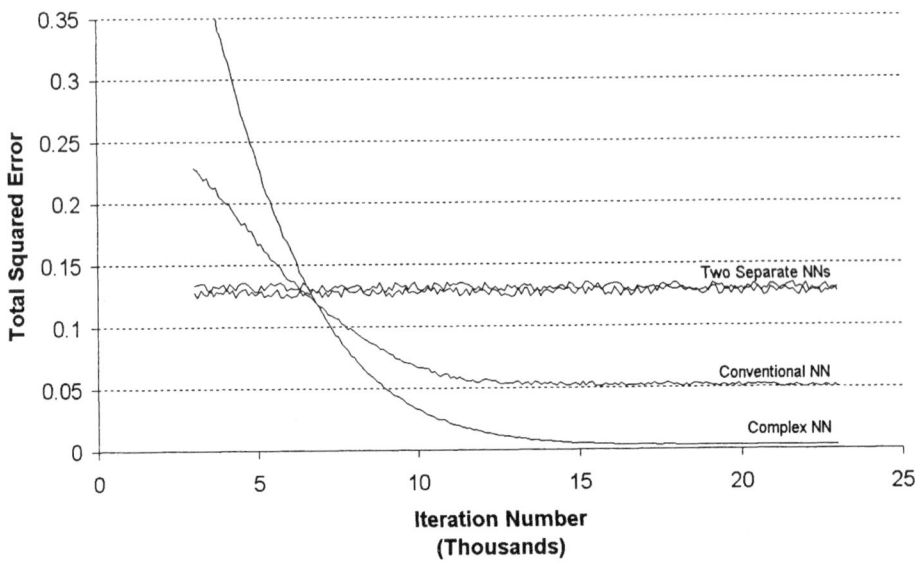

Figure 11.21 Error history of three ANNs for comparison

11.5.4 Application of "Complex" ANN to Load Flow Analysis

The Power Network and ANN Learning

It was shown that a multi-layer feedforward NN with one or more hidden layers is sufficient in order to approximate any continuous non-linear function arbitrarily well on a compact interval, provided sufficient hidden neurons are available [53]. The power load flow problem is by itself a non-linear problem and, hence, it can be analysed with the help of an ANN. A six-bus network, as shown in Figure 11.22, has been used to test the performance of the newly developed 'complex' ANN. Bus-1, bus-2 and bus-3 are generator buses while bus-1 is the swing bus.

Bus-4, bus-5 and bus-6 are ordinary load buses where the P (active power) and Q (reactive power) are to be specified. The training example is generated by ordinary load flow software using Newton-Raphson algorithms. This is just an illustrative example and therefore, during real application, the 'complex' ANN will continuously monitor the real-time state of the network in terms of voltage, P and Q. The details of the network parameters are shown in Tables 11.7(a) and 11.7(b) below:

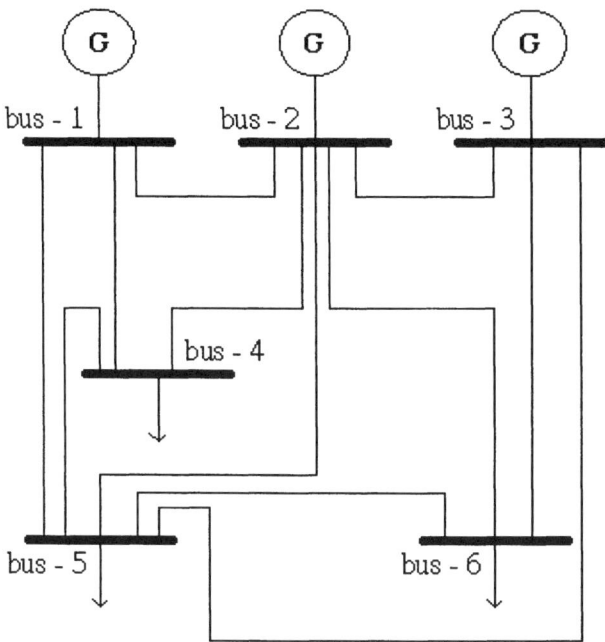

Figure 11.22 The six-bus network for load flow computation

Table 11.7(a) Busbar power for load flow study

Bus	P_{load}	Q_{load}	P_{gen}	V_{spec}
bus-1	0	0	---	1.05
bus-2	0	0	0.5	1.05
bus-3	0	0	0.6	1.07
bus-4	P_4	Q_4	---	---
bus-5	P_5	Q_5	---	---
bus-6	P_6	Q_6	---	---

Table 11.7(b) Network parameters for load flow study

From	To	R (p.u.)	X (p.u.)	B (p.u.)
bus-1	bus-2	0.1	0.2	0.02
bus-1	bus-4	0.05	0.2	0.02
bus-1	bus-5	0.08	0.3	0.03
bus-2	bus-3	0.05	0.25	0.03
bus-2	bus-4	0.05	0.1	0.01
bus-2	bus-5	0.1	0.3	0.02
bus-2	bus-6	0.07	0.2	0.025
bus-3	bus-5	0.12	0.26	0.025

bus-3	bus-6	0.02	0.1	0.01
bus-4	bus-5	0.2	0.4	0.04
bus-5	bus-6	0.1	0.3	0.03

14 training examples, shown in Table 11.8, have been generated by the software package for learning by the two ANNs. In this case, the voltage at bus-5 is to be estimated while the three generators maintain constant voltages at the corresponding busbars.

Table 11.8 Training examples for the neural networks

$P_4 + jQ_4$ (p.u.)	$P_5 + jQ_5$ (p.u.)	$P_6 + jQ_6$ (p.u.)	V_5 (p.u.)
0.7+j0.7	0.7+j0.7	0.7+j0.7	0.975-j0.089
0.9+j0.9	0.9+j0.9	0.9+j0.9	0.864-j0.137
0.9+j0.7	0.7+j0.7	0.7+j0.7	0.969-j0.101
0.7+j0.9	0.7+j0.7	0.7+j0.7	0.960-j0.088
0.7+j0.7	0.9+j0.7	0.7+j0.7	0.962-j0.115
0.7+j0.7	0.7+j0.9	0.7+j0.7	0.944-j0.084
0.7+j0.7	0.7+j0.7	0.9+j0.7	0.964-j0.108
0.7+j0.7	0.7+j0.7	0.7+j0.9	0.951-j0.086
0.9+j0.7	0.9+j0.9	0.9+j0.9	0.883-j0.137
0.7+j0.9	0.9+j0.9	0.9+j0.9	0.872-j0.125
0.9+j0.9	0.9+j0.7	0.9+j0.9	0.903-j0.142
0.9+j0.9	0.7+j0.9	0.9+j0.9	0.882-j0.108
0.9+j0.9	0.9+j0.9	0.9+j0.7	0.894-j0.140
0.9+j0.9	0.9+j0.9	0.7+j0.9	0.880-j0.116

Therefore, inputs to each ANN consist of P_i and Q_i, $i = 4, 5$ and 6, i.e. six inputs to the 'real' ANN and three inputs to the 'complex' ANN. The output of each ANN is V_5, i.e. two output nodes for the 'real' ANN and one output node for the 'complex' ANN. Here, the subscript refers to the number of the load bus. All other parameters associated with the network remain unchanged during the trial test.

The Results
The power load flow network is learned by the 'real' and 'complex' ANN under 14 combinations of $P_4, P_5, P_6, Q_4, Q_5, Q_6$ and V_5. The 'real' ANN is a 6-2-2 network while the 'complex' ANN is a 3-2-1 network. The initial values of all weights are randomly selected. Since the output of an ordinary sigmoid function for 'real' ANN has a range [0, 1] and it is not suitable for this application, the sigmoid function was slightly modified to the following form:

$$\frac{2}{1+e^{-x}} - 1 = \frac{1-e^{-x}}{1+e^{-x}} = \tanh\left(\frac{x}{2}\right) \qquad (11.19)$$

Information Technology Application

The limit of iterations for both ANNs is set to 230000 as in the case of Section 11.5.3. Figure 11.23 shows the variation of the total squared error of the two ANNs with respect to the number of iteration.

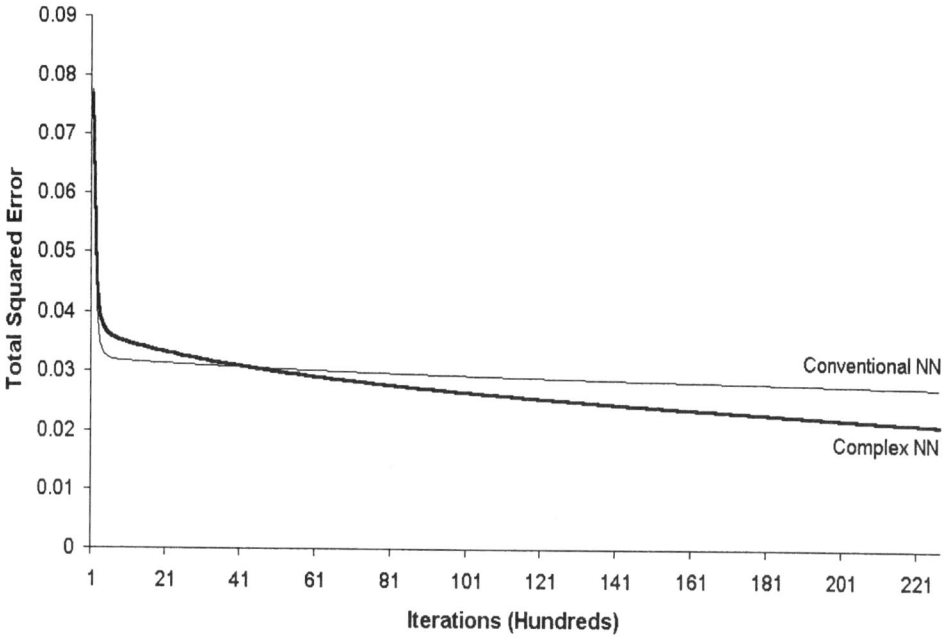

Figure 11.23 Training errors of two ANNs for power load flow

After the two ANNs have been trained up, they are used to estimate V_5 under different testing samples of P_i and Q_i, $i = 4,5$ and 6. There are two categories of testing samples, first set (Cases 1 to 7) being those P and Q randomly selected in between the limits of P and Q included in Table 11.8. Another set (Cases 8 to 12) is randomly selected outside the limits to test the ability of generalisation of the two ANNs. The P_i and Q_i under test are shown in Table 11.9 while the results are shown in Table 11.10.

Table 11.9 Test cases for the neural networks

Case	$P_4 + jQ_4$	$P_5 + jQ_5$	$P_6 + jQ_6$
1	0.77+j0.82	0.75+j0.79	0.84+j0.73
2	0.72+j0.76	0.88+j0.81	0.77+j0.80
3	0.83+j0.87	0.72+j0.79	0.82+j0.89
4	0.75+j0.77	0.82+j0.89	0.80+j0.76
5	0.84+j0.81	0.71+j0.77	0.79+j0.82
6	0.88+j0.81	0.83+j0.87	0.75+j0.82
7	0.80+j0.80	0.80+j0.80	0.80+j0.80

8	0.61+j0.69	0.92+j0.95	0.78+j0.67
9	0.58+j0.69	0.76+j0.94	0.97+j0.88
10	0.79+j0.87	0.61+j0.57	0.94+j0.68
11	0.60+j0.60	0.60+j0.60	0.60+j0.60
12	1.00+j1.00	1.00+j1.00	1.00+j1.00

Table 11.10 Comparison of the neural networks

Case	V_s/Correct	V_s/Real NN	V_s/Complex NN
1	0.935-j0.109	0.920-j0.111	0.932-j0.108
2	0.924-j0.115	0.921-j0.108	0.922-j0.116
3	0.912-j0.103	0.923-j0.115	0.919-j0.102
4	0.917-j0.110	0.922-j0.109	0.919-j0.114
5	0.932-j0.101	0.923-j0.114	0.931-j0.101
6	0.907-j0.113	0.923-j0.118	0.910-j0.113
7	0.924-j0.112	0.922-j0.113	0.921-j0.112
8	0.923-j0.113	0.919-j0.101	0.922-j0.114
9	0.897-j0.105	0.920-j0.105	0.920-j0.120
10	0.974-j0.108	0.923-j0.112	0.970-j0.106
11	0.993-j0.064	0.914-j0.077	0.960-j0.062
12	0.785-j0.169	0.928-j0.141	0.889-j0.160

Discussions

According to Figure 11.23, although there are only 14 training examples, it can be seen that, initially, the total squared error of the 'real' ANN is smaller than that of the newly developed 'complex' ANN. Actually, after 500 iterations, the total squared error of the 'real' ANN has already attained a steady-state value, around 0.032. After 4300 iterations, the 'complex' ANN catches up with the 'real' ANN and the total squared error is improving. After 23000 iterations, the error is only 0.019. Actually, if we compare Figure 11.23 with Figure 11.20, one very interesting result can be noted. It seems that the 'real' ANN can easily get itself into a minimum after a small number of iterations, where as the 'complex' ANN can continuously improve itself during the learning process. The CPU time for training is about 90 s and 150 s for the real and complex NNs respectively when a PII 300 PC is used under Windows 98. Although the training error is less, the prediction accuracy is similar. It is suspected that it is quite easy for this NN to over-fit the data.

Other random initial guesses of weights have been tried for both ANNs and arrived at similar results. This is one merit of the newly developed 'complex' ANN. Next, Tables 11.9 and 11.10 are referred to where we want to test the power of prediction of the two ANNs. The first seven testing samples have been randomly selected in between the limits of [0.7, 0.9] of both the P and Q. The 'complex' ANN behaves better in all cases. However, when five alternative testing samples, selected outside the limits, are tried, the 'complex' ANN behaves better except in case 9. It appears that, in general, the 'complex' ANN is more preferable for the present application.

11.6 Virtual Reality

Since the 1980s, information processing has exploded. Processing power has doubled every two years. Today, the Intel Pentium III runs at a clock speed in excess of 700 MHz. It is expected that by 2002, the chip could run at a clock speed of 3 to 5 GHz. At present, the technology could create, store, search and process vast amounts of information, but we have yet to advance the technology further to access and interface this information more easily. Traditionally, interaction with a computer has involved the use of a keyboard, mouse or joystick/trackball device to input information and the use of a visual display unit (VDU) to receive output from the system. With the development of virtual reality (VR) systems, new interaction methods have been developed that allow the user to 'step into' computer-generated, or virtual, environments (VEs). VR can be considered an extension of ideas which have been around for some considerable time, such as flight simulation and wide screen cinema. Using such systems, the viewer is presented with a screen, which takes up a large portion of the visual field giving a powerful sense of presence or 'being there'. VR refers to a suite of technologies which permit human interaction with real-time, three-dimensional (3-D) representations of information held in computer databases. A VR system may consist of visual, auditory and tactile stimuli, each of which is transmitted to sense organs. VR is a significant extension to the way the users interact with computer systems, greatly improving individual and shared understanding, leading to competitive business advantage. In VEs it is possible to simulate inaccessible or risky experiences, allowing the user to extract the lessons to be learned without the inherent risk. This allows the user to interact in real-time with a computer-generated environment in a simple, 'natural' manner, without the need for extensive training. Presentation to the user may be in many different forms depending on available budget and the needs of the application. There is a general perception that VR requires high levels of computing power implying high costs. However, recent developments in low-cost desktop PC-based VR and a greater appreciation of its value have made the technology more readily available within the budgets of smaller companies. The strength of VEs is in achieving this through their exploitation of the natural interactive skills of the human. As the cost of ownership of VEs technology decreases, 'immersive' and 'desktop' VEs systems, integrating novel display and interactive peripherals, will be more widely used. The potential applications of VR technology are vast and a great deal of research is currently under way in many commercial and academic institutions to develop these technologies into effective useable systems. Today, VR may be displayed cheaply on a conventional desktop PC, or through a large-screen display for multi-user participation. Although not always relevant to business use, immersive representations can involve the use of head-mounted displays (HMDs), tactile gloves, and other devices to enhance the effect. Applications range from simulations of physical items (ranging from buildings to molecular structures) to more abstract concepts such as the display of large amounts of time-varying data (e.g. analysis of world stock markets, complex databases) or illustrating intangible concepts [54].

11.6.1 Types of VR systems

Although it is difficult to categorise all VR systems, most configurations fall into three main categories and each category can be ranked by the sense of immersion, or degree of presence, it provides. Immersion or presence can be regarded as how powerfully the attention of the user is focused on the task in hand. Immersion presence is generally believed to be the product of several parameters including the level of interactivity, image complexity, stereoscopic view, field of regard and the update rate of the display. For example, providing a stereoscopic rather than monoscopic view of the VE will increase the sense of immersion experienced by the user. It must be stressed that no one parameter is effective in isolation and the level of immersion achieved is due to the complex interaction of the many factors involved.

11.6.2   Non-immersive (Desktop) Systems

Non-immersive systems are the least immersive implementation of VR techniques. This includes mouse-controlled navigation through a 3-D environment on a graphics monitor, stereo viewing from the monitor via stereo glasses, stereo projection systems, and others. Interaction with the VE can occur by conventional means such as keyboards, mice and trackballs or may be enhanced by using 3-D interaction devices. The non-immersive system has advantages in that it does not require the highest level of graphics performance, no special hardware and can be implemented on a high-specification PC. This means that these systems can be regarded as the lowest cost VR solution, which can be used for many applications. However, this low cost means that these systems will always be outperformed by more sophisticated implementations, provide almost no sense of immersion and are limited to a certain extent by current 2-D interaction devices. Additionally, these systems are of little use where the perception of scale is an important factor. However, one would expect to see an increase in the popularity of such systems for VR use in the near future. This is due to the fact that virtual reality modelling language (VRML) is expected to be adopted as a standard for the transfer of 3-D model data and virtual worlds via the Internet. The advantage of VRML for the PC desktop user is that this software runs relatively well on a PC. Furthermore, many commercial VR software suppliers are now incorporating VRML capability into their software and exploring the commercial possibilities of desktop VR in general.

11.6.3   Fully Immersive Head-mounted Display Systems

In immersive VR, the user becomes fully immersed in an artificial, 3-D world that is completely generated by a computer. These systems are probably the most widely known VR implementation where the user either wears an HMD or uses some form of head-coupled display such as a binocular omni-orientation monitor [55]. Fully immersive VR systems tend to be the most demanding in terms of computing power. Consequently the cost required to achieve a satisfactory level of realism is high. Major areas of research and development include field of view, resolution, reducing the size and weight of HMDs and reducing system lag times.

11.6.4   Semi-immersive Projection Systems

Semi-immersive systems are a relatively new implementation of VR technology and borrow considerably from technologies developed in the flight simulation field. A semi-immersive system will comprise a relatively high-performance graphics computing system, which can be coupled with:

- a large-screen monitor;
- a large-screen projector system; and
- multiple television projection systems.

Using a wide field of view, these systems increase the feeling of immersion or presence experienced by the user. However, the quality of the projected image is an important consideration. It is important to calibrate the geometry of the projected image to the shape of the screen to prevent distortions and the resolution will determine the quality of textures, colours, the ability to define shapes and the ability to the user to read text on-screen. The resolutions of projection systems range from 1000 to 3000 lines but to achieve the highest levels it may be necessary to use multiple projection systems, which are more expensive [56].

Semi-immersive systems therefore provide a greater sense of presence than non-immersive systems and also a greater appreciation of scale. In addition, images can be provided that are of a far greater resolution than HMDs and this implementation provides the ability to share the virtual experience. This may have a considerable benefit in educational applications as it allows simultaneous experience of the VE, which is not available with head-mounted immersive systems. Additionally, stereographic imaging can be achieved, using some types of shuttered glasses in synchronisation with the graphics system.

11.6.5 Comparison between Different VR Systems

Kalawsky [56] provides a good comparison between the various VR systems (Table 11.11). It is also important that these systems are not regarded as distinct boundaries for implementations. For example, it is possible to turn a desktop system into a semi-immersive system by simply adding shutter glasses and the appropriate software, or a fully immersive system by connecting an HMD.

Table 11.11 Performance of different VR systems (adapted from [56])

Main Features	Non-immersive VR	Semi-immersive VR	Full-immersive VR
Resolution	High	High	Low–Medium
Scale	Low	Medium–High	High
Sense of situational awareness	Low	Medium	High
Field of regard	Low	Medium	High
Lag	Low	Low	Medium–High
Sense of immersion	None–Low	Medium–High	Medium–High

Vision is the main sense that VR designers have concentrated on and many sophisticated techniques have been developed such as 3-D graphics, variations in light and shade, etc. Vision is very complex and concentrates on interpretation of information that is highly subjective (i.e. what is seen varies from person to person). Immersive VR is generally better at simulating vision than non-immersive VR as it cuts out the 'background' forcing concentration on the virtual landscape. However, immersive VR tends to be non-portable. There are also three related VR technologies and they are listed below.

11.6.6 Cave

Cave is a small room where a computer-generated world is projected on the walls. The projection is made on both the front and side walls. This solution is particularly suitable for collective VR experience because it allows different people to share the same experience at the same time. It seems that this technological solution is particularly appropriate for cockpit simulations as it allows views from different sides of an imaginary vehicle.

11.6.7 Telepresence

Telepresence systems immerse a viewer in a real world that is captured by video cameras at a distant location and allow for the remote manipulation of real objects via robot arms and manipulators. Telepresence is used for remote surgical operations and for the exploration/manipulation of hazardous environments such as space and underwater.

11.6.8 Augmented

The technologies of 'augmented reality' allow for the viewing of real environments with superimposed virtual objects. As a matter of fact the user's view of the world is supplemented with virtual objects and items whose meaning is aimed at enriching the information content of the real environment.

11.6.9 Applications

While VR has been used by the military and by space scientists for the last decade, pharmacologists, molecular biologists and theoretical physicists are beginning to venture into its domain. Simply speaking, the technology provides heightened representation of the real, physical attributes of scientific models. It is a tool that takes visualisation and interpretation of data to a new dimension, to the point at which the user, in a sense, touches, interacts with, or is engulfed by the model that has been created. Indeed, it promises to accelerate scientific understanding by enabling the users to refine their hypothesised models.

Using VR to visualise and prototype imaginative projects can shorten the lifecycle of their development and make them available much earlier than would otherwise be the case. VR is an interactive media, where a wide variety of experiences can be created for users to explore at their own pace, choosing their own pathways.

As the technologies of VR evolve, the applications of VR become literally unlimited. It is assumed that VR will reshape the interface between people and information technology by offering new ways for the communication of information, the visualisation of processes, and the creative expression of ideas. Note that a VE can represent any 3-D world that is either real or abstract. This includes real systems like buildings, landscapes, spacecraft, sculptures, crime scene reconstructions, solar systems, and so on. Of special interest are the visual and sensual representation of abstract systems like magnetic fields, turbulent flow structures, molecular models, mathematical systems, auditorium acoustics, stock market behaviour, population densities, and any other artistic and creative work of abstract nature. These virtual worlds can be animated, interactive, shared, and can expose behaviour and functionality.

Though still relatively new, VR has already been put to use in a number of different, innovative ways. In the world of industrial design, engineers are using computer-generated simulations of prototypes to speed up the time required to take a new product from the drawing board to the production line. In the world of science and medicine, doctors are using computer-simulated pathologies to determine the outcome of potentially risky surgical procedures before these procedures are actually performed on patients. In the world of civil engineering, architects and interior designers are using VR systems to create realistic, computer-generated simulations of proposed environments. These environments can then be modified in real-time based on client input, zoning ordinances, aesthetic concerns and budgetary considerations.

In the world of weather forecasting, VR is being used to predict weather patterns and to create hurricane models which can accurately determine where a storm will make landfall and when. In the world of higher education, VR lets astronomy students tour distant galaxies and physiology students tour the innermost workings of the human body. A VR simulation of a complex pipework layout, for example, could allow access, maintenance and safety aspects to be examined at the design stage, more effectively than by modelling. It immediately permits the evaluation of routing and accessibility, thereby avoiding expensive, time-consuming correction during or even after construction. There are many more opportunities that have yet to be explored.

In the information age, VR has been identified as one of the most promising development areas. There is a constant improvement in marketing perspective of both quality of applicative VR systems and receptiveness of potential customers. This is due to mainly three reasons: (1) the decrease of the cost of VR systems and devices, (2) the constant improvement of performance reliability of the technology, (3) the extremely valuable economic benefits derived from VR use in its various forms and purposes such as design [54]. Although the technology is mature enough to have different applications, there are key issues to be resolved for its use for practical applications.

The sensational press coverage associated with some of these technologies has led many potential users to overestimate the actual capabilities of existing systems. Many of them must actually develop the technology significantly for their specific tasks. Unless their expertise includes knowledge of the human-machine interface requirements for their application, their resulting product will rarely get beyond a 'conceptual demo' that lacks practical applications. Current VR products employ proprietary hardware and software. There is little doubt that incompatibility between different systems is restricting market growth at present. It is probable that as the market matures, certain standards will emerge.

The premise of VE seems to be to enhance the interaction between people and their systems. It thus becomes very important to understand how people perceive and interpret events in their environments, both in and out of virtual representation of reality.

Fundamental questions remain about how people interact with the systems, how they may be used to enhance and augment cognitive performance in such environments, and how they can best be employed for instruction, training and other people-oriented applications.

11.7   3-D Thermal Imaging for Power Equipment Monitoring

It is well known that most faults in a power system appear as localised hot spots in their early stages of development and gradually grow into accidents that are not remediable, such as total insulation breakdown or even explosion. Through the detection of hot spots and abnormal rise in working temperature of conductors or insulators, infrared thermography [57] is one of the most useful tools in identifying potential failures so that preventive measures can be taken to avoid occurrence of any hazard. Thermography has also been employed in medical diagnosis [57,58] and diagnosis of advanced electronics and/or electrical equipment [57,59,60,61]. Another major reason for the numerous applications of infrared thermography in the electricity power supply industry is that many points of interest requiring temperature measurement are either difficult to access or hazardous. The plant can be large, remote, electrically live, rotating or extremely hot and a remote sensing technique capable of inspecting large areas quickly has obvious attractions. The infrared imager used in our project has an overall temperature range from -40°C to 950°C with a maximum resolution down to 0.01°C, making it very sensitive to even a slight imbalance in heat dissipation. However, though thermography bears so many advantages for power system applications, it has not drawn popular attention worldwide.

The major reason behind this is the extreme difficulty in analysing thermal images of power equipment and very often, only skilful maintenance engineers are able to make correct judgements based on the images themselves, thus hindering the widespread applications of this intrinsically useful technique. The fact is that, on a thermal image, almost all 3-D geometrical features of a physical object are lost. It is therefore very difficult for a non-skilful engineer to recognise the object, not to mention the colour of any point on the object that represents the surface temperature instead of the reflected light. With the help of techniques in VR, the visualisation problem mentioned above can be solved. This section describes the development of a 3-D thermal imaging technique so that the user can easily understand the geometrical structure of the equipment and at the same time analyse the temperature distribution on the surface of it by swapping between the geometrical mode and the thermal mode. Furthermore, the user can freely control the viewing angle of the object as if he/she was moving it with the feeling of total immersion in the environment. This technique, besides aiding condition-based maintenance, can be a very useful tool for designing new power plants and installation. The proposed system will also be useful for training maintenance engineers as they are able to visual the real object under a 3-D environment and to have a feeling of the variation of surface temperature which is always a good indication for health of the equipment. The key techniques involve a true regeneration of the 3D surfaces of the equipment using absolute dimensioning based on on-site measurements and the integration of thermal images with the re-generated 3-D surfaces. Finally, the processed data will be displayed on the screen for 2-D projection or viewed through head-mounted helmets for true 3-D visualisation. For the sake of simplicity, we shall use i and j as common running variables throughout the whole section, their meanings varying from subsection to subsection.

11.7.1 The Hardware

The thermal system consists of an infrared camera, shown in Figure 11.24, and a processor, shown in Figure 11.25. The infrared detectors inside the camera are cooled by compressed argon to -186°C and they sense the infrared spectrum in the range between 3 and 5.4 µm. Image scanning speed is 30 frames/s while floppy disks and high-speed RAM are two media for saving the images for further processing on a PC.

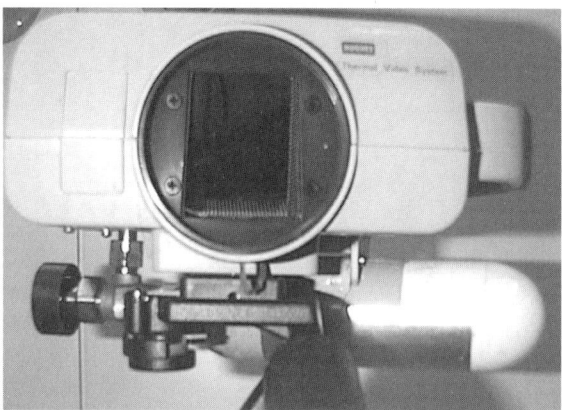

Figure 11.24 Infrared camera

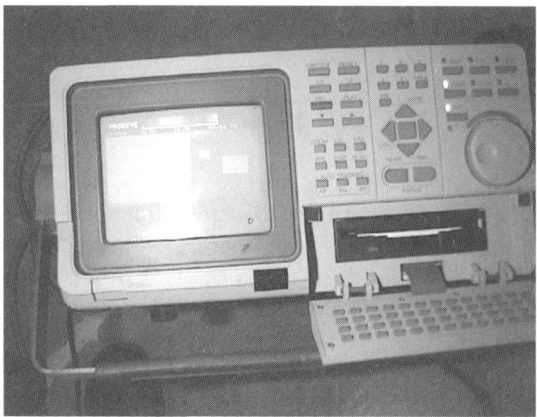

Figure 11.25 Processor of the thermal system

The surface of the object under measurement is divided into grid points whose absolute coordinates are measured with the aid of a laser-based distance-measuring kit, the DMK, shown in Figure 11.26. The length of black wire is the RS-232 communication cord for communication with a PC. The resolution of DMK is 5 mm over a distance of 10 m. Panning and tilting functions are offered to the DMK by an assembly of two stepper motors. The horizontal and vertical angles of the assembly are controlled by the PC, with a resolution of 0.36 degrees, while the values are updated on a real-time basis. This assembly is mounted on top of the infrared camera, which is stationary. Therefore, the

DMK with panning/tilting functions can give the absolute coordinates of each grid point on the power equipment while the thermal system can give the real-time surface temperature of that grid point. When this information is fed into a tailor-made VR-based software package, a 3-D thermal image can be displayed and manipulated. The major problem here is with the correspondence between the DMK and the thermal system, i.e. matching every point sensed by the DMK to a corresponding point on the thermal image.

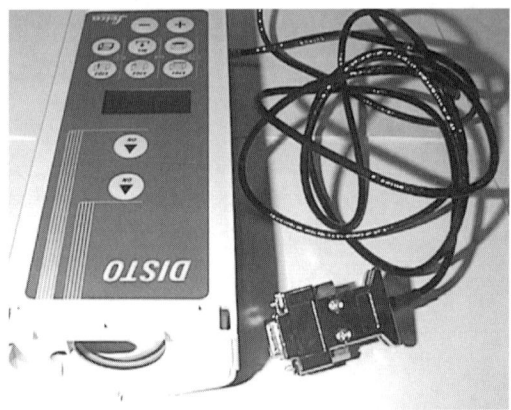

Figure 11.26 Laser-based distance-measuring kit

11.7.2   The Correspondence

Infrared Camera Modelling
Camera modelling defines the imaging geometry of the system and the procedure of finding out the model of a camera is known as camera calibration. Sobel [62] and Gennery [63] used full-scale non-linear optimisation for camera calibration. A comprehensive survey of the literature and discussion of methods for electronic cameras was presented by Lenz and Tsai [64]. It has been found that the perspective transformation model [65] is the simplest camera model and its calibration procedure is straightforward as it only involves linear optimisation. At the same time, its level of accuracy is up to a satisfactory level, suitable for our application. A pin-hole camera model has been the most popular representation of a standard electronic camera. Let (x_w, y_w, z_w) be the 3-D coordinates of an object point **P** in the universal 3-D world coordinate system. (X_f, Y_f) are the corresponding coordinates of the image point on the thermal image. The transformation from $\mathbf{P}(x_w, y_w, z_w)$ to $\mathbf{I}(X_f, Y_f)$ is modelled by a standard approach involving a linear transformation **T** [64] in projective coordinates. **T**, a 3×4 matrix, is known as the 'perspective matrix' of the camera where the element, t_{34}, is assigned a value of 1 by proper scaling. By proper scaling, all the t inside the transformation, **T**, are divided by t_{34} and, naturally, t_{34} will automatically be scaled to '1'. Let $\mathbf{I}^* = (U, V, S)^T$ be the projective coordinates of **I** and equation (11.20) will hold. The element '1' in the homogeneous coordinate vector, together with the two parameters t_{14} and t_{24}, take care of the linear translation of the transformation.

$$I = \begin{pmatrix} X_f \\ Y_f \end{pmatrix} = \begin{pmatrix} \dfrac{U}{S} \\ \dfrac{V}{S} \end{pmatrix} \quad I^* = \begin{pmatrix} U \\ V \\ S \end{pmatrix} = T \begin{pmatrix} x_w \\ y_w \\ z_w \\ 1 \end{pmatrix} \quad (11.20)$$

Therefore,

$$X_f = \frac{t_{11} x_w + t_{12} y_w + t_{13} z_w + t_{14}}{t_{31} x + t_{32} y_w + t_{33} z_w + 1}$$

(11.21)

$$Y_f = \frac{t_{21} x_w + t_{22} y_w + t_{23} z_w + t_{24}}{t_{31} x_w + t_{32} y_w + t_{33} z_w + 1}$$

In theory, six non-coplanar calibration points with known, (x_{wi}, y_{wi}, z_{wi}) and (X_{fi}, Y_{fi}), for $i = 1,...,n$, where $n = 6$ are enough to determine the 11 unknown t by the method of least-squares estimation. In practice, it is better to have more points so that the accuracy can be enhanced since least-squares optimisation is an averaging technique that can provide a more accurate solution when n is larger. In other words, the following n sets of linear equations have to be solved:

$$\mathbf{A\,a = b}$$

where

$$\mathbf{a} = \begin{pmatrix} t_{11}\, t_{12}\, t_{13}\, t_{14}\, t_{21}\, t_{22}\, t_{23}\, t_{24}\, t_{31}\, t_{32}\, t_{33} \end{pmatrix}^T$$
$$\mathbf{A} = \begin{pmatrix} A_1, ..., A_n \end{pmatrix}^T$$
$$\mathbf{b} = \begin{pmatrix} b_1, ..., b_n \end{pmatrix}^T$$

$$\mathbf{A}_i = \begin{pmatrix} x_{wi} & y_{wi} & z_{wi} & 1 & 0 & 0 & 0 & 0 \\ 0 & 0 & 0 & 0 & x_{wi} & y_{wi} & z_{wi} & 1 \\ -X_{fi}\, x_{wi} & -X_{fi}\, y_{wi} & -X_{fi}\, z_{wi} \\ -Y_{fi}\, x_{wi} & -Y_{fi}\, y_{wi} & -Y_{fi}\, z_{wi} \end{pmatrix} \quad (11.22)$$

$$\mathbf{b}_i = \begin{pmatrix} X_{fi} \\ Y_{fi} \end{pmatrix}$$

With optimisation of the least sum of squares over the n number of calibration points, the solution is given by

$$a = (A^T A)^{-1} A^T b$$

From experience, eight calibration points are good enough for calibrating the infrared camera up to an acceptable accuracy.

Correspondence of Grid Points with the Aid of the DMK

Before the infrared camera is calibrated, the PC controls the two stepper motors and takes readings of the grid points on the surface of interest of the power equipment, say a power transformer in this case. A range of vertical angles, θ_V, and horizontal angles, θ_H, has to be specified and the two stepper motors are activated in turn to scan through the whole range. θ_H is measured in an anti-clockwise direction as viewed from the top. θ_V is measured from the zenith, i.e. 0 degrees, downwards until 90 degrees meaning that the DMK is pointing in the horizontal direction and furthermore 180 degrees means that the DMK is pointing vertically downwards. The number of grid points available, M, is equal to [range of θ_V/0.36 degrees] × [range of θ_H/0.36 degrees]. The higher the number of grid points, the finer the resultant image is and the slower the whole process is. After scanning, the following list of coordinates is produced, $(\theta_{Vj}, \theta_{Hj}, d_j)$ where $j = 1, \ldots, M$, and d_j is the distance of the jth grid point from the origin of the DMK coordinate system. The center of the DMK's motor assembly has been chosen to be the origin of the 3-D coordinate system. As shown in Figure 11.27, the coordinates of each grid point within the 3-D system, (x_{wj}, y_{wj}, z_{wj}), can also be given by:

$$z_{wj} = d_j \cos \theta_{Vj}$$
$$y_{wj} = d_j \sin \theta_{Vj} \sin \theta_{Hj} \quad (11.24)$$
$$x_{wj} = d_j \sin \theta_{Vj} \cos \theta_{Hj}$$

All dimensions are measured in metres. As all the t in equation (11.21) have been identified after the calibration of the infrared camera, the coordinates of each grid point in x_{wj}, y_{wj} and z_{wj} are fed into equation (11.21) and the corresponding (X_{fj}, Y_{fj}) can be estimated, producing a new list of grid points, i.e. $(x_{wj}, y_{wj}, z_{wj}, T_j)$ where T_j is the surface temperature of the jth grid point, where $j = 1, \ldots, M$.

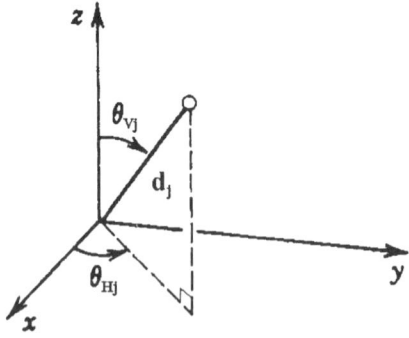

Figure 11.27 3-D world coordinate system

11.7.3 Display with VR

Recently, VR techniques have been widely used in industry [66]. VEs are being produced to simulate real environments, the application described in this session being one of them. The whole piece of electric power equipment was modelled to include not just the geometrical details but also the distribution of surface temperature. VEs can be distinguished by their relation to a real environment in three aspects. The VE is actually a projection of a real environment of a very different scale [66] or at some distance from the viewer [67]. The VE does not exist but is fairly realistic. In this case, the VE is the power plant where the designer can manipulate existing and real objects, the power equipment. In order to enhance the feeling of immersion which is the feeling of a VR user that his/her VE is real, surface rendering and controllable angle viewing are built into the system. The user can select two different modes of display, the geometrical mode where the grid of the object's surface is shown, or the thermal mode where the surface of the object is coated with a pseudo colour scale to reveal the surface temperature of each point. Three transformations have been built into the program, namely translation, scaling and rotation.

Any grid point with coordinates (x_{wj}, y_{wj}, z_{wj}) can be translated to $(x'_{wj}, y'_{wj}, z'_{wj})$ by a translation vector (t_x, t_y, t_z) in accordance with equation (11.25). Hence, the whole object can be moved about the VE simply by modifying the translation vector. The scaling transformation alters the size of the object by scaling all of its coordinates relative to a reference point in accordance with equation (11.26) based on a scaling vector (S_x, S_y, S_z).

$$\begin{bmatrix} x'_{wj} \\ y'_{wj} \\ z'_{wj} \\ 1 \end{bmatrix} = \begin{bmatrix} 1 & 0 & 0 & t_x \\ 0 & 1 & 0 & t_y \\ 0 & 0 & 1 & t_z \\ 0 & 0 & 0 & 1 \end{bmatrix} \begin{bmatrix} x_{wj} \\ y_{wj} \\ z_{wj} \\ 1 \end{bmatrix} \quad (11.25)$$

$$\begin{bmatrix} x'_{wj} \\ y'_{wj} \\ z'_{wj} \\ 1 \end{bmatrix} = \begin{bmatrix} S_x & 0 & 0 & 0 \\ 0 & S_y & 0 & 0 \\ 0 & 0 & S_z & 0 \\ 0 & 0 & 0 & 1 \end{bmatrix} \begin{bmatrix} x_{wj} \\ y_{wj} \\ z_{wj} \\ 1 \end{bmatrix} \quad (11.26)$$

Scaling is different from translation in that the resolution of the object's surface may change significantly after the scaling process in terms of texture, size and shape. For translation, it is just relocating the object from its original location to a new location, the whole object being unchanged. For scaling, if the object is enlarged, we may come across the appearance of cracks or trap holes. Hence, some sort of correction in terms of an interpolation process needs to be carried out to form a continuous surface of the object, i.e. surface rendering before any form of scaling is executed. New points are generated from the interpolated surface and are scaled according to equation (11.26) to make sure the scaled object has the same spatial resolution with respect to the original one. Interpolation of the surface temperature is by means of a similar process.

The grid points are generated in appropriate sequence by the two PC-controlled stepper motors. For each of the n number of θ_V within the specified range, there are m numbers of θ_H within another specified range. Hence, the grid points can be viewed as elements of three $n \times m$ matrices where $n \times m = M$, each representing the x, y and z coordinates of each grid point respectively. For the (i,j) grid point where $i = 1, ..., n-1$, and $j = 1, ..., m-1$, three more grid points, namely $(i+1,j)$, $(i,j+1)$ and $(i+1,j+1)$, are considered. Two planes are generated, one having (i,j), $(i+1,j)$ and $(i,j+1)$ and the other having $(i+1,j)$, $(i,j+1)$ and $(i+1,j+1)$. The equation of the first plane is given by the following equation:

$$Ax + By + Cz - 1 = 0$$

$$\begin{bmatrix} A \\ B \\ C \end{bmatrix} = \begin{bmatrix} x_{w,i,j} & y_{w,i,j} & z_{w,i,j} \\ x_{w,i+1,j} & y_{w,i+1,j} & z_{w,i+1,j} \\ x_{w,i,j+1} & y_{w,i,j+1} & z_{w,i,j+1} \end{bmatrix}^{-1} \begin{bmatrix} 1 \\ 1 \\ 1 \end{bmatrix} \quad (11.27)$$

During scaling, points are created on relevant planes. The surface temperature, T, of any point on the plane in equation (11.27) can be found by the following equation, if the distances of this point from the three vertices are known to be $d_{i,j}$, $d_{i+1,j}$ and $d_{i,j+1}$:

$$T = \frac{T_{i,j}\, d_{i+1,j}\, d_{i,j+1} + T_{i+1,j}\, d_{i,j}\, d_{i,j+1} + T_{i,j+1}\, d_{i,j}\, d_{i+1,j}}{d_{i+1,j}\, d_{i,j+1} + d_{i,j}\, d_{i,j+1} + d_{i,j}\, d_{i+1,j}} \quad (11.28)$$

The temperature of any point on the three sides of the triangles is produced by linear interpolation of the two vertices, i.e. the two end points of the side, only. Regarding rotation, a rotational matrix consisting of nine r must be specified and the transformation is shown in the equation (11.29).

$$\begin{bmatrix} x'_{wj} \\ y'_{wj} \\ z'_{wj} \\ 1 \end{bmatrix} = \begin{bmatrix} r_{11} & r_{12} & r_{13} & 0 \\ r_{21} & r_{22} & r_{23} & 0 \\ r_{31} & r_{32} & r_{33} & 0 \\ 0 & 0 & 0 & 1 \end{bmatrix} \begin{bmatrix} x_{wj} \\ y_{wj} \\ z_{wj} \\ 1 \end{bmatrix} \quad (11.29)$$

$$\sum_{i=1}^{3} r_{ij}^2 = 1\ \forall j\ \text{and}\ \sum_{j=1}^{3} r_{ij}^2 = 1\ \forall i$$

11.7.4 Implementation Example

The competitive electricity market raises utility cost consciousness. Huge expenditures are normally associated with equipment investment and continuous maintenance for the power system. Power transformers are one of the most expensive elements in the system. The identification of any hot spots, i.e. potential faults, could provide benefits including extended transformer lifetimes, reduction in risk of failures and improvement in maintenance strategies. A transformer room, shown in Figure 11.28, housing three 1500 kVA 11 kV/380 V transformers, was used for implementing the developed system.

Figure 11.28 Three 1500 kVA transformer in a typical transformer plant room

Transformer No. D3 was used as the object for 3-D thermal imaging. The optical image of a portion of the surface is shown in Figure 11.29. The thermal image of size 256 × 100 pixels, 256 grey level per pixel of the portion, is shown in Figure 11.30. For the DMK, θ_V is selected within the range of [88.2 degrees, 96.84 degrees] while that of θ_H is [33.84 degrees, 52.2 degrees], all in steps of 0.36 degrees. Hence, all grid points are selected within the range [2.50 m, 3.25 m] in the x direction, [2.00 m, 3.50 m] in the y direction and [-0.60 m, 0.15 m] in the z direction, while the surface temperature varies from 24.55°C to 44.8°C with a resolution of 0.2°C. It takes about 3 seconds to record the distance and orientation of a grid point by the PC, thus needing more than 1 hour to scan through the whole portion by the infrared camera.

There is not enough space here to list the coordinates and surface temperature of each of the 1300 grid points. A typical data list of eight points belonging to a constant θ_V = 91.44 degrees is shown in Table 11.12 below as an illustration while the full graphical plot is shown in Figures 11.31 and 11.32. The x, y and z coordinates measured in metres, are the absolute coordinates of an image point with respect to the coordinate system of the DMK. In the geometrical mode, the 3-D surface of the transformer is as shown in Figure 11.31 without any information on temperature. It should be noted that the 3-D surface is not totally identical to the real surface in this situation. The reason is that part of the

Figure 11.29 Transformer No. 3 under imaging

Figure 11.30 The thermal image of portion of transformer D3

object's surface is hidden from the line of sight of the DMK and hence no geometrical or thermal information is available for those points on the surface of this hidden part. However, the PC has no knowledge about which part is hidden and which part is not hidden as the whole process of distance measurement and thermal image grabbing is fully automatic. The program just produces a continuous surface based on the 25 × 52 grid points in proper sequence, thus forming a slightly different virtual surface with respect to the real object. For a full implementation in the future, the DMK and the thermal imager will be manoeuvred around the power equipment so that all hidden surfaces can be retrieved into the database for display. By superimposing the information on temperature on Figure 11.31, Figure 11.32 can be obtained where the user can identify the surface temperature at any non-hidden grid point on the surface of the portion. From Figure 11.29, we can see three main parts of the surface, namely the cooling fins on the right, the incoming cable cover in the middle and the cooling fins on the left. The cooling fins, normally with a higher temperature except for the labelling strips, are protruding outwards from the transformer and, hence, they are nearer to the origin of our coordinate system.

The cable cover has a comparatively lower temperature. All these features are clearly shown in Figure 11.31 and Figure 11.32.

Table 11.12 Coordinates versus surface temperature

x(m)	y(m)	z(m)	T(°C)
2.5	1.7	0.1	43.8
2.5	1.8	0.1	37.0
2.5	1.9	0.1	31.4
2.5	2.0	0.1	41.2
2.8	2.3	0.1	27.2
2.8	2.4	0.1	27.2
2.8	2.5	0.1	27.2
2.8	2.6	0.1	27.5

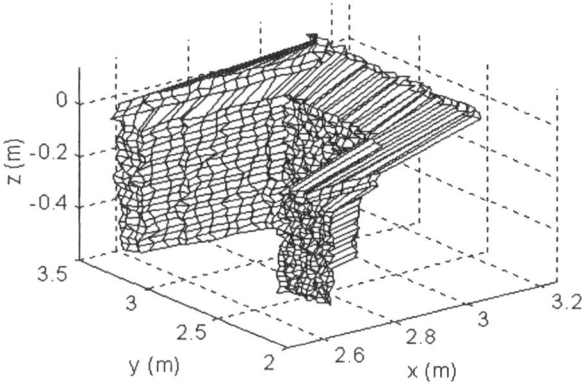

Figure 11.31 3-D surface of the transformer in the geometrical mode

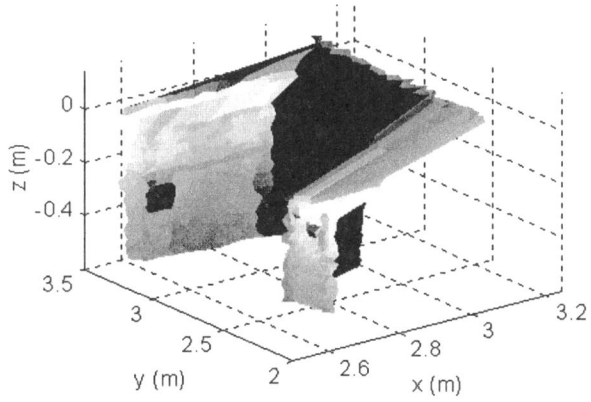

Figure 11.32 3-D thermogram of the transformer in the thermal mode

Furthermore, the user can freely adjust the viewing angle to concentrate on any particular part of the surface, thus fulfilling the full objective of producing a VR environment for electric power equipment design and maintenance. This 3-D thermography can serve two purposes. The first one is for designing power plants in the geometrical mode. The 3-D information of all components is recorded on the PC for proper 3-D display. The designer can thus fly around the power plant to check any obstruction and improper placement of equipment. Such checking is more powerful than using the 2-D drawings conventionally supplied by the manufacturers. After the power plant has been commissioned, regular thermal imaging procedures can be carried out so that any hot spots in the equipment can immediately be identified manually and the physical locations of these hot spots can be cross-checked against the geometrical 3-D surface. A point to be noted is that a skilful engineer in thermography is no longer required because any technically proficient person in power engineering can fully understand the 3-D thermograms.

11.8 Conclusions

In this chapter, we have considered four hot topics in information technology and their applications, namely, intelligent agents, evolutionary programming, virtual reality and neural networks.

The use of derivative financial instruments such as futures and options will be an important and useful tool for managing risk in the deregulated electricity industry. To test a genetic algorithm's applicability to valuing electricity options, we simulated an electricity market and used Black-Scholes to set the market valuation of put options on electricity. The genetic algorithm was then used to evolve agents whose fitness was measured by their ability to make a profit trading put options. The genetic algorithm was able to earn a profit over the course of a trading run.

An evolutionary-programming-based optimal power flow algorithm (EP-OPF) has been developed. The performance of the EP-OPF algorithm has been demonstrated by its application to the IEEE 30-bus test system. More difficult generator input/output curves such as piecewise quadratic and sine-wave/quadratic curves have been used to provide a non-convex solution surface. The algorithm has converged accurately and reliably to the global optimum solution in each case, while the steepest descent method has failed in the more difficult cases. The algorithm is also capable of producing a more favourable voltage profile while still maintaining a competitive cost.

The advantages of the application of thermography in power equipment diagnosis have been highlighted. Owing to the difficulty in analysing thermal images by most unskilled technical personnel, three-dimensional thermograms are generated so that all engineering staff can easily understand them within a virtual reality environment for more efficient power plant designing and planning as well as condition-based maintenance. When engineers need to gain knowledge in understanding thermograms for maintenance, this system will become a very useful tool for training because both the optical and thermal images of the same object are on the screen with precise geometrical correlation.

The structure of a new 'complex' artificial neural networtk has been developed. Operational elements, such as addition, subtraction and multiplication, are shown graphically. A corresponding sigmoid function utilising a complex function has been devised, resulting in a new gradient function for back propagation. In order to demonstrate that the newly designed 'complex' ANN is superior to the conventional 'real' ANN, a simple application of this novel technique is carried out for load flow analysis of an electric power network consisting of six buses. It concluded that the 'complex' ANN is superior to the conventional 'real' ANN in two aspects. Firstly, the 'complex' ANN will not easily be trapped in a local minimum. Secondly, it seems that there is an improved ability to evaluate cases not falling within the training zone.

11.9 Acknowledgements

The authors would also like to thank IEE and IEEE for granting permission to reproduce the materials contained in references [4,61] and [9,11] respectivley.

11.10 Appendix: System Data and Parameter Settings

System data: The load flow data for the system is that of the standard IEEE 30-bus test system. Branches 11,12,15 and 36 are in phase tap-changing transformers with allowable tapping ranges of ±10% with a step size of 1%. The lower voltage magnitude limits for all busses is 0.95 p.u. while the upper limit is 1.05 p.u. for node 1 and all load nodes; all other generation nodes have an upper limit of 1.10 p.u. This data may be found in [47]. The load flow convergence tolerance is 10^{-3} p.u.

Algorithm parameter settings: The population size is set at 20 and the total number of generations or iterations is 50. The number of tournaments N_t was set at 15. The scaling constant M in the fitness function in (11.3) is determined by the set of cost curves being used. The value of a in (11.5) was 0.9. Q-limit checking was started at iteration 2 of the load flow when a previous solution was not used in the load flow initialisation and in iteration 1 when it was. For all cases the weightings within the fitness function (11.3) if relevant to the case are $K_v = 1000$, $K_q = 10,000$ where Q_{slack} is in p.u. and $K_f = 1000$. The percentage of candidates undergoing gradient acceleration is 50 % in all cases.

Both the SD algorithms used in EP and in comparison studies include penalties identical to those described in [43] for voltage violations with weighting of 20. Penalties are also included for active and reactive power limit violations at the slack node of this form with weightings of 30 and 10 respectively. For case (d) above, the penalties for voltage violations are replaced by penalties of the form of VF_j in (11.11) with a weighting K_f of 10. The SD step size for all cases (a)–(c) is 2.0 for active power, 0.1 for transformer tap and 0.001 for voltage; for case (d) the step sizes are 0.005 for active power, 0.005 for transformer tap and 0.001 for voltage. Q-limit treatment for all nodes other than the slack node is handled by switching within the load flow routine.

11.11 References

[1] K.P. Wong, 'Artificial intelligence and neural network applications in power systems', Invited Paper, *Proceedings of the International Conference on Advances in Power System Control, Operation & Management*, IEE, 1993, pp.37-46.

[2]   S.B. Lau and K.P. Wong, 'An artificial neural network approach to transient stability assessment', *Australian Journal of Intelligent Information Processing Systems*, Vol.3, No.1, 1996, pp.75-85.

[3]   K.P. Wong and S.B. Lau, 'An artificial neural network approach to modelling generator fuel cost characteristics', *Journal of Institution of Engineers*, Singapore, Vol.36, No.6, November 1996, pp.71-77.

[4]   W.L. Chan, A.T.P. So and L.L. Lai, 'Initial applications of complex artificial neural networks to load-flow analysis', *IEE Proceedings – Generation, Transmission and Distribution*, Vol.147, No.6, November 2000, pp.361-366.

[5]   K.P. Wong and E. Tsoi, 'Genetic algorithms approach for the evaluation of trade-off between economic and environmental costs in power dispatch with multiple fuels and pollutants', *Proceedings of the International Conference on Advances in Power System Control, Operation & Management*, IEE, 1995, pp.553-558.

[6]   K.P. Wong and S.Y.W Wong, 'Combined genetic algorithm/simulated annealing/fuzzy set approach to short-term generation scheduling with take-or-pay fuel contract', *IEEE Transactions on Power Systems*, Vol.11, No.1, 1996, pp.112-118.

[7]   K.P. Wong and S.Y.W. Wong, 'Hybrid genetic/simulated annealing approach to short-term multiple-fuel-constrained generation scheduling', *IEEE Transactions on Power Systems*, Vol.12, No.2, 1997, pp.776- 784.

[8]   L.L. Lai and J.T. Ma, 'Genetic algorithms and UPFC for power flow control', *International Journal on Engineering Intelligent Systems*, Vol.4, No.4, CRL Publishing Ltd, UK, December 1996, pp.237-242.

[9]   J. Yuryevich and K.P. Wong, 'Evolutionary programming based optimal power flow algorithm dispatch', *IEEE Transactions on Power Systems*, Vol.14, No.4, 1999, pp.1245-1250.

[10] K.P. Wong, A. Li and T.M.Y. Law, 'Advanced constrained genetic algorithm load flow method', *IEE Proceedings – Generation, Transmission and Distribution*, Vol.146, No.6, November 1999, pp.609-616.

[11] Derek W. Lane, Charles W. Richter, Jr. and Gerald B. Sheblé, 'Modeling and evaluating electricity options markets with intelligent agents', *Proceedings of the International Conference on Power Utility Deregulation, Restructuring and Power Technologies 2000*, City University, London, IEEE, April 2000, pp.203-208.

[12] L.L. Lai, H. Subasinghe, N. Rajkumar, E. Vaseekar, B.J. Gwyn and V.K. Sood, 'Object-oriented genetic algorithm based artificial neural network for load forecasting', *Lecture Notes in Computer Science*, LNCS, Springer-Verlag, Xin Yao et al. (Editors), May 1999.

[13] K.P. Wong and C.C. Fung, 'Development of a fuzzy-logic-based control algorithm for the commitment of energy sources in an integrated energy system', *IEEE Conference Proceedings First Australian and New Zealand Conference on Intelligent Information Systems* (ANZIIS-93), December 1993, pp.432-436.

[14] P.C.K. Luk, L.L. Lai, T.L. Tong, 'GA optimisation of rule base in a fuzzy logic control of a solar power plant', *Proceedings of the International Conference on Power Utility Deregulation, Restructuring and Power Technologies 2000*, City University, London, IEEE, April 2000, 221-225.

[15] L.L. Lai, 'An expert system used in power system protection', *IFAC Symposia Series*, 1990, No.8, Pergamon Press, Oxford, pp.489-494.

[16] J Bradshaw, ed., *Software agents*, MIT Press, Cambridge, Mass., 1997.

[17] J.S. Rosenschein and G. Zlotkin, *Rules of Encounter: Designing Conventions for Automated Negotiation among Computers*, Cambridge: MIT Press, 1994.

[18] Barbara Hayes-Roth, Robert van Gent, Rembert Reynold, M. Vaughan Johnson and K. Wescourt, 'Agents application', *IEEE Intelligent Systems & their Applications*, March/April 1999, pp.23-27.

[19] R.A. Brooks, 'A Robust Layered Control System for a Mobile Robot', *IEEE Journal of Robotics and Automation*, Vol.2, 1986, pp.14-23.

[20] J. Ferber, 'Simulating with Reactive Agents', in E. Hillebrand and J. Stender (Eds.), *Many Agent Simulation and Artificial Life*, Amsterdam: IOS Press, 1994, pp.8-28.

[21] R.A. Brooks, 'Intelligence without representation', *Artificial Intelligence*, Vol.47, 1991, pp.139-159.

[22] M. Wooldridge and N. Jennings, 'Intelligent agents: Theory and practice', *The Knowledge Engineering Review*, Vol.10, 1995, pp.115-152.

[23] H.S. Nwana and M. Wooldridge, 'Software agent technologies', *British Telecommunications Technology Journal*, Vol.14, October 1996.

[24] J. Bates, 'The Role of Emotion in Believable Characters', *Communications of the ACM*, Vol.37, 1994, pp.122-125.

[25] A. Newell, A. (1982), 'The Knowledge Level', *Artificial Intelligence*, Vol.18, 1982, pp.87-127.

[26] P. Maes, (ed), *Designing Autonomous Agents: Theory and Practice from Biology to Engineering and Back*, MIT press, 1991.

[27] A. Chavez and P. Maes, 'Kasbah: An agent marketplace for buying and selling goods', *Proceedings of the First International Conference on the Practical Application of Intelligent Agents and Multi-Agent Technology (PAAM 1996)*, London, April 1996, pp.75-90.

[28] P. Wayner, 'Free Agents', *Byte*, March 1995, pp.105-114.

[29] M.R. Genesereth and S.P. Ketchpel, 'Software agents', *Communications of the ACM*, Vol.37, 1994, pp.48-53.

[30] G. Wiederhold, 'Mediators in the architecture of future information systems', *IEEE Computer*, Vol.25, 1992, pp.38-49.

[31] G. Sheblé, 'Electric energy in a fully evolved marketplace', *North American Power Symposium*, Kansas State University, KS, 1994.

[32] G. Sheblé, 'Priced based operation in an auction market structure', Paper presented at the 1996 IEEE/PES Winter Meeting. Baltimore, MD, 1996.

[33] J. Marshall, *Futures and Option Contracting: Theory and Practice*, South Western Publishing, USA, 1989.

[34] J. C. Hull, *Options, Futures, and Other Derivatives*, Prentice Hall, USA, 1997.

[35] D. Goldberg, *Genetic Algorithms in Search, Optimization & Machine Learning*, Addison-Wesley Publishing Company, Inc., 1989.

[36] R. Ristanovic, 'Successive linear programming based OPF solution', *Optimal Power Flow: Solution Techniques, Requirements and Challenges*, IEEE Power Engineering Society, 1996, pp.1-9.

[37] S.M. Shahidehpour and V.C. Ramesh, 'Nonlinear programming algorithms and decomposition strategies for OPF', *Optimal Power Flow: Solution Techniques, Requirements and Challenges*, IEEE Power Engineering Society, 1996, pp.10-24.

[38] J.A. Momoh, S.X. Guo, E.C. Ogbuobiri and R. Adapa, 'The quadratic interior point method solving power system optimisation problems', *IEEE Transactions on Power Systems*, Vol.9, August 1994, pp.1327-1336.

[39] K.P. Wong, and Y.W. Wong, 'Genetic and genetic/simulated-annealing approaches to economic dispatch', *IEE Proceedings – Generation, Transmission and Distribution*, Vol.141, No.5, 1994, pp.507-513.

[40] IEEE Committee Report: 'Present practices in the economic operation of power systems', *IEEE Transactions on Power Apparatus and Systems*, Vol.PAS-90, 1986, pp.1768-1775.

[41] D.B. Fogel. *Evolutionary Computation: Toward a new Philosophy in Machine Intelligence*, IEEE Press, 1995.

[42] K.P. Wong, and A. Li, 'A technique for improving the convergence characteristic of genetic algorithms and its application to a genetic-based load flow algorithm', *Simulated Evolution and Learning*, J.H. Kim, X. Yao, T. Furuhasi (Eds), Lecture Notes in Artificial Intelligence 1285, Springer-Verlag, 1997, pp.167-176.

[43] H.W. Dommel and W.F. Tinney, 'Optimal power flow solutions', *IEEE Transactions on Power Apparatus and Systems*, Vol. PAS-87, 1968, pp.1866-1876.

[44] K.P. Wong and J. Yuryevich, 'Evolutionary-programming-based algorithm for environmentally-constrained economic dispatch', *IEEE Transactions on Power Systems*, Vol.13, No.2, 1998, pp.301-306.

[45] K.P. Wong, A. Li and M.Y. Law, 'Development of constrained genetic algorithm load flow method', *IEE Proceedings – Generation, Transmission and Distribution*, Vol.144, No.2, 1997, pp.91-99.

[46] D.C. Walter and G.B. Sheblé, 'Genetic algorithm solution of economic dispatch with valve point loading', *IEEE PES Summer Meeting*, 1992, Paper No.92 SM 414-3 PWRS.

[47] O. Alsac and B. Stott, 'Optimal loadflow with steady state security', *IEEE Transactions on Power Apparatus and Systems*, Vol.PAS-93, 1974, pp.745-751.

[48] J.M. Zurada, Eds. *Introduction to Artificial Neural Systems*, Info Access and Distribution Pte Ltd., Singapore, 1992, pp.1-3.

[49] L.L. Lai, *Intelligent System Applications in Power Engineering - Evolutionary Programming and Neural Networks*, John Wiley & Sons, Chichester, 1998.

[50] T.T. Nguyen, 'Neural network optimal-power-flow', *Proceedings of the Fourth International Conference on Advances in Power System Control, Operation & Management*, IEE, Pub No 450, November, 1997, pp.266-271.

[51] T.T. Nguyen, 'Neural network load-flow', *IEE Proceedings – Generation, Transmission and Distribution*, Vol.142, No.12, January 1995, pp.51-58.

[52] W.L. Chan and A.T.P. So, 'Development of a new artificial neural network in complex space', *Proceedings of 2nd Biennial Australian Engineering Mathematics Conference*, Sydney, July 1996, pp.225-230.

[53] J.A.K. Suykens, J.P.L. Vandewalle and B.L.R. De Moor, *Artificial Neural Networks for Modelling and Control of Non-linear Systems*, Kluwer Academic Publishers, Boston, 1996.

[54] Virtual reality: personal, mobile and practical applications, *IEE Colloquium*, Digest No. 1998/454, October 1998.

[55] M.T. Bolas, 'Human factors in the design of an immersive system', *IEEE Computer Graphics and Applications*, Vol.14, 1994, pp.55-59.

[56] S. Kalawsky, Exploiting Virtual Reality Techniques in Education and Training: Technological Issues, SIMA Report Series, 1996.

[57] S.G. Burnay, T.L. Williams and C.H. Jones, Eds, *Application of Thermal Imaging*, Adam Hilger, 1988.

[58] A.T.P. So, F.H.Y. Chan and A.W.C. Kung, 'A real time system for the diagnosis of thyroid diseases using computerized thermography', *Biomedical Thermology*, Vol.14, No.4, 1994, pp.27-35.

[59] F.H.Y. Chan and A.T.P. So, 'Application of thermography in advanced consumer electronics', *Proceedings of the International Symposium on Consumer Electronics*, Beijing, CIE & IEE, October 1992, pp.337-340.

[60] Niancang Hou, 'The infrared thermography diagnostic technique of high-voltage electrical equipments with internal faults', *Proceedings of POWERCON 1998*, IEEE, 1998, pp.110-115.

[61] W.L. Chan, A.T.P. So and L.L. Lai, 'Three-dimensional thermal imaging for power equipment monitoring', *IEE Proceedings – Generation, Transmission, and Distribution*, Vol.147, No.6, November 2000, pp.355-360.

[62] I. Sobel, 'On calibrating computer controlled cameras for perceiving 3D scenes', *Artificial Intelligence*, Vol.5, 1974, pp.185-198.

[63] D.B. Gennery, 'Stereo-camera calibration', *Proceedings of Image Understanding Workshop*, 1979, pp.101-108.

[64] R.K. Lenz, and R.Y. Tsai, 'Techniques for calibration of the scale factor and image center for high accuracy 3D machine metrology', *IEEE Transactions on Pattern Analysis and Machine Intelligence*, Vol.10, No.5, 1988, pp.713-720.

[65] O.D. Faugers and G. Toscani, 'The calibration problem for stereo', *Proc. of CVPR'86*, Miami, 1986, pp.15-20.

[66] R.M. Taylor, W. Robinett, V.L. Chi, F.P. Brooks, W.V. Wright, R.S. Williams and E.J. Snyder, 'The nanomanipulator: a virtual reality interface for a scanning tunnelling microscope', *Computer Graphics*, Vol.27, 1993, pp.127-134.

[67] G.M. Herb and C.A. Shaffer, 'A real-time robot arm collision avoidance system', *IEEE Transactions Robotics and Automation*, Vol.8, No.2, 1992, pp.149-160.

12

Application of the Internet to Power System Monitoring and Trading

Dr Harald Braun
City University, London
UK

Dr Loi Lei Lai
City University, London
UK

12.1 Introduction

Utility companies presented themselves on the Internet but many are hesitant to move their businesses towards the Internet as quickly as possible. Determining the amount of technology available for Internet applications is difficult, but delaying the move of business opportunities towards the Internet will reduce a company's competitive advantage. The utility industry has always been waiting for technology products to mature so that they can be purchased easily. Waiting for the Internet technologies to settle down could take a long time and can only result in loss of revenue and market share.

12.2 The Internet

12.2.1 What Is the Internet?

There are many interpretations of what the Internet is or what it represents. Loosely speaking, the Internet is a connection between computers all over the world, allowing users to access shared documents and exchange messages. It represents the connectivity between different types of computers and operating platforms. One of the major uses of the Internet is the distribution of static information [1]. Static information can be read by an Internet user but cannot respond to user requirements or allow the user to interact with the information provided. The information available on the Internet can be split into two categories: commercial and non-commercial. Examples of commercial use are publications of company profiles, financial data, product advertisements and information. Non-commercial examples are publication of papers, references, on-line tutorials and information about events. The Internet is not only able to distribute static information: non-static information can also be distributed in the form of active Web pages, which change depending on information requested, or in pages such as search engines, where the page content is changed in response to queries from the Internet user. Other non-static Web pages are pages in which changing data is constantly received. Such pages can contain on-line music, radio stations, video or real-time data updates.

12.2.2   How Does the Internet Work?

The Internet allows computers to talk to each other via a cable or wireless connection. In order to allow computers running different operating systems to communicate, a common language, or transmission protocol, is required. The most common transmission protocol on the Internet is TCP/IP. The use of a protocol ensures that a user can access the information on the Internet regardless of the computer, operating system and software being used. For the information on the Internet to be universally accessible in a meaningful way, it must be provided in a format that can be displayed successfully on any computer. For static pages, HTML was developed to allow data to be received in a formatted and presentable layout [2,3]. HTML documents are plain text documents containing tags that allow software to display the text in a formatted layout. For active pages, programming languages such as JavaScript or Java allow software to be included in a Web page for added interactivity. The software products used to display Web documents are called browsers because they assist the user in browsing or surfing the Internet. At present, the most common Internet browsers are Internet Explorer by Microsoft and Netscape Navigator by Netscape.

12.2.3   What Would Happen Without the Internet?

Without the Internet most computers would be standalone or connected to a smaller, local network. These computers would only be able to access information that is available within the local area network (LAN). This information would have to be created by the users themselves or transferred from a physical medium such as a CD-ROM to a computer within the LAN.

It would not be possible to access the latest news or obtain up-to-date information using a computer. Whenever a software component on a computer required an update, a physical medium containing an updated software version would need to be available at its location. The need for a physical medium to be supplied to the location of the computer would result in delays and additional costs when compared with providing information and software over the Internet.

The Internet provides multiple types of electronic information, such as newspapers, manuals, magazines, tutorials, handbooks, frequently asked questions (FAQs), expert help on computing or programming problems and many more which would otherwise not be available.

12.2.4 How Can the Power Industry Benefit from the Internet?

The IT industry has been utilised by the power industry for streamlining plant management and pricing productivity. The best power plants are not the plants with the most IT tools on their computers. The best power plants will be the ones which are using the right IT tools and using them appropriately.

There are many benefits to the power industry by accessing the largest resource of IT tools, the Internet and some of them are listed below:

- Deregulation of energy market
  - General information on power privatisation available for customers
  - Presentation of private energy supply companies
  - On-line price comparison for energy customers
  - On-line sign-up to electricity supplier
  - A revolutionised supply chain by reducing supply chain costs
  - Establishment of remote e-partnerships
  - Improved customer relationship management
  - Broadcast load control for managing peak demand energy pricing
- Power station monitoring
  - Remote power systems component monitoring and component control
  - Real-time expert advice for problems which have been experienced on other sites
  - On-line consultancy (e-knowledge) improving knowledge management.
  - Real-time distribution automation for continuous energy supply monitoring to optimise power system operation, e.g. in case of point failure
- Digital marketplace in the energy sector
  - On-line energy trading floors or NetMarkets for electricity suppliers
  - Real-time energy price auctions and negotiations between energy producer, distributor and supplier
  - Ability of governmental regulators to monitor energy companies on-line
  - Purchase of machinery or spare parts from a wider range of suppliers
  - Online trading of raw materials such as oil, coal or gas
  - Independent on-line marketplaces control inventories
  - Quick sales of energy surplus in the marketplace
- Power industry services
  - Using the power grid for telecommunications
  - Offering Internet service provider (ISP) services
  - Advertising outsource services
  - Selling of commodities and equipment
  - Offering specialist consultancy and training
  - Publishing power benchmarking information
  - Advertising electricity prices for end users

12.2.5 How Can I Find the Information I Need?

In order to find the information required within millions of Web sites a search engine can be used. Most ISPs provides the Internet user with a simple search facility to search by category or keyword. The number of Internet search engines is constantly growing. Companies which are providing a free Internet search facility are generally using advertising as a source of income. Search engines are constantly combing through the vast amount of accessible Web pages trying to index the information they contain. This indexing job is done by a part of the search engine called a *Web crawler*. If specific keywords are used for searching through the accessible Web sites, these keywords are matched against the index and lists of pages containing the keywords are displayed.

12.3 Usability of the Internet

Generally, the Internet can be used to gather or publish information on all topics, which can be captured in electronic format. But finding the right information with an acceptable quality describes the problem of the usability of the Internet. One of the major problems with connecting to the Internet is slow technology. If Internet users have slow connections, old computers or old software, the usability of the Internet is not high. Keeping technology updated requires investment in hardware and software upgrades. Commercial usage of the Internet can only be effective if such investments are met. But there are more parameters which affect the usability of the Internet and which cannot be influenced by investment on the client side parameters like: which search engine to use, what keywords will give the best search results and which Web pages contain the information required? Furthermore, broken links, which are connections between Web sites where the target site has been removed, fragmentation and repetitive or duplicated contents will reduce the usability.

Increasing Internet usability is the ultimate objective for many commercial users. Therefore, keeping a cooperative database of practical keywords, places of interests and bookmarks on an internal Web page will increase productivity and reduce Internet surfing.

12.3.1 Scientific Use for Researchers

Originally, in addition to the US military effort, universities created the Internet to share information on research programmes. In other words, the Internet itself has been a research programme between universities in the USA. With the Internet in place, research projects can be continued where other research projects have stopped. This is particularly true for open governmental and university projects, avoiding duplication of research effort. Most private research projects are executed behind closed doors for economical and competitive reasons although there are exceptions.

Research software projects which are sponsored by universities or the public sector for product development on the frontier of technology are often *open source* and accessible to others. Such projects often benefit from the input of hundreds of contributing programmers from all over the world. One example of such a collective effort is the Linux operating system. It has been developed by an uncountable number of contributors and matured into a very stable and reliable system. Most importantly, its source code is freely available on the Internet.

The Internet is the ideal medium for publishing information without having to pay high rates to commercial publishers. Everybody with Internet access and Web space can publish research results or join newsgroups to exchange research information. Almost any imaginable topic is available within the never-ending lists of newsgroups. Newsgroups allow researchers to publish and discuss their results with an interested and generally competent audience. Whenever research problems accrue which might seem to stop the project, help can be found in the Internet newsgroups. Scientists can use the Internet to query and collaborate with colleagues and access or share software and database information made available on remote machines across the Internet.

12.3.2 Educational Use

The Internet can be used to access information on schools, universities, scholarship, fellowships, and others. It is able to improve interactions between institutions by sharing information about events, projects, timetables, resources and activities, which may improve the usability of resources, such as sharing transportation or avoiding overcrowding in the local swimming pool. The Internet can even help reduce the running costs of schools, e.g. by allowing equipment to be bought in bulk and shared by several institutions.

Teaching materials can be made available to students on-line, which saves on material costs, cannot be lost or left at home and allow students to get prepared. Furthermore, they allow potential students to gather more detailed information about a course prior to signing up for it.

There are several on-line training courses available on the Internet. They allow people who live in remote locations to continue their education after leaving school. With the help of an on-line tutor, which monitors the progress of students remotely, queries can be sent and answered via e-mail within minutes. On-line examinations and on-line multiple-choice questions generally follow such studies, including the publishing of examination results.

12.3.3 Internet Products

Prior to making a decision on which product to purchase, extensive information such as fact sheets and general opinions can be analysed. The Internet enables users to compare product or component performance by being able to access directly information on competing companies. Good starting points for obtaining lists of competing companies are specialised on-line magazines, virtual exhibitions or virtual shopping centres.

12.3.4 Business Competition

Businesses can compete with on-line quotes for services and goods to attract possible customers. They can show detailed statistics on their business performance to attract potential investors and shareholders. Businesses can publish information and communicate via a secure Internet connection and firewalls to improve communication between remote departments.

12.3.5 Multimedia Access

Multimedia means the simultaneous use of more than one medium. A single medium can be text, image, video and sound. Multimedia devices are able to play music, animated images, motion pictures and videos. But multimedia technology is not just about playing multiple media, it also includes storing, transmitting and presenting information from multiple sources. Uses for such technology include entertainment, video conferencing, video on demand (VOD), close circuit television (CCTV) and distance learning.

There are many different formats in which multimedia contents can be stored. The most common ones found on the Internet are listed in Table 12.1.

Table 12.1 Common multimedia types

Category	Extension (MIME type)
Audio, Sound, Music	AIF AVI M3U MID MP3 SND WAV
Movie, Video	AVI DV DVD M1V MOV MP2 MPE MPEG MPG RA
Images, Photos	BMP GIF JPEG JPG TIF PCX WMF

This list summarises only a fraction of available file types. There are many more file types for images, sound or movies and new ones emerge constantly.

If the Internet browser receives a multimedia file, it identifies its contents by the MIME type, which is related to the file extension. Once the content is identified, the browser executes software to use the file as intended by the originator. In cases where the browser does not include software for opening a file type, e.g. MOV (Windows Movie), the browser invokes a helper application, e.g. a movie player. For an unknown or new file type a browser plug-in or an external helper application may be required.

With the power and ability of the Internet to distribute multimedia content, copyright problems with pirated music and video occur. Without copy protection it is very easy to convert music or video tracks into an Internet distributable format. But there are obvious advantages for selling music electronically over the Internet, e.g. no record company, no intermediaries, low cost, large audience and many more.

12.3.6 On-line Services

Internet on-line services such as Internet banking and account managing, shopping in virtual shopping malls, live news and trading floors, just to name a few, have become very popular. On-line shopping has become very popular for light goods (small postage cost) such as music, videos and books. Several large supermarket chains are trying to push Internet shopping for food and groceries by introducing a fixed delivery fee and guaranteed delivery times. Such business is not time critical and can be accomplished without a continuous connection to the Internet.

Time-critical on-line services, such as trading floors, require a continuous connection to the Internet in order to follow and react to market changes. They can only be successful if the used IT infrastructure can handle real-time data transmissions and if contingency plans are in place in case of technical failure.

Trading floors and on-line auctions are a very promising development on the Internet. They allow multiple sources and end users to meet in a common virtual place, conducting business without verbal communication or travelling.

12.3.7   Support for Professionals

Referencing information plays an important part in many professions, particularly the legal, medical, scientific, financial and information technology professions. Before the Internet was established, these professions relied on an extensive amount of published papers, e.g. books, journals and reports. The production and distribution of paper reports can be expensive and slow, resulting in information being unavailable when it is required or only available to those who can afford it. Since the introduction of the Internet, such information is readily available to everybody. People who are working in a fast moving environment such as the IT sector require frequent updates. The Internet provides a medium in which updates can be made available to everybody quickly and cheaply; therefore, the latest technology and manuals can only be found on the Internet.

The Internet user can be his/her own doctor or lawyer. But there is a danger. Using the information without the necessary experience can sometimes lead to wrong conclusions. This is especially true for self-analysis of illnesses. Some information on the Internet should only be used for the purpose of improving information for a specific group of professionals. It might be useful to know which illness matches the symptoms but a professional should compile the final conclusion.

Internet-based business analysis solutions for the utility marketplace provide utilities and energy companies with large-scale sophisticated analyses of their data and allow them to extend access to these analyses to a larger number of users. As a result, users will have point-and-click access to energy and power plant data to support their energy trading and power plant management and for improved business decision-making capabilities.

Internet forums have been successfully used as an information source for the power utility industry with regards to IT-related questions, such as the year 2000 (Y2K) issues.

There is a vast amount of power utility related data already available on the Internet but some of it is poorly organised and difficult to find. Energy information companies address this shortfall because they specialise in the collection of information related to power utility research.

Internet companies have created virtual conference room, where visitors can review case studies, survey results, white papers, reports and studies, meet with staff consultants, and participate in an on-line survey.

12.3.8   The Power Industry and the Internet

Can the Internet really stand up to its promise to increase productivity and profitability and lower the running cost for the power industry?

At present, there is not much evidence to support this statement, but with continuous reduction of Internet connection ownership, this will change in the near future. The aim is to increase productivity by using the Internet to find the right information easily and quickly and to attract potential clients to commercial Web pages to increase profitability.

Efficient Web page design should start from the need of the clients (the community) to find the information or products they require (intermediate content) and move to the actual Web page functionality (e-commerce tool). Making money via e-commerce requires the client to have confidence to conduct real business on-line. Increasing such customer confidence, removing the psychological barrier to purchase goods on-line, is one of the highest priorities.

An Internet site must provide more than just industry news. It must offer a complete package of low- to high-value services tailored specifically to the power industry. It should go beyond the stage of offering information to the average Internet *surfer* by helping professionals to satisfy a need, or to solve a problem. For example, lowering emissions by optimising the combustion process or reducing shipment delivery time by finding spare shipload capacity using other Web sites.

Increased traffic to Web pages will increase its popularity and therefore its value. To increase the rate of recurrent visitors, the Web page should contain added context and perspective. Power industry Web sites should provide easy access to electrical industry resources for manufacturers, distributors, contractors, engineers, purchasing and facilities managers, consultants, and other electrical industry professionals.

Energy applications across the Internet can support the core competitive market functions. An energy-trading platform can be designed to receive bids/offers for energy from participating organisations and use these bids/offers to schedule and control energy resources locally. Additional functions provided to the market include submission of meter data, settlement, billing and publishing of pricing and trading information through an information portal. Such a platform can be designed to give market participants the ability to access the market 24 hours a day, seven days a week.

Utility companies are developing electronic billing and payment services over the Internet. Internet billing will reduce the utility company's overheads by creating such an on-line service for residential and commercial customers. Publishing common questions and their answers on the Internet can reduce call centre capacity. Joint ventures between utility companies and large computer manufacturing and software companies can result in an ISP network by utilising existing power cables. This will allow power companies to offer a connector service by providing continuous Internet connections using existing wiring.

One of the most promising Internet technology advances for specialised industries are application service providers (ASPs). The idea is to provide software functionality across the Internet, where the software is installed on a remote server which is accessible from any computer with Internet access. This avoids the problem of having an expensive software product installed and maintained or upgraded on fixed location computers. Furthermore, renting access to expensive and highly specialised software on a time share basis only on demand reduces the standby time and cost of ownership extensively. Software packages, for example, for training purposes can be leased just for the training day and for the number of people attending. This is one example just where the Internet can really reduce costs.

12.3.9 Recent Improvements on the Internet

Since the computer manufacturing industry and the Internet are directly related, faster and cheaper computers will constantly cause expansion of the Internet network. This, in itself, is a positive development as long as the transportation network is at least expanded at the same rate. Therefore, a constant improvement the Internet infrastructure is necessary to ensure a continuous quality of service.

With more and more users using and publishing information on the Internet, more and more information becomes available. It is certainly not the case that every Web site on the Internet on a specific topic contains valuable information. Sometimes information is duplicated or even wrong. This makes it sometimes difficult to find *quality content* for serious researchers without wasting time visiting different Web sites containing effectively similar information. Consequently, the more information there is on the Internet the more diluted the quality of content on a common topic and the more difficult it becomes to find quality content. But the Internet is constantly improving its search engines, which are now using information-refining processes with artificial intelligence (AI).

The processing power of PCs has been doubling almost every two to three years and with the new generation of multimedia extended processors (MMX), Web pages now include images, sound or video clips. This development has increased the appeal of Internet *surfing*, sometimes caused just by the graphical design and functionality (interactivity) and not the content of a Web site. People visiting interactive Web sites can interact with their contents for fun or for business purposes. The Internet has dramatically improved the appearance and interactivity of Web sites and has created a new job title, the Internet Graphical Designer.

Security is the most important issue surrounding the Internet. A system which is not secure cannot be used for serious business. When the Internet was established, security was not built in, since its inventors did not intend to use it for business. Because of its cheap advertising capabilities, businesses started to emerge on the Internet and subsequently the need for security arose. There are several reasons why *hackers* can gain access to remote computers. Some of the most common ones are client/server configuration errors or bugs. A simple security-related bug in browsers can allow hackers to gain access to private information. With the latest 128 bit encrypted Secure Socket Layers (SSL) http connection protocol, hackers will have at least to spend a serious amount of time in gathering information, if it is within their ability.

Accessing the Internet while on the move is one of the major improvements of the IT industry for the future. Portable hand-held devices such as WAP phones, palmtops or even wristwatches are currently available for accessing the Internet.

12.4 Internet Technology

Currently, the PC is still the most common way to access the Internet. But within the next few years, Internet access will be dramatically increased through mobile phones, palmtops, TV set-top boxes, game consoles or video telephones. Such devices will have a built-in Internet browser enabling Internet access at any location.

WAP or *Internet-enabled* telephones are currently being pushed as a PC alternative. So far, there is still a lot of convincing and improvements to be done until WAP phones will take a serious market share. They suffer from bandwidth restrictions and small display size. Palmtops in contrast look quite promising, since their display is reasonably sized, but they too suffer from bandwidth restrictions. Currently, WAP phones and palmtops are suitable for plain text display, which is comparable to mobile phone text messaging.

One of the most promising technologies for accessing the Internet in the future are TV sets. They can be used as Internet display units, if connected to the Internet via a telephone connection Set-top boxes will sit between the TV and the telephone connection and will deliver services such as digital TV or video on demand.

Since the great success of game consoles in the late 1990s, game consoles are another alternative for accessing the Internet. They need to be equipped with a similar technology as set top boxes and will require a telephone connection. Already some game console manufacturers allow Internet connection from game consoles for playing against other contestants but do not offer full Internet browsing.

Video telephones already have a display, are connected to a telephone line and subsequently are Internet ready. They also have QWERTY keyboards and the ability to send and receive e-mails. So far the costs of ownership and maybe the awkward location of telephones in some households (e.g. in fridge doors) have dampened their use to some respect.

12.4.1 Access to the Internet

Access to the Internet for the general public or small/medium-sized companies (the first tier) is generally given through a telephone connection to an ISP. The ISP (second tier) channels the users' data queries, e.g. a Web address, to the Internet (third tier) via a fast fiber-optic connection, sharing the expense of a fast connection with all its users. This positions the ISP between the user and the Internet, making it the intermediary or middle tier in this three-tier connection (Figure 12.1).

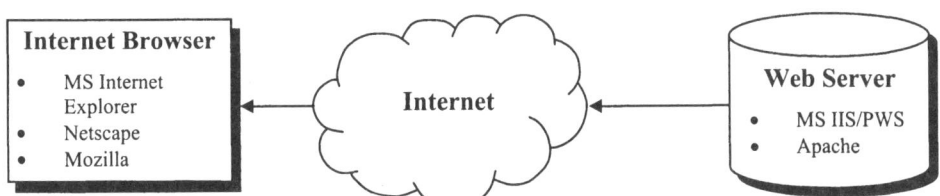

Figure 12.1 The Internet as a three-tier connection

If the Internet user wants to access information about a specific subject, the user needs to know to which Internet location to connect. Each Internet location is uniquely identifiable by a uniform resource locator (URL) which points to a single Web page or Web directory. The URL is dependent on the type of protocol required to access the information. The most common one is hyper text transfer protocol (HTTP), which results in the familiar-looking URL: http://www.wiley.com.

12.4.2 Operating Platforms on the Internet

The Internet is building on a client-server relationship model, where the client's Internet browser connects to an Internet Web server. For the browser to function properly, an operating system or platform needs to be installed on the client's PC. The same applies to Web servers. Operating platforms supply the basic structure of the computer environment such as convenient access to all peripheral devices installed on the computer. While it is possible to create an Internet browser for a specific computer hardware layout, the multitude of computer hardware combinations would require a browser for every possible option. Therefore computer software is generally written for operating systems. The most common operating systems for client machines are Microsoft's Windows, Linux and Apple Macintosh's MacOs. The more applications these are available for any platform, the more popular this platform becomes. Therefore, most home or office-based computers will have one of the previously mentioned operating systems.

Web servers have different criteria for choosing the platform on which they reside. While broad application support is necessary for home and office computers, security and reliability are the major criteria for Web servers. Operating platforms such as UNIX, Sun Solaris, Windows NT and Hewlett Packard are focused on secure access restrictions and secure memory management. Access restrictions incorporate multi-user capabilities permitting different levels of access, e.g. Web users can only access directories dedicated to Web publishing. Secure memory management incorporates multi-tasking capabilities permitting different levels of memory access, e.g. every operating-system-dependent program is functioning in its own address space and will not conflict with other programs or the operating system in case it crashes.

The really hot issue for Web servers is security. While Web servers allow access to Web pages for everybody, access restrictions apply to all other areas on the server's hard disk.

12.4.3 Web Clients

A Web client is a piece of software that is able to receive information for display or storage purposes. Web clients are used to access information published on an Internet-enabled Web server via a URL. Web clients do not exist in isolation since they have to access a server for information retrieval. They are part of the client-server model, which is shown in Figure 12.2. The Internet utilised the client–server model because of its globally distributed nature.

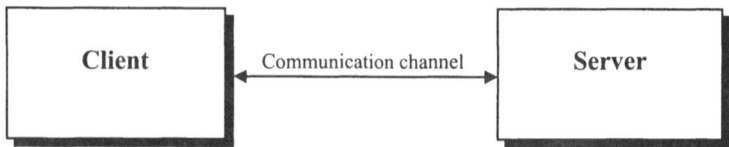

Figure 12.2 The client-server model

The most common Web clients used for displaying information on a computer screen are Web browsers such as Internet Explorer or Netscape Navigator. Web clients for information storage can be found in Internet search engines. A Web client must be able to understand the format of the remote information accessed for successful processing. If, for example, a Web site containing Chinese writing is accessed, the browser must have the required fonts installed. If real-time data should be displayed in a Web client, the client has to have the capability to receive data updates and display changes accordingly. The multitude of data types and constantly emerging new technologies and standards forces companies building such clients to release frequent updates. Users of Web clients should always try to update client software in order to access new Internet technologies.

12.4.4 Web Servers

Accessible URLs must be located on a dedicated Web server. Only Web servers which are enabled for Internet access are accessible by the Internet user. Basically, a Web server is a computer with Web server software such as Apache Web Server, Internet Information Server (IIS), Personal Web Server (PWS) or any other Web server software. This software allows other computers to connect to a specific port (normally port 80) and display the contents via a Web browser.

12.4.5 Web Protocols

Web servers are able to understand several protocols. A protocol is a method computers use to communicate with each other. There are several types of protocols. The different types of protocols are required for different tasks, e.g. Web page access or file transfer.

The most common protocol used over the Internet is a combined protocol called TCP/IP. The TCP part is responsible for the communication and the IP part is required for identification of computers. In order to address uniquely any Web server on the Internet, a unique token is required. This has been realised with telephone numbers in mind. Therefore, a Web server can be addressed by a set of numbers, the IP number. It can be used in the browser as a hexadecimal, octal or decimal number. Its most common appearance is decimal and it looks like this: 123.456.789.012.

Since such numbers are difficult to remember, a more friendly way has been developed, called a domain name. The domain name allows the use of friendly names such as http://www.wiley.com instead of 199.171.201.14. Good domain names are limited and most of them have already been occupied. Some of them are available on the market for bidding, which is very similar to personalised car number plates. Recent court rulings have tried to discourage *domain name hogging* by forcing individuals to release branded and trademarked company domain names so that the companies can represent themselves on the Internet without paying millions of dollars.

12.4.6 E-mail

The Internet owes parts of its popularity to the e-mail system. E-mail is an electronic means of sending a message from one computer to another in an organised fashion. E-mail services are offered by an ISP. Mail accounts can be created from ISP e-mail providers such as CompuServe or AOL. E-mail is the fastest and cheapest way of sending messages to any location in the world. There are specific protocols for sending and receiving e-mail messages. The protocol used to send e-mail messages across the Internet is the Simple Mail Transfer Protocol (SMTP). The protocol used to receive e-mail messages is the Post Office Protocol (POP). Newer versions of these protocols have been improved in robustness and safety and are called POP2 or POP3.

If e-mail contains more than just text, e.g. attachments, another protocol is required. This protocol allows downloading or uploading of files on remote machines and is called File Transfer Protocol (FTP). It is automatically invoked if an attachment is copied to a hard disk. If, for instance, the graphics adapter driver software requires updating, it is more than likely that it is available on the Internet. Generally, there will be more than one location, called FTP site, for downloading.

The most used protocol for Web browsing is the HTTP. This protocol carries information about the originator of the information and the information itself. It is able to tell the browser of which type (e.g. plain text or compressed) and format (e.g. HTML, JSP, ASP) the information is so that the browser can display it correctly. Free Internet-based e-mail services are available, e. g. from HotMail or BT.

12.4.7 Internet Security

Internet security is necessary to protect computer resources against the risks and threats that arise as a result of a connection to the Internet.

The design of the Internet originated from the idea of connecting computers between universities etc. for communication and knowledge-sharing purposes. There was no reason for anybody to consider sabotaging the connections, since only a selection of trustworthy people had physical access to the computers connected to the Internet, sharing sensitive research information. Therefore, security issues were not been part of the initial Internet design. Since more and more users have access to the Internet and its utilisation for business and financial transactions has grown, Internet security has become a primarily concern. The reasons for exploiting or sabotaging the Internet are manifold.

One of the major security concerns is caused by the fact that data is transported as plain text, allowing easy access for third parties. This risk is mostly acceptable for non-sensitive information such as data intended for the public domain. Since the Internet is a public place, sensitive information should not be transported across these lines if avoidable. Transporting information across a private network is the safest option for sensitive data but also the dearest. Most companies are not in a position to afford a continuous cable connection between all their offices, which in fact is very uneconomical. In addition, companies have to support their workers with Internet access and allow travelling and home workers to have access to sensitive company data from any location. These requirements have persuaded many companies to open up their private Intranet to connect to the public Internet.

Protecting a private network and shielding it from hackers without restricting communication to remote users can be achieved with a firewall. A firewall is software written to combat unauthorised access to files or underlying operating systems. Depending on the company policies, only selected services are granted access to the outside world. Figure 12.3 shows how an Intranet can be protected with a firewall. Local computers are able to connect to each other and to the Internet, but remote computers with Internet access can only access information behind the firewall which has been published on port 80 (the default port used by a Web server). Every attempt to connect to any other computer behind the firewall is prevented and logged. It is the responsibility of the IT personnel to analyse such logged files and detect unauthorised connection attempts.

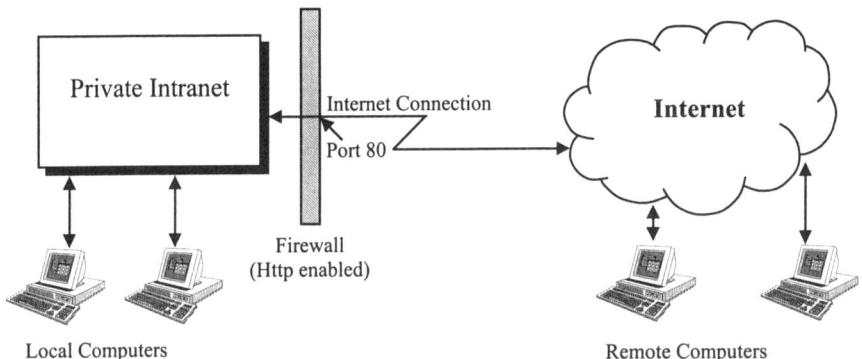

Figure 12.3 Intranet protection with a firewall

Protecting an organisation's sensitive information from unauthorised access without restricting accessibility to authorised users requires a system of identification. Such access restriction can be accomplished with password-protected logins. Once a user has logged on to the protected area, the data can be accessed. Passwords should be easy to remember but difficult to guess. These include words which can be found in dictionaries, songs or movie titles or slang. A safe password should consist of at least eight characters with mixed upper and lower case letters and numbers, e.g. 'My2001HolsInSpaiN'.

E-mail has become one of the most important communication channels in a modern office environment. Sending e-mail is very easy and therefore very easy to abuse. If, for instance, somebody is using a false name pretending to be somebody else with instructions for selling shares, say how can this be prevented? The answer is to use a digital signature. A digital signature can be purchased from reputable sources enabling e-mail users to protect their e-mail from forgery and encryption of the contents.

Digital signatures and encryption should strongly be considered when it comes to power station monitoring and controlling. Such systems should be designed with security in mind from the first day. Designing a Web application without security and bolting security onto the finished product can lead to incomplete or insufficient safety. A safe way of communicating between a Web client and a Web server is via the secure sockets layer (SSL). This is a protocol developed by Netscape Communications Corporation to provide security and privacy for transmissions over the Internet by using both data encryption and authentication of the server and client. SSL uses an encryption key, which can be up to 128 bits long.

There are basically two ways of encrypting information by using an access key. The first method involves a pair of private keys (Figure 12.4) and the second method a public and private key pair (Figure 12.5). In the case of data encryption with a private key, data can only be decrypted if the same private key is known. The major disadvantage is that the recipient needs to receive the private key, which can be intercepted. Private key generators produce only one key (A) for encryption and decryption of data.

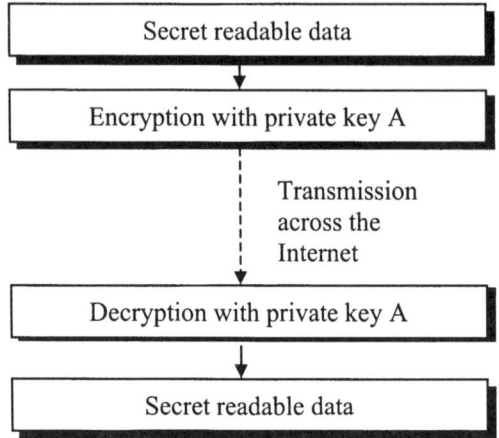

Figure 12.4 Data encryption using a private key

The alternative is to use a private/public key pair. In this case, a message can be encrypted with the public key, but only decrypted with the private key. This allows publishing the public key to many people, who are able to send messages back to the publisher of the public key. Since the publisher of the public key is the only person who has access to the private key matching the public key, the messages can be decrypted. Once a message is encrypted with a public key, only the private key can decrypt the message back into a readable format. Public key generators always produce a public and private key pair (key A and B), where A is used for encryption and B for decryption.

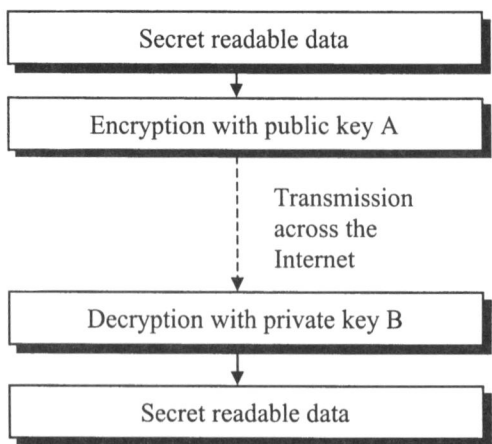

Figure 12.5 Data encryption using a public and private key pair

12.4.8 Internet Bandwidth

Internet bandwidth is the amount of information that can be transmitted via a given physical transmission line in a given period of time; it is usually measured in bits per second (bps or baud). In other words, the larger the bandwidth available, the faster information can be sent.

Bandwidth is the most important attribute of the Internet since without it nothing can be transmitted. On a network, bandwidth is shared between all network users, regardless if it is the Internet (wide area network, WAN) or an Intranet (LAN). The more users use a network the less bandwidth there is available for each user; the less the bandwidth available for a single user, the lower the data transfer rate and the longer the time for receiving information. Users transferring large files across a network connection are sharing bandwidth resources with other users. Resources are shared only with active users which cause network traffic and are released until completion of any transfer operation. Subsequently if data transfers are slow, the release of bandwidth will accelerate the transactions of other users. Figure 12.6 shows how multiple users share bandwidth.

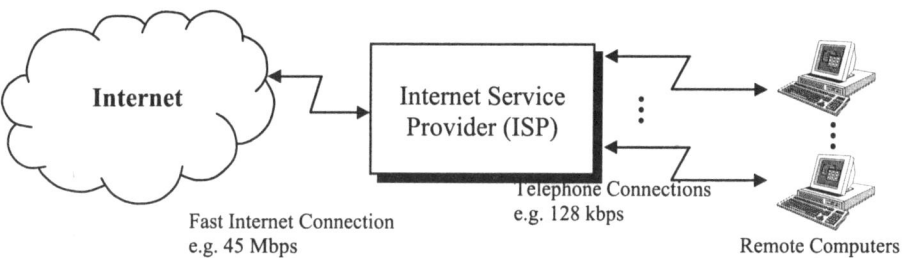

Figure 12.6 Bandwidth shared by multiple users

Connecting to a Web site on the Internet requires time to download all the components of the Web site. If the Web site contains text only, the download time for a constant bandwidth is shorter compared with a Web site containing text and images. Connecting to the Internet with a modem with a low bandwidth, e.g. 14.4 kbps/baud compared to 128 kbps/baud, will result in longer download times for the same page.

Bandwidth must be the primary concern if real-time data is transmitted across the Internet. If the Internet is congested and data transmission is delayed contingency plans must be in place. Critical power station monitoring or control over the Internet is relatively safe as long as sufficient contingency plans are in place in case the ISP or telephone connections are not functioning. Considerations to install wired or wireless dial-up modems are highly recommended. If real-time data is transmitted across the Internet it is generally sent in small packets. In order to reduce the size of the packets, they can be compressed with a range of compression algorithms freely available on the Internet. But the best solution is to keep transmitted data to a minimum.

Web page designers have to ensure that their pages are relatively small, not exceeding 50 kB; otherwise page loading will take too long and hit rates will fall dramatically. If a Web page contains too much information it should be divided into smaller pages. If a Web page contains images, they should be optimised for Web downloading. Optimising images could result in using less server resources so the page is loaded faster. The right choice of image formats is important if bandwidth is to be saved, e.g. the use of compressed image formats such as Graphics Interchange Format (GIF). The GIF image format can be reduced in size by increasing their image compression ratio. One of the basic rules is the higher the compression, the lesser the quality. The best compression ratio can be found by experiment until the optimal balance of image quality and compression is found. The GIF images format does not allow control of the compression ratio. Its compression algorithm takes the image size and colour depth (number of colours, e.g. 256) into account to optimise the image size. In order to reduce the file size of a GIF image, the image size or colour depth must be reduced. Most images do not use all the unique colours available, therefore wasting a lot of space. Such images can easily be converted into a lower colour depth for file size reduction. Most graphics packages allow image analysis by a colour histogram. A colour histogram determines how many colours are used by the image and can be used to decide the best colour depth for the image.

The *experienced load time* of a Web page can be different to the *real load time* if a Web page is loaded in stages and displayed to the user while loading. Large images on Web pages can be displayed once they have been received completely or line by line until the image is complete. There is a third and more professional-looking option, which is to divide the image into sections, called tiling. Each section (tile) is received as a small piece of the entire image and displayed instantly, as shown in Figure 12.7.

 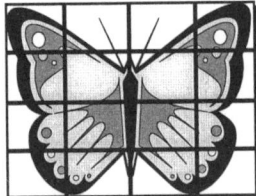

Figure 12.7 Images can be tiled to reduce Web site loading time

Another bandwidth problem is that most Internet traffic goes through the USA. A large volume of Internet traffic passes through the USA before reaching its destination. This is mainly caused by the fact that high-speed lines to the USA are much cheaper than high-speed lines to neighbouring countries. Since the TCP/IP of the Internet is optimising its path efficiency it will not always take the shortest route. If the available bandwidth to a neighbouring state is comparably small to that in the USA, the probability of using the channel to the USA is higher.

There are methods to increase the available bandwidth by diverting network traffic to servers used in peak times. Such methods increase the scalability of network structures and are known as network load balancing. Figure 12.8 shows that in network load balancing, a client request is distributed within a group or cluster of servers if network traffic is increasing, thus redirecting network traffic to other servers for balancing purposes. This technology represents the servers within a cluster as one *virtual* server.

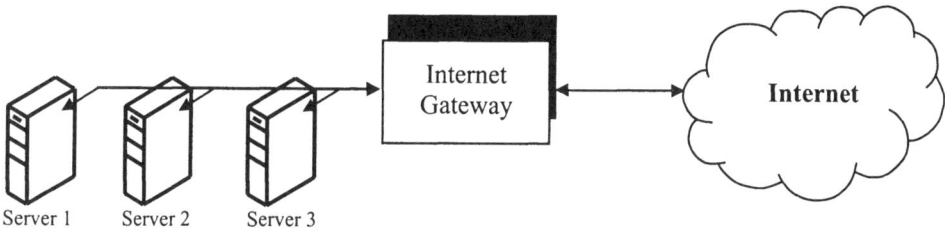

Figure 12.8 Network load balancing with three servers

12.5 Internet Programming Languages

12.5.1 HTML

One of the most popular programming language for the Internet is the Hyper Text Markup Language (HTML) [4]. In reality its more of a textual formatting markup language as opposed to a software development programming language. HTML is good for designing static Web pages with textual, graphical and multimedia contents. Most Internet programmers start with HTML until they reach the limit of HTML and move on to JavaScript or Java. The Internet owes HTML quite a lot since it made it possible for almost everybody to create Web pages before Web site graphic design packages were introduced.

HTML is a plain text document which contains tags understandable by the software used to display the document. An HTML page can be created with a plain text editor, e. g. notepad or simpletext, by writing the page content in between markups or tags. Tags in HTML documents tell the browser how to display the following information. If, for example, tags such as <TITLE>*This is a Title*</TITLE> are within an HTML document, the browser will display the text, encapsulated within the tags, in the browser's title bar. There is an overwhelming number of 'build-in' tags for formatting and specifying a document layout. The list of HTML tags is dependent on the version of the browser used. It is constantly changing and can be found on the Internet.

HTML pages are not bound to contain only textual information. Later versions of Internet browsers understand HTML tags which identify various types of images, music, and video or generally speaking multimedia content.

12.5.2 Interpreted Versus Compiled Languages

There are quite a few programming languages today. Programming languages can be differentiated into interpreted and non-interpreted languages. Interpreted languages are compiled during execution into commands that the processor can understand. Non-interpreted languages are compiled before execution into processor-specific commands. Generally, non-interpreted languages are executed faster than interpreted ones since the interpretation step during execution is not required. The price for higher execution speed is the loss of platform portability. Non-interpreted languages are platform independent as long as a matching interpreter for the platform is available. Especially on the Internet, where different types of computers are connected, fast executable platform-independent software is important. The answer is Java. Java has been developed with the Internet in mind. It is not exactly interpreted or non-interpreted, but somewhere in the middle. This is because the source program code is compiled into byte-code, a half-cooked compilation process. Java byte-code is interpreted by a Java Virtual Machine (JVM) and translated into processor-specific instructions during run-time. This mechanism allows execution of the same Java byte-code on different platforms if an appropriate JVM is installed.

Internet programming languages are enabling Web pages to be dynamic and interactive. Dynamic Web pages are important for on-line and real-time information services. Interactive Web pages are required if feedback from the Web user is relevant.

12.5.3 What Is JavaScript?

JavaScript is a programming language which is executed in the Web browser. It is one of several languages that provide a means of adding dynamic elements to an otherwise static environment. As shown in Figure 12.9, JavaScript can be included in a static HTML page in order to enhance its interactivity. It is a hybrid language that lies somewhere between HTML and the programming language Java. JavaScript shares some of its syntax with Java, allowing Java programmers to program in JavaScript and vice versa. On loading an HTML page from a Web server, the browser interprets the HTML tags for page formatting and executes JavaScript code for page interactivity. If, for example, the page content is a selectable list of products and prices, JavaScript can keep track of the running of all products selected from the list. Such interactivity cannot be accomplished with basic HTML.

JavaScript versatility allows Web pages to be created without HTML formatting. This means that JavaScript does not necessarily need to be embedded in an HTML document. Writing entire Web pages in JavaScript might defeat the purpose of JavaScript, which should enhance HTML pages by adding interactivity instead of replacing it.

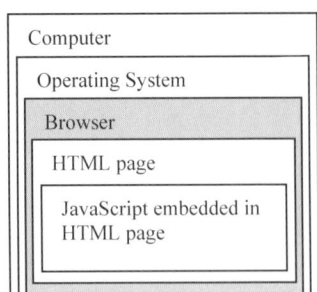

Figure 12.9 Environment for JavaScript applications

Writing code in JavaScript is very similar to writing code for a Java application. The major difference is how events for executing code sections are triggered. JavaScript enables HTML objects, e.g. a button, to trigger code, which might calculate a subtotal of display fields and show the result in a pop-up message window.

But whenever unknown programming code is executed caution must be taken. The Internet browser executes the JavaScript code in an encapsulated environment, preventing access to system resources, e.g. the hard disk. Theoretically it should be just as safe to execute JavaScript as it is to execute applets. But holes have been found in some browsers' Java security implementation, allowing cleverly written JavaScript code to access files with known location and name. Nevertheless, the execution of Java code in the form of JavaScript or an applet within the Internet browser can be turned off when browsing untrusted Web sites.

12.5.4 What Is Java?

Java is a semi-interpreted software development programming language optimised for use on the Internet [5]. As with many programming languages, instructions for the computer are written in plain text, readable to humans. These plain text instructions, called source codes are stored in a text file with the extension '.java'. Java source code can be produced with any text editor which is able to store the source code in plain text format. After the creation of a Java source file, this file can be compiled into Java byte-code via a Java compiler as shown in Figure 12.10.

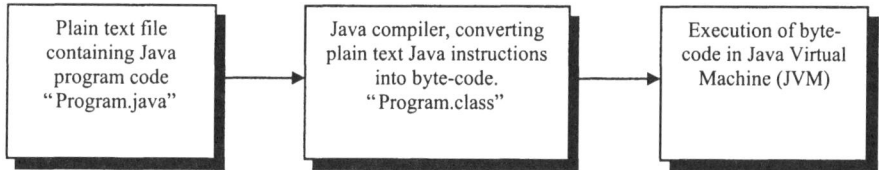

Figure 12.10 Java code compilation and execution

Java compilers can be downloaded from the Internet from various locations. There are numerous commercial Java development platforms on the market, which allow faster development and easier debugging than non-commercial ones. Since Java has been introduced by Sun Microsystems, it is one of the most reliable sources for free Java tutorials, compilers and other Java resources. It can be accessed via the following URL: http://www.java.sun.com.

Java byte-code can be executed on many different computers. Therefore, Java is a real multi-platform programming language. Any computer which has a Java virtual machine (JVC) installed is able to execute java byte-code. This means that software programs need only be designed, written and compiled once. This is a real advantage for its flexibility in terms of distribution on the Internet.

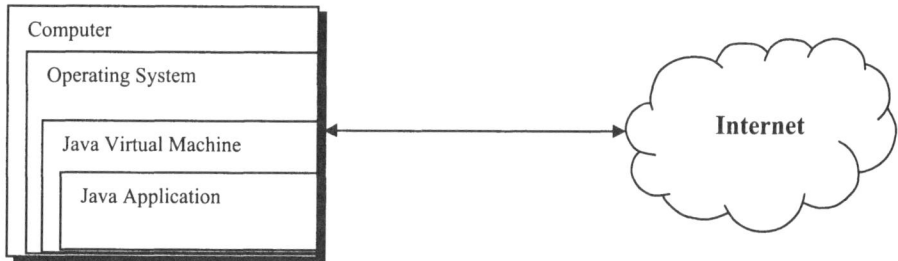

Figure 12.11 Environment for a Java application

Applications written in Java have the advantage of cross-platform portability. But how can this advantage be of use over the Internet? As shown in Figure 12.11, the Internet is linked by a combination of different kinds of computer platforms, most of which have a JMV for the execution of Java byte-code. Java applications can be made accessible for downloading on the Internet by anybody.

Java has, simply speaking, two execution modes. One mode is outside a browser as an application, and the other mode is inside a browser as an applet. The major difference between a Java application and a Java applet is that applications are executed within the boundaries of the operating system (e.g. Windows), while an applet is executed within the boundaries of the Web browser (e.g. Netscape). In order to define the boundaries of execution for a Java application, a dedicated operating system and system administration knowledge are required. Definition of the boundaries for a Java applet executed in a browser is much easier. Firstly, no dedicated operating system is required, since the browser is able to control the applet itself. Secondly, the user of the browser has nothing to do in terms of administration, since all common browsers control applets by default. In most platforms, execution of Java applets is enabled by default, i.e. no intervention after installation of the browser. Browsers execute applets in a 'sandbox'. The term 'sandbox' has been introduced to visualise the boundaries in which an applet can work within a browser. Browsers do not allow an applet to access any directories or data files on the computer's hard disk, thus making it safe to run applets from untrusted sites. The only connection an applet can perform if it is live on the Internet is back to the server from which it was loaded.

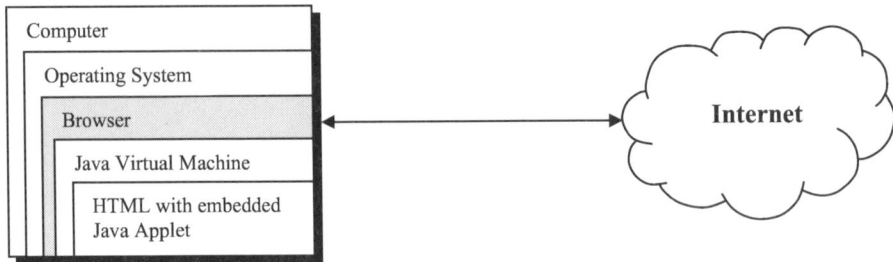

Figure 12.12 Environment for a Java applet

The purpose of Java applets is to add dynamic content to a Web page. If a Web site containing Java applets within an HTML Web page is accessed, the Web page together with the applet is temporarily downloaded to the computer's local hard disk and displayed in the Web browser. Dynamic content can range from a very basic graphical animation to a very comprehensive graphing tool. Applets are related to applications, they can be designed to perform as fully functional desktop applications or even act as an operating platform itself. Figure 12.12 shows the environment for a Java applet.

12.6   Web Pages

Web pages generally speaking are shared documents accessible to everybody on the Internet. The Internet has made it possible to access information on any computer connected to it. Restricting access to confidential information is the sole responsibility of the administrator.

The Internet is also called the World Wide Web, WWW or 3W. It is often described to be a web because Internet documents have hyperlinks, which allow links to other pages. It can be seen as a web if lines representing the connections between pages are visualised. People who are spending time on the Internet, following connections from one page to another, are *browsing* or *surfing* the *Web*.

12.6.1   Setting up a Web Page

Setting up a simple Web page for information distribution purposes requires access to the Internet. Internet access can only be gained via an ISP. ISPs invest into networking technology permitting private users or small businesses access to the Internet via a telephone connection at a competitive rate. There are alternatives, such as Internet access via Internet cafés, libraries, universities, and future plans include post offices and other public sector institutions.

Creating a simple Web page is very easy. There are several word processing packages on the market, which allow the user to store documents not only in the native word processor format but also in HTML.

Once the page(s) are designed, written and saved as HTML, they need to be copied onto the Internet. Placing information on the Internet is very easy. All that is required is some disk space on a Web server called *Web space*. Web space can be bought, if commercially used, or it is free. Locating free Web space can be accomplished by using a search engine, e.g. http://www.google.com, by using a search phrase containing the words *free* and *Web space*.

If dynamic or interactive web pages are required, the pages have to become more complicated as a database content is to be displayed. In order to use a database for information retrieval, Common Gateway Interface (CGI), servlets or Enterprise Java Beans (EJB) might be required. Most Web space providers will allow users to use CGI and a light database, e.g. MySQL, free of charge. But if servlets or EJB in conjunction with a SQLServer or Oracle database is used for communication between the Web browser and the database, commercial Web space is required.

12.6.2   Difference Between a Static and a Dynamic Web Page

Static Web pages are created once and published on the Internet. If the content is likely to change, the Web page requires manual updating. Creation tools for HTML documents range from simple text editors that edit raw text, such as Notepad or EasyText, to What You See Is What You Get (WYSIWYG) editors, such as FrontPage or HotMetal, for editing of sophisticated HTML. Many of these HTML design products incorporate wizards or assistants that help produce a complete Web site with multiple pages. Most word processing packages (such as Word or Lotus) are also capable of generating static Web pages.

In contrast, the Web server creates dynamic Web pages during run-time. Such Web pages are built depending on information available or request by the client. Generally, an HTML Web page is divided into static and dynamic parts, where on access of the Web site, all parts are combined into a single HTML page. The static part can include headings, general information and logos. The dynamic part can be a table where data is queried from a database, formatted and enclosed by HTML tags. Such dynamic creation of HTML pages can be achieved via a CGI or servlets. Therefore it is not unlikely that Web pages are initially being developed as a set of templates with the contents added via dynamic HTML creation. Generation of a dynamic Web page is illustrated in Figure 12.13.

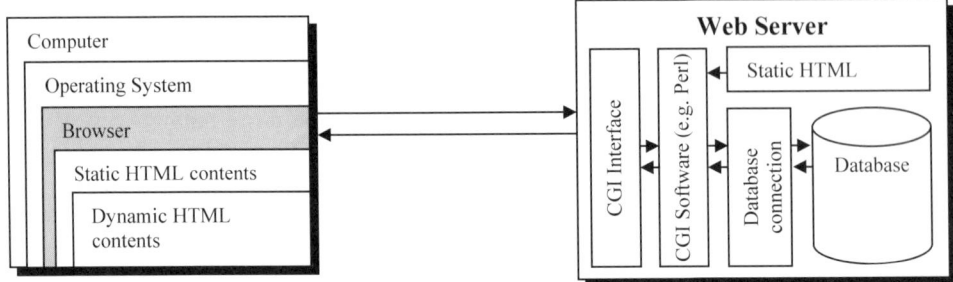

Figure 12.13 Generation of a dynamic Web page

12.6.3 Displaying Database Content

Databases are the most used data structure for storing information electronically. They all contain the most basic functions such as inserting, updating, deleting, sorting and selecting data records. Databases are categorised by their data storage methodology. The most common databases are hierarchical, relational and object-oriented databases. Their main difference is how the data is stored and therefore this effects how the most basic database functions can access the data. Examples of a hierarchical database are file systems and XML data files as shown in Figure 12.14 [6,7].

With hierarchical databases, data is stored in a tree-like format, where more generalised data can be found near the root and more specialised data can be found further down the branches, nodes or directories.

Relational databases are the most commonly used databases because of their simplicity. In a relational database information of the same type is stored in a spreadsheet, for example, in a table. Such tables can roughly be categorised as tables containing business information and tables containing lookup information. If a business record needs to refer to the lookup table, a matching key must be presented in both tables.

Object-oriented (OO) databases combine tables and functionality specific for the type of information stored in the table. For example, a table may contain a list of daily share prices and a function could be the calculation of moving averages.

Once the data is stored in a database, information needs to be retrieved depending on the question in hand. A typical way of communicating with database software is through a programming language called Structured Query Language or SQL for short (often pronounced 'sequel'). SQL is implemented in almost any available commercial databases. The SQL standard said to be the 'most common denominator' is SQL-92. Its implementation is not always guaranteed because different database vendors need specific SQL commands to allow for features specific to their database.

Application of the Internet to Power System Monitoring and Trading

```
<TRANSFORMER>
    <TIME = 09.00>
        <TEMP UNIT= CENTIGRADE>53</TEMP>
        <ANGLE UNIT = DEGREE>16</ANGLE>
        <RATING UNIT = KVA>200</ RATING>
    </TIME>
    <TIME = 10.00>
        <TEMP UNIT= CENTIGRADE>56</TEMP>
        <ANGLE UNIT = DEGREE>18</ANGLE>
        <RATING UNIT = KVA>240</RATING>
    </TIME>
</TRANSFORMER>
```

Hierarchical   Relational

Figure 12.14 Most common database types

Once the SQL query has been defined and coded, it needs to be sent to the database manager for execution. Database vendors have their own version and implementation of their database manager and query optimisers. Therefore, a common cross-platform database connectivity standard for Java has been introduced called Java database connectivity (JDBC). JDBC drivers have been developed from JDBC's predecessor, ODBC, and are available for almost every database. JDBC comes in different levels of database accessibility. For example, Level 1 JDBC drivers are JDBC-ODBC bridges for databases where only an ODBC driver exists and Level 4 JDBC drivers can access a database directly and are generally written in pure Java.

When the database and the SQL application reside on the same computer, and no server exists, the database model is called a two-tier model with the first tier being the application and the second tier the database as shown Figure 12.15.

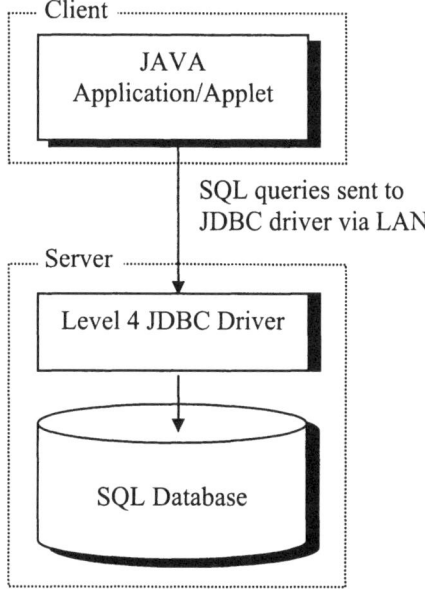

Figure 12.15 Two-tier JDBC driver connection

If the database is located on a serve the application accesses the database via server software. Such server software can be accessed via an ordinary http request. It can be written in any CGI executable language, e.g. Perl or C++, or as a pure Java application, e.g. servlets or EJBs).

Server-side software generally contains parts of the business logic of the database. Business logic is, for example, pre-programmed SQL methods for accessing a database or invoking transaction scripts as shown in Figure 12.16.

In most Web applications the third tier is to be regarded as the connection to the database, since applications cannot be granted direct access to the database across the Internet for reasons of security. Therefore, whenever a database is accessed across the Internet, an appropriate CGI, a servlet or EJBs must be coded. There are several software companies creating 'off-the-shelf' client-server software for data presentation for the client and database access on the server.

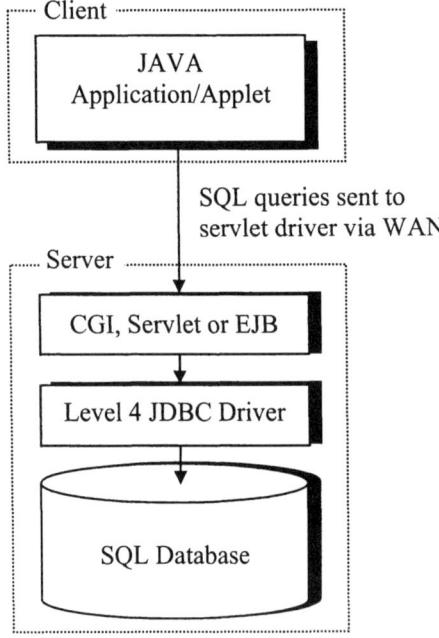

Figure 12.16 Three-tier JDBC driver connection

12.6.4 Web Pages with Functionality

Web pages can include functionality, e.g. collecting data typed in by users and its validation using JavaScript or VBScript. DHTLM is a collective description of mixing the functionality of a scripting language with Web page interactivity.

12.6.5 Web Pages with Integrated Applications

A new trend into leased applications on the Internet can be noted. Many small companies, which cannot afford to develop an application on their own, have used this service, for example, Web pages for e-commerce, for shopping on the Internet and trading of shares.

12.7 XML

This section aims to give readers who are not Web developers a quick background in eXtended Markup Language (XML). XML is primarily used to define data contents without formatting information.

HTML is the most supported formatting language by browsers on the Web. HTML lacks extensibility, since the tags which are used must be defined within the browser. When the race began between major Web browser manufacturers, style definition tags were introduced as a matter of competitive advantage.

Another HTML weakness is that the tags are used for formatting and only little about what the information is. XML can describe the stored information clearly.

12.7.1 Why the Need for XML?

The shortcomings in HTML have accelerated the introduction of XML. One of the major differences between HTML and XML is that XML does not contain tags which relate to the formatting of the data. The formatting of XML documents is assigned to individual data elements, representing specific elements with a constant format throughout the document. By introducing new data elements and representing them in a pre-defined format, entire industries are able to interchange information in a suitable format. Since XML documents contain the data elements, a new type of document is required to store information about its representation. Such documents are called stylesheets and further information will be given later. Stylesheets can change the way XML is displayed in the browser. If the formatting needs to be changed, only the stylesheet requires modification, separating the maintenance between data and formatting or content and layout.

12.7.2 Reasons for XML

Plain text:
If an XML document is accessed, a plain text editor can be used to access the data. The advantage is that in years to come everybody will be able to read/write the files. Plain text files are platform and application independent. This means that it is not necessary to use the same program in which the files have been created in order to read the information. Data losses or expensive data conversion can be saved if data creation software is discontinued, e.g. try to open a document file created in the 1980s (WordStar or MS Word 2.0) 50 years later. This gives XML a truly universal and timeless data structure solving cross-platform data archiving and compatibility problems.

Data identification:
Information can be split by type of data and subsequently displayed depending on meaning. This is because different parts of the data can be identified which enables different applications to utilise it in different ways, e.g. searching or summarising. A data element starts with a tag describing the meaning of the data, e.g. <NAME>, and ends with a terminating tag, e.g. </NAME>. XML data structured in such a way is referred to as being well formed.

Hierarchical data format:
XML data is presented in a hierarchical format. Hierarchical formats have the advantage of faster drill-down for more specialised information or move-up for more generalised information. One of the major disadvantages is that they suffer from data duplication.

Formatting:
XML data can be formatted for display by using a stylesheet. Stylesheets define how a specific element is displayed, e.g. on a screen or printer. This enables the user to reuse the XML data for different views or presentations by applying different stylesheets. As well as displaying XML data, stylesheets can be used to convert XML data into different formats such as LaTeX or PDF.

Inline contents:
XML allows the inclusion of other files containing XML. This results in manageable chunks of XML data. Files containing XML data chunks can then be included in one or more XML documents, reducing the amount of data duplication.

XML sample:
XML allows users to define a tag set of their own. Some rules with regards to its layout are listed below:

- XML requires one large container element, which encapsulates sub-elements.
- All open tags must have a corresponding closing tag, e.g. <H1></H1>.
- All sub-elements within a hierarchy must be closed in reverse order. Outer elements containing sub-elements can only be closed if all sub-elements belong to the outer element are closed, e.g. <H1><H2><H3></H3></H2></H1>.
- Attribute values for tags must be in quotes, e.g. <H1 colour="blue"></H1>.

The same data can be formatted in different ways by introducing different ways of representing elements. Once the data has been generated in well-formatted XML, it can be reused by different industries.

12.7.3 Separation of Content and Layout

Information contained in static Web pages may change form time to time, challenging Web page designers for fast and reliable update mechanisms. Maintaining the flexibility of static Web pages is therefore one of the major design issues driving the introduction of new strategies and technologies. HTML pages contain content and layout within one document. Content is the information displayed on an HTML page; it can be in form of plain text, tables, charts, graphics or others. Layout is the presentation of the HTML page; it is embedded as HTML markup tags and is not explicitly displayed to the viewer since the browser translates the markup tags into positioning information.

Classic HTML pages contain both content and layout in the same file, causing difficulties since common layout needs to be replicated for all pages if changes are needed. For example, if a large company changes the layout of its Web pages a modification of each Web page is required if they were written in static HTML markup language. This problem can be avoided if content and layout are separated. The technology used for the separation could be achieved with XML for content and eXtended Stylesheet Language (XSL) for layout. Figure 12.17 shows the relationship. Details on XSL will be given later.

By separating content from layout, Web design can be split among specialised teams such as graphical experts, script programmers and site managers. This allows each component to be reused and versioned, reducing maintenance complexity.

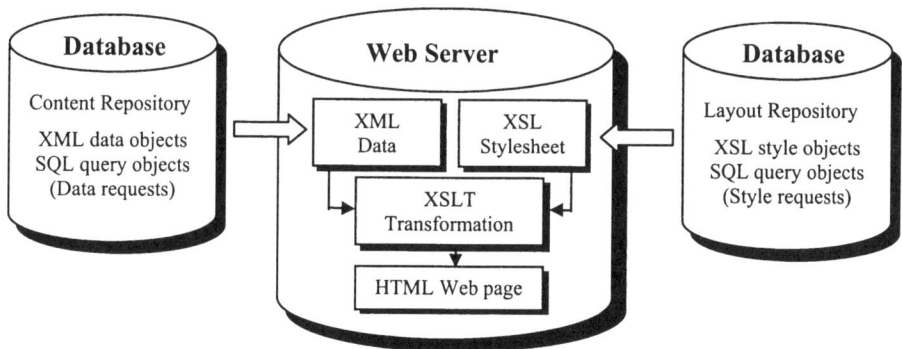

Figure 12.17 Rendering of XML data with an XSL stylesheet for HTML display

Depending on the XSL stylesheet output formats such as WebTV, WAP, PDF or others could be constructed. Since XSL stylesheets are in principle XML documents, they can be converted by another XSL stylesheet into a new XSL stylesheet as shown in Figure 12.18.

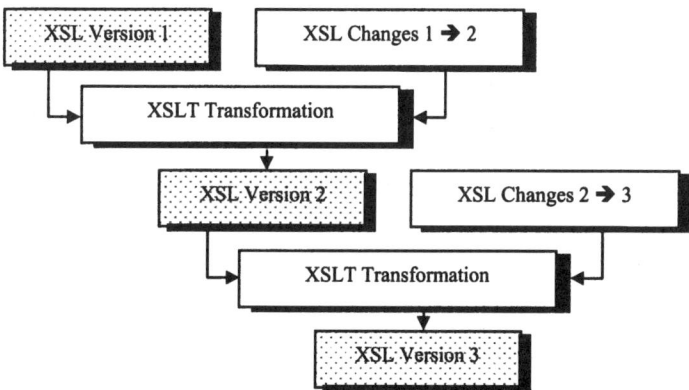

Figure 12.18 Change management system based on XSL stylesheets

XML and XSL can be transferred between multiple platforms, applications, databases and programming languages. This protects the technology investment since information is stored in plain text and therefore always accessible. This makes XML and XSL future proof and allows new emerging technologies to rely on a simple but comprehensive data structure.

12.7.4 XML Layout Validation with DTD

A document type definition (DTD) file contains layout rules written in plain text XML. DTD is used to validate XML to avoid invalid or incomplete XML data files. This is important since XML files require to be well formed so they can be read by an XML parser. A DTD document contains rules, with which an XML document must comply in order to pass a DTD validation test. Such rules consist of tag hierarchy, node and parameter data types, mandatory or optional tags or data. Using DTDs is optional but their utilisation can assist error analysis within an XML document.

12.7.5 Stylesheets

The purpose of a stylesheet is to display XML data in a format specified by the stylesheet. Stylesheets allow the same XML data to be displayyed in different ways. Since a data view is dependent on which information about the data is required, stylesheets offer an enormous flexibility not matched by HTML. If an HTML document is displaying data, e.g. in table format, and a transposed view is required, a new HTML document needs to be created and this will cause a duplication of data. If the data is changed, both HTML documents need to be updated. By using a set of stylesheets, this problem could be avoided. Regardless of which view is required by the user, there will be only one XML document containing the data. This arrangement offers greater maintainability for data since only one data source is involved. If the user needs to change the view, a different stylesheet can be used. One XML document can be formatted in many different ways just by changing the stylesheet.

There are several types of stylesheets, which can be used with the most common browsers. The two most frequently used are cascading stylesheets (CSS) and extensible style language (XSL). Stylesheets allow modification of thousands of XML Web pages concurrently and consistently, and this makes the redesign of Web sites much simpler.

Cascading Style Sheets (CSS)
CSS is a basic language for applying font styles to XML elements. Some font styles are **bold**, *italic* or combinations ***bold and italic,*** Arial or Courier. A font can be described by its attributes, such as name, weight, sizes, foreground colour, background colour, character, paragraph spacing, and many more. CSS stylesheets were introduced in 1996 as a standard means of extending the style properties of HTML and XML documents.

HTML has a set of pre-defined elements, such as <H1></H1> for headings. If a CSS document contains the same element names and tries to redefine these tags, ambiguities will result which might not result in the desired formatting. XML documents, which do not have any pre-defined elements in a browser, will not experience these problems and are therefore more reliable.

CSS gets its name from the fact, that the stylesheets can be cascaded. This means that more than one stylesheet can be applied to a data source.

eXtensible Style Language (XSL)
XSL is a data formatting and translation language. Therefore, XSL consists of two languages. The data formatting part, on one hand, is similar to CSS in terms of formatting XML elements for, say, display or printing. On the other hand, the translation language is able to transform an XML document into any other document types such as HTML, LaTeX, PDF or XML itself. Document translation from one XML representation into a different XML representation of the same data is required for data exchange on all levels.

XSL allows transformation of XML documents and XML element formatting at the same time. For example, it is possible to filter only selected information from an XML document and format it for the purpose of creating a static HTML document for publishing on the Internet.

eXtensible Linking Language (XLL)
XLL defines how one document links to another document. This is accomplished via XPointers and XLinks. XPointers define how individual parts of a document are handled. XLinks build a link to a URL resource such as other XML, XSL or XLL documents.

XLinks are similar to HTML hyperlinks with the difference that they are more flexible. Whereas HTML hyperlinks allow a connection to entire documents, XLinks allow a more powerful tool for linking between documents. XLinks allow multi-directional links where the links allow running in more than one direction. They allow every element to become a link not just pre-defined elements.

XPointers allow links to arbitrary positions in an XML document, allowing cross-referencing, footnotes, end notes, interlinked data, connections between documents and parts of remote documents and other more complex document navigation. In short, XLL is designed to link by reference rather than by exact location or resource name. By establishing a series of relationships among information held in XML documents, XLL allows pinpointed links to other XML documents. XLL links can be more complex than HTML since they allow for one-to-many, giving the user more choice.

12.8 Case Study1: Power Station Monitoring [8]

The increasing complexity of large electric power systems has resulted in a greater need for maintenance to ensure a reliable supply of power. Condition-based maintenance and distributed on-line HV condition monitoring have been the current trend. In Hong Kong, with the construction of an international airport, new power substations have been built to meet the huge energy demand. The capacity of the existing distributed monitoring system, which is based on one-to-one communication, was considered inadequate and therefore a completely new design concept was tried. The schematic block diagram of the newly developed system is shown in Figure 12.19.

12.8.1 Requirements of Airport Substation

An international airport, currently the largest in Southeast Asia, was constructed and was opened in 1998. A number of electric power substations for the new terminal building and associated infrastructure have been constructed. A detailed study into one of the numerous substations revealed the shortfalls of the existing distributed on-line monitoring system because the substation there had been too remote from the maintenance centres. The engineers in charge of the transmission network in China Light & Power Company Ltd (CLP) very often need to know not only the real-time status of power equipment but also the security and fire safety of the substation. Furthermore, in consideration of a more efficient operation of the system in the future, personnel in other organisations, such as the Airport Authority, Fire Services Department and other operation and maintenance departments within CLP, may need to gain access simultaneously to the important information within the substation.

The original information system needed to be enhanced and extended to tackle the fire safety and security requirements. Therefore, the idea of remote vision for substation monitoring has been employed. This enabled engineers and relevant staff to see on their remote display monitors the real-time scene of the indoor environment of the substation at different office locations or at home during standby duty. Intruders and fire outbreak in terms of smoke emissions can be detected immediately. To allow simultaneous access to information by all parties concerned, the old method of using modem-based peer-to-peer communication has been abolished and replaced with an Internet-based client-server concept.

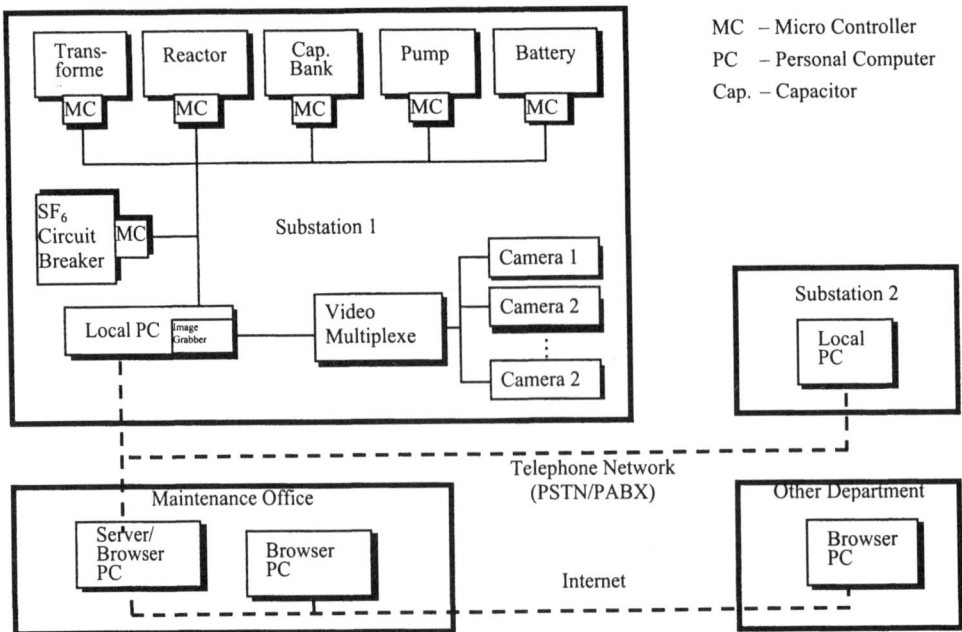

Figure 12.19 The whole Internet-based monitoring system

12.8.2 System Implementation

The substations, though having great impact on the integrity and normal operation of the whole airport, are normally unmanned. Existing substations are equipped with fire alarm panels that retrieve signals from smoke and heat detectors. False alarms are frequently encountered and this leads to wasting resources as the fire services are only able to discriminate them when they arrive at the remote sites. Illegal intruders must be detected and prohibited from entering such substations at any time. To accomplish the aims mentioned above, a remote vision system was developed.

Remote Vision System
Eight off-the-shelf CCTV cameras are installed at different locations in each substation. Figure 12.20 shows the structural schematic diagram of the remote vision system. The aim is to cover all internal areas as completely as possible. For example, the eight locations of the airport substations being monitored are the fire panel, control room, 11 kV switchgear room, 132 kV switchgear room, substation entrance, 132/11 kV transformer bay, cable basement 1 and cable basement 2. Each camera is equipped with the functions of panning, zooming and tilting. The video signal from each camera is wired back to a tailor-made 'remote control and multiplexing box'. The on-site PC controls each box via the printer port. Through this box, the lighting contactors of the eight locations can be energised and de-energised based on commands from a remote server. This is to ensure an adequate illumination level for each camera to grab a satisfactory real-time image of each location. Via this box, the video signal of any one camera can be selected by an image grabber card on a time-multiplexing basis. Furthermore, the PC is communicating with all of the microcontrollers in the existing distributed monitoring system. In addition, control signals for panning and tilting each camera can be output from the box. Communication between the PC and the CLP maintenance centre is accomplished by a modem.

On the software side, the on-site PC has two modes of operation, namely the regular mode and the real-time mode. The regular mode is active during normal operation. The on-site PC sequentially grabs images from the eight cameras at a frequency of 5 seconds per frame.

The value of the average grey level can be used to assess the overall illumination level of the site and the lighting system of the site can be switched on and off accordingly. The average grey level of this updated image is further compared with that of the previous image, which was grabbed and saved onto the hard disk 40 seconds ago. If there is a significant change in the average grey level, the two images cannot be compared directly and the system will regard it as an error and wait for another 40 seconds. Otherwise, the updated image is subtracted from the previous image so that any significant change in the scene can be identified.

If the change is considered significant, the on-site PC will first of all save the two relevant images onto the hard disk for later reference and then inform the maintenance centre by producing an alarm at the server. On top of analysing the images, the on-site PC saves the real-time images onto the hard disk at a frequency of two sets per hour.

There are two levels of operation being selected by the server, namely the coarse level and the fine level. Under the coarse level, images of size 320 pixels × 200 pixels are transmitted, resulting in a transmission cycle of only 48 seconds for the eight images from the eight respective cameras. If the user finds anything unusual, the fine level can be switched in, resulting in a transmission rate of around 35 seconds for each image of size 640 pixels x 400 pixels. The user is able to fix any camera 'on-line' and pan/tilt/zoom that particular camera. The compression algorithm for these images is 'standard JPEG format' with the quality factor set at 15 % so that the file size of coarse-level images is around 5 kb while that of fine-level images is around 30 kb. There are two factors that govern the transmission rate, namely the quality factor and the speed of the connection; 15 % for the quality factor is the optimal value based on experimental trials and thus room for improvement is limited. If an ISDN link is provided from the server to the airport substation, the transmission rate will be substantially improved.

This remote vision system requires neither spare contacts nor additional transducers. It can be used to prevent theft as well. General inspection of the substation can be carried out, such as checking cleanliness and quality of maintenance work. The alarm indicators on the fire panels can be grabbed as images so that the user at the central operation and maintenance centre can confirm whether the alarms are false or genuine by selecting the relevant camera to see the existence of smoke or fire in the activated zone. Furthermore the remote vision system can be used to monitor external contractors working on site. No CLP staff is deemed necessary in the substation. Equipment in hazardous areas or areas without adequate clearance, such as confined spaces or equipment rooms with live conductors, can be monitored by this system. During major overhauling or fault handling, the maintenance manager is able to visualise the equipment status through the display monitor to give direct instructions to the site engineers. Site problems encountered can be efficiently solved with the cooperation of the site staff and central management personnel.

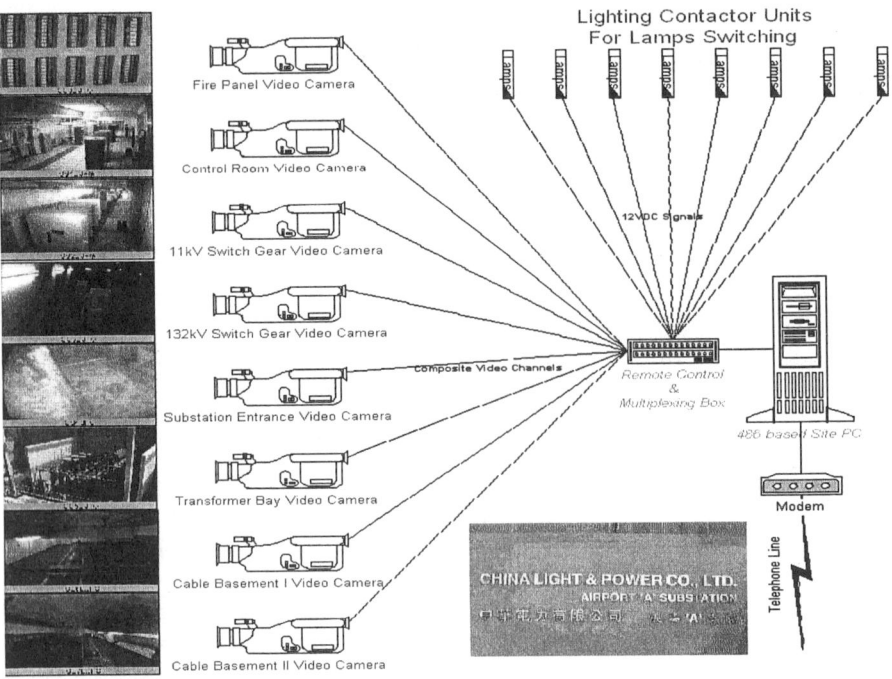

Figure 12.20 Remote vision system

Connection from Substation Components to Local PC

Equipment used for the transmission of energy requires constant monitoring and maintenance in order to ensure a permanent power supply. Monitoring of such equipment is necessary for those occasions in which vital changes of important parameters can be used to predict a potential problem with the equipment. Such monitoring can prevent sudden loss of sub-station equipment leading to unexpected power cuts. Furthermore, uncharacteristic behaviour of one element in the supply chain can influence the performance of other equipment or might even cause damage. This is even worse if sporadic malfunction of one device leads to damage in another device. Such sporadic malfunction of devices can only be detected if continuous monitoring is practised.

Therefore constant monitoring of equipment will improve power system reliability and reduce maintenance costs because devices can be replaced before they cause further damage.

In practice, it is impossible to monitor every unit since they are geographically distributed. Furthermore, it would be too expensive to keep qualified personnel in remote locations for 24 hours a day, 7 days a week. Therefore, remote monitoring of power equipment with the help of computers has been practised for some time. Computers are the perfect alternative to monitoring personnel since they are able to monitor constantly and accurately, detecting even the smallest changes in critical parameters.

Computers can be placed at important points of a substation. The price of computer hardware has been falling continuously for a number of years, making remote monitoring with computers effective and economical viable. Most of the hardware components required for real-time data collection are well established and robust. Once the hardware is connected and configured, device drivers for each hardware component make it possible to access and control its functionality on an abstract level. Device drivers for display adapters, data sampling cards or other hardware, for example, are generally written in a low-level programming language such as Assembler, C or C++ to increase processing speed. Using low-level programming languages will increase processing speed, since they are optimised, compiled and linked for a specific type of processor and operating system. Such processor and platform dependency plays an important role when it comes to deciding which components to buy. Not all companies can afford to update their device drivers in good time if a new or changed processor or operating system is introduced. Expensive hardware components such as I/O sampling cards can be rendered useless if device drivers are not compatible with the latest processors and operating systems. Therefore, care should be taken if hardware is purchased in large numbers from a manufacturer with no proven record of a continuous supply of device drivers.

Once the monitoring computers collect data, access to the data by users must be granted. Computers can distribute data in many different ways. Basically, there are two distinct network constellations. LANs, where local computers are connected via a local network, and WANs, where remote computers are connected by means of long-distance connections. Granting access to substation data via the Internet requires an Internet connection.

In this case study, substation computers act as data collection points for a remote maintenance office. Data collection from power devices is accomplished by sequentially converting their analogue signals into digital information via analogue/digital (A/D) converters. Such A/D converters can be found on standard PC I/O sampling cards. The typical range of an A/D converter can be ±12V or ±5V, which requires appropriate conversion of the analogue signal to match the A/D converter's input range.

The A/D conversion sample frequency depends on how many data sources are converted and on the required data accuracy. It can be quite low (<1 kHz) if no complex data transformations are planned. If, for example, spectral analysis or other data-intensive transformations are part of the overall monitoring process, the sampling frequency must satisfy the mathematical constraints of the transformations used.

In order to avoid loss of accuracy or injections of harmonics into the analogue signals, A/D converters should be placed as close as possible to the source. Once the analogue signals are converted into digital information, transmission will not cause loss of accuracy. Figure 12.21 shows how substation components exchange data via a LAN.

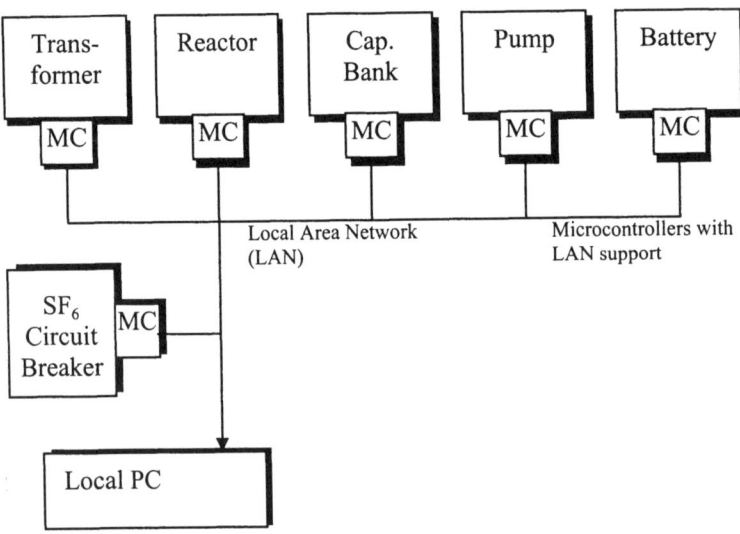

Figure 12.21 Collection of data in a local PC

The raw data received via the LAN from the microcontrollers needs to be converted in such a way that it can be sent to the maintenance centre. It requires a format that is easily extendable, in case new components are added to the monitoring requirements. Data received from different sources needs to carry additional information such as the name of the source, its location, date and time, scaling factors, units and many more. There are different possibilities on how to encode this additional information. The most configurable and extendable formatting standard, which is widely accepted, is XML. It is compatible with all operating platforms since it is contained in a plain text file, for example:

```
<TRANSFORMER>
        <Temperature Unit = Centigrade> 60   </Temperature>
        <PowerAngle  Unit = Degree>     20   </PowerAngle>
        <PowerRating Unit = kVA>        200  </PowerRating>
</TRANSFORMER>
```

Connection from Substation to Maintenance Office

Once substation data has been collected and stored in local PCs, it needs to be published. The purpose of the case study is to grant access to the substation data for all responsible parties. Such parties may be the personnel of the electricity and security companies, the fire brigade or other remote experts and advisers. In order to publish information over the Internet, a Web server connected to the Internet is mandatory.

In this study, there are several different ways of distributing information on the Internet, such as:

- Static but frequently updated Web pages.
- Dynamic Web pages which are updated if data changes.
- Static Web pages with a dynamic applet and data polling.
- Static Web pages with a dynamic applet and data streaming.

As mentioned previously in brief, static Web pages are not really suitable for constantly changing data since they require the data to be embedded within the document. If the data is continuously changing, the document needs to be changed manually. Therefore, fast data changes cannot be represented by static Web documents.

Dynamic Web pages are one way of publishing changing data over the Internet (Figure 12.22). Generally a new static Web page is generated and transmitted to the client each time a data update takes place. Such generation of Web pages can be done via a common gateway interface (CGI) and an executable program located in the Web servers' CGI-bin directory. The task of the CGI interface is to instantiate the CGI program and pass information sent by the user. A program used for the CGI can be written in any programming language but must be compiled or interpretable by the server. This way of updating information on the Internet is suitable for slow-changing data such as daily or weekly events.

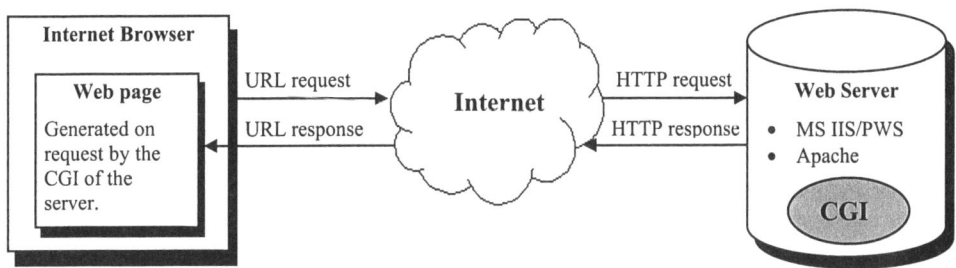

Figure 12.22 Dynamically created Web page via server CGI

Dynamic Web pages are not limited to the CGI. They can be generated with servlets. Servlets are written in Java and executed on the Web server. They function in a similar way to CGI-based programs but are more flexible and reliable in terms of robustness and security.

If a Web page is generated dynamically on a server with frequent data changes, e.g. every 10 minutes, it is likely that the client browser might display obsolete information. The problem with server-side-generated Web pages is that the browser does not know when new data becomes available on the server. Browsers have no functionality of testing if new data has become available on the server. Therefore, the user is required to reload the Web page continuously by selecting the refresh or reload option in the Web browser.

There are even more options to generate dynamic Web pages. Internet programming languages such as active server pages (ASP), server side include (SSI) or JavaScript and VB Script pages allow web pages to update themselves. Their underlying functionality is to execute code inside the browser in which the page has been loaded and use the browser functionality to connect back to the server and check for new data updates at set intervals. Continuous requests to the server to test for new data updates is called data polling (DP).

Static Web pages can contain a dynamic program called an applet. They have full graphical functionality to a designated area within the browser's display area or client area. Applets can be written to represent data in graphs, charts or any other way specified by the applet developer. They allow full interactivity with the mouse and other pointing devices. The technology used to send data to an applet can be categorised into DP and data streaming (DS).

DP, as shown in Figure 12.23, is the simplest implementation of receiving data updates into Web browsers. With data polling, the applet connects back to the server from which it was loaded originally and tests if new data has become available. The connection and new data tests are repeated at set intervals. If new data becomes available on the server, a data update is sent to the applet and the applet will display the new data. Data polling is not restricted to applets. It can almost be implemented in any programming language which is executed in the client browser, and supports repetitive loops.

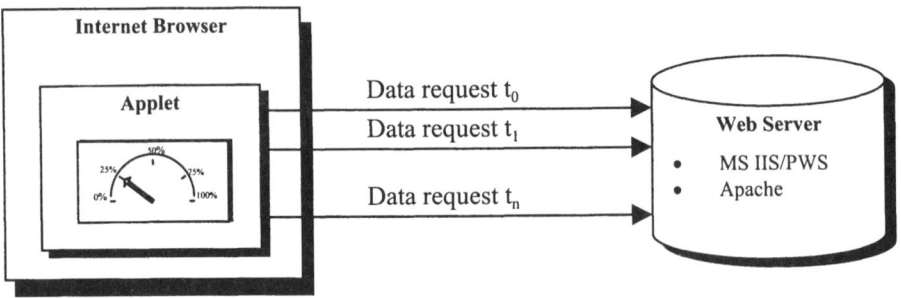

Figure 12.23 Real-time data update request via repeated requests to the server (data polling)

DS, as shown in Figure 12.24, on the Internet can mostly be found in Web applications, which require continuous data updates such as Web TV, Web radio, MP3/MP4/Flash video and/or music. It is based on keeping the connection between the client browser and server open. Generally, a connection is closed as soon as a Web page has been downloaded to a client browser. But with data streaming, the connection remains open.

A data streaming connection can be established by keeping the HTTP connection initiated by the browser. It can be compared to a telephone conversation in which the called person keeps the phone off the hook by talking continuously.

Application of the Internet to Power System Monitoring and Trading 453

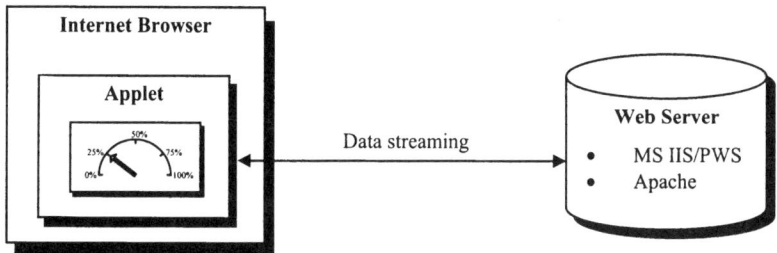

Figure 12.24 Real-time data updates via continuous connection to the server (data streaming)

The maintenance office is connected to the remote power substations using a standard telecommunication connection, as shown in Figure 12.25. Depending on which parameters are monitored in the maintenance office, different data update technologies need to be considered. In the case when all measurements taken from the appliances within the substations are within their set tolerances, transmission of averaged measurements might be sufficient. In case a fault occurs, all measured and locally stored data from a defined point could be transmitted. In order to receive continuous data transmission, data streaming is required for fast real-time data updates.

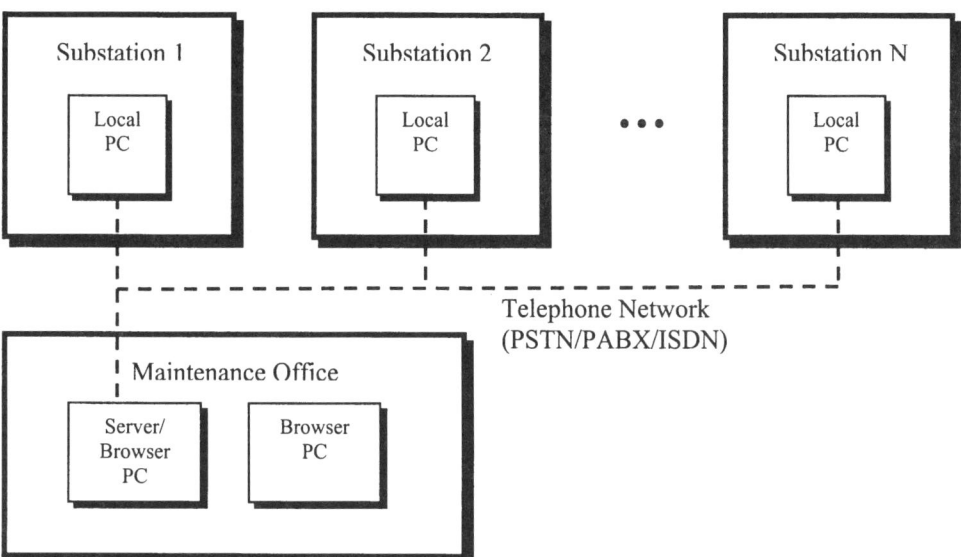

Figure 12.25 Connection between power substations and the maintenance office

If remote expert advice or an on-line reporting is required, data transmissions of real measurements can be transmitted across the Internet. There are several ways of displaying real-time data, e.g. as numerical values in an analogue or digital display or time series graph, as shown in Figure 12.26. At present, browsers do not have a built in functionality for supporting graphical representations of data. Therefore software extending the browser's display capabilities must be used. Applets could use the browser's client area for drawing lines, shapes or colours. Such shapes offer the basic functionality required for designing controls capable for displaying real-time data.

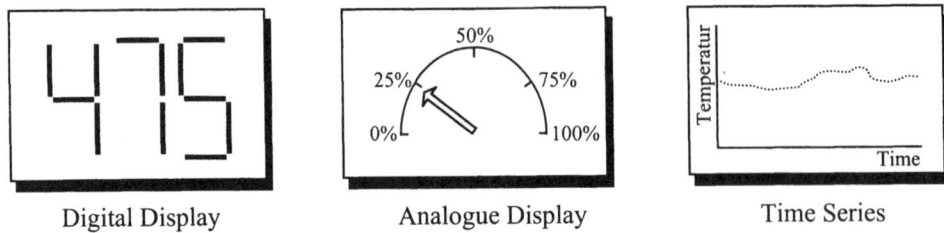

Figure 12.26 Different controls to display real-time data

Another important aspect of working with applets is that they are able to connect back to the server from which they were loaded to retrieve new data updates, regardless of whether data polling or data streaming is used.

Displaying real-time data in an applet is roughly a two-step process, as shown in Figure 12.27. The first step is to transmit the applet from the Web server to the browser. The second step is to transmit data to the applet.

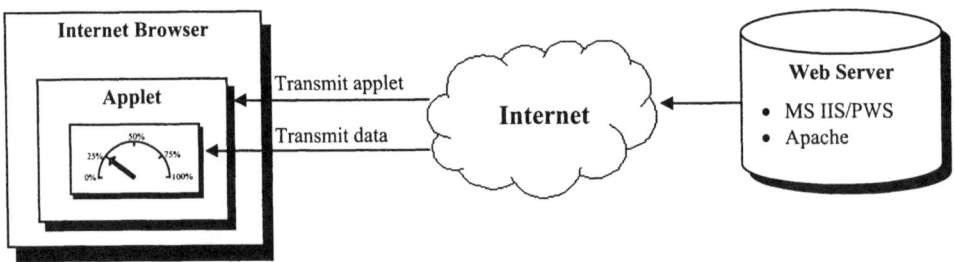

Figure 12.27 Real-time data display in a browser as a two-step process

12.8.3 Monitoring Power Station Equipment

A few examples will be given on the monitoring of power station equipment such as circuit breakers for the prevention of major faults and supply interruptions.

The SF_6 gas pressure measurement history over four months, from October 1995 to January 1996, was presented for a circuit breaker (CB) and the trend is shown in Figure 12.28. It can be seen that there has been a very serious SF_6 gas leakage problem with the CB and the system was successful in giving a warning to the maintenance team on 17 December 1995. The gas topping exercise was completed on 18 December 1995 to avoid a major failure of the CB.

A method was developed to measure the travelling of CBs based on looking at the current waveforms. Figures 12.29 and 12.30 show the measurement results for a typical 132 kV CB which is used to switch a 132 kV, 80 MVA reactor. From the figures, it can be seen that the closing time for the CB is 125 ms while the tripping time is 50 ms.

Application of the Internet to Power System Monitoring and Trading   455

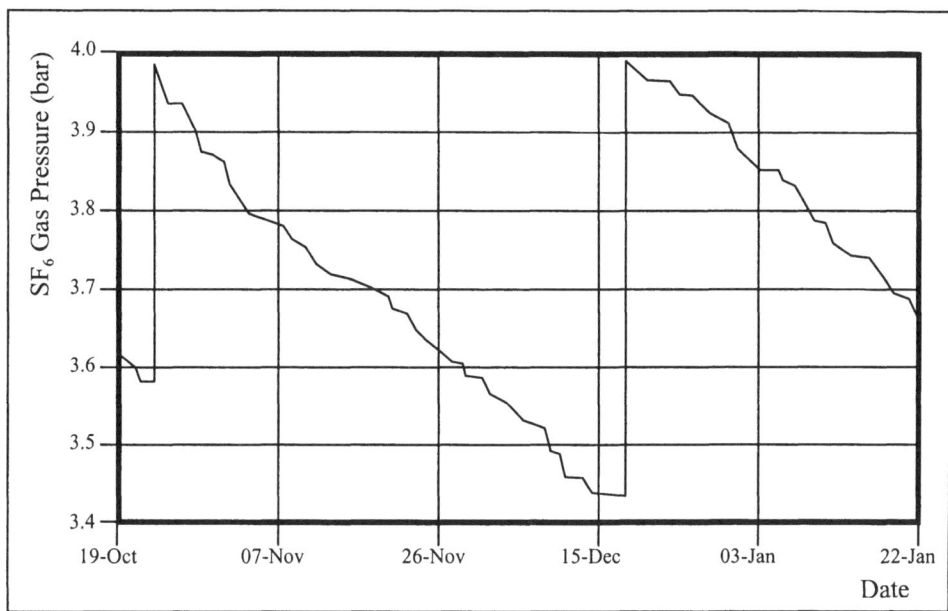

Figure 12.28 SF$_6$ gas pressure variation in the CB

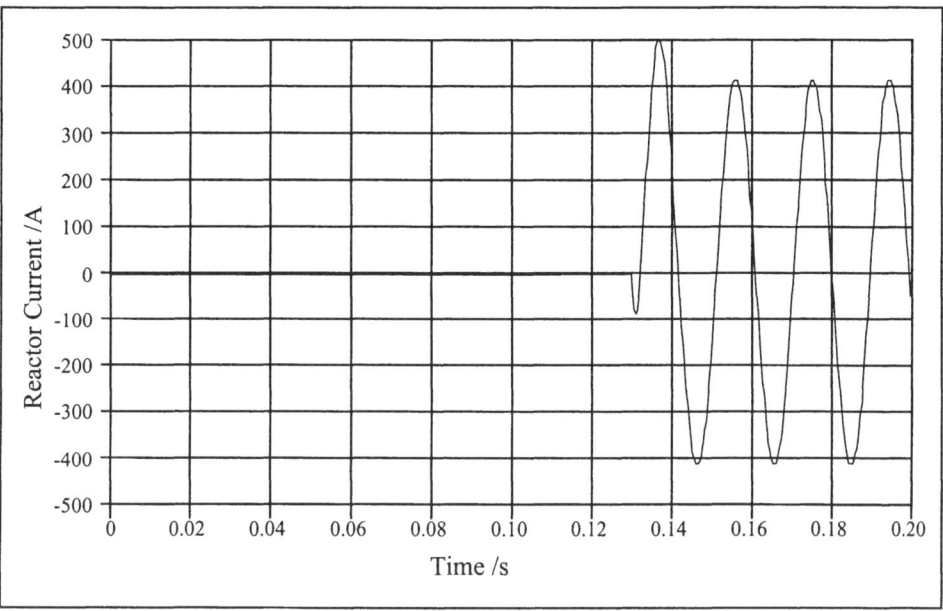

Figure 12.29 Current waveform for closing of reactor CB

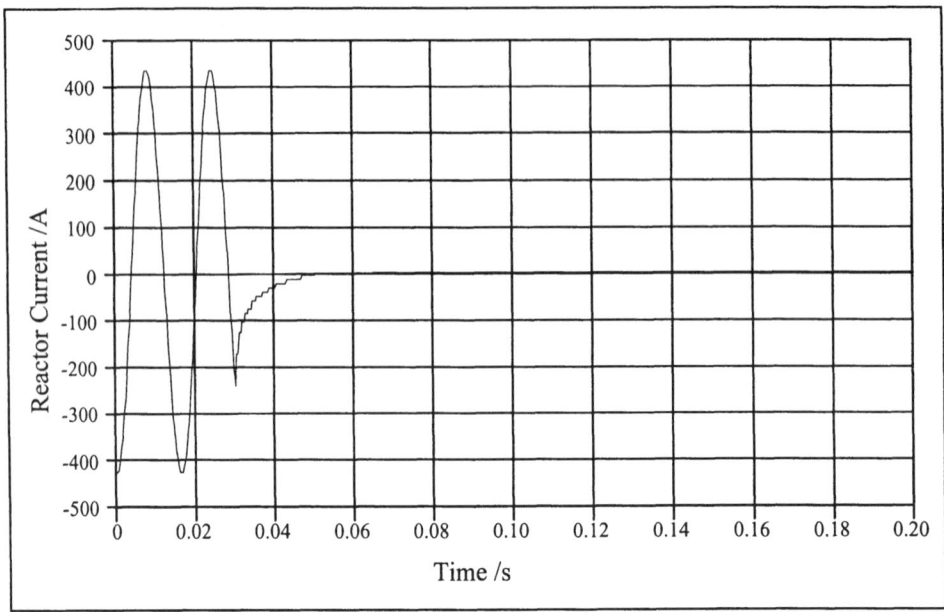

Figure 12.30 Current waveform for tripping of reactor CB

The air compressor operating time (Ton) and idling time (Toff) can be monitored and are shown in Figure 12.31. If Toff is short, this indicates that the pressure of the compressed air will meet the lower limit very quickly. It may be caused by air leakage in the piping system or the storage tank. If Ton is long, this may indicate that the efficiency of the air compressor has become poor as it takes a longer period of time to charge up the storage tank. As a result, precautions could be taken at an early stage.

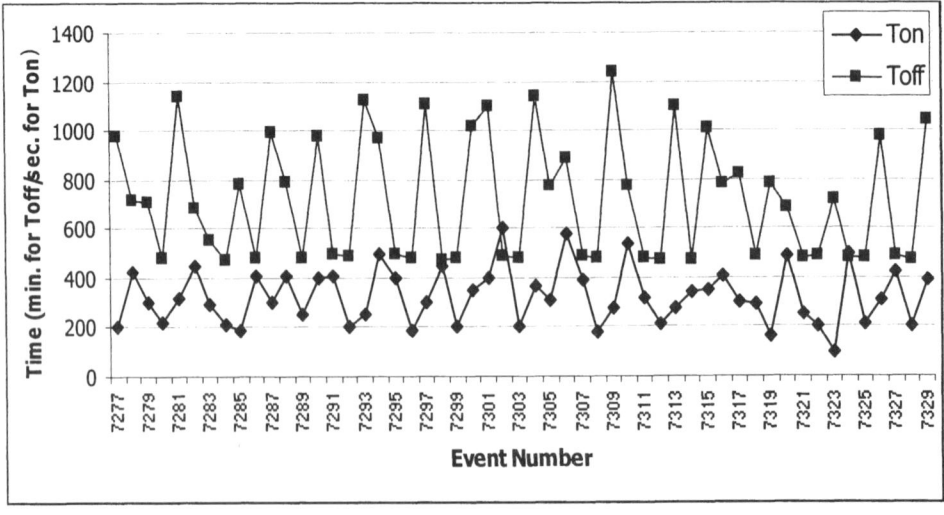

Figure 12.31 Air compressor on/off timing

12.9 Case Study 2: Power Trading Application

Electricity deregulation is creating a free electricity market which is different from country to country. For each restructured utility, the market operator provides the essential service of exchange function. Electricity trading in Europe will change dramatically as the wholesale and retail markets open up to competition. Competition between utility suppliers will bring benefits to end users only if each competitor has the same access to information regarding power pricing and distribution. Tp keep the energy marketplace competitive, it needs to be non-discriminatory, transparent and easily accessible for each competitor.

Energy trading is not confined within a country's borders. Many countries are exporting energy to neighbouring countries so that a centralised operated power exchange can have a key role [9]. That kind of power exchange will have to offer a reliable and efficient exchange information between the market participants by operating a reliable, highly distributed and low-cost information network.

If the open energy market is to succeed, all participants must be wired into a standard data exchange infrastructure that must be platform and language independent. Therefore the Internet, with its platform and language independence, is the choice for hosting on-line trading platforms.

Power traders require fast reaction to market changes. They need to control their trades across all current bids, offers and negotiations by means of a mouse-click and require real-time market information, including market depth as well as vital news and weather information. Furthermore, anonymity during negotiations and tools for the immediate analysis of market condition are the relevant requirements.

The complexity of the power exchange with its large numbers of variables makes the prediction of market trends difficult to predict. Therefore, participants must be aware of the effects of market parameters to support decision making in the daily offering process.

The ideal power exchange can schedule enough capacity to meet all requirements for demand. Auction mechanisms describe different kinds of auctions such as the discriminatory and uniform auction system [10]. An ideal power exchange can control energy, regulation and reserve in parallel based on auctions. The dispatched energy will match the demand, the regulation is the capacity to maintain real-time generation/load balance and the operating reserve is the provision that can respond to imbalances and outages.

The bidding process is based on the market situation and follows an offer of a generation unit. Market participants have to ensure high profitability before agreement. It is generally difficult to find a clear relationship between the value of a generation unit and the real market price. Therefore, participants can use an *agent* with a specific behaviour to help in the decision-making process [11,12].

Internet-based electronic scheduling and trading systems could provide all transmission services and new tools and technologies for controlling, scheduling, buying and selling electricity. Therefore, intelligent agent technology has been developed for the electric power market as described in Chapter 11. Complex distributed systems for the electricity power market could benefit the interaction between intelligent software agents in helping the buying and selling of electricity. Intelligent agents perform as buyers, sellers, auctioneers or bidders in an on-line auction [13].

As mentioned previously, agents for buying or selling electricity are primarily representing either generators or consumers. In order to use agents to gain market advantage, each agent needs to present a unique economic and strategic behaviour model. These models are based on human behaviour with respect to different trading environments. For example, agents can show an anxious buying and selling behaviour, greedy behaviour or relaxed behaviour to emulate market participants.

There are several Internet-based simulation environments for experimentally testing the various power exchange mechanisms available on the Internet [14]. Such simulation tools allow participants from different locations to compete in the open market.

This is advantageous for the training of personnel, who are able to try different buying and selling strategies under changing market conditions without causing interference on a real trading floor. With the help of more advanced trading platform models, different auction types, e.g. uniform price, single and double-sided auctions, and different constraints, e.g. transmission losses, line capacity and stability limits and congestion situations, can be explored. The ultimate objective for each simulation will always be to maximise profits from trading energy.

12.9.1 Trading Platform Architecture

The first step in building a trading platform over the Internet is to gain quality Internet access with enough bandwidth to serve all clients at a reasonable speed. High-quality Internet access cannot be achieved by telephone. It is necessary to rent or buy a dedicated server with a reliable ISP, which offers a 24-hour, 7-day customer service.

Once a reliable Internet connection is established, server software must be purchased for running Web services. Currently, the most common Web servers are IIS from Microsoft, Apache Web Server from Apache and Web Logic. There are many software companies offering competitive Web server solutions, which can also integrate e-commerce packages.

Running a reliable trading platform across the Internet is not trivial. A primary objective must be to ensure data security and data integrity. Data security across the Internet has constantly been improved by the introduction of better and faster security algorithms. The most used and trusted method is secure sockets layer (SSL). SSL is relatively simple to implement and does not require changes to any existing Web pages. Data integrity can be achieved by buying a database from a major vendor. Such systems may include startup consultancy and customer support. It is important to design the database in such a manner that the database structure will deliver optimal performance. If, for example, data updates are sent to clients in XML format, conversions from table format to XML can increase database response times. Therefore, the choice of database layout should match the distributed data format if possible [15].

Figure 12.32 shows a simplified block diagram of a communications architecture between clients connected to a trading platform. On connection to the trading platform, the client receives an HTML page containing all the required fields to perform an auction-related transaction. On submission of a transaction, the Web server will receive the transaction details, which should be validated for correctness prior to storage in the persistent database. If invalid data is contained in the transaction, the transaction can be interrupted and changes will be rolled back to restore the database to its previous state. Specialised *transaction servers* can be purchased for keeping up data integrity.

As with many real-time auction and trading platforms, data updates are sent to the client in XML data format. Received data updates via XML allow faster data updates, since the changes can be directly drawn to the browser client area to avoid the generation of dynamic HTML pages. Furthermore, more clients can be synchronously updated because small portions of XML data are sent across the Internet, saving precious bandwidth.

It will take an entire programming team to create a real-time auction platform from start to finish. There are several software companies offering complete solution packages for e-commerce and on-line auctions. Internet applications have different operational requirements unknown to desktop applications. Requirements such as scalability and continuity are of great importance for Web applications. Web-based software development for highly scalable products require a great knowledge of multi-threaded environments and parallel processing architectures.

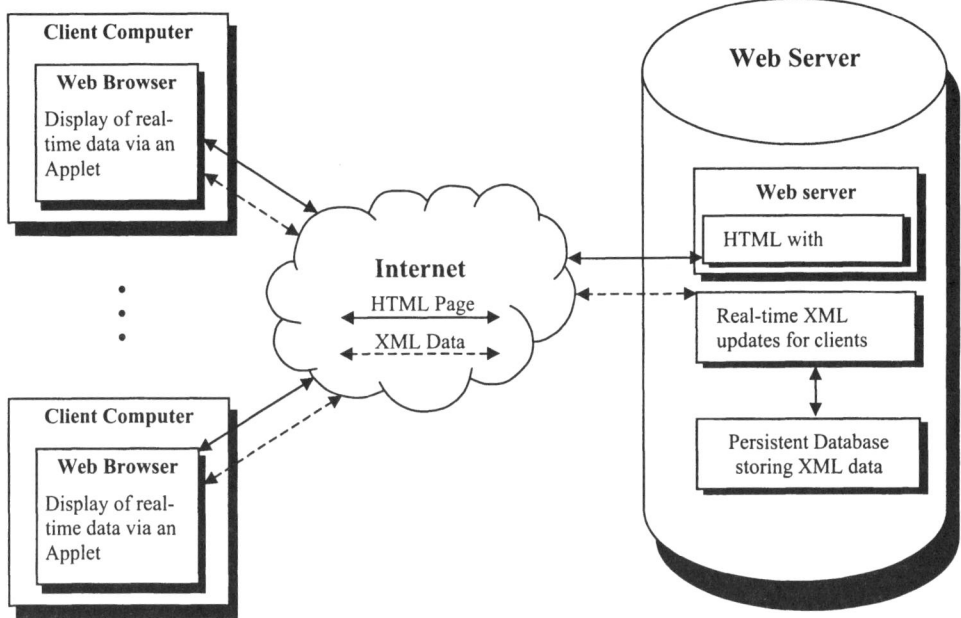

Figure 12.32 Communications architecture

12.10 Conclusions

The application of Internet to power system monitoring and trading is a very exciting area. Some examples have been given and the benefits derived from it are obvious. However, it can be seen that much work remains to be done. One area is system security in the open-access environment.

12.11 Acknowledgements

The authors would also like to thank IEEE for granting permission to reproduce the materials contained in reference [8].

12.12 References

[1] Raymond Greenlaw, *Introduction to the Internet for Engineers*, McGraw-Hill, November 1998.
[2] Danny Goodman, *Dynamics HTML*, O'Reilly UK, August 1998.
[3] Dick Oliver, *Sams Teach Yourself HTML 4 in 24 Hours*, Sams Publishing, September 1999.
[4] Elizabeth Castro, *HTML 4 for the World Wide Web*, Fourth Edition: Visual Quick Start Guide, October 1999, Peachpit Press.
[5] Brett McLaughlin and Mike Loukides, *Java and XML*, O'Reilly & Associates Inc, June 2000.
[6] James K. Tauber, *Teach Yourself XML in 21 Days*, Sams. Net, March 1998.
[7] Frances M. Cleveland, 'Information exchange modelling (IEM) and eXtensible markup language (XML) technologies', *IEEE Power Engineering Society Winter Meeting*, Singapore 2000.
[8] W.L. Chan, A.T.P. So and L.L. Lai, 'Internet Based Transmission Substation Monitoring', *IEEE Transactions on Power Systems*, Vol.14, No.1, February 1999, pp.293-298
[9] Marta Marmiroli and Hiroshi Suzuki, 'Web-based framework for electricity market', *Proceedings of the International Conference on Power Utility Deregulation, Restructuring and Power Technologies 2000*, IEEE, April 2000, pp.471-475.
[10] G.B. Sheblé, *Computational Auction Mechanisms for Restructured Power Industry Operation*, Kluwer Academic Publishers, 1999.
[11] G. B. Sheblé, Chapter 6: Agent based Economics, in M.Ilic, F. Galiana and L. Fink, (Editors), *Power Systems Restructuring and Economics*, Kluwer Academic Publishers, 1998.
[12] Chen-Ching Liu, Haili Song, Jacques Lawarrée and Robert Dahlgren, 'New methods for electric energy contract decision making', *Proceedings of the International Conference on Power Utility Deregulation, Restructuring and Power Technologies 2000*, IEEE, April 2000, pp.125-129.
[13] L.L. Lai, T. Motshegwa, H. Subasinghe, N. Rajkumar, and R. Blach, 'Feasibility study with agent on energy trading', *Proceedings of the Fifth International Conference on Advances in Power System Control, Operation and Management*, Vol.2, IEE, Publication Number CP478, October/November 2000, pp.505-510.
[14] POWERWEB User's Manual, Cornell University, NY, USA, 1999, http://www..pserc.cornell.edu/powerweb/
[15] E.Z. Zhou, 'XML and data exchange', *IEEE Power Engineering Review*, April 2000, pp.66-68.

Index

3-D thermal image, 402
access fees, 162, 165, 167
active reserves, 25
ageing assets, 295
air pollution, 234, 256
allocation factor, 248, 249, 251
amplitude modulation, 344
ancillary service, 25, 46, 61, 65, 76, 78, 79, 80, 86, 87, 92, 93, 94, 96, 97, 98, 99, 100, 101, 104, 106, 107, 139, 179, 180, 186, 191, 192, 193, 194, 195, 198, 199, 218
ancillary services markets, 79, 80, 86, 92, 93, 94, 97, 99, 107
aperiodic component, 346
Application Service Provider, 423
arcing, 299, 323, 334
artificial intelligence, xix, xxi, 121, 354, 413, 414, 415, 424
artificial neural network, xix, 107, 121, 151, 353, 412, 414
asset governance, xx
asset manager, 111, 112, 127, 296, 306, 313, 322
asset owner, 111, 289, 293
asset replacement, 116, 117, 122, 124, 125, 130, 290, 307
asset utilisation, 129, 281, 287
asynchronous interconnector, 331
auction mechanisms, 61, 96, 109
auto-change-over devices, 127
automatic generator control, 76
autonomous generation schemes, 27
autonomy, 355, 356, 359
auto-reclosers, 127, 128
auto-regressive moving average, 121
available transfer capability, 138

back-to-back thyristors, 269, 273
balancing market, 68, 78, 85, 113
battery charging, 28
benchmark, 116, 125, 128, 157, 163
bid prices, 23, 98, 176
bilateral contracts, 24, 61, 68, 71, 73, 74, 91, 107, 154, 155, 158, 167, 168, 184, 197, 209, 257
bilateral model, 96
binding day-ahead market, 86
biomass, 9, 12
black-start capability, 93, 194

C++ language, 136
capacity payment, 23, 100
capacity reserve, 240
capacity rights, 215
catastrophic failure, 100, 296, 314, 315
cellular phone, 134, 144
central auction, 82, 83, 84, 86, 89, 90, 91, 95, 96
central control systems, 128
central utility model, 52
CGI, 139, 437, 438, 440, 451
chromosomes, 40, 41
clean coal technologies, 20, 45
client server, 426, 446
climate change, 11, 19
co-generators, 2, 3
collaborative agents, 354, 355, 356, 357
combined heat and power, 5, 15, 45
combined-cycle, 10, 65
common gateway interface, 139, 437
communication systems, 127, 128, 146, 148, 259
competition, xii, 1, 2, 4, 5, 8, 9, 11, 15, 16, 20, 45, 47, 50, 51, 52, 53, 54, 55, 56, 57, 60, 63, 67, 72, 73, 76, 77, 78, 79, 100, 110, 111, 114, 125, 126, 127, 142, 145, 151, 153, 154, 155, 163, 167, 173, 174, 183, 185, 226, 228, 229, 236, 254, 258, 262, 290, 293,

304, 329, 330, 332, 334, 347, 356, 360, 373, 377, 420, 457
competitive bidding, 1, 65
competitive framework, xi, 110, 353
competitive generation, 2, 3, 4, 107
competitive metering, 114
competitive trading, 24
computational intelligence, xxi, 353
condition monitoring, 129, 132, 295, 300, 304, 312, 313, 320, 322, 328, 445
congestion management, xiii, xxi, 58, 69, 70, 71, 75, 78, 79, 86, 88, 89, 90, 92, 93, 94, 95, 97, 99, 104, 178, 180, 188, 195, 198, 200, 209, 215, 216, 217, 218, 219
congestion management markets, 93, 94
contract market, 10, 61, 68, 179
contract path allocation, 57
corporate restructuring, 293
cost/benefit analysis, 310
crossover, 40, 41, 273, 274, 370

damper, 273, 274
data polling, 451, 452, 454
data security, 458
data streaming, 451, 452, 453, 454
database, 136, 137, 319, 321, 408, 419, 420, 437, 438, 439, 440, 458
day-ahead, 61, 69, 79, 86
day-ahead market, 71, 78, 90, 176, 178, 191
delivery time, 86, 421, 423
demand growth, 123
demand side management, 115
demand-side bidding, 68
deregulation, xii, xiii, xiv, xviii, xix, 1, 2, 5, 6, 7, 9, 10, 15, 19, 45, 48, 50, 51, 52, 55, 57, 58, 64, 70, 71, 73, 108, 111, 116, 119, 133, 140, 153, 161, 167, 171, 173, 175, 202, 217, 218, 232, 259, 283, 316, 330, 331, 332, 334, 348, 353, 360, 457
deregulation of energy market, 418
desalination plant, 38, 49
discrete wavelet transform, 338

dissolved gas analysis, 296, 323, 329
distributed generation, 13, 16, 17, 20, 21, 22, 23, 25, 26, 46, 48, 99, 108, 144, 164
distributed generation technologies, 13
distribution automation, 127, 128, 147, 148, 151, 418
distribution companies, 4, 63, 64, 110, 111, 113, 115, 116, 117, 119, 154, 175, 302, 316, 318, 353, 361
distribution loss, 63
district heating, 21
disturbance recognition, 341, 350

economic dispatch, 53, 77, 78, 82, 109, 121, 133, 374, 414
eddy currents, 325, 326
elasticity, 59, 192, 195, 196, 209, 215, 220
electrical discharge, 296, 300
electricity and gas networks, 111
electricity distribution industry, 111
electronic auction markets, 10
e-mail, 354, 420, 425, 427, 428, 429
embedded cost, 57, 58, 186, 187, 189, 190, 194
embedded generators, 112
embedded systems, 128
emissions-free electricity, 19
energy function, 206, 385
energy mix, 6
energy policy, 16, 48
energy purchase cost, 113
energy storage, 5, 13, 259, 264, 269, 270, 285
English auction, 55, 56
equilibrium point, 68, 69, 84, 97, 206, 207
ethernet, 348
evolutionary computing, xvi, 353
evolutionary process, xiii, 98, 163
evolutionary programming, 49, 353, 410
ex ante market, 61
ex post market, 61, 73
excitation capacitance, 27, 28, 32, 35
expert systems, 353, 355

facilitators, 359
fiber optic communication, 147
fiber-based transmission, 142
file types, 421
financial markets, 78, 88, 94, 97, 171
financial transmission rights, 95
first rejected offer, 55
flexible AC transmission system, 162
flicker, 266, 331, 342, 346, 347, 352
force-commutated converters, 278
forward markets, 71, 86, 95, 106, 178, 361
fossil fuel, 3, 4, 6, 45, 53
Fourier transform, 336, 347
free space lasers, 141
frequency modulation, 144
fuel cells, 10, 12, 13, 20, 21, 26, 99, 330
full graphics interface, 134
futures market, 8, 68, 74, 104, 186, 361, 362, 364
fuzzy diagnosis, 323, 325, 328
fuzzy logic, 38, 49, 341, 412

gaming, 50, 78, 83, 88, 91, 92, 95, 98, 99, 107
gas industry, 165
gas turbine technology, 173
generation companies, 22, 67, 72, 73, 99, 175, 361
generation mix, 11
generation model, 8
generation scheduling, 53, 104, 109, 156, 180, 412
genetic algorithm, xix, 49, 360, 362, 364, 365, 367, 370, 410, 412, 414
GIF image, 421
government intervention, 16, 45
graph theory, 246, 251
green certificates, 17
green energy, 20
greenhouse gas reduction, 5
GTO, 262, 263, 264, 268, 270, 274, 275, 278, 279, 280, 283

harmonic distortion, 13, 26, 331, 346, 348
harmonic instabilities, 344
head-mounted displays, 395
hedging, 65, 95, 360
hedging contracts, 65
hidden nodes, 384, 386, 388, 389
hot spots, 297, 400, 407, 410
hour-ahead market, 158, 176, 178
HVDC, xvii, 73, 260, 263, 264, 266, 274, 277, 278, 279, 280, 281, 286
hybrid agent, 355, 358, 359
hydro, 3, 5, 6, 12, 13, 20, 68, 72, 73, 105, 174, 229, 259, 280, 330

IGBT, 262, 263, 264, 269, 278, 280
immersion, 395, 396, 397, 400, 405
incipient faults, 296, 323, 329
incremental cost, 53, 57, 83, 84, 85, 88, 90, 91, 92, 99, 196
incremental cost allocation, 57
Independent Power Producers, 2, 65, 330
independent system operator, 25, 51, 104, 121, 175, 217
inelastic load, 65, 92
inequality constraints, 198, 211, 212, 373
information technology., 2, 54, 59
infrared detectors, 297, 401
infrared imager, 400
infrastructure planning, 125, 126
installed capacity, 22, 25, 115, 122, 222, 223, 231, 266
intelligent electronic devices, 139
interface agents, 355, 356, 357, 358
interharmonics, 342
international financing agencies, 124
Internet, xiv, xvii, xix, 114, 118, 140, 141, 143, 144, 145, 358, 416, 417, 418, 419, 420, 421, 422, 423, 424, 425, 426, 427, 428, 429, 431, 432, 433, 434, 435, 436, 437, 440, 441, 445, 446, 449, 451, 452, 453, 457, 458, 459, 460
  auction, 55, 56, 60, 61, 65, 67, 82, 84, 90, 91, 95, 96, 98, 105, 108, 109,

178, 193, 194, 195, 362, 413, 457, 458, 459
  bandwidth, 431
  education, 143, 354, 399, 420
  energy trading, 418, 423, 457
  multimedia, 421
  on-line training, 420
  power trading, 418
  real-time data, 138, 417, 421, 427, 431, 449, 453, 454
  usability, 419
Internet Service Provider, 418, 423, 427, 431, 458
inter-provincial power market, 242, 244
inter-temporal constraints, 77, 91, 96
inter-zonal congestion, 88
investment planning, 307

Java Virtual Machine, 434
JDBC, 439, 440

LAN/WAN technology, 134
last accepted offer, 55
liberalisation, 47, 73, 144
load forecasting, xiii, 120, 121, 122, 151, 383, 412
load levelling, 21, 270
load management, 21, 135, 145, 150
load pocket, 55
load profiles, 26, 63
load shedding, 80, 145, 334
loading vector, 158
locational marginal prices, 88, 96, 98
locational marginal pricing, 88, 99
long term planning, xii, 122
long-term contract, 8, 18, 93
long-term contracts, 18, 93
loop flow, 8
loss allocation, 246, 247, 248, 249, 251, 253, 254
loss of load probability, 238

magnetising reactance, 30, 31, 32, 37
mandatory system operator model, 158, 159, 160, 162
marginal costs, 3, 53, 58, 210, 240, 242

marginal pricing, 23, 57, 99, 164, 165, 187
market clearing, 65, 71, 85, 87, 89, 90, 91, 96, 177
market clearing price, 177
market mechanism, 53, 155, 160, 167, 219, 231, 234, 236
market operator, 457
market reform, 1
market regulation, 21, 46, 48
market transparency, 22
maximalist ISO, 96, 97, 103
megawatt mile allocation, 57
merit order, 70, 82
meter asset management, 114
meter assets, 111, 114
microeconomic principle, 99
micro-gird, 20
MIME type, 421
minimalist ISO, 96, 97, 103
mitigation, xiii, 6, 10, 98, 107, 125, 231, 335
mobile agents, 354, 357, 358
mother wavelet, 337, 338, 339
multi-converters, 278
multilateral, 56, 59, 158, 160, 161, 179, 184, 185, 192, 195, 196, 197, 198, 199, 200, 212, 214, 361
multilateral trades, 185, 361
multilateral trading, 184
multilateral transactions, 179, 184, 192, 197, 198, 199, 212
multilayer feedforward network, 121
multiple linear regression, 121
multi-tiered structure, 159
mutation, 40, 41, 372, 373, 375, 376, 377, 378

negative generator, 115
negative sequence, 344
network automation, 119
network management system, 132, 138
newsgroups, 420
no load bid, 96
nodal pricing, 59, 73, 88, 166, 167, 187, 188

Index   465

non-discriminatory auction, 55
non-immersive systems, 397
nonparametric regression, 121
non-stationary voltage, 336
nuclear, 3, 4, 5, 6, 9, 19, 45, 55, 64, 72,
   154, 231

object models, 143
object oriented, xvii, 438
ODBC, 136, 439
on-line load forecasting, 122
open access, xiii, 2, 3, 57, 60, 73, 79,
   108, 109, 140, 163, 174, 176, 183,
   192, 198, 206, 208, 216, 218, 219,
   372, 459
operating platform, 416, 436, 450
opportunity costs, 83, 170, 186
optimal power flow, 52, 57, 59, 77, 109,
   159, 160, 188, 219, 371, 373, 410,
   412
options contract, 362
options markets, 412

parallel processing, 139, 348, 459
partial discharge, 300, 311, 323
pattern search method, 32
peak load time, 241
peer-to-peer protocols, 134
penalty payments, 291, 317
phase balancing, 33
phase-locked loop, 272
photovoltaics, 9, 12
planning horizon, 58, 122
plant expansion, 120
pollution, 9, 15, 18, 19, 221
pool selling price, 191
poolco model, 96, 108
positive sequence, 344
postage stamp method, 187, 188
power exchange, 21, 22, 46, 60, 106,
   175, 177, 215, 256, 275, 331, 457,
   458
power injections, 79, 210, 211
power line carrier, 134, 144, 145, 148,
   150

power pool, 4, 22, 82, 86, 87, 98, 100,
   109, 159, 176, 179, 182, 183, 184,
   185, 192
power quality, xiv, 21, 25, 117, 127,
   143, 258, 259, 270, 330, 331, 332,
   333, 335, 339, 346, 347, 348, 349,
   350, 351
power station monitoring, 418, 445
power swing, 139, 260, 269
PQ monitoring, 348
PQ standards, 348
price cap regulation, 165, 167
price-cap, 72
pricing scheme, 3, 8, 188, 221
priority insurance service, 167, 168, 169,
   170
privatisation, 1, 2, 4, 50, 51, 52, 54, 63,
   64, 65, 67, 72, 110, 117, 119, 120,
   125, 144, 174, 287, 304, 418
probabilistic curve, 308
probabilistic models, 77
probabilistic production simulation, 238,
   239, 243
programming language, 417, 433, 434,
   435, 438, 444, 449, 451, 452
   HTML, 138, 417, 428, 433, 434, 436,
      437, 441, 442, 443, 444, 445, 458,
      459, 460
   Java, xiv, 417, 433, 434, 435, 436,
      439, 440, 451, 460
   JavaScript, 417, 433, 434, 440, 452
   XML, DTD, XSL, 438, 441, 442, 443,
      444, 445, 450, 458, 459, 460
programming language
   Java applet, xiv, 435, 436, 451, 452,
      454
   Java servlet, 440
public electricity suppliers, 111
pulverised coal combustion, 17, 18
put options, 360, 365, 367, 370, 410

quality of supply, 116, 117, 119, 126,
   127, 129, 130

ramping rates, 77, 91, 96
reactive reserves, 25

real time pricing, 218
real-time energy market, 97, 158
real-time markets, 78, 86
regression analysis, 116
regulatory body, 110, 334
regulatory incentives, 293
reliability benefit, 189, 190
reliability constraints, 78, 79
remote meter reading, 134, 147
remote terminal units, 127, 129, 132
remote vision, 447
renewable, xiii, 2, 4, 5, 9, 11, 13, 16, 17, 19, 20, 21, 27, 45, 53, 231, 330
reproduction, 41
reregulation, 1
retail competition, 2, 51, 63, 73
right-of-ways, 169
risk assessment, xiii, 115, 117, 125, 316
risk control, 313
risk management, 106, 292, 311, 314, 316, 317
rule base, 38, 40, 42, 49, 412

SCADA, 119, 127, 134, 135, 136, 137, 138, 144, 145, 146, 297
scheduled wheeling transaction, 121
scheduling co-ordinators, 175, 179
search engine, 417, 419, 424, 427, 437
  web crawler, 419
second-tier suppliers, 72
security, 24, 47, 117, 139, 140, 152, 202, 205, 219, 320, 358, 414, 424
  digital signature, 429
  encryption, 429, 430
    private key, 429, 430
    public key, 430
  firewall, 428, 429
  password, 429
  Secure Socket Layers, 424, 458
security limit, 163
selection, 40, 70, 100, 104, 149, 150, 174, 177, 282, 296, 338, 370, 372, 377, 428
self-excited induction generator, 27, 48
sequential electricity market, 86

series compensation, 260, 261, 271, 272, 275, 276, 282, 285, 286, 331
service provider, xiii, 111, 156, 162, 163, 164, 170, 288, 289
settlement, 55, 63, 69, 71, 79, 177, 290, 423
shadow prices, 96
shareholder, 4, 112
shoulder load time, 241
sigmoid function, 384, 386, 387, 392, 410
simultaneous electricity market, 87
single-phase loads, 27, 46
smart agents, 355
smart metering, 63
social welfare, 54, 84, 85, 87, 91, 92, 94, 107, 163, 192
software functionality, 423
solar, 5, 9, 12, 13, 14, 21, 22, 37, 38, 39, 42, 46, 49, 330, 399, 412
solar collectors, 38
solar power plant, 37, 38, 46, 49, 412
solar radiation, 37, 39, 42
spinning reserve, 61, 71, 93, 94, 105, 108, 121, 194, 195, 271
spinning reserve markets, 71
spot markets, 3, 168
spot purchases, 63
spread spectrum, 144, 149
SQL, 136, 437, 438, 439, 440
stand-alone generation, 20
start-up bid, 96
state-owned monopoly, 2
stator winding, 28
stock markets, 79, 395
storage heating, 28, 115
stranded costs, 6, 9, 46, 51, 54, 63, 71
strike price, 154, 362, 365, 367, 369
structured query language, 136
sub-synchronous resonance, 271, 285
super-conducting material, 259
supply curve, 56, 68, 69, 84, 98, 159
swing curve, 269
switching element, 264, 265
symmetrical components, 30, 335, 350
system dynamics, xxi, 80, 101

Index

system marginal price, 23, 66
system operator, xiv, 51, 53, 56, 59, 60, 61, 65, 69, 73, 103, 115, 120, 121, 139, 154, 157, 158, 166, 168, 177, 178, 192, 193, 194, 195, 210, 331
system-wide blackouts, 155

take-or-pay, 412
tap-changer, 261, 277
telecommunication industry, 153, 154
telephone network, 114
thermal heating technology, 37
thermal limit, 58, 59, 66, 259, 283
thermography, 400, 410, 415
thermovision cameras, 297
thyristor controlled reactors, 266
thyristor controlled series capacitor, 271, 285
tier supplier, 112
time of use, 135, 190
tournament scheme, 377
transient energy margin, 206
transient stability, xvii, xx, 139, 206, 219, 285, 412
transmission access, xvii, 51, 175, 184, 191, 197, 200, 216
transmission channels, 141
transmission charge, 58, 90, 95, 165, 168, 199, 211
transmission loss, xiii, 57, 60, 65, 72, 105, 120, 165, 186, 191, 192, 196, 197, 198, 204, 214, 247, 257, 373, 374, 376, 458
transmission model, 8
transmission open access, xiii, 216, 371
transmission pricing, xxi, 58, 105, 168, 169, 187, 191, 218, 221, 246
transmission protocol, 417, 427
   FTP, 428
   TCP/IP, 143, 417, 427, 432
transmission revenue, 162, 164
transmission system expansion, 162, 163, 170
triplen harmonics, 344
two-tier system, 120

UHF radio, 144, 149, 150
unbundling, xii, 50, 52, 53, 73, 194, 371
unconstrained schedule, 65
unified power flow controller, 275, 331
uniform price auction, 55, 68
Uniform Resource Locator, 425
unit commitment, 77, 78, 82, 91, 96, 97, 103, 104, 108, 109, 177, 180
UNIX, 136, 426
uplift charge, 55
usage charges, 162, 164, 165, 166, 167
use of system charges, 27, 72, 111, 115

valley load time, 241
vertically integrated, 8, 50, 58, 64, 72, 77, 153, 155, 156, 157, 163, 164, 172, 178, 210, 360
vertically integrated utilities, 77, 153
virtual environments, 395
visual display unit, 395
voice activated messages, 114
voltage collapse, 140, 260
voltage control, 14, 26, 80, 93, 148, 193, 194, 284
voltage dip, 117, 333, 350
voltage sags, 13, 331, 332, 334, 335, 348, 349, 350
voltage source converter, 280
voluntary system operator model, 158, 160, 161, 162, 163

WAN, 134, 139, 358, 431, 449
wavelet transform, 336, 337, 339, 350
weather forecasting, 399
Web browser, 426, 429
Web page, xiv, 417, 419, 422, 423, 424, 425, 427, 431, 432, 433, 434, 436, 437, 440, 441, 442, 444, 451, 452, 458
   design, 423
   dynamic, 417, 437
   static, 417, 433, 437, 442, 451
Web server, 426, 427, 429, 434, 437, 451, 454, 458
web space, 420, 437
website, 75, 114

wheeling, 7, 54, 57, 58, 74, 188, 189, 198, 246, 247, 249, 254, 257
wheeling costing, 246, 249
wholesale competition, 2
wide-scale power outages, 114
willingness-to-pay, 59, 178, 198, 199, 200, 201, 208, 215
wind, 5, 9, 12, 13, 14, 17, 20, 21, 22, 26, 45, 49, 53, 147, 259, 280, 330, 349

wireless connection, 417
  WAP phone, 424, 425
wires business, 174

XLPE cables, 313

zonal price, 71, 166, 167, 168, 188, 189
zonal pricing, 90, 166, 167, 188